And the wolf finally came

The Decline of the American Steel Industry

John P. Hoerr

D0169115

University of Pittsburgh Press

HD
9517
.M85
H64
1988

Published by the University of Pittsburgh Press, Pittsburgh, Pa., 15260
Copyright © 1988, University of Pittsburgh Press
All rights reserved
Feffer and Simons, Inc., London
Manufactured in the United States of America

Second printing 1988

Library of Congress Cataloging-in-Publication Data

Hoerr, John P., 1930–
  And the wolf finally came: the decline of the American steel industry / John P. Hoerr.
    p. cm.—(Pittsburgh series in social and labor history) Includes index.
  ISBN 0-8229-3572-4. ISBN 0-8229-5398-6 (pbk.)
  1. Steel industry and trade—Monongahela River Region (W.Va. and Pa.) 2. Steel industry and trade—Pennsylvania. 3. Steel industry and trade—West Virginia. 4. Iron and steel workers—Monongahela River Region (W.Va. and Pa.) 5. Iron and steel workers—Pennsylvania. 6. Iron and steel workers—West Virginia. 7. Collective bargaining—Monongahela River Region (W.Va. and Pa.) 8. Collective bargaining—Pennsylvania. 9. Collective bargaining—West Virginia. I. Title. II. Series.
HD9517.M85H64 1988
338.4'7669142'09—dc19   87-24932
                                                          CIP

And the Wolf Finally Came

One of the problems in the mills is that no union man would trust any of the companies. To the average union man, they're always crying wolf....

Quotation from Joseph Odorcich,
Vice-President, United Steelworkers of America

For Joanne

Contents

Foreword

My association with the author of this book has been unusual. For many years, until 1979 (the year of my retirement as an official of the United Steelworkers of America), John Hoerr was an observer and I was one of those being observed as I took part in labor-management negotiations in the steel industry.

Hoerr was (and is) highly respected by both labor and management. He is a superb journalist, with an insider's perception of the economics, management systems, labor relations techniques, and social policies of steel. His book is notable for its breadth and insight, and it has a wealth of information about steel labor relations, especially in the recent years of the restructuring of the industry. *And the Wolf Finally Came* reveals more about steel and some of its management and union personalities than has ever been written. It provides an intimate look at cause and effect in the decline of one of America's great basic industries, and it shows the importance of steel's labor relations experience to the country as a whole.

There is an added dimension. Hoerr grew up in the Monongahela Valley, once the heart of the American steel industry. This has added to his understanding of steelworkers and the unique steel environment. Hoerr's background is reflected in his concern about the disruptions experienced by many thousands of people as the industry closed plants. It is inevitable that such disruptions take place in a free economy. If one is looking for a villain, one must look to the society, to our government, to our own sense of priorities as a people. The virtually hands-off policy taken by government toward the victims of disruption is a less humane policy than that pursued by most other countries.

I do not always agree with Hoerr's interpretation of events; and other people, on both sides of the bargaining table, are certain to disagree with some. But a history of labor relations is not a chronicle of precise scientific or statistical data. If there is one thing that history teaches us as a certainty, it is that change is constant and that events look very different as we look back at them. Indeed, time and events have changed my own evaluation of what was accomplished during the nearly four decades that I served with the Steelworkers.

History will thank John Hoerr for *And the Wolf Finally Came*. It will help the participants see themselves and their activities in a broader perspective, and other readers will be more appreciative of the realities, rather than the countless myths, about the steel companies and the union of their employees.

And the Wolf Finally Came will be remembered.

Ben Fischer
Director, Center for Labor Studies
Carnegie-Mellon University

Acknowledgments

This book could not have been produced without the help of many friends and colleagues. Although I had reported on steel labor for twenty years, I needed extended time off my job to do more research for the book. Stephen B. Shepard, editor-in-chief of *Business Week*, generously granted leaves of absence in 1986–87 under McGraw-Hill's policy of unpaid leaves. My mother, Alyce Clark, and my sister and brother-in-law, Lynn and Ronald McKay, graciously gave me lodgings at their home in White Oak, Pennsylvania, during two long periods of research in the Pittsburgh area in 1986. The Pittsburgh Foundation provided a research grant.

Early versions of my manuscript benefited from criticisms by Thomas N. Bethell, who encouraged me to use the somewhat unorthodox personal approach. Thomas A. Kochan, professor of industrial relations at the Sloan School of Management, Massachusetts Institute of Technology, contributed good ideas about my setting of the industrial relations scene in the early chapters. Ronald W. Schatz, professor of history at Wesleyan University, advised me on the labor history portions of the book. Ben Fischer, director of the Center for Labor Studies at Carnegie-Mellon University, who is mentioned often in these pages, gave me valuable opinions drawn from his encyclopedic knowledge of collective bargaining in the steel industry from 1946 to the present.

Other persons who read parts of the manuscript were Charles C. Robb, who provided thoughtful analyses of several chapters; my son Peter, who curbed my tendency to stylistic excesses; Lawrence Delo, who kept me straight on steelmaking techniques; Peter Gall, a colleague at McGraw-Hill; Bob Arnold and Seymour Zucker, senior editors of *Business Week*; Lynn and Ron McKay; David and Eleanor Bergholz; and my wife Joanne, who is an exceedingly good copy editor (and breadwinner). Matt and Gladys Lillig helped financially. John DeGregorio and Darlene S. Payne supplied material and commentary about the recent history of McKeesport; my son David drove me to various airports at odd periods of the day; Jim McKay and John P. Moody, the current and retired labor writers for the *Pittsburgh Post-Gazette*, and Henry P. Guzda, historian at the U.S. Labor Department, provided me with old files; and Judy Adel gave technical advice. Bill Deibler of the *Post-Gazette* and Peggy O'Neill of the *Pittsburgh Press* located many of the photographs.

The United Steelworkers of America allowed me access to certain files, and USW leaders gave freely of their time for extended interviews without restrictions. Most steel executives that I approached were willing to do the same. Others were not. A number of people have written extensively and well about labor in the steel industry, and I have benefited from their beginnings: David Brody, historian of early steel workers; John Herling, author of the only political history of the USW; A. H. Raskin, former chief labor writer of the *New York Times*; and Jack Stieber, an industrial relations scholar.

I want to thank especially Frederick A. Hetzel, director of the University of Pittsburgh Press, who encouraged and aided me in many ways; Catherine Marshall, the managing editor, who edited the manuscript and helped me shape the book; and other members of their staff who provided an exceptional backup service by clipping local newspapers. Finally, I would like to thank all the men and women in the Monongahela Valley who have helped me, often unknowingly, over many, many years. I have not been able to mention them all by name, but I pay homage to them all.

John P. Hoerr

And the Wolf Finally Came

Chapter 1

Collapse of the Steel Industry, 1982

It is a trip weighted with shock and nostalgia. I am driving east on Second Avenue in Pittsburgh, heading out of the city and up the Monongahela River. Behind me stand the eminences of steel and glass, bunched in the heart of the city, where management makes decisions for its far-flung steel empire. Ahead lie the mill towns and steel plants, strung along the winding river artery, where labor produces molten iron and steel and finished steel products. Once vital parts of Andrew Carnegie's wondrously profitable linkage of mines and mills, most of these plants now sit idled and empty, soon to be churned into rubble.

Times are hard in the Monongahela Valley. The most devastating business slump since the Great Depression has eviscerated the steel industry, causing plant shutdowns and widespread unemployment. On my right, the black metal siding of the cavernous old plant of the Jones & Laughlin Steel Corporation looms over the street, as intimidating as ever, forcing me instinctively to lean away from it as I drive past. A large "For Sale or Lease" sign is posted on one end of the building. It is a bold stroke on someone's part, advertising a mile-long steel plant for sale as one would a corner grocery store. The plant had been expiring slowly for years, the victim of foreign steel's relentless attack, as well as capital starvation and an ill-suited merger of J&L with the LTV Corporation. Across the river, J&L's Southside plant is still working, though at a low level of operations.

I have flown in from New York to take a reporter's measure of this destruction. But I can make no pretense of objectivity. This is my home. I was born in the valley, grew up here, went to school, played, worked, and was part of a family here, and I

1

dread the sight of silent mills and dying mill towns. I have made this trip up the valley from Pittsburgh hundreds of times over the years, in automobiles, commuter trains, trolley cars, and once or twice by boat. It has always been for me the prototypical journey home, coming from the harsh outer world into the mysterious inner world of earliest memories and first thoughts. The further I penetrate the valley and move into the shadows of the river bluffs, the more I shed journalistic indifference.

The J&L plants are the only steel mills remaining in the city. Pittsburgh itself shows little visible evidence of the recession which started in the summer of 1981 and is now, in October 1982, approaching its low point. The visitor's eye tends to focus on the impressive skyscrapers which house the corporate offices of such giants as U.S. Steel, Alcoa, Mellon Bank, Westinghouse Electric, Rockwell International, Koppers, Jones & Laughlin, National Intergroup (formerly National Steel), Consolidation Coal, Joy Manufacturing, and Wheeling-Pittsburgh Steel. It is an enormous concentration of power, an island remaining high and dry while the rest of the industrial Midwest struggles in neck-deep debt and jobless misery. The appearance is somewhat deceptive, for thousands of white-collar employees of these companies also have lost their jobs. Hundreds of offices also stand vacant in the newer buildings, thrown up to accommodate a business expansion that hasn't occurred.

Nevertheless, Pittsburgh always puts forth its best front, aided by the spectacular juxtaposition of hills, rivers, bridges, and architecture. The city lies at the "forks of the Ohio," as early settlers called it, where the Monongahela and Allegheny rivers join to form the Ohio. A visitor driving in from the airport first sees the city upon leaving a tunnel that cuts through a high ridge on the south side of the Monongahela. At night, emerging from the tunnel is like bursting into a dazzling new world of light and form. The skyscrapers, with row upon row of floors lit for the night cleaning crews, give off a white glow that shimmers on the dark rivers. This spectacle, you might think, must surely radiate outward, projecting strength and prosperity into every corner of the surrounding countryside.

To cross the river, I must turn off Second Avenue, drive up the hill into the Greenfield section of Pittsburgh, and descend on the other side to the Homestead High-Level Bridge. The bridge affords a broad view of the Monongahela as it curves up the valley between stiff-backed ridges and disappears around a hairpin bend. The river is about a half mile wide, greenish brown and slow-moving. As I gaze up the valley, I am struck by the absence of smoke. Not even a suggestion of a wisp hangs over the Homestead Works of U.S. Steel, which sits to my left at the end of the bridge. But something else is lacking—a sense of life, the teeming, active, energetic life the valley once knew. I dislike exaggeration, but this is what I feel: Death is in the air.

The October day is cool and bright. There is still plenty of green foliage on the steepest parts of the river bluffs, where not even the craziest house can safely perch. For all that we have done to the Monongahela Valley over the past century, the wooded hills and bluffs retain the primitive look of the dense forests that once blanketed the region. When times are bad, and no plumes of sooty smoke drift across the bluffs, that appearance of rusticity distracts one from what man has made of the valley's narrow flatlands by the river. The industrial detritus of a fading culture stretches for mile upon unrelieved mile on these riverbanks: abandoned furnaces, mill buildings, railroad tracks and bridges, storage yards, pumping stations, pipelines, transmission towers. The American steel industry lies dying in its cradle.

I am a labor journalist, and I have come to report on relations between labor and management in the steel industry. As of now, the United Steelworkers union (USW) has refused to grant wage concessions that the industry says it needs to become competitive. There is a standoff. Both sides have displayed an obduracy to change that is all too common in American industrial relations.

Indeed, the steel industry is the best example of what has gone wrong with union-management relations in the United States. Such poor relations are one reason why organized labor is declining in most industries, and why American companies have lost their competitive edge in the new international marketplace.

I can see that an immense tragedy is unfolding in the "Mon Valley," as it is known to the people who live here. The industry is now operating at only slightly more than 30 percent of capacity, close to the record low of the Depression. Some twenty thousand steelworkers, or about two-thirds of the work force in the Mon Valley, are laid off. Most will never work again in a steel plant. The almost legendary mill towns in the valley—Homestead, Rankin, Braddock, Duquesne, McKeesport, Clairton, Monessen—are being drained of their life's blood. The same is true of many other mill towns: The Southside and Hazelwood sections of Pittsburgh, Aliquippa on the Ohio River, Johnstown on the Conemaugh, Youngstown on the Mahoning, Bethlehem in eastern Pennsylvania, Lackawanna in New York, Cleveland, Gary, South Chicago, and others.

I grew up in McKeesport and have vivid memories of the way it used to be. In the late 1940s, when I was a teenager, a dozen great steel plants lined the banks of the Monongahela, extending forty-six miles up the valley from Pittsburgh. The mills worked twenty-four hours a day and provided jobs for nearly eighty thousand men and women, not counting employees in the companies' Pittsburgh offices. They were enormous steaming vessels, clanging and banging, spouting great plumes of smoke, and searing the sky with the Bessemer's reddish orange glow. The narrow brick streets of the mill towns were filled with streetcars, automobiles, workers going to and from the plant, and shoppers carrying big brown paper bags. There were two or three saloons to a block on the main street near the mill, and almost as many churches scattered through each town.

And, yes, noondays were often as dark as night—as awed visitors usually reported, when inversions trapped great clouds of smoke close to earth, and the downtown sidewalks were so thick with ferruginous dust from the open hearth and Bessemer furnaces, that they gave off a metallic sheen. Smoke seemed to seep out of the very pores of the mill buildings. Every morning housewives all over town put on babushkas and swept clouds of dust off their front porches.

The activity in the plants was never frenetic, but always intensive. You would see diesel engines hauling long strings of tank cars filled with molten iron along the riverbanks; heavy trucks heaped with smoking slag grinding out of the plant gates to dump their loads on the huge, incessantly burning slag piles; skip cars laden with iron ore crawling up cable hoists on the 100-foot tall blast furnaces to feed the vessels from the top; crane buckets dipping into mounds of coke on barges tied up along the river wall. Crane sirens were always screaming out in the depths of the mill, diesels were honking, pipes were banging against pipes with a resounding, hollow noise. Workers moved about everywhere. From outside the plant you could see them strolling on the plant roadways, or hooking up loads on cranes in the storage yards, or striding out of plant offices with blueprints under their arms, or lined up just inside the plant gates, awaiting the siren for shift change.

Now those giant sprawling places of enormous energy have become rusting hulks: silent and lifeless, like obsolete dreadnoughts sunk to their stacks in shallow water. This image occurs to me as I turn left off the bridge and drive south through Homestead on East Eighth Avenue, the town's main street. The Homestead Works, the largest plant in the valley, is a few blocks to my left.

Twelve thousand people once were employed here; a few thousand are left. There is little traffic on Eighth Avenue, where it once took fifteen minutes to drive four or five blocks. Gone are the retail emporiums, the movie theaters, most of the bars, the crowds of shoppers. Amity Street, which leads down to the main plant gate, is deserted. Farther south, where the plant crosses into the Borough of Munhall, the mill buildings extend right up to Eighth Avenue. Their open ends face the street, but they are dark inside. Nothing moves. At one of the half dozen entrance gates, I see a sign: "This gate closed. Use Amity St. gate." Now there are so few working that all gates except one are closed.

I drive on through Munhall and up the steep four-lane highway that cuts across the face of a high bluff. At the top is Kennywood Park, where three generations of steelworkers, and everybody else in the valley, have spent the traditional end-of-school holiday riding roller coasters and picnicking. Just before getting to Kennywood, I pull into a cinder parking lot and walk to the edge of the cliff overlooking the Monongahela.

The cliff is at least a hundred feet high and makes a sheer plunge down to railroad tracks on my side of the river. Directly across the river is Braddock and the Edgar Thomson Works of U.S. Steel. In 1872 Carnegie and a group of investors bought 107 acres of farmland in what was known as Braddock's Field (where the French and Indians ambushed and slaughtered General Edward Braddock's army in 1755) and built Carnegie's first steel plant. Its two Bessemer furnaces and rail mills, representing the latest in steelmaking technology, began operating in 1875. The early success of this plant launched Carnegie in an enormously lucrative steel business which set the pattern for all other steel companies in terms of commercial and production practices, formed the core of the future U.S. Steel Corporation, and forever changed the face of America.

The plant extends about a mile along the riverbank and consists of a score or more of different structures—furnaces, tanks, metal buildings, brick buildings, all of varying sizes. It is dominated at the northern end by a huge, pale blue building housing the basic oxygen furnaces. Nearer the river is a two-story, red brick building with high arched windows that have been bricked over, which looks like a turn-of-the-century structure. I see nothing moving in the plant. No trucks, no trains, no people, no smoke. The river locks adjacent to the plant, where strings of coal barges would be lined up awaiting passage over a five-foot change in water level, are vacant. After a while I hear a faint sound of metal beating on metal. It seems to come from the old brick building, as if someone were pounding a sledge on an anvil, pounding for no reason but to proclaim his existence in that vast and empty place.

I drive on, glancing wistfully at the arching curves of Kennywood's three famous roller coasters, the Rabbit, the Racer, and the Thunderbolt. The Thunderbolt, which used to be called the Pippin, was the most exciting. It dived over the side of the cliff toward the muddy river and zoomed back up again as our cries of thrill stretched out behind like a flyer's scarf. The park is closed for the winter. Across the street, showing more life than anything I've seen so far, is a large McDonald's where once stood a row of brick homes. A big red sign in the parking lot says: "McDonald's. Billions and billions served. Drive through." Cars are queued up at a red light, waiting to drive through.

Now I'm going down the hill into Duquesne, where another U.S. Steel plant dozes in the bright sunlight. The pride of the Duquesne Works is Dorothy Six, the largest blast furnace in the valley, 125 feet high and swathed in piping that looks like rippling muscle. As I drive past, a cloud of steam envelopes Dorothy's cast house at ground level. Is she tapping iron or just blowing steam? A number of friends from high school days work in the Duquesne plant—if they still have their jobs, I remind myself.

A half mile beyond the Duquesne Works, I turn onto the bridge that crosses the Monongahela to McKeesport. Now I am really coming home. McKeesport's National Works lies on the riverbank just south of the bridge. From my viewpoint on the bridge, I note that the plant gives off a pale blue glint hinting of fresh paint, and the dashing USS logo is prominent on one of the mill buildings. This neat, fresh look is an optimistic sign.

At the end of the bridge, I turn south once again on Lysle Boulevard (named for a mayor who for years banned union meetings) and drive past the plant. On the town side, it looks completely different, old and worn out. Nothing is painted. The blast furnaces, long unused, are piles of dirty rust. The buildings housing the finishing mills expose to McKeesporters ancient, yellow-brick facing and rows of broken windows. It's as if the plant promoted itself as a sleek, modern mill to the outside world across the river but revealed its obsolete, abandoned face to its own community.

The mill buildings look the same as they did thirty years ago. Like most young men of my generation who grew up in steel towns, I worked in the mill during summers and other time off from school. One job lasted a year. Going to and from work every day, I walked past the buildings on the mill roadway and took a short cut through a dark storage building where electric motors, steam engines, pumps, generators, and unopened crates of forgotten equipment lay heaped in dusty rows. Great age was stamped on everything in the mill, including the lined faces of older men who still wore narrow-brimmed hats with grease spots in the finger hollows near the peak. I dreamed of finding footprints of an ancient steelworker in the dirt of the mill grounds, like the arrowheads still reported to be buried in Braddock's Field. I worked then for a construction company that was building a new boiler house in the plant, but I felt as much a part of the mill as a steelworker. The boiler house is still there, looking weary and depressed but still spouting little puffs of steam now and then.

It is close to three o'clock, and so I park the car and walk to the corner of Locust Street and Third Avenue to watch the shift change. Workers stand inside the Locust Street gate as always, waiting to burst through at quitting time. When the siren sounds, everybody in town will look at their watches and think, "Mmmm, three o'clock. The men'll be coming out now."

But I am wrong. There is no siren. Now it is only a faint tinkling sound. The mill doesn't want to embarrass itself by proclaiming an enormous change of shifts, I think, when in fact only a fraction of a normal day turn is now working. In the old days, indeed only a year ago, it would have taken ten or fifteen minutes for the entire day turn to file through the gate. Now the exodus takes a total of one and a half minutes. Only 150 hourly employees, out of 4,200, are still working.

I begin to walk away. Someone calls my name. I turn and see a straggler coming out of the gate. He is Manny Stoupis, an acquaintance since the 1940s and now chairman of the grievance committee of USW Local 1408. I always remember Manny (nobody used his full name, Manuel) as a slight, wiry kid who was good in solid geometry. He has grown a bit thicker since I last saw him a few years ago, and his black hair is graying above the ears. Manny has the same earnest look, though now it is more knowing and verges on the bitter. He has been a grievance committeeman for the better part of twenty years. Anyone who survives this long as an elected local official will have experienced more than the normal share of human conflict.

After we shake hands, I ask what it is like to work in the huge plant with only 150 fellow workers. "It's eerie," he says. "You don't see anybody. You don't hear anything. There's no security. The company closed the two gates at the lower end of the plant and laid off the guards. They canceled the roving guard who used to drive around in a jeep. People are breaking in at the lower end, climbing over the fence, and stealing brass and iron scrap." He laughs. "Two foremen down at the lower end lock themselves in their office all day."

Manny was an ambitious young man who, like so many back in the early 1950s, got caught in steel's good times. For young men in the mill towns of those days, there was a very tangible sense of having to make an implicit bargain with life from the outset. There were two choices. If you took a job in the mill, you could stay in McKeesport among family and friends, earn decent pay, and gain a sort of lifetime security (except for layoffs and strikes) in an industry that would last forever. You traded advancement for security and expected life to stick to its bargain. Or you could spurn the good pay and long-term security, leave your family and the community, and take a flyer on making a career in some other field. In a sense, everybody growing up in America must decide whether to stay or move on. In the Mon Valley, however, the very presence of the mill on the riverbank, its gates flung open to young men (though not women and not blacks in the better jobs), forced you to make this choice.

Manny could have moved on, but his ambition trapped him in McKeesport. He started working part-time in the mill in 1947 during his junior year in high school and never stopped. His steelworker wages carried him through college and graduate school, and by the time he received a master's degree in mathematics from the University of Pittsburgh in 1958, he was making too much money as a steelworker to leave the mill. He had accumulated seniority by then and was grossing $4,000 more a year than he would have received if he had gone into biophysics. Manny stayed in the mill and got into union politics. I had talked to him many times in the 1960s and 1970s when he held a variety of local union offices.

We walk slowly toward the Local 1408 hall. Manny talks in hard, curt sentences. He doesn't complain, but I can guess. The way things are going, he will have to take early retirement before he is fifty-five . . . and then what? There is plenty of anger in him, but it comes out only a little at a time. "Most of our people will never see the inside of this plant again," he says. There is a long pause. "National Works is just about done for."

The Local 1408 hall is a two-story, pale green building fronting on Fifth Avenue, McKeesport's main business street. Several men in the lobby are discussing recent layoffs; somebody with forty-two years of service has just lost his job. I leave Manny and walk down a narrow corridor, passing a committee room where more men sit around a table and talk about layoff benefits. I find Dick Grace, the local president, sitting behind a desk in his tiny office. There are three chairs and a bookcase containing union brochures. On the wall are color photographs of John F. Kennedy, the five current International officers of the USW, and Grace himself competing in various road races. In his late forties, Grace took up jogging and has run in several ten-kilometer races. Now fiftyish, he is trim around the middle and wears a brush moustache that is turning gray.

I've known Grace for several years. Sometimes he is evasive about union business, but not today. He pulls his baseball cap advertising "Local 1408" more firmly down on his forehead and throws his feet up on the desk, ready to talk. He is angry. The company wants to padlock the unused locker rooms in the plant and has been calling laid-off workers and telling them to come in and pick up their clothing. "They call them at home and just say, 'Come in, empty your locker out.' That scares the hell out of them. These bastards have no finesse."

We talk about local negotiations. The company has asked the union to approve changes in the local agreement which would enable management to move workers from one department to another inside the plant. Grace doesn't like the idea. I ask a demurring question: If the union wants National Works to continue in business, doesn't it make sense to help management reduce labor costs?

Grace drops his legs behind the desk and straightens up in the chair. He has called off negotiations, he tells me angrily, upon discovering that the company has sent a letter to all employees ordering them to take their vacations in January so the plant can shut down for two weeks. It's not the policy that angers him so much as the fact that management negotiators failed to tell him about the mailing (purposefully, he suspects) even as they were asking him for concessions. "I haven't talked to them for two weeks," Grace says. "My relations with them are at the worst level I've ever seen."

Tragedies are made of such things.

I drive back to Pittsburgh after dusk. There are no stars out, and the valley is dark and silent. I feel as if I am in another country, or another time. The Mon Valley that I have known, a socioeconomic system that for a hundred years provided the labor for America's industrial explosion, is slipping into the past. Perhaps it should. Perhaps it is time to discard that culture and build another someplace else. What's one century in the sweep of progress?

The Mon Valley, 1987

By the middle of 1987 the death whose presence I had felt five years earlier had moved in on bulldozer's treads. The steel depression of 1981–82, followed by a continuing dismantling of the industry in a worldwide economic upheaval and a 184-day shutdown of U.S. Steel plants in a labor dispute, had left the Mon Valley reeling and helpless. The Homestead, Duquesne, and McKeesport plants of U.S. Steel were closed, permanently. (In 1986, after acquiring two large energy companies, U.S. Steel changed its name to USX Corporation.) Farther up the valley, Wheeling-Pittsburgh Steel shut the gates of its historic plant in Monessen in 1986. The old Jones

& Laughlin mills in Pittsburgh were being torn down. Dozens of related manufacturing plants had also closed.

The number of people who derived their income from steel had declined to less than four thousand, down from over thirty-five thousand in 1981, and from eighty thousand in the late 1940s. The mill towns, once so alive with the heavy throb of industry, now gave off the weak pulse of welfare and retirement communities. The degree of suffering caused by lost jobs, mortgage foreclosures, suicides, broken marriages, and alcoholism was beyond calculation. Many people, especially the young, had left the valley, but middle-aged and older workers, unable or unwilling to migrate from the only home they had known, went through the anguish of trying to start new careers. The standard of living, boosted to a high level by the bargaining strength of the United Steelworkers, was falling steadily.

More than might have been suspected, families held together. Many steelworkers *did* start new lives, discarding forever the hard hat and accepting employment as clerks and low-skilled technicians in service industries. The mill towns, instead of giving up, kept searching for new employers. Grass-roots organizations dedicated to self-help served as advocates of the jobless before legislative bodies and government agencies. Some groups urged a revolutionary new role for labor, as employee-owners of abandoned enterprises. Other activists turned to physical confrontation with bankers and corporate executives, the symbols of uncaring wealth. But the real struggle in the Mon Valley was waged by individual people who, in countless cases of quiet heroism, prevailed over difficulties they had never before experienced.

They were left with bitter memories, however. From 1981 on, U.S. Steel acted with a callousness that will not soon be forgotten. The economics of the world steel trade dictated that the corporation shut down plants and reduce its work force. There was no pleasant way to do that. U.S. Steel, however, seemed to go out of its way to turn unpleasantness into nasty displays of power. It fired thousands of supervisors and managers—its most loyal employees—on no more than a moment's notice: "Clean out your desk, this is your last day," they were told. In fairness, it must be said that U.S. Steel poured out enormous amounts of money to support retirees and their families; in 1985 alone, it paid $123 million in pension and medical benefits to recipients in the Mon Valley.

Nonetheless, the company left a legacy of bitterness, even hatred, that will last for many years in the valley.[1]

The times clearly called for a partnership between steel management and the USW to meet the new competitive challenges. At companies such as National Steel, leaders on both sides managed to break out of old patterns of adversary relations to forge such a partnership. This did not happen at U.S. Steel, partly because of union resistance to change, which was more pronounced at some plants than others. But the company's confrontational style of management also impeded cooperation, with unhappy consequences. The corporation practically invited the 1986–87 work stoppage through bad faith dealings with the union.

What were the comparative dimensions of the disaster in the Mon Valley? The life expectancy of single-company towns in the fast-changing American economy is counted in decades, not centuries. New England's Merrimack River Valley, with its acres of abandoned textile factories (some now restored as historic sites) is another example of an industrial valley that went bust. Starting in the 1820s, this birthplace of America's Industrial Revolution grew into a mighty textile and apparel-producing area. The introduction of electricity eliminated the valley's water power as a competitive advantage, and the textile industry moved to the South, which offered an abundance of raw materials and cheap labor. After a generation of economic depression, the Merrimack Valley is now experiencing a resurgence as an electronics center.

The Merrimack Valley's major period of decline extended from the 1920s to the 1950s. The economic destruction of the Mon Valley was even more swift and brutal. The valley's manufacturing base had been eroding, it is true, for twenty-five years, but the final paroxysms would not be those of a moribund creature. A vibrant forty-six-mile stretch of river valley, providing primary jobs for over thirty-five thousand steel employees, and subsidiary jobs for nearly three times as many people, would be devastated and expunged from economic memory in less than five years.

An entire industrial culture overthrown and trampled into the dirt within a few years! There would be no gradual change here, no slow dying of one technology and the simultaneous growth of another. This isn't the Ruhr Valley, where old steel communities are kept alive even as their mills are torn down and replaced with other industry. This is wide, broad-shouldered America, where there is always room someplace else for people abandoned by their livelihood. Are you an unemployed steelworker in the Mon Valley? Well, move on, brother! The first hill is the hardest one to cross. After that, the opportunities are limitless ... Texas, Arizona, or anyplace from here to there where McDonald's needs someone to serve the one-trillionth burger.

For more than a century the mill towns had sustained an economic system that provided labor and social stability for America's industrial explosion. In the 1980s these towns—Homestead, Braddock, Duquesne, McKeesport, and Clairton—lost industries, people, pride, and their tax base. And not just the mill towns themselves were affected, but also dozens of other suburbs and river communities that form part of the same system: Munhall, East Pittsburgh, Turtle Creek, Dravosburg, Glassport, West Mifflin, White Oak, Elizabeth, Wilmerding, among others. Schools, roads and highways, transportation systems, churches, sewerage, business districts, and shopping centers—the entire infrastructure of the region was in danger of being swept away, as if a huge wave were surging down the valley from the headwaters of the Monongahela.

How would the region survive without its manufacturing base? Well, it was said, Pittsburgh would become a center of high technology and financial service industries, transmitting its knowledge and expertise around the world via satellite and electronic networks. Instead of exporting steel and machinery, the region would export services and research (medical research already was well established) and employ people as office workers, technicians, and operatives in high-tech manufacturing. So said the bankers and regional development experts. This forecast had some merit. While the five-county Pittsburgh metropolitan area lost 127,500 manufacturing jobs between 1979 and 1987, it gained 75,800 service-type jobs. Many of the latter, however, were low paid, demanded little skill, and existed largely in Pittsburgh and surrounding communities, not in the decaying depths of the Mon Valley. The truth was, nobody knew what to do about reviving the mill towns and their suburbs.[2]

What This Book Is About

This book tells two stories. The main story is about the decline of the labor movement and the deterioration of the

American system of industrial relations. The second, related story is about the wastage of the Monongahela Valley. The valley was *not* primarily a victim of labor-management conflict. However, I believe, and hope to demonstrate, that the singular use of the valley as a factory system for the steel industry produced a unique social and economic environment. Both the conduct of work in the mills and the course of labor-management relations were profoundly influenced by this environment.

The two stories, therefore, are intertwined, and the telling of both helps illuminate the major theme of this book, which is that the American system of organizing and managing work is obsolete.

The precipitous decline of the steel industry in the 1970s and 1980s provides the most dramatic evidence of the need for reforms. The industry's shrinkage resulted from many complex, interrelated forces (see chapter 4). In some part, it was inevitable, given powerful structural shifts in the international steel industry: the growth of foreign steel and the resulting excess of steelmaking capacity over demand by 1986; the increasing use of plastics, aluminum, ceramics, and other materials as a substitute for steel; and the rise of American minimills with modern equipment and low labor costs which enabled them to capture high-volume products such as rod, wire, and bars from the integrated steelmakers.[3]

The USW's resistance to wage cuts and work-rule reforms constituted only one element in this array of causes. But it was a critical one. And for every barrier thrown up by the union, higher barriers were raised by management. Management was too arrogant to see that labor could contribute ideas and commitment that might arrest the decline. That these antagonisms remained intact fifty years after unionism came to the steel industry is proof of fundamental flaws in the union-management relationship and, more importantly, in the shop-floor organization of work. The USW and the steel industry failed to work together to make the plants more productive, even though a mechanism existed for doing this, the voluntary Labor-Management Participation Teams (LMPTs) which could be established at the department level within plants (see chapter 6).

Aside from LMPTs, which dated from 1980, the steel companies had engaged in little innovation, social or technological, for most of the postwar period. Led by a management that ranged from mediocre to poor, the American steel industry simply failed to make the right business decisions at the right time. Most important, it failed to change its authoritarian style of managing people in order to gain cooperation in a common endeavor.

As a result, the conflict between union and management on the shop floor, which was necessary and healthy to some degree, degenerated into pointless battles for supremacy between rival bureaucracies. The union's one-time fight for a decent standard of living and better working conditions became the captive of a bureaucratic industrial relations system. By the 1980s, the Mon Valley plants—indeed, most American steel plants—were no longer competitive.

In the six years covered by this book, practically everything that could happen to an industry and a union *did* happen in the steel industry: massive unemployment, wage-cutting, national and local bargaining, internal union strife, rank-and-file rejection of labor pacts bargained by their leaders, bitter disagreements between companies, the collapse of an employer bargaining group, the breakdown of industrywide wage patterns, plant shutdowns, chaotic market conditions and bankruptcies, and—in some instances—the beginnings of new labor-management relationships.

One of the more dramatic of these events had occurred a few months before my October, 1982, trip to the Mon Valley.

Union Drama in a Ballroom

At 10:30 A.M. on July 30, 1982, a great shout of defiance rang out in the Grand Ballroom of the William Penn Hotel in Pittsburgh. Some four hundred local union presidents, who constituted the ratification body of the United Steelworkers, had rejected by voice vote a steel-industry proposal calling for wage concessions. As the echo of the vote receded, many of the delegates leaped to their feet, shaking their fists and shouting.

It was an extraordinary demonstration. In the 1970s the same body, called the Basic Steel Industry Conference (BSIC), had ratified new wage agreements once every three years. Elated cheering had accompanied the votes—and for good reason. Each new agreement in those years was described as the "best ever," and cumulatively they made steelworkers the highest paid industrial workers in the United States, if not the world.

In 1982, the circumstances were reversed. The steel companies wanted to modify the wage terms they had agreed to in 1980. Although the USW's contracts with eight large steelmakers would not expire until August 1, 1983, the companies had proposed replacing those pacts with new, lower cost contracts. The companies were U.S. Steel, Bethlehem, Jones & Laughlin, National, Republic, Armco, Inland, and Allegheny-Ludlum. In 1982 they were still bargaining as an industrywide unit, a practice that began in 1956 and ended in 1985. A bargaining committee named by the companies negotiated an industrywide labor agreement, which set forth wage and benefit terms, as well as general provisions on matters such as hours of work and the grievance procedure. The companies, therefore, paid identical rates of pay and levels of benefits. Each firm also had its own labor contract which spelled out other noneconomic provisions, which differed to some degree from company to company.[4]

Because the wage terms were identical across the industry, this method of negotiating met the union's goal of "taking wages out of competition." The companies gained the protection of unity; the USW couldn't strike them singly and leverage one firm against another. The uniform wage policy also suited the industry's oligopolistic pricing structure. But by 1982 industrywide bargaining was an anachronism. It was not flexible enough to cope with the industry's changing problems. Despite the strong arguments of some executives to disband the committee, however, the companies hadn't mustered the nerve to take that bold step.

The industry had been falling deeper into financial trouble since the brief recession of 1980. Now, in the midst of the 1981–82 recession, the nation's mills were operating at only 43.8 percent of capacity. Japanese steelmakers, along with American minimills, had captured more than 25 percent of the U.S. steel market. In the Big Eight companies, the average hourly cost of providing wages and benefits (pensions, medical care, vacations, and holidays, etc.) had risen to between $22 and $23 in 1982, roughly twice the employment cost of Japanese steelmakers and $5 per hour above the minimills' cost. One of the Big Eight, Republic, was said to be close to failure, and the industry negotiators told Lloyd McBride, the Steelworkers' president, that all eight companies would vanish by the end of the decade unless the steep upward trend of labor costs were leveled off.

About 107,900 hourly workers, or 37 percent of steel's 1981

work force, were laid off at the time of the BSIC meeting, and many of the local presidents realized the business slump was no ordinary recession. Largely for this reason, the conference had reluctantly authorized McBride to "discuss" labor-cost issues with industry negotiators. McBride and two union aides, Vice-President Joseph Odorcich and General Counsel Bernard Kleiman, met throughout July with the industry bargainers, J. Bruce Johnston of U.S. Steel and George A. Moore, Jr., of Bethlehem. It was a peculiar set of negotiations, the first in which the USW had contemplated breaking John L. Lewis's old dictum of "no backward step" in bargaining.[5]

Nothing came of the July talks. The industry asked the union to cancel a wage increase scheduled for August 1, 1982, and accept a pay freeze for the next three years, with limited cost-of-living adjustment (COLA) payments. McBride would have accepted a wage freeze, but he rejected the COLA limit. The negotiations ended in a deadlock.

The USW president called the July 30 BSIC meeting merely to ask the local presidents to ratify his rejection. The delegates' response indicated in part the depth of their anger with management. But in part it resulted from an internal union problem. McBride had bred suspicion among the presidents by refusing to admit that he actually was discussing possible wage cuts with the industry men. A few dozen local presidents had taken a stand against any wage concessions, at any time, for any reason. McBride's reticence made the militants' campaign all the easier. They spread the word that the USW president had already "cut a deal" with the industry bargainers and intended to ram it down the throats of the BSIC delegates on July 30.

So it was that the presidents who assembled in the William Penn for the 10 A.M. meeting were in a rebellious mood. Many were annoyed that the companies seemed to have placed all the blame for the industry's predicament on labor. The "no concessions" group included unionists from each of the major companies, but especially from Bethlehem and U.S. Steel, where relations had deteriorated in recent years. The delegates from U.S. Steel plants were furious that the company had bought Marathon Oil instead of investing the $6 billion in steel. As they filed into the ballroom on the seventeenth floor, their faces grim and angry, one might have thought they were about to do battle with a hated foe rather than to meet among themselves to decide the fate of their industry, and their jobs.

It has been an abiding irony that the Steelworkers use the William Penn ballroom for many of their large meetings. Gaudily dressed in late Renaissance decor, the ballroom today differs little from the original of sixty-five years ago. Gold draperies cascade down over high, domed windows, three enormous chandeliers with tiers of glass candles hang from concave pits in the ceiling, and maidens carrying baskets of fruit stand out in bas-relief on wooden panels. The walls are covered with white paper flocked with undulating gold curls and lined with dozens of candelabra that give off a dull glow. A kind of ornate gloom permeates the place—a curious setting for a latter-day class struggle.[6]

Today, the ballroom is an interesting museum piece. But when the William Penn opened in 1916, the ballroom's Old World dazzle represented the pretentious aspirations of the businessmen who built the seventeen-story building. One of their leaders was Henry Clay Frick, the coal and steel magnate who in 1892 sent bargeloads of Pinkerton goons up the Monongahela River from Pittsburgh to attack striking steelworkers at Carnegie's Homestead Works. Those were heady days for capitalism in Pittsburgh. It must have seemed that the iron and steel furnaces would never stop pouring out molten metal, belching smoke and dust, and employing endless streams of immigrants.

By July 30, 1982, everything had changed in Pittsburgh and the Mon Valley. In one sense, however, nothing had changed: Management-labor relations in the plants were still implacably hostile.

McBride told the BSIC meeting that the companies had asked for "more than was reasonable" in negotiations and that he had turned down their proposal. When he called for a vote on a motion that the BSIC reject the companies' demands, an overwhelming shout of "aye" resounded in the ballroom. The local presidents jumped up from their chairs and cheered and clapped. In the balcony, scores of nonvoting union "observers" joined in the demonstration. It was loud and boisterous.

Why did the steelworkers react this way, knowing even then that their vote might mean a further loss of jobs? Some of the delegates applauded out of relief that they had not been called upon to vote yes for concessions. For others, particularly from U.S. Steel and Bethlehem plants, the demonstration was the release of a collective anger, a challenge to a hated enemy, "the company." McBride stood grimly at the podium. At that time he was a relatively young-looking sixty-five, a chunky, solid man of medium height with a balding head and a professorial face. (Thirteen months later, he would die of a heart ailment probably aggravated by the strains involved in trying to steer the union through a perilous time.)

Most union presidents would have turned such a showing of militancy to political advantage; an obvious tactic would have been to make a stemwinding speech attacking the "greedy" companies. But it was not in McBride's style, or character, to use political rhetoric. Believing that the survival of the industry and the union was at stake, he had tried to reach an accommodation with the companies and had failed only because their price, in his view, had been too stiff. When the noise finally subsided, he told the delegates in a stern voice: "I say to you, the decision we have made today is not cause for celebration. The problems are mutual [between labor and management], and they have to be solved. This is a mutual failure."

As the delegates streamed out of the ballroom, one group went whooping down the corridor, clapping backs and saying: "We did it! We did it!" They were claiming credit for having forced McBride to reject the industry's proposal, a dubious claim but one that indicated the temper of the more militant delegates. Although relatively few in number, they were louder and more passionate (either from serious intent or demagoguery) than the delegates in the opposite camp and tended to pull the undecideds in their direction. McBride's rebuke at the end of the meeting had annoyed rather than sobered the militants and added to their suspicions that McBride sided with the companies in favoring wage concessions. "Sometimes you wonder which side he's on," one of them said to me.

The events of the next seven months, however, confirmed McBride's assessment of the economic picture. Indeed, it was even bleaker than he suspected. The steel industry had been badly hurt by the 1981–82 recession, and it continued sliding downhill through the late summer and fall of 1982. Steel unemployment soared; by the end of the year, 153,000 hourly workers, or 53 percent of the work force, were laid off. Many thousands of the jobless had run out of unemployment benefits, and hundreds had lost, or were on the verge of losing, their homes and cars in mortgage foreclosures.

Nevertheless, in November 1982 the ratification committee once again rejected an industry proposal, this time repudiating McBride by voting down a package of concessions that he negotiated and recommended. Once again, the local presidents' anger at the companies, particularly U.S. Steel, played a large role in the rejection. Not until March 1, 1983, would the Basic Steel Industry Conference finally approve a settlement accepting an immediate 9 percent wage cut.

Those seven months of negotiating impasse became one of the most momentous periods in the history of steelmaking in the United States. This book will tell the story of what happened during that time and afterward, from 1981 to 1987. The story unfolds in many places, in negotiating sessions between top-level union and industry officials, on mill floors and picket lines, in turbulent union meetings and staid company conferences, and in the homes of workers and managers in the Monongahela Valley.

The Procedures Obscured the Reality

As a journalist, I have been reporting on and writing about labor since 1960. By far the most excruciating rounds of collective bargaining I have covered were those in the steel industry in 1982–83 and again in 1986–87. I became convinced that the problem in steel was rooted in forty years of poor management of people and a misdirected union-management relationship. For most of that period, steel management regarded hourly workers as an undifferentiated horde, incapable of doing anything more than following orders and collecting the paycheck. Suddenly, in 1982, that horde was expected to turn into an enlightened group of rational individuals with implicit trust and faith in the company and a willingness to accept whatever management said was necessary to put its financial affairs in order.

Some companies had better relations with their employees than others, and the atmosphere differed from plant to plant within companies. Generally, however, those decades of adversary relations on the shop floor had created an atmosphere of suspicion and hostility that could not be overcome by the normal methods of communication.

This was not evident on the record. There had not been a major strike since the 116-day national steel walkout of 1959. Such a long streak of peaceful bargaining was extraordinary in a major U.S. industry. But it obscured the real problems in the steel industry, largely because the negotiating procedure did not reflect the reality of everyday life in the plants. Industrywide bargaining put control of the negotiating process in the hands of top-level industry and union leaders who were determined to avoid strikes, a worthwhile goal but one which, by itself, could not contribute nearly as much to solving steel's competitive problems as improving relations on the plant floor.

This misplaced emphasis had its beginnings in the 1950s and 1960s, when the collective bargaining procedures that had evolved during and after World War II became ever more procedural and bureaucratic in form. Management wanted to maintain stability at almost any cost, and avoiding unnecessary strikes became a primary mission of industrial relations professionals in American industry. Company negotiators accumulated power in the company, and were compensated and promoted, on their ability to hold the union at bay while eschewing conflict. This was especially true in the steel industry, which went out of its way to buy labor peace in the 1970s with its no-strike agreement.[7]

The USW's contract ratification procedure inserted another blurred lens between the leaders' eyes and what was happening down at the mill-and-furnace level. Industrywide pacts were approved by a vote of the BSIC, local union presidents from all companies that were construed, however remotely, to be part of the basic steel industry. While the BSIC in its later years became an increasingly independent creature, still it was essentially a committee whose votes were subject to some degree of leadership control. Ratification by committee had been used in the USW since its inception in the 1930s, an outgrowth of its top-down organization. It was one of those traditions—of which there were many on both sides in the steel industry—that seemed to survive all evidence that it was outmoded.

By the early 1980s, the neglect of shop-floor relations had adversely affected attitudes, productivity, and product quality. On one side a managerial bureaucracy consisting of the plant manager, department superintendents, general foremen, and line foremen, ran the plant in the old authoritarian fashion. Orders came down through the hierarchy and, right or wrong, were to be followed to the letter. On the other side, a union bureaucracy consisting of the local president, chairman of the grievance committee, and his committeemen (few women held these posts) protected the workers' "rights," as spelled out in the contract. Management made no attempt to involve the union officials, much less the workers, in cooperative efforts to improve the work process; such an effort would have violated management's unilateral "right" to run the plant. The union officials spent most of their time reacting to management decisions and protecting work rules which established crew sizes and work jurisdictions, even though those rules might be obsolete. In this climate, there was much yelling and puffing of chests on both sides, and ordinary workers were caught in the middle.

This stultifying relationship was particularly true of the old plants in the Monongahela Valley. Not much had changed since the USW organized them in the 1930s. It was as if a regiment of paratroopers (the union) had been dropped into hostile territory and remained surrounded all these decades by management, the two sides accommodating to an uncomfortable balance of power. The degree of hostility waxed and waned over the years as managers and union leaders came and went. But management usually sets the tone of a relationship, and in the Mon Valley that tone was at its best one of polite belligerence.

The local unions became highly politicized; any leader who made peaceful overtures to the company was promptly denounced as a "sell-out artist." Union officials and supervisors conducted daily warfare over such matters as discipline, job assignment, and the handling of worker grievances. It was as though each side had to prove to itself, day after day, that it had the capacity to hurt the other.

Over the years, this relationship hardened and became impervious to change. Many people tried to reform the system, only to run into barriers of tradition and bureaucracy— on both sides. At U.S. Steel's Duquesne plant, for example, Local 1256 President Mike Bilcsik went to management in 1983 and proposed that the union and company begin an LMPT program. He was repulsed. Plant management wanted no part of participation.

On the management side, many supervisors had to suppress their participatory tendencies and instead act as disciplinarians. One of these was Lawrence Delo, who worked for more than twenty years in several Mon Valley mills as a first-line foreman, one of the toughest jobs in any industry. In late

1983, when he was working at U.S. Steel's Homestead Works, we talked about the corporation's management style. "I really think a lot of Homestead's labor problems are caused by management," Delo said. "They are so hard-assed! That's been the history of the steel industry in the Mon Valley. You have to manage by force and by setting examples, rather than by reason and persuasion. You can't get ahead if you don't scream and yell at people."[8]

By the early 1970s, I began to realize that the "mature collective bargaining relationship" that the academics were ascribing to the steel industry existed only between top-level union and company negotiators. But even then I didn't perceive the depth of the hostility in the plants. From the summer of 1982 on, however, it became clear that the union and industry hadn't advanced beyond primitive stages in their relationship. Each side sought only to advance its own interests. But the decline of the steel industry—its poor profit showing, decreasing productivity, and lack of investors—should have galvanized both sides to focus on their mutual interests. Instead, each side blamed the other and stood aside as jobs dwindled.

The result of all those years of neglect was best summed up by Joseph Odorcich, a blunt-spoken USW vice-president. Odorcich started work in the coal mines of western Pennsylvania at the age of fifteen and went on to a colorful career in the Steelworkers, first as a local officer at a McKeesport foundry during the late 1930s. Later, as a union reformer in the Mon Valley, he had to buck entrenched union leadership to make his way to the top of the union.

One day in the fall of 1983, we sat in his Pittsburgh office, reviewing events of the past year. The bargaining failures of 1982 had embarrassed the union leadership, and Odorcich conceded that serious errors in strategy and tactics were partly responsible. The largest problem, however, was that local union leaders and rank and filers distrusted management, and even the union to some extent. "One of the problems in the mills," he said, "is no union man would trust any of the companies. To the average union man, they're always crying wolf. And the wolf finally came."[9]

Chapter 2

Steelworkers were not the only group to suffer in the 1980s. The collapse of the once mighty American steel industry occurred at a time of vast economic restructuring that affected every industry and virtually every person. After a long, seemingly stable period following World War II, a series of upheavals sped along the fault lines of the world economic system, turning a familiar landscape into an unrecognizable jumble of upended industries and overlapping trade borders.

During a period of only ten years, from the mid-1970s to the mid-1980s, the world saw the rise and fall of the OPEC oil cartel, the emergence of Developing World countries as manufacturing powers, the fading of American technological supremacy. Perhaps the most significant development was the growth of a new technology—based on the silicon chip, microelectronics, and the computer—which enables capital, that volatile generator of economic growth and corrupter of men, to fly about the world on electronic air waves and create new assets in the twinkling of a second.

In the United States, meanwhile, several trends and events of a more parochial nature produced major economic changes for Americans: the deregulation of the trucking and airline industries, the election of a president bent on disengaging the federal government from any role in the relationship between corporations and people, the expansion of the federal deficit to proportions that reversed the U.S. balance of trade, and the rise of a new breed of financial speculators skilled at raiding (and sometimes looting) the decaying domains of old-time corporate America—and paying for it with the assets of the pirated companies and their employees.

These forces had disastrous consequences for two income

24

groups in the United States, the poor (who sometimes benefit from political upheavals but rarely from economic eruptions) and middle-class wage earners once employed in the shrinking goods-producing and exporting industries. These industries include the auto, steel, tire, leather, apparel, textile, mining and other sectors that made up the core of American manufacturing. Not only the employees and their families suffered, so did suppliers, retail businesses, professionals, and all of the supporting institutions—municipal government, schools, churches, social agencies, and, not least, trade unions.

From one vantage point, the great economic changes of the 1980s can be viewed as a continuation of the U.S. economy's massive shift away from union labor over a period of three decades. This shift was barely visible for many years. Finally, two events of overwhelming importance forced recognition upon the labor movement. The first was the 1980 election to the presidency of Ronald Reagan, whom organized labor regarded as a bitter enemy. But it was a "crisis of capitalism," a cyclic business slump of giant proportions, that really brought labor to heel.

When the devastating recession of 1981-82 finally began to recede in early 1983, organized labor found itself stranded on an eroding beach. It had lost 3.4 million members, 11 percent of its 1979 membership. There had been plant closings and business failures in numbers matched only during the Depression. In the past, labor had rebounded from business slumps when the economy recovered. That was not to be in this cycle. Manufacturing employment did not return to the previous level, and the unions continued to shrink as business returned to normal, losing another 328,000 members by the end of 1984.

In the Pittsburgh region, one of the hardest hit by all of the above trends, ninety-five thousand manufacturing jobs disappeared between 1980 and 1983 in a labor force of slightly over 1 million. The region includes Allegheny County, sometimes called the birthplace of organized labor, where the predecessor of the American Federation of Labor (AFL) was founded in 1881 and the forerunner of the USW in 1936. According to one estimate, union membership here dropped from 30 percent of the work force to 21 percent in the 1970s and to less than 17 percent in the 1980s.[1]

These losses forced the unions into a retreat that exposed a vulnerability to attack on many fronts—economic, political, and legal—that hadn't been tested since before the New Deal. American unions lost jobs, members, dues income, and respect, but the biggest blow to their pride and standing was a reduction in living standards through wage cuts. Although Europe's recession was every bit as deep as, and even longer than, the U.S. downturn, European unions avoided a similar fate by winning larger government subsidies for underemployed members. Lacking that kind of political clout, American unions were forced to swallow embarrassing wage reductions. In 1983, the peak year for wage concessions, about 28 percent of all newly negotiated major union contracts contained either wage freezes or cuts, and the trend diminished only slightly in the next two years (27 percent in 1984 and 25 percent in 1985).

The concessions often enabled financially troubled companies to survive. But in the anything-goes, free markets atmosphere encouraged by the Reagan administration, many healthy firms exploited the concessionary trend without fear of being called to moral account. As the wave of "concession bargaining," as it came to be called, swept through industry, the old practice of "scabbing" reappeared. High rates of joblessness created what AFL-CIO President Lane Kirkland termed "an army of long-term unemployed" who were willing to cross picket lines and take strikers' jobs.[2]

In Washington, an unfriendly administration controlled the White House and the U.S. Senate (during Reagan's first term) and sealed off all channels of government decision making from the AFL-CIO. Labor's influence on Capitol Hill diminished markedly, even in the Democratic-controlled House. The National Labor Relations Board (NLRB) had come under the control of conservative members appointed by President Reagan and soon began reinterpreting labor law in ways that would further weaken unions.

All of this sent an unmistakable message to the unions: In the eyes of the administration, many employers, and even many workers, they were expendable.

Indeed, the tide seemed to be running out on the American labor movement. It had been around for well over a hundred years, waxing and waning erratically. It had been harassed, suppressed, and incarcerated, but it had always fought back and always persisted. From the earliest days of the Republic, the movement had given life and hope to a long procession of tradesmen and craftsmen—coal miners, carpenters, riggers, masons and bricklayers, ship builders, teamsters, molders, locomotive engineers, machinists, glass bottle blowers, printers, hatters, boot makers, cigar makers, garment makers, boiler makers, and many other kinds of "makers" who formed the base of American industrialization.

In the 1930s the movement added a young and militant wing, one molded by the Depression out of the industrial masses who crammed the nation's factories. Laboring in faceless swarms at giant machines, they became known as "workers"—auto workers, rubber workers, steel workers, electrical workers, oil and chemical workers, aluminum workers, smelter workers, dock workers, glass workers, and later, when the industrial ethic penetrated the service trades, hospital workers, clerical workers, and food and commercial workers. As an invading force in the 1930s, this wing led the movement in a major assault on American-style capitalism in the mining, manufacturing, construction, and transportation sectors and sent advance columns into retail stores and the white collar area.

With passage of the National Labor Relations (Wagner) Act of 1935 and the creation of the NLRB, the movement was given legal title and guaranteed the right to exist. For meritorious service during World War II in observing a no-strike pledge, organized labor was given turf and protection by a grateful government acting on behalf of the people. Aided by the country's rapid postwar growth, the movement entrenched itself—deeply, it appeared at the time, although this was illusory—in the American economic system. Its constituent unions amassed enormous economic and political power. The unions advanced in the political area by rewarding their friends and punishing their enemies in Congress, state legislatures, and city councils. Through collective bargaining they raised the standard of living of millions of members, as well as nonmembers whose employers paid higher wages to avoid unions.

There was corruption in labor's ranks, no more than in any commercial walk of life, though more publicized; there were tyrants who ruled with an iron hand and paid little attention to the needs of their members, but they were far fewer in number than those on the corporate side of the field; there were strikes and slowdowns, but not much more—and perhaps even less—than might have been expected of any industrial democracy. Business rallied to limit labor's power in the late 1940s with passage of the Taft-Hartley Act, and union organizing slowed down. Nonetheless, the movement had extended into every part of American life by this time. The most prominent labor economist of the time predicted in 1948 that the United States would become a "laboristic" society.[3]

Barely three decades later, however, the movement was in deep trouble, for many reasons. Fundamental structural changes occurred in the economy that were not foreseen in the 1940s. Few predicted the great shift of employment from the Northeast and Midwest to the South and Southwest, where employers were stronger in their opposition to unions. It was inconceivable then that Japan, West Germany, and other European nations would rebuild so well after the war, with American aid and technology, that they would become major competitors of our basic industries. No one in the 1940s or 1950s foresaw the rise of the Developing World as an industrial power.

Who could have foreseen that American women, though they had always represented a comparatively large proportion of the U.S. labor force, would in the 1970s constitute half of that force? There were good reasons for women to enter the workplace in increasing numbers, including the corroding effect of inflation on the family budget and women's desire to move out of the kitchen and into a wider world. But if the husband remained the primary breadwinner, the wife had less reason to embrace unionism. By the mid-1970s, as more and more low-income women held jobs, polls began to show a positive attitude toward unions. The unions, however, hadn't found a way to translate this change into increased membership.

One fundamental change was already under way in the 1940s: the long-term decline of employment in goods-producing industries and the corresponding rise in service industries where unions are weak. The change had been occurring gradually for most of this century, and this very gradualism seemed, to labor leaders unused to long-term planning, to permit infinite opportunities to exploit the trend—somewhere down the road. As always, gradual change occurred much faster than the human mind will admit. Suddenly, in the 1960s and 1970s, the United States became a fast food addict, and McDonald's—the prototypical employer of young part-timers who think they have no need for unions because they are bound for better things—became one of the largest employers in the nation. From 1959 to 1986, service-producing jobs increased by 42.4 million, to 75 percent of nonfarm employment, while the goods-producing sector added only 4.5 million jobs and fell from 38 percent to 25 percent of employment.[4]

Labor's own mistakes and lack of aggressiveness contributed to the slowing of momentum. In highly organized industries like steel and autos, unions felt little need for further organization and did not branch out rapidly to other occupations in anticipation of shrinkage in their core industries. The unions spent $1.03 (in wage-deflated dollars) per nonunion worker for organizing in 1953 but only $0.71 in 1974, a drop of 30 percent. Spending less and encountering greater employer resistance, the unions had proportionately less success recruiting new members as the population and labor force grew.

It may well be, as one sociologist contends, that Americans began backing away from organized labor as their dislike for union behavior grew. Other students of the U.S. labor movement demonstrate the growth of "an alternate nonunion system of industrial relations" since the 1960s. Companies involved in this trend take great care to establish compensation and other programs that preempt the unions' appeal to workers. At the same time, some companies have increasingly resorted to illegal methods, such as firing union activists during organizing campaigns, to dampen workers' enthusiasm for the union. By the 1980s, one analyst calculated, for every twenty workers who voted for a union in a representation election, one was illegally fired—an astonishing indictment of employer behavior.[5]

Whatever the reasons, labor's bright promise of the 1940s began to fade. Sometime during the 1950s, organized labor in the United States reached a turning point, although few union leaders or labor observers recognized it at the time. Even when observers began reporting on declining membership, in the early 1970s, labor leaders—especially AFL-CIO President George Meany—belittled its importance.

Statistics on union membership show a remarkable rate of decline. The greatest surge in unionization in the United States occurred in the twenty years between the depths of the Depression and the mid-1950s, when millions of workers in the mass production industries poured into the newly organized industrial unions as well as the older craft unions. In 1953, the peak year, unions represented 32.5 percent of all nonagricultural workers, but only 18 percent in 1987. (In private industry, the density fell from 35.7 percent to 13.4 percent in 1987.) During that thirty-year period, union membership grew in absolute numbers but not nearly as fast as the labor force.

And the decline continues. Two Harvard University economists, Richard B. Freeman and James L. Medoff, describe it as the "slow strangulation" of organized labor. In the early 1950s, unions each year organized roughly 1 percent of the private, nonagricultural work force through elections conducted by the NLRB. Twenty years later, in the early 1970s, the proportion had dropped to 0.3 percent, producing fewer new members annually than are lost through attrition. At that rate, Freeman and Medoff calculate, the union share of the work force tends to decline by roughly 3 percent per year.[6]

A movement with 17 million people—the total union membership in 1986—cannot be said to be moribund—yet. But the prospect of a slow wasting away, fighting losing defensive battles in retreat, does not augur well for organized labor. The movement needs to reshape itself on the lines of a radically different model, and it is questionable whether major reform is possible short of a crisis—that is, the prospect of instant death. A union derives its effectiveness from economic leverage—which increases in proportion to the percentage of employees it represents in a company or industry—and from the morale of its members. Declining membership reduces leverage and makes it difficult for workers to hold to the old belief in the collective strength of numbers. The sweeping dislocations caused by the recession have altered that rank-and-file view in once-strong union centers such as Detroit, Pittsburgh, and Chicago. "I don't need to belong to a union to take a wage cut or lose my job," became a common saying.

The point is that when the members have little faith in a labor movement, it may in fact be dying. Union leaders scoff at this notion. They point out that the death knell for organized labor has been sounded many times, only to be followed by a resurgence. However, the last time this happened, in the 1930s, the unions were aided by a new political environment accompanying the Depression and new federal laws protecting unions.[7]

Such a turnaround is not out of the question in the 1980s or 1990s, although people who study labor's situation are hard

put to say how it could come about. The political environment shows no sign of significant change, and if any new
labor legislation is enacted, it is likely to be unfavorable to
unions. There have been suggestions that the crowding of
labor markets by well-educated men and women of the "baby
boom" generation might eventually lead to pressures for
unionization. Young white-collar workers are likely to be
jostling one another, fighting for the limited number of jobs
with salaries and the potential for advancement that they had
been led to expect.

On paper, white-collar workers seem ripe for unionization,
especially those who labor in the huge clerical factories of
insurance companies and banks, processing claims, checks,
and whatnot on electronic assembly lines. A few unions have
organized some of these workers. But there is no mass movement toward unionization. Still, the hopeful contend, growing discontent in these areas could seed the ground for
another great spurt in unionization. This is possible.

Few stars fall from the sky, however. Rather than wait for
such an epochal event, the labor movement would better
serve itself, and the national interest, by seeking new ways to
do what it alone can do: represent the interest of workers in a
capitalist society. The old ways are no longer sufficient for
this purpose.

Management's Strategic Failures

It is not only organized labor, however, that has lost its way.
So has management, with even greater repercussions for the
nation's competitive ability. The dimensions of management's failure can't be clicked off on the same counter that
measures declining union membership. But one can look at
America's dwindling competitive ability in the following
light. It is management that sets the tone of a relationship
between supervisors and employees in a workplace. It is management that determines how much is going to be produced,
in what amount of time, with how much emphasis on quality.
And it is management that constantly looks to the future
and plans ahead for shifting patterns of consumption, the
growth of competition, and the development and use of new
technologies.

That American management generally has performed less
than adequately in these functions has been amply docu-

mented in recent studies. Steel management, in particular, failed to develop and carry out competitive strategies in the 1960s when Japanese steelmakers began threatening the United States' world leadership in this industry. As one study points out, managers of the domestic steel companies lacked the worldwide business outlook that drove Japanese firms to seek raw materials in all parts of the world and to build modern plants next to deep-water harbors in order to export steel. The American companies did not develop resources abroad and followed an inconsistent approach in modernizing plants at home. As a result, the domestic industry lost its long-standing advantage in raw material costs and wound up with partially rebuilt plants scattered across the country (see chapter 4). The Japanese, concludes one study, "had put together raw materials with modernized large-scale facilities in strategic locations. They had gone ahead; we had not. Their steel industry had been well managed; our steel industry had been badly managed."[8]

Managements in industries such as autos, rubber, agricultural implements, and semiconductors made similar strategic mistakes. This poor performance is one reason for the decline of American manufacturing as a source of high-income employment. Other reasons include unfavorable currency exchange rates, which aided imported goods in the early 1980s; the industrialization of low-wage nations such as South Korea, Taiwan, Brazil, and Mexico; relatively high labor costs in the United States; and automation. As a result of these and other factors, manufacturing employment fell by 1.9 million from the prerecession 1979 level to 1986.[9]

In some ways, the most troubling failure of American management occurred in the area of organizing work and managing people in the workplace. Contrary to the notion that the workerless factory is just around the corner, people are still the most critical element in production. The structure of the industrial relations system used by most American companies, however, fails to take full advantage of the ingenuity of employees. Loosely defined, the industrial relations system is comprised of employees (union or nonunion), managers, and government interacting in a system that provides the rules, ideology, and economic framework for everything that affects the employment relationship. Included are wage decisions, work organization, working conditions, and enforcement of labor legislation.

Among the problems that emerged within the system were these: The traditional mode of "pattern" bargaining produced excessive wage increases in many cases; the organization and management of work on an old-style basis—narrow, functionally defined jobs and autocratic management style—resulted in alienated workers, poor product quality, and lagging productivity growth; a decades-old political hostility between labor and management organizations at the national level prevented progress on needed legislation.

Americans had lived with these problems for a long time. But in the increasingly competitive world markets of the 1980s, they assumed larger importance. Reforming the industrial relations system had become imperative, and so had the need for unions to alter their roles in the economy. Their function in the workplace, for example, is based on concepts developed in the 1930s and 1940s.

To show how this role has become largely irrelevant, the next section sketches the evolution of labor relations after World War II.

The Turning Point After World War II

Before the war, the hostility of management forced the unions to operate as outsiders. They were interlopers trying to batter down the doors of capitalistic greed. This role conferred upon the unions considerable ideological force which kept alive the possibility that they would lead the nation down the road to socialism. However, the needs of wartime production forced a compromise. Management reluctantly accepted the unions and joined them in devising peaceful procedures for resolving disputes over wages and working conditions.[10]

In the glow of patriotism, some industrial employers even agreed to set up labor-management committees within the plants and to listen to workers' ideas for improving output and quality. Philip Murray of the Steelworkers, Walter Reuther of the United Automobile Workers (UAW), and a few other labor leaders pressed insistently for a greater union voice in company decisions. Murray's concept of "industrial democracy" involved joint management through "industry councils."

When the war ended, employers needed labor stability to reap the huge profits promised by a pent-up demand for consumer goods. Instead of reviving the prewar battles against unions, management decided to live with them, although largely on management terms. The companies, for example, proclaimed the existence of a broad class of matters (production and work scheduling, operational decision making, pricing, etc.) that were covered by "management prerogatives" and thus protected from union interference. The corporate community also lobbied for, and won, passage of the Taft-Hartley Act, which severely limited union actions in many areas (for example, it outlawed the secondary boycott).

Unions grumbled about this bargain but accepted it. The prosperous postwar economy presented them with an unparalleled opportunity to raise the living standards of their members. They focused their energies on bargaining for wages and the then-new fringe benefits, such as pensions and health insurance. The idea of industrial democracy was thrown out the factory window, where it lay unused for more than a generation.

This was a turning point of immense importance and one that I shall address later in greater detail. Had the USW and the UAW taken the road of participation at this crucial juncture, the histories of their two major industries would have been vastly different. Of course, there is something illusory about dealing with what might have been. In this case I believe it is justified, especially with regard to the Steelworkers. Murray, the USW's first president, did not choose the road that the union would travel for the next thirty-five years by stumbling into it blindfolded. He made a purposeful decision, reacting to the closed-mindedness of industry leaders of the time. As I shall demonstrate, the deterioration of work and life in the mills of the Monongahela Valley—and their eventual closure—flowed partly from the management and labor decisions of the 1940s.

The point is that when organized labor became an insider, so to speak, in the economic system, it lost interest in what was happening in the work process. The union presence remained strong on the factory floor in the person of the steward or grievance committeeman. But he or she fulfilled a narrow, quasi-legal function of filing grievances when members complained about management decisions such as disciplinary actions or job assignments. The union took virtually no part or responsibility in making the workplace more productive, or in gaining a direct voice for its members in operations.

Management insisted on this arrangement as long as prosperity lasted, for it kept the factories humming and the profits flowing. Moreover, it kept the union out of management's business. Achieving labor peace and preserving management control were perceived to be the key labor problems that industrial relations executives were expected to solve. The union had no concern for competitiveness and rejected the idea that it bore any responsibility for seeing that the plant operated efficiently. The old saying that "management manages the business and the union 'grieves'" concisely described the explicit, and artificial, division of responsibilities.

While unions were no longer the ideological outsiders they had been in the prewar days, they kept up the pretense. They crouched just outside the corporate boundaries, driving in to the attack during contract negotiations, but then pulling back with their gains and returning to a sated, watchful rest. Through this pretense the unions maintained a mythical independence, apparently standing aside from capitalism even as they rooted in the corporate fruit cellar.

American labor and management constructed a rigid, legalistic industrial relations system that, in a sense, ignored the outside world. It tended to alienate workers and could not adjust readily to changes in technology, geopolitics, and international trade.

For twenty or twenty-five years after the war, U.S. industry had virtually no competition from abroad. Wages in the unionized sector of the private economy jumped far out of line in the late 1970s just at the time when foreign and domestic competitors were becoming strong. According to data compiled by Freeman and Medoff, the union wage advantage over nonunion workers rose to about 25 percent in the late 1970s, some 10 to 15 percentage points above the normal premium.[11]

The economic crises of the late 1970s and early 1980s exposed the failures of this system. Management by and large had not shared information with the unions, had not attempted to establish a feeling of mutual trust, had not dealt with workers as adults. Nor had unions, at most companies, demanded such treatment. Lacking the employees' trust, management often found it almost impossible to convince workers that wage cuts and changes in work practices were necessary.

The tragic loss of jobs and the decline in union membership detailed in the early pages of this chapter resulted in part from these failures. The dislocations undoubtedly would have been severe even with ideal union-management relations. A certain amount of strife was inevitable; people were being asked—or forced—to reduce their standard of living. But the extent of the misery and turmoil was aggravated by the inherent weaknesses in the system.

The economic strength of unions in the United States has depended largely on the role they choose to play, or, given the restrictions of labor law, are able to play in industrial relations. That the unions failed to occupy a central, unassailable position in the corporation—and thus in the economy—goes a long way toward explaining their decline.

Many thoughtful labor leaders and union supporters are now coming to see the need to adopt a new role. One of these is Ben Fischer, director of the Center for Labor Studies at Carnegie-Mellon University in Pittsburgh. A white-haired, slow-talking man with a sharp wit and not much patience for conventional wisdom, Fischer worked for more than thirty years on the USW staff. An arbitration expert, he was an innovative negotiator and served as an "idea man" for four Steelworker presidents. He sees the future of labor this way: "What the labor movement really must do is update the definition of its objectives. It must become part of the management structure, to help secure the success and position of the firm as the thing most meaningful to the worker. There are two reasons for the union to have a role in management. One is that the union has a better capacity, or should have, to know what's best for workers. The other is that it gives management a good channel for relating to the work force. Management and the union can manage the work force more effectively than management alone can."[12]

These are hard words to swallow for a labor movement used to acting as the outsider. It is difficult for unionists at all levels to put aside fears of being co-opted in the corporate vortex. Militant leftists and the espousers of traditional "business unionism" argue that this is the wrong road to travel. However, growing numbers of labor leaders are advocating shop-floor collaboration with management as a means of making work more satisfying and the firm more competitive. In the 1970s only two high-ranking leaders, Douglas A. Fraser, president of the UAW, and Irving Bluestone, a vice-president, actively supported worker participation, or Quality of Worklife (QWL), as the concept was known at the time. By the mid-1980s, the list had grown substantially and included such strong advocates as Lynn R. Williams, who became president of the USW in 1984; Donald F. Ephlin, the UAW's chief negotiator at General Motors; William Bywater of the Electronic Workers; and Morton Bahr of the Communications Workers. Even the AFL-CIO, which had for years either ignored the issue or spoke of it in skeptical terms, officially urged its affiliated unions to "accelerate" participation efforts.

In the beleaguered steel industry, the USW's Williams began pushing his union aggressively toward gaining a major voice in management decisions. "A lot of trade unionists seem to be frightened of [participation]," Williams said at a 1984 conference. "[They are] frightened of it as an antiunion tool, frightened of it as a way of defeating the labor movement.... I'm inclined much more to see it as a way in which to build the labor movement."

There was increasing evidence that rank-and-file workers themselves wanted to be more deeply involved in their work, assuming some responsibility for the success of the business. The "baby boom" generation's demand for more challenging jobs converged with the need to become more competitive, and participation spread significantly in the 1980s. Thirty-six percent of two thousand workers surveyed in 1985 reported that their companies had formal involvement programs, and 23 percent said they were personally involved.[13]

Management generally, however, was not willing to share power to the extent necessary. Many involvement programs did not give workers significant voice in company affairs. Nevertheless, a real movement—with leaders and grass-roots backing—was under way. In a minority of cases, worker participation had moved to higher levels than cooperation on the shop floor, involving, for example: union representatives on boards of directors; workers consulted on technological change, plant location and even product design; unions with a voice in hiring and training decisions; rank and filers virtually bossing themselves on semiautonomous work teams. Slowly, inexorably, workers were becoming involved in running the business, on the shop floor, in the plant manager's office, and on the corporate board.

I will deal in greater detail with crucial issues posed by participation for both management and labor. The point of this chapter is that the labor movement has been severely weakened by its failure to adapt to changing circumstances. Of course, some adverse effects of the changing international economic order were unavoidable. But by the mid-1980s, the balance of power in labor relations had changed fundamentally. Three examples will suffice.

The Teamsters, once a national leviathan that held most of the nation's truck-freight system under one contract, could no longer call a nationwide strike. In the auto industry, the weakening of the UAW led to the defection of its one-hundred-thousand-member Canadian branch. This would greatly complicate the ability of the two unions to coordinate bargaining across the border at U.S. and Canadian units of GM, Ford, and Chrysler.

The final example involves the Steelworkers. In 1959 the USW shut down the entire steel industry for 116 days, causing such economic damage that a federal court ordered a halt to the strike so a settlement could be negotiated. But in the 1980s a USW "industrywide" work stoppage would have barely shut down 50 percent of production, given the rise in nonunion producers and foreign imports. When the union embarked on a pivotal strike in August 1986 against the company that had become its arch-enemy, a vastly shrunken U.S. Steel, the nation displayed almost total indifference.

Union Weakness in the Reagan Era

It wasn't only the unions' refusal to adopt a new role that endangered them. During all their days of glory from the 1940s on, they occupied a highly vulnerable position in the American economy. In the rest of this chapter, I shall explore the peculiar status of an organized labor movement in a society that both dislikes unions and yet stands up for their right to exist. I begin in the watershed year of 1981.

Ronald Reagan became the thirty-ninth president on January 21, 1981. He embarked on an economic program that would, with the help of Congress and the Federal Reserve Board, dramatically reduce inflation, establish a tax policy that favored the rich over the poor, and lead to unemployment rates of close to 11 percent in 1982. In federal agency after agency, Reagan's appointees began dismantling machinery that had been set in place under laws lobbied into existence by the unions. The Labor Department, the NLRB, the Equal Employment Opportunities Commission, and the Occupational Safety and Health Administration—all of which have a major impact on labor's rights in the workplace—came under the control of people whose main intention was to reduce or eliminate the regulatory impact of their agencies.

The crowning blow for organized labor, however, came in the middle of summer, from an unexpected direction.

Monday, August 3, 1981, was an uncomfortably muggy day in Chicago. At 7 A.M., Lake Michigan was placid and pond-like. Just north of the Drake Hotel, vapors rose from wavelets that barely ebbed ashore on the sandy Oak Street beach. Low-lying fog trapped this warm mist, turning the lake front into something like a vast steam room. As I jogged north along the concrete walk skirting the beach, feeling sticky and out of breath, I could hear approaching runners before they suddenly materialized. Fortunately, they were all more agile than I and swerved to avoid a collision, leaving me bucking their slip streams.

As I meandered along the lake front, strike pickets were forming in front of the control tower at O'Hare International Airport, some twenty-five miles north and west. The Professional Air Traffic Controllers Organization (PATCO) had called a national strike for 7 A.M., local time. By 9 A.M. Chicago time, more than eleven thousand three hundred air controllers were engaged in a walkout that, from the beginning, was destined to become a long, tragic march to the graveyard of lost strikes—and broken unions.

Eleven thousand three hundred men and women—"strong, emotionally normal, dominant, independent, highly motivated" people with higher intelligence levels than the national average, who held elitist notions about themselves ("a breed apart," they boasted), who worked under stressful conditions (and loved their jobs), who earned considerably more than the average American ($31,000 a year in 1981), and who thought they were indispensable (and history may show that they were)—had taken on the federal government to better themselves and the air traffic control system. When they lost, they lost everything.[14]

That same morning, hundreds of conventioneers and conference attendees of various faiths and disciplines assembled in the many "conference facilities" of Chicago's Hyatt Regency Hotel. Among them were the thirty-two members of the executive council of the AFL-CIO. Made up of the AFL-CIO's two principal officers and thirty presidents of major

unions, the council is the ruling body of the federation be
tween biennial conventions. Since the merger of the AFL
and the CIO in 1955, it had been the custom of the council to
hold two major meetings each year, a mid-winter gathering in
Bal Harbour, Florida, and a mid-summer assemblage in Chicago. That the 1981 summer meeting started on the same day
as the PATCO strike was an unfortunate happenstance, for it
revealed with painful clarity just how powerless organized
labor is in the United States.

Although PATCO was an AFL-CIO affiliate, its officers had
failed to keep the federation informed about the status of
negotiations. The failure was especially galling to the unions
whose members would be affected immediately by the walkout, including the Machinists, the Air Line Pilots, the Railway and Airline Clerks, the Teamsters, and various flight
attendant unions.

Shortly after noon, Lane Kirkland, the AFL-CIO president,
recessed the council meeting and walked down the corridor to
hold a news conference. A native of South Carolina, Kirkland
had been a member of the Masters, Mates & Pilots Union
while serving in the Merchant Marine. But he had spent most
of his career in labor as an intellectual and staff aide (he was
George Meany's right-hand man for years). Portly and solemn,
he looked less like a picket-line unionist than he did an aging Shakespearean scholar with rumpled hair and a crooked
tie.

Kirkland sat behind a cluster of microphones in a crowded
room and read a statement deploring Reagan's 30 percent,
"supply-side" tax cut (recently approved by Congress) as a
"trickle-down" program that benefited the rich over the poor.
When the questioning started, a reporter noted that Reagan
had issued an ultimatum to the striking controllers: either
return to work within forty-eight hours or be fired. Kirkland
characterized the threat as "harsh and brutal overkill directed
against a relatively small number of loyal and responsible
American citizens." But there was little more he could say.[15]

It became obvious during the four-day council meeting that
the Reagan administration had no intention of seeking the
AFL-CIO's help to end the walkout. The federation was cut
off from any important channel of communication with the
administration. Although Kirkland was the titular head of
some 18 million union members, he would have to stand by

powerless as the administration broke the air controllers' union across its knee.

Like all federal employee unions covered by Civil Service laws, PATCO was prohibited from striking or engaging in real wage bargaining. Nevertheless, since its inception in 1968, PATCO had staged six slowdowns or "sickouts." But the union's past success in disrupting air traffic led its leaders into serious mistakes of tactics and judgment. In 1981 they planned to use this power to highlight the faults of the Federal Aviation Administration (FAA), an agency notorious for its militaristic, unresponsive management. The union asked Congress to establish an independent FAA with authority to bargain contracts. Despite the no-strike rule, PATCO set up a strike fund to flaunt its strength.

In June 1981, PATCO President Robert Poli accepted an FAA contract offer with pay increases more than double those received by other federal employees. However, the package fell short of PATCO's high demands, and the membership rejected it in a mail ratification vote.

The FAA refused to increase the offer and prepared to implement a strike contingency plan that had been devised by the Carter administration. It called for staffing the control towers with retirees, army personnel, trainees, and controllers who were willing to cross the picket lines. PATCO leaders dismissed the plan. They were confident that they could paralyze air traffic, a disastrously mistaken judgment that was shared by local leaders as well. At the FAA's Air Traffic Control Center in the Chicago suburb of Aurora, Local 301 President John Schmitt was quoted as saying: "It would take a minimum of ten to twelve years to replace thirteen thousand of the most highly skilled air traffic controllers in the world." Eventually, the FAA scraped together ten thousand people to replace the fifteen thousand five hundred who had operated the air traffic control system (about forty-two hundred of the original controllers kept working) and ordered airlines to cut about 25 percent of their flights.[16]

On August 5, the third day of the walkout, President Reagan made good his threat and fired all striking controllers. Neither the executive council nor unions representing other airline employees—pilots, mechanics, baggage handlers, and ticket clerks—had taken any strong actions to support the strike. The most likely leader of a sympathy strike was

William Winpisinger, president of the International Association of Machinists (IAM). A big, likable man and a capable union president, Winpisinger was an avowed socialist and tended to talk in more militant terms than most other council members. The IAM represented some fifty-five thousand machinists and other ground personnel at the airlines and, second only to the Air Line Pilots' Association (ALPA), had the occupational leverage to plunge the nation into a transportation emergency.

Winpisinger had told the executive council he would call out his airline members if the other involved unions did so. But the council held back because of two formidable problems. One was legal in nature. A work stoppage by the airline unions would be viewed as an illegal secondary boycott against employers (the airlines) who were not directly involved in the dispute. Moreover, the union presidents feared that to call out their members to support the air controllers would be to put themselves at the head of phantom columns. Only a tiny minority of airline employees regarded the PATCO battle as a "working-class struggle" that all must join. Indeed, like the general public, many machinists probably viewed Reagan's threat to fire the controllers as an overdue disciplinary action against a badly behaved child—a child who was already paid far more than most machinists, ticket clerks, and flight attendants.[17]

On the first day of the strike, reporters cornered Winpisinger and asked whether he would order his locals to walk out. "I'm not prepared to tell anybody anything," he snapped. "The locals will make their own decisions." But I saw the rage and frustration in his eyes as he wheeled and walked away.

ALPA, the conservative pilots' union, was the least likely to stage a sympathy strike. There was no love lost between pilots and air controllers who often accused one another of air-traffic foulups. During the strike, ALPA repeatedly issued statements that flying was safe, an action that reassured the public but undermined PATCO's position. In the end, labor's support of PATCO was limited to contributions to a relief fund, symbolic demonstrations, and a temporary boycott of air travel by some union officials.

When a reporter pressed Kirkland on why the federation didn't call for a "general" strike of all workers, he replied: "I am not prepared to declare, nor do I have the power to declare, or order, things to be done by great masses of working people in this country."

Kirkland went on to note that the AFL-CIO was a body of autonomous union affiliates with no power to compel its member unions to do anything. Moreover, the illegality of the secondary boycott posed the possibility of "enormous and unlimited fines" for violators. Given those circumstances, Kirkland added, only a "midnight-gin militant" would talk seriously of calling a general strike.

PATCO quickly lost public support. The administration dramatized the illegality of the walkout, based on the no-strike oath that all federal employees are required to make as a condition of employment. Meanwhile, comments by PATCO leaders created the false impression that the controllers were striking for a $10,000 salary increase.

The real strike issues involved demands for shorter work hours (a four-day, thirty-two-hour schedule) and a voice in choosing equipment to help relieve excessive job-related stress in the control towers. Controllers also wanted a liberalized pension plan, contending that "burnout" forced many controllers to retire before accumulating full pension rights. "The $10,000 demand killed us in the media," writes David Skocik, a fired controller and coauthor of a book on the strike. "Of primary importance to most was a *reduced work week* and *achievable retirement*—which meant hiring more controllers."

PATCO leaders committed other errors. They had supported Ronald Reagan for the presidency in 1980 and apparently expected him to respond in kind. They failed to consider that an illegal strike would present Reagan with an ideological issue that was ripe for conservative plucking. Herbert R. Northrup, a labor scholar at the Wharton School, concludes: "Rarely has such an amateurish performance by a union been displayed so publicly or dealt with so decisively."[18]

Although PATCO pickets continued to appear at airports into November, their strike probably had been lost by about mid-August. Unable to escape the tightening grip of injunctions and criminal and contempt actions brought by the administration and the airlines, PATCO was decertified. But the administration's victory came at a considerable cost. A congressional study group estimated that the cost of training new controllers, along with the losses suffered by the airlines, would total $12 billion. Five years after the firings, the air traffic control system still had not recovered from the loss of experienced personnel. Despite a 10 percent increase in air traffic by 1986, the FAA was operating with 13 percent fewer controllers than in 1981. Near collisions in midair rose alarmingly, from 395 in 1981 to 777 in 1985.

Moreover, a number of studies later vindicated Robert Poli's charges about working conditions under the FAA. A congressional report in 1984 said that "human relations in the FAA have not improved since the strike." The truth of this was borne out in a 1987 vote by air controllers to form a new union.

In the final accounting, 11,345 controllers were fired. A few hundred eventually won reinstatement through the courts. As of spring 1986 roughly 10,000 had found other jobs, but most would have returned to the FAA if offered reinstatement. The tragedy for them was not merely one of losing a job. They had been "traumatically cut off from the most invigorating and rewarding jobs they could ever imagine filling," writes Arthur B. Shostak.[19]

It was a matter of note that these white-collar technical workers, without much background in unionism, were able to sustain the strike as long as they did. Some labor intellectuals viewed this, with hope, as the emergence of a new working-class. "The PATCO strike's importance," wrote Stanley Aronowitz in 1983, "lies in the fact that it was the first major internationally noted strike among the new 'class.'" But the "new class" had much to learn about the uses of power.[20]

Before the 1930s, the federal and state governments had crushed many strikes and destroyed countless lives by serving as enforcement arms of the courts. Only once before, during the 1894 strike of Eugene V. Debs's American Railway Union at the Pullman Company, had the federal government broken a national union. Such an act would have been unthinkable in France, Italy, Sweden, West Germany, and the Low Countries, where organized labor was entrenched. In Britain, however, Prime Minister Margaret Thatcher had substantial public support in her 1984 stand against a violent strike by coal miners.

President Reagan had good reason to fire the air controllers and decertify their union. They violated the law and broke an oath. But the refusal of the administration, years after the strike, to grant amnesties to the controllers seemed to visit more vengeance on them than their actions called for. As Shostak reports, the controllers paid a very high price for their mistake "in terms of suicide, divorce, substance abuse, nervous breakdowns, and downward job-skidding."

Within months of the controllers' strike, evidence began to mount that the administration's decisive action was encouraging employers to take a harder line in contract negotiations. The practice of hiring strikebreakers to defeat legitimate strikes was used increasingly in the ensuing years, by companies such as Greyhound, Phelps Dodge, Continental Airlines, TWA, Danly Machine, Hormel, and the National Football League. Whether these employers took heart from Reagan's victory over PATCO is not known. But one thing was certain after the PATCO affair: Union-busting had received the official sanction of the U.S. government.[21]

A Train Back East

Although the Reagan administration dismissed Kirkland's views on the PATCO strike, I took one of his comments very much to heart. "The one thing that I want in that control tower," Kirkland had told reporters, "is people who are reasonably happy in their work, whose morale is good, who are satisfied with their working conditions." Having no confidence that this would be the case on August 5, the third day of the strike, I went, along with hundreds of other frightened travelers, to Chicago's Union Station and boarded the Broadway Limited at 7 P.M. Through the night, we jolted across the Midwest, stopping frequently in pitch darkness for no discernible reason.

Dawn broke on a misty Pittsburgh as we crossed the Allegheny River. That was when I discovered that the Broadway Limited no longer stopped in Pittsburgh but roared through as if the city were a whistle stop. I waited an hour in the dining car for a breakfast of mushy scrambled eggs, read everything I could find to read, and saw innumerable cows on innumerable hills as we lumbered across Pennsylvania. We arrived at Penn Station in New York at 4 P.M. I may have been the last journalist in the United States to spend twenty hours on a train to avoid a two-hour plane trip.

Union-Busting Precedents

The AFL-CIO's inability to protect, or help in any way, 11,300 of its own members in the PATCO strike stripped away all pretenses about organized labor's vaunted strength in the United States. Legally and economically, the unions occupy a tenuous position in American life, and always have. From the first efforts of artisans in the early nineteenth century to form trade associations, American capitalism has regarded organized labor as an intruder in its domain and has fought to destroy it when times were propitious. The coming to power of Ronald Reagan in 1981 was one such time. The creation of the United States Steel Corporation in 1901 was another.

In both situations, the general atmosphere was one of vast economic restructuring, strong antiunion sentiment, and legal restraints on labor's power. In both cases, a mortal blow dealt to one union by a powerful employer caused shuddering consequences for the entire labor movement.

Eighty years, almost to the week, before Reagan presided over the demise of PATCO, the new U.S. Steel Corporation broke the back of the Amalgamated Association of Iron, Steel & Tin Workers, which predated the USW. Before the turn of the century, the Amalgamated was strongly entrenched in the iron and steel industry, representing at its peak in 1891 some twenty-four thousand workers, or two-thirds of those eligible for membership. Its members were mostly skilled workers such as iron puddlers and steel rollers and heaters, and their semiskilled helpers. The companies could not produce raw or finished iron or steel without them. Gradually, however, the union lost its leverage. It failed to change with the times, while the steel companies became more aggressively antiunion.[22]

The precedent was set in 1892 when Henry Clay Frick sent boatloads of Pinkerton guards up the Monongahela to dislodge Amalgamated strikers from Andrew Carnegie's plant at Homestead. In the ensuing battle, the workers won their famous but short-lived victory. The strike was crushed by the state militia, and Homestead—the union's largest plant—turned nonunion.

Later in the 1890s, two growing trends undermined the Amalgamated's position. One was steel's increasing displacement of iron as America's basic metal, thus making obsolete the skills of the iron foundrymen who formed the core of the Amalgamated. At the same time, steelmakers were introducing new machinery which could be operated by relatively unskilled laborers, with some training (and a very high accident rate). Despite this movement away from skilled labor, the union—like most other AFL unions—remained elitist. It refused to admit the growing numbers of immigrant laborers in the mills.

The Amalgamated also failed to adapt its policies to a great wave of corporate mergers starting in the late 1890s. Up to then, manufacturing had been characterized by cutthroat competition between large numbers of relatively small companies. To reduce competition, corner the market, and stabilize profits, companies began a frenzied merging which eventually produced huge corporate combinations. By 1904 some 318 industrial firms controlled 5,288 separate plants. In the steel industry, the merger trend produced larger and larger combines, raising the threat of massive price-cutting wars which would reduce profits and endanger the investments of J. P. Morgan and other Wall Street financiers.

To prevent their elaborate financial structure from toppling, Morgan and other money men put together the largest of all mergers and formed a "steel trust" which was incorporated on February 1, 1901, as the United States Steel Corporation. It was a holding company of subsidiaries, including Carnegie's firm (Carnegie retired to devote the rest of his life to philanthropy), two other basic steel producers, five leading fabricators, and a host of smaller companies. This brought 60 percent to 70 percent of the steel industry under one corporate roof.

Almost immediately, "the Steel Corporation," as it was known for years, adopted an antiunion policy. Some of the subsidiary presidents who sat on the executive committee were violently antilabor. "I have always had one rule," said one of the presidents at an early meeting. "If a workman sticks up his head, hit it." On June 17, 1901, the committee ordered the subsidiary companies to oppose "any extension of unions in mills where they do not now exist."[23]

The purpose of this policy was to squeeze the union out of the plants it already held; the consolidation enabled the new company to switch work from the union to the nonunion plants. The Amalgamated demanded that U.S. Steel pay the union scale at all plants, and when the corporation refused, the union went on strike at some of the subsidiaries. Union leaders rejected one compromise offer that would have installed the union in eighteen of twenty-three mills. This was a grave mistake, for U.S. Steel hardened its demands, forcing the union to expand the strike in August 1901 to all U.S. Steel plants.

In addition to this tactical blunder, according to David

Brody, a labor historian, the union made "a final miscalcula-tion" that (in a remarkable instance of historical parallelism) would be repeated eighty years later by PATCO: The Amalga-mated "assumed the indispensability of its experienced men." Unskilled laborers, refused admittance to the union, had re-fused to strike at some plants. With these workers and im-ported strikebreakers, the company managed to open plants and produce steel. The Amalgamated appealed to Samuel Gompers, president of the AFL, for help, urging him to make the steel walkout "the central fight for unionism" in the United States. Gompers, however, refused to get other unions involved. Mounting sympathy strikes was an extremely risky business in the early 1900s, a time when courts routinely issued antistrike injunctions that were enforced by the police power of local and state governments.[24]

In September, with the strike breaking up, the union caved in and accepted a settlement that was worse than the com-pany offer of ten days before. The Amalgamated had to give up fourteen formerly unionized plants. The union limped along until 1909, when U.S. Steel declared that it would no longer recognize the union at any of its plants. A hopeless strike followed, and the union was ousted from U.S. Steel.

Although the Amalgamated remained in existence for an-other thirty years, it was little more than a name in the offi-cial files of the AFL. Astonishingly, its superannuated officers continued to deny union membership to unskilled, immi-grant, and black workers, who thereupon helped the steel industry break a national steel strike in 1919. At its founding convention in 1942, the USW mercifully ended the Amalga-mated's interminable death throes by declaring it part of the new union.

U.S. Steel's victory over the Amalgamated in 1901, wrote the noted labor historians Selig Perlman and Philip Taft in 1935, "created an antiunion pattern of conduct amongst the large industrial interests." They added: "If unionism in America has never been taken as a matter of course in the big industries, in contrast with England and the industrial nations of the Conti-nent, not the least of the causes was the antiunion attitude set by the United States Steel Corporation."[25]

It is reasonable to ask whether the death of the Amalga-mated at U.S. Steel in 1909 really mattered to anybody except the union officers whose income depended on members' dues. One way of approaching this is to consider what the union might have accomplished had it remained. The assumption is that it would have addressed labor's historic concerns, wages, working conditions, and hours of work. For at least a decade after the Amalgamated was kicked out of U.S. Steel, steel-workers made little progress in these areas.

For example, the two-shift, twelve-hour day was the norm in the steel industry for many years. In 1912, 50 percent to 60 percent of U.S. Steel's employees worked six days a week, or seventy-two hours; some workers labored seven days, or eighty-four hours a week. Although efforts to reduce the work week began at U.S. Steel in 1907, the corporation fended off all critics of the system and all campaigns for shorter hours until 1923. Finally, embarrassed publicly by Secretary of Commerce Herbert C. Hoover, U.S. Steel gave in and adopted the eight-hour day. During the sixteen years of delay, the long work week undoubtedly cut short the lives of an incalculable number of steelworkers.

In addition, steel wages remained unconscionably low, resulting in depressed living standards. Even a former vice-president of U.S. Steel wrote of his "disgust at the squalid living conditions" in the mill towns of the Monongahela, Allegheny, and Ohio valleys. Whose fault was this, he asked rhetorically, and answered: "Who maintained working conditions which tended to brutalize the body and soul? Answer—Carnegie Steel Company and U.S. Steel Corporation."[26]

Would the Amalgamated have made any difference? One must assume that it would have tried.

The Vulnerability of Unions in America

The purpose of the preceding section is not to suggest that labor strife at U.S. Steel in the 1980s flowed directly from the corporation's antiunion policy of 1901. What I mean to demonstrate is that American history is replete with examples of union insecurity in the corporate world. The breaking of the Amalgamated at the beginning of the century and of PATCO near the end demonstrate that management antipathy to giving workers an individual or representative voice in the workplace survived more than eighty years, despite growing evidence that this was the wrong way to run a company, or an economy.

American union leaders may understand better than anybody that their position is precarious. Yet after World War II they did little to carve a secure niche for themselves in the American economic system. They failed to make themselves indispensable to workers—and managers—in leading the way toward work reforms that would increase productivity, raise the level of "democracy" in the plants, and keep the companies competitive.

From the mid-1970s to the mid-1980s, the unions rediscovered that they are highly vulnerable to economic slumps and corporate campaigns to achieve a union-free environment, and that they could no longer count on the law to protect them from their enemies. Having digested this unpleasantness, the labor movement cast about for another solution to their growing dilemma. One remedy, in a democracy, is what the unions call "political action," and they gave it their best effort in 1984. For the first time, the AFL-CIO endorsed a presidential candidate before he was nominated by a party convention. The experiment turned out badly, both for Democrat Walter F. Mondale (although he could not have beaten Reagan in any case) and organized labor. More than ever, the unions were accused of being a "special interest" group outside the main current of American politics. The AFL-CIO strategy, perversely, demonstrated that organized labor could not deliver on its claims of voter power.

And so, by the early 1980s, American unions were in deep trouble. They could no longer work the political magic that had made them a powerful lobby as recently as the 1970s. They rated low in public esteem. Legal embroidery was hemming them into a small corner of the economic quilt. They were ignored by a federal government that once trembled when coal miners, auto workers, or steelworkers shut down their industries. The protections once afforded unions and their industries by domestic regulatory laws and American dominance in the world economy were vanishing. This is the context in which my story takes place.

Of all the setbacks suffered by organized labor in the 1980s, none was more painful than the forced retreat from high wage levels that had been won in a more prosperous time. Collective bargaining was the one activity that labor had emphasized above all others for most of its history. Now, its power in that area would be challenged on a massive scale.

This was nowhere more true than in Pittsburgh, home of the large and powerful United Steelworkers, a union that had always prided itself for its ability to negotiate "top dollar" wages. Pittsburgh also was the home of U.S. Steel, the titular head of the nation's steel industry, which had the most authoritarian of industrial managements. It soon would be the scene of a clash between these old-style titans, pointing up all that was wrong with traditional industrial relations in the United States.

Chapter 3

The Life and Style of Lloyd McBride

An old black and white photograph stood on a shelf in a conference room adjoining Lloyd McBride's office in Pittsburgh. It showed about seventy men lined up in neat rows outside the Foster Brothers Manufacturing Company in St. Louis. They were mostly young and confident-looking and, except for the coarse shirts, could have been mistaken for a graduating class at a small college. In reality, this was a snapshot of the American working class, Depression Era, 1940. Despite their obliging smiles for a photographer, the future held little promise for these men as long as the Depression lasted. There was, however, one cause for optimism: The union had come to Foster Brothers. Standing in the middle of the front row with folded arms, the sleeves of his denim shirt rolled up above the elbows, was the president of Lodge 1295 of the Steel Workers Organizing Committee (SWOC), a husky, smiling Lloyd McBride at the age of twenty-four. He projected a sense of directness, competence, and responsibility. This was the quintessential McBride.

He had become the family breadwinner at fourteen, dropping out of school to work at Foster Brothers for 25¢ an hour when the firm laid off his ill father. In 1936 SWOC organizers came to St. Louis, and McBride helped sign up his fellow workers in the bedspring manufacturing shop. When the first president of the local failed to stand up to management, McBride led a sitdown strike and took command of the union. Stepping up to a challenge, assuming responsibilities that others shirked, taking hard knocks for others—it was this character trait that, paradoxically, would make political enemies for McBride when he served as USW president. It

52

drove him in 1982 into a political quicksand of concessionary bargaining and, perhaps, endangered his health.

McBride was one of a legion of young, able men and women who were tossed up by the Depression to become local leaders in the burgeoning ranks of the new industrial unions. It was the chance of a lifetime for McBride, who had little education and no special skills beyond operating a punchpress. Yet he didn't fit the popular image of the young firebrand unionists of the 1930s who rose up and dealt capitalism a stunning blow. For example, Walter Reuther, a well-educated tool and die maker who had been steeped in socialism and trade unionism by family background, actually struck fear into the old-time industrial leaders with his fiery speeches and challenging ideas. Although he later dropped socialism as a political doctrine, Reuther never abandoned a personal vision of an ideal society with centralized economic planning and world government. McBride had a more limited vision. Unions existed to protect workers from ill treatment, to bargain a fair share of the employer's profits, and to influence the political process in a positive way for union members, consumers, and the poor. McBride had no ambition to lead the labor movement in pursuit of an ideological Utopia.

In 1940 McBride left the plant and joined SWOC as a staff man in St. Louis. His rise in the union was less than meteoric. When SWOC became the USW in 1942, the union was divided geographically into some thirty-five districts (the number has declined over the years as districts were merged), each headed by a director who also was a member of the union's executive board. In 1981 the board had thirty members, including twenty-four district directors, five top officers, and the national director in Canada. McBride worked on the staff of District 34, covering Missouri, southern Illinois, and neighboring states. The USW's structure, borrowed from the UMW, was based on a feudal distribution of power. The top-level tier of leaders included the president and two to four other officers (also changing over time). The president decided important policy matters and spoke for the union on national and international issues. He ruled the kingdom, while the directors ruled the regional duchies. The latter were allowed considerable autonomy, as long as they paid fealty to the president.

Local union offices were filled through vigorously contested elections. But one couldn't advance to district director without the blessing of the USW's "official family," which consisted of the current directors and top officers on the International's executive board. Anyone who ran for director without being anointed was labeled a "rebel" or "dissident" and almost invariably defeated (until the 1960s). To become a director, a unionist had to get on the district staff, demonstrate an allegiance to the establishment, and wait patiently until the incumbent director quit, retired, or died.[1]

McBride waited for nineteen years, carrying out the duties of a field staff representative (organizing, negotiating contracts, and handling arbitration cases). Far from becoming impatient, he was grateful to the union. "It took me from a job that was dull and routine and put me into an exciting job," he once said. When District 34's first director retired in 1965, McBride was elected to succeed him and served twelve years in the post.

I met McBride in the late 1960s and was baffled by him. He went about quietly, in a businesslike manner and, in fact, looked like a moderately successful businessman, the owner of a small machine shop, say. He had a high forehead and a round slightly fleshy face that seemed to look upon life with an alert skepticism. Unlike most ambitious unionists with thoughts of high office, McBride made no effort to woo the press. I never saw him make a strong or flamboyant speech at a USW convention for or against anything. His speeches were practically devoid of polemics and rhetoric, and he seemed constitutionally unable to talk in terms that even hinted at a class struggle.

During these years, McBride (called "Mac" by friends and staffers) became known as one of the more competent directors in the union, a man of impeccable integrity. He worked hard and spent most of his free time with his family. He had converted to Catholicism when he married his wife Delores and remained deeply religious. Indeed, when he became president, a number of staffers noted a "religiosity" in the way he pursued some goals (wage concessions eventually became one of them). His hard early life may have contributed to his tendency to take moral offense at unionists who made excessive money demands. "I suspect he really thought that everybody was paid too much—steelworkers, staff members, union officers, everybody," said one staff man. McBride felt the same way about corporate executives. "When you leave employers to their own devices," he told me in 1983, summing up his feelings about the continuing need for unions, "greed quickly comes on the scene, and they take more than their share. Many of them, left to their own devices, will really—as they have in other years—organize the workers for the union movement."[2]

McBride's long years in District 34 did not prepare him well politically or administratively for the problems he would face as International president. Although it had a few steel plants, the majority of the district's twenty-eight thousand to thirty-four thousand members worked in small fabricating shops, foundries, and lead and iron ore mines. McBride negotiated contracts at the small Granite City Steel Company and served as secretary of the union negotiating committee at Armco. This experience gave him little background for the industrywide steel negotiations that were so important to the USW. But it did influence him in developing a management style that would make him vulnerable as USW president. "Lloyd is a very tough-minded person," noted Bernard Kleiman, the USW's general counsel, in 1983. "He worked as a staff representative for many years. The really good staff representatives are very lonely people. They do it themselves, or not at all."

District 34 commanded less interest at the International headquarters in Pittsburgh than the dominant steelmaking districts. Operating far from Pittsburgh's prying eye, McBride developed an inclination to do things himself rather than ask for staff help from headquarters or delegate authority to his own staffers. Subordinates gathered information for McBride to act on. "I learned very early in my union career that you never went to Mac without having all the i's dotted and the t's crossed," said George Becker, a USW vice-president who worked under McBride in St. Louis. "He had an uncanny ability to sift out all the arguments and emotions and get right down to exactly what the issue was. I've seen him take letters that had been written by lawyers and correct spelling and punctuation, much to everybody's embarrassment." In staff meetings, McBride often expressed his displeasure with a technician's opinion or report in insulting terms.

McBride's district also was remote from the union politics that kept the larger steelmaking centers in turmoil. Districts 15 (Monongahela Valley), 31 (South Chicago, Gary, and northern Indiana), 20 (Aliquippa and the upper Ohio Valley), 26 (Youngstown), 7 (Philadelphia), and 8 (Baltimore) were always churning with internecine battles for control of local and district offices. Starting in the 1950s, rank-and-file campaigns to overthrow the top union leadership also drew most strength from these regions. This dissent was healthy, but it also produced demagoguery and intrigue. District 34, by comparison, was a placid gulf, unswept by political tides. McBride always ran unopposed for office. As a consequence, he was ill-prepared for the hothouse political atmosphere that he encountered as president.

Indeed, McBride's election in 1977 to the top post resulted in part from deep political divisions in the big steel districts. When McBride announced his candidacy in late 1975, there was no obvious heir apparent to I. W. Abel, the USW's third president, who had served since 1965 and was due to retire in 1977. One likely candidate was Edward Sadlowski, the thirty-six-year-old chief of District 31 who had created a sensation in the union by bucking the "official family" to rise to director. He had criticized Abel publicly and fought bitterly with the president in executive board meetings. Abel hated the younger man and worried because leftist factions formed one source of Sadlowski's support, although Sadlowski also could be expected to draw a large vote from younger, independent unionists and from among the perennially disaffected steelworkers in basic steel plants.

But Abel lacked enthusiasm for other prospective candidates, including Vice-President John Johns and Joe Odorich, then the head of District 15. Abel contended that he would have no part in picking a successor. But top staffers and directors close to Abel began to push McBride on the assumption that the president preferred him. Eventually, most of the union establishment rallied to defeat Sadlowski by electing McBride. Curiously, McBride seems to have had some misgivings at this point because of health problems and other concerns. According to Buddy Davis, a close friend and union colleague, McBride had a strong sense that if he became president, he wouldn't have a "normal retirement," a euphemism for dying in office. But the "establishment man" in McBride triumphed over these worries. After a long, rancorous campaign, in which the establishment did all it could to crush Sadlowski, McBride won by a vote of 328,861 to 249,281 and took office on June 1, 1977.[3]

A Threatening Trend

From his office on the twelfth floor of the Steelworkers' building in Pittsburgh, Lloyd McBride could take in one of the most stunning scenes in industrial America. His desk faced two floor-to-ceiling windows on the western side of the building. He had only to raise his eyes to see, spread across his view, what used to be known as "the place where the West begins"—where the Monongahela and Allegheny rivers flow together and form the Ohio. That the president of an American union in the early 1980s could routinely take in this view, from a skyscraper owned by the union, showed how far the USW had advanced since its founding. Almost fifty years earlier, Philip Murray had launched the SWOC organizing campaign from a rented office in Pittsburgh's Grant Building. The union soon moved and for more than thirty years occupied several floors in the old Commonwealth Building on Fourth Street, hemmed in by other tall structures. The offices were small and shabby, but many union staffers rather liked the idea of working in a building that had ancient, cage-type elevators operated by old men in faded brown uniforms.

In 1973 the USW's fortunes changed (or seemed to change) dramatically. President I. W. Abel signed an agreement with the steel industry which *guaranteed* steelworkers a 3 percent annual wage increase, plus cost-of-living adjustments, in return for pledging not to strike the industry. The union no longer had to put its economic strength on the line; the gains would come automatically. It was like having a lifetime contract, and the future seemed secure. Later that year, the USW purchased a thirteen-story office building on the Bouvelard of the Allies in downtown Pittsburgh. The timing of the two events, while coincidental, was symbolic.

The new building had automatic elevators, and an odd exterior. Criss-crossing steel beams formed diamond-shaped windows from top to bottom, resembling a diagram of interlacing genes in a chromosome. One couldn't look at it too long without blinking away the diamond patterns. It was in Gateway Center, where office buildings stood around a pedestrian plaza and tree-shaded walkways led into the neat lawns of Point State Park.

From his window, McBride had an unobstructed view of the park and its centerpiece, a fountain which in the summertime spouted a fifty-foot geyser of water. On the left, a steep bluff known as Mt. Washington towered five hundred feet over the south shore of the Monongahela, its brow crowned by old frame houses and high-rise apartment buildings. Below the ridge, McBride could see the gold-painted superstructure of the Fort Pitt Bridge curving across the Monongahela. On the far side of the Allegheny, Three Rivers Stadium, a large, wart-shaped structure, sat at the corner of the three rivers, a fat paperweight preventing Northside Pittsburgh from floating away.

And in the middle, dominating the view, flowing northwest out of Pittsburgh like a great, broad highway, was the Ohio River. Farther on, it would loop back on itself and begin the long southwesterly descent to the Mississippi. Another golden bridge crossed the Ohio before it wound out of sight. There was often a towboat on the river, pushing a string of coal barges. On the south side of the Ohio, the Mt. Washington escarpment folded into forested hills that rolled down the valley. Except for periods of inversion, the valley air was clear of the smog and dust that once had given Pittsburgh its bad name. But a solitary puff of steam often hung over the farthest ridge, signaling the site of Jones & Laughlin's huge steel plant twenty-six miles down the Ohio at Aliquippa.

It was a scene of immense power. A century ago, iron and steel from the mills of the three Pittsburgh valleys had flowed down the Ohio into the sprawling domain of smokestacks and factories that came to be known as the industrial Midwest. There it was used to build the railroads, bridges, factories, and office buildings that formed the backbone of mid-America.

In 1981, however, an economic trend was flowing in the opposite direction, back up the Ohio. The business slump that began in the summer of that year was growing worse, forcing increasing numbers of small companies to seek wage cuts. It was a phenomenon that hadn't occurred on such a widespread basis since the early Depression. Then, it had been called "wage-cutting," implying the unilateral nature of the act when it is performed by managements that do not have to contend with a union. In the 1980s the trend picked up various names, none of them entirely satisfactory and all of them implying that the workers were to blame for the recession and the loss of industry's competitive ability. Thus, the press and television began talking about "give-backs" and "wage concessions" that union members were granting to employers in a process that became known as "concession bargaining." If workers refused to "give" voluntarily, employers either threatened to, or did, shut their plants to force wage reductions. In these cases, employers were seen as winning "take-backs" or "take-aways." The old, familiar term in labor circles, "sell out," was dredged up by the more disaffected union members to describe the behavior of leaders who negotiated "give-backs."

Whatever it was called, the trend had begun in late 1980 among small manufacturing firms in the industrial Midwest. High interest rates made it impossible for them to borrow money to buy needed equipment and perform maintenance work, or to pay the interest on previous loans. Union-represented workers in these companies agreed to accept pay cuts or freezes, reductions in benefits such as vacations and holidays, and changes in work practices and customs, all in an effort to cut labor costs, keep their employers solvent, and retain their jobs. The trend received little national attention except in the case of Chrysler. The UAW first acceded to small concessions at the third largest car manufacturer in December 1979. Eventually, the auto union swallowed two further rounds of cuts to keep Chrysler afloat; the final one in early 1981 was mandated by the federal government as part of its agreement to issue $1.3 billion in loan guarantees to Chrysler.

Encouraged by the UAW's acquiescence at Chrysler, other companies, particularly firms that supplied equipment and parts to the auto industry, began to ask for wage cuts. Some UAW regional offices were inundated with such requests. In January 1981, UAW President Douglas Fraser sent a memo to his regional directors, warning that some employers would try to test the union's resolve. The directors should respond to these requests by pointing out that the government had forced the UAW to cut labor costs at Chrysler as the price of keeping it in business, Fraser wrote, and the union should renegotiate contracts only if a company were willing to open its books and prove that it was in financial trouble. Increasing numbers of companies did just that.[4]

The trend began broadening in the manufacturing sector as a slump in consumer spending put a further squeeze on cash flow. Dissatisfied with the quality and price of small domestic cars, the American public was turning in droves to Japanese Toyotas, Datsuns, and Hondas. As the dollar rose in exchange value against foreign currencies, other manufactured goods poured into the United States, displacing domestically produced steel, agricultural implements, machine tools, and many other products. By July 1981, the United States was in a recession. At about the same time, President Reagan's firing of the PATCO strikers set a hard-line labor policy that many companies began emulating. Even healthy employers began demanding give-backs.

Concessions were now being negotiated by unions in the airline and trucking industries, both of which had recently been deregulated by Congress. In both industries, new companies sprang to life and, with the national unemployment rate approaching 10 percent, they had little trouble hiring workers at well under the union rates paid by the established, unionized firms. Companies like People Express and New York Air began taking business away from the major carriers by offering low fares. Pan American, American, and United, among the big airlines, managed to persuade their unions to accept pay freezes and change costly work rules.

The phenomenon of wage concession was hardly a new one; since the founding of organized labor, many unions, in many places, had accepted a reduction in compensation to save a failing firm. However, this latest episode was not confined to small, isolated companies whose coming or going would have little effect on general wage movements. By early 1982, powerful unions in trucking, airlines, retail food, rubber, automobiles, agricultural implements, meatpacking, and metals firms on the fringes of the basic steel industry had negotiated some form of wage deceleration. As it spread through these major industries, concession bargaining (for want of a better term) undermined industrywide wage patterns, severed wage linkages between individual firms, and eroded the concept of "wage comparability." This term refers to the tendency of unions in related businesses, though sometimes in unrelated industries, to base their contract demands on terms won elsewhere. It was the blue-collar version of keeping up with the Joneses. The concept had served during most of the post–World War II era as a major determinant of wage-setting and, carried to its extreme, linked wage rates between industries that had no economic linkage.

To the degree that this "pattern bargaining" produced wage levels which did not reflect the health of a company, moving away from the practice was good for the economy. For unions, however, a general retreat from past gains in collective bargaining would damage them severely in the eyes of their members. They could not afford to fall senseless under the heel of the corporation. For this reason, many of the manufacturing and airline unions demanded, and won, trade-offs for wage revisions that could, properly nurtured and built upon, change power relationships in industry.

In many concession cases, for example, the companies had to provide financial data to prove the extent of their economic misery. Wage concessions often were accompanied by an increased worker voice in management, ranging from representatives on the board of directors (at Chrysler and Pan Am) to rank-and-file participation on shop-level committees with a role in increasing efficiency and improving the work environment and product quality. In many cases, it was management that actually demanded the shop-floor committees. Some unions also gained strengthened job security provisions. But as important as these changes were, they could not gainsay the fact that unions and their members were taking a hard pounding.

In late 1981 and early 1982, concession bargaining came marching out of the Midwest toward Pittsburgh. Growing numbers of small employers in steel-related businesses were experiencing significant financial problems and asking for concessions. The volume of requests suggested a spreading weakness in the economy that would undermine the entire steel industry, forcing even the major companies to seek wage reductions.

McBride may have appreciated the breadth of the problem, but he made no attempt to formulate a comprehensive policy for dealing with it. This was understandable at the beginning. During its half-century of existence, the union had been engaged in negotiating the largest possible wage increases, always moving forward, never backward. The possibility of a major retreat was too painful to consider. Moreover, the union seemed impregnable to a corporate attack. In terms of size, prestige, influence, and internal solidarity, it still appeared as strong as the day I. W. Abel led the move to the new building in Gateway Center.

The USW's total dues-paying membership at the end of 1981 was down slightly from the 1,042,730 of June. This was to be expected because of increasing layoffs in industry (laid-off members were exempted from paying dues). The number of members also had decreased from close to 1.3 million in the peak year of 1975. But the USW was still the fifth largest union in the country, exceeded in size only by the Teamsters (1.9 million), the National Education Association (1.5 million), the Auto Workers (1.3 million), and the Food & Commercial Workers (1.2 million). If measured by legislative and bargaining strength and general influence in the labor movement, only the UAW rivaled the USW.

Contrary to the union's name, Steelworker members were spread through the economy. In addition to about 313,000 working members in the basic steel industry, the USW also represented thousands of employees in the aluminum, nonferrous metals, and container manufacturing industries. Its members mined iron ore in Minnesota, Michigan, and Missouri, and copper, lead, zinc, uranium, silver, and gold in the West; they served as seamen on Great Lakes ore boats and built ocean-going ships at Newport News. The USW had become the largest union in the chemical industry and counted over 200,000 members in hundreds of fabricating concerns which produced pumps, valves, forgings, and a large variety of metal products. As the result of mergers and organizing efforts, Steelworker members also were employed in businesses as diverse as supermarkets, limestone quarries, municipal government offices, furniture-making, and construction. The members belonged to 5,300 locals. To provide services to the locals, the International employed nearly 900 staff representatives, technicians, and secretaries in its twenty-four geographical districts in the United States and Canada.[5]

At the center of this empire, a staff of about two hundred persons worked in the International headquarters. The building itself, girded by the diamond-shaped steel shell, seemed impervious to intrusions from the outside world. It was not difficult for rank-and-file critics to imagine (or contend, as they did) that McBride and the other officers were isolated in their tower offices from work-a-day concerns in the shops.

In 1981 the five International officers—McBride; two vice-presidents, Joseph Odorich and Leon Lynch; the treasurer, Frank S. McKee; and the secretary, Lynn R. Williams— occupied the twelfth floor, along with various aides and secretaries. Their quarters could not properly be described as plush, but each office was expensively furnished with walnut paneling, large wooden desks, and enough chairs and couches to hold a meeting of a small local.

Each officer had a suite of interconnecting offices that could be barred from the officer in the adjoining suite. It was a "power" arrangement abetting the old idea that election to union office carried with it the right to build a little political empire. In this setting, days, weeks, or even months could go by without the officers seeing one another. Presumably, each officer could shut himself in his quarters, surrounded by friendly aides and secretaries, and scheme the downfall of his fellow officers. Something very much like this happened, in fact, after McBride died in late 1983. Three of his colleagues— Odorcich, McKee, and Williams—began scrambling to succeed him. They met occasionally, and by chance, on the elevator. Williams eventually won out.

At the end of 1981, however, there was no such divisiveness. The five officers were still unified as the "McBride Team," now serving its second term. The union was strong and outwardly healthy. But the appearance was deceiving.

Concession Bargaining Comes to Steel

It came in a sudden rush. Starting in late 1981, McBride and his aides began receiving phone calls from field representatives who reported requests from small employers for wage relief. The companies were mainly metal fabricators, but they included a few small steel producers. For instance, producers of stainless steel and other high alloy steels, faced with an 18 percent falloff in orders for their products, had laid off 25 percent of their twenty-six thousand workers.

For the entire steel industry, including the plants that produced "carbon" steel, unemployment was about 21 percent at the beginning of 1982. (Carbon steels were used in the manufacture of cars, appliances, and other high-volume products.) There was little demand for steel, and foreign producers were taking more than 20 percent of the puny market that existed. Indeed, McLouth Steel of Detroit, the nation's eleventh largest steelmaker, had filed for protection from its creditors under chapter 11 of the federal bankruptcy code in December 1981.

The question immediately arose, what should the union negotiators look for in determining whether the company needed help? James Smith, head of the USW's research department, counselled some of the negotiators by phone. At his request, McBride on January 12, 1982, wrote a letter to all district directors advising them that Smith's research department could help in evaluating renegotiation proposals. By late January the research department was analyzing data from five companies that claimed to be in trouble, including the Crucible Stainless & Alloy Division of Colt Industries (an important producer of stainless steel), Wheeling-Pittsburgh Steel, and Penn-Dixie Steel of Kokomo, Indiana. McLouth had already obtained a wage cut.[6]

A tall, slow-speaking Texan with a shock of white hair, Smith had been a boilermaker and a local union president before joining the USW staff in 1953. In 1970, while on leave from the union, he graduated from the University of Texas at the age of forty-five with an economics degree. He had a quick, penetrating mind and a good knowledge of economics. An intensely political man, Smith involved himself in practically every facet of union affairs. He had a reputation for hatching schemes, in furtherance of union goals, that sometimes went awry.

This was not the case when Smith suggested in early 1982 that the union develop a policy for dealing with the concession requests. He discussed the problem with McBride and Bernard Kleiman, the union's chief counsel, and formulated an approach consisting of three elements:

1. Standards for determining whether a company really needed financial help, and how much. The International notified the districts that they should not enter negotiations unless the company opened its books. The research department would analyze sales volume, selling prices, production costs, debt, earnings (or losses), and the company's projections for the future. On the basis of this data, Smith and his staff would determine whether bankruptcy was likely if the company didn't receive relief on labor costs. A research technician would take the final report to the affected district and local, explain it and, frequently, help in negotiations.

2. A procedure for membership ratification of the terms of a concession agreement. Most of the local unions involved were considered part of the steel industry, broadly construed, and belonged to the USW's Basic Steel Industry Conference (BSIC). In the past, the BSIC ratified only the industrywide steel agreement negotiated with the industry leaders. The terms of this master agreement would be applied (more or less intact) to many scores of "me-too" firms without membership approval. Smith, McBride, and Kleiman decided that a concession agreement should be ratified by the members at that company.

3. A "system of trade-offs" that the union would demand in return for cutting wages and benefits. Each contract was to

have provisions enabling workers to recoup the money they gave up when, and if, the company became healthy. In many cases, the mechanism used was stock ownership.

These informal guidelines were an ad hoc response to a growing problem, not a formal statement developed through comprehensive study and discussion by the policymaking executive board. For the most part, it described *how* to negotiate wage reductions and did not address the issue of *whether* the union should grant concessions even at troubled firms. Still, the new procedures in themselves were important. Smith thought they should be publicized to indicate to members and companies under USW contracts that the union had adopted a coherent approach to a potentially chaotic situation. But McBride rejected the idea on grounds that publicity would only increase the flood of companies seeking relief.[7]

On January 8, 1982, McLouth's Steelworkers local agreed, under the stern eye of a bankruptcy judge, to take a 79¢ per hour wage cut, suspend a cost-of-living allowance (COLA) provision, and give up some benefits. The agreement reduced hourly labor costs to $17, about $6 below that paid by the eight major firms headed by U.S. Steel. This gave McLouth a competitive advantage over U.S. Steel and other companies that produced flat-rolled sheet steel for the auto market. In another chapter 11 proceeding, Penn-Dixie Steel (later renamed Continental Steel) on January 29 won concessions from its USW local that would reduce labor costs by $2 per hour. In this case, the company agreed to remit the cost-savings to workers after it returned to profitability and before it paid dividends to stockholders.

Two USW districts had now granted large, dissimilar wage reductions that undercut the master agreement. Although the McLouth and Penn-Dixie decisions had been directed by bankruptcy courts, they nonetheless set a precedent that other districts would follow in the months ahead. The union had, in effect, given the signal that each district could make its own decision on whether to cut wages (after complying with the open-the-books procedure). The International made no attempt to minimize the competitive effect on other firms and the USW members they employed. As late as March 31, McBride told me: "We don't have a policy on it. We're reacting to the problems that come to our attention." In other words, the USW president acted on instinct rather than on the basis of a strategy. "He looked upon steel negotiations as something that you run 'by the seat of your pants,'" USW economist Edmund Ayoub later told me. "Those were his words to me, which appalled me. He expected team-playing among the rest of us, but he went off on his own toot."

Within a few months, the USW found itself dealing with many such situations, including large companies like Colt and Wheeling-Pittsburgh Steel. In April, the USW agreed to reductions that cut Wheeling-Pittsburgh's labor costs by $1 per hour. But the union balked at Colt's demands, and the company sold its Midland, Pennsylvania, plant to Jones & Laughlin.

Subsequently, in 1982, the USW reopened contracts and negotiated concessions at a number of smaller, yet fairly substantial me-too steel firms, including Guteryl, Laclede, Roblin, Colorado Fuel & Iron, Connors, and Northwestern. Many other, tiny companies also asked for relief and submitted financial data. Jim Smith had six people, including himself, analyzing these "poverty cases" almost on a full-time basis. Over a period of twelve months, four hundred such cases poured in, and four years later the data and analyses were still stacked in cardboard boxes in Smith's office.

The signs of market chaos were everywhere in early 1982. "The company people we see are terribly frustrated, frightened by the economic events they're caught up in," Smith said in May. "As soon as the suppliers and customers discover that a company has been in to see the union about concessions, it sends shock waves through the markets. Competitors jump in and offer lower prices, warning customers that their regular supplier is near death. What's happening in the steel business can only be compared to a whole big school of sharks in the ocean that's running red with blood.[8]

If the union continued to react on an ad hoc basis to each request for wage cuts, a proliferation of wage rates would replace wage uniformity in the basic steel industry. From the USW's earliest days, one of its primary goals was obtaining common wage rates for each of hundreds of different mill jobs across the entire industry. It was both a moral and a political necessity to "take wages out of competition." Moral, the union believed, in that workers should not be treated like other raw-material commodities that went into the cost of making steel. Political, in that the union should not be perceived by its members as bargaining harder for one group of workers than for any other.

For the most part, the Steelworkers had achieved this goal. Pay levels were similar, if not identical, throughout the unionized segment of the industry. The few non-USW plants, such as National Steel's Weirton Works, historically paid higher wages to keep the USW out. Pension, vacation, insurance, and other benefit plans also were nearly identical, although the cost of providing the benefits varied substantially. This was because average age and length of service of the work population differed from plant to plant. Generally, however, total employment costs ranged from $20 to $25 per hour. Granting each troubled company a different wage-and-benefit reduction based on individual company circumstances would throw wages back into competition. The firm with the lowest labor rate theoretically would be able to reduce prices and take customers away from competitors.

Once wage-cutting got under way, it would pick up speed. High-rate firms would appeal to their union employees to cut back to the level of competitors, and so forth. In the absence of a unionwide policy, such wage fighting was especially likely in the Steelworkers, because of its division into autonomous districts. Indeed, before the end of 1982, the USW had set a pattern of departure from the uniform wage policy that caused repercussions for years in the steel industry. It seriously eroded the union's position as the industry's enforcer of a common labor rate, and it was a factor in the long work stoppage against U.S. Steel (USX) in 1986. In December 1983 the USW's executive board adopted a policy preventing districts from undercutting one another, but it was not entirely successful.

Within the union, two arguments were advanced against case-by-case concession bargaining. The first, put forward by some of the more militant local unionists, was quite simply that concessions, however badly needed by an individual firm, weakened the union and the labor movement. In addition to this ideological objection, militant leftists contended that granting concessions at a time of low demand was no guarantee against ultimate company failure.

A purely economic argument against wage-cutting at the me-too firms was made by the USW's chief economist, Edmund Ayoub. In 1982, he contended, U.S. mills could produce annually about 15 million more tons of steel than could possibly be consumed, given the level of demand. The excess capacity had to be taken off the books one way or another, either by the weaker companies going belly up or the stronger companies closing plants, he said. The McLouths and Penn-Dixies had got into trouble earliest because they were the most inefficient companies. Ayoub put his argument in these words at a USW staff meeting in November 1982: "What the union is doing when it gives a concession to a less efficient company is giving that company a competitive advantage over an efficient company. We have no business doing that. We shouldn't penalize companies for being efficient. We have no business doing what the marketplace should be doing. It's not for the union to decide who survives and who doesn't survive."

How would Ayoub have handled this problem? By early 1982, he later said, it was apparent that the entire steel industry eventually would need wage relief. By granting wage reductions on a company-by-company basis, the Steelworkers created a situation in which the companies could whipsaw the union, by asking for ever greater cuts. What the union should have done was develop a policy for the entire industry, including the major integrated firms as well as the me-too producers, and freeze or cut wages uniformly for all companies. "We took the entire industry up the wage scale with a uniform policy, and we should have taken it down the same way."

Ayoub, however, was not involved in formulating a response to the renegotiation requests of early 1982, partly because McBride compartmentalized union staff functions and kept decision making to himself. When he needed advice on negotiating matters, McBride talked to Bernie Kleiman, an experienced bargainer and legal counselor but not an economist. McBride relied on Ayoub and Smith for technical expertise, not strategic advice.

It is not clear whether McBride was aware of Ayoub's line of reasoning early in 1982, but it is doubtful—given his later expressions on the issue—that he would have accepted it. Implicit in Ayoub's argument was the assumption that an inefficient plant would close and take its marginal capacity out of the marketplace after the union rejected concessions. But Smith and McBride anticipated a different reality. Whether such a company closed shop or went into bankruptcy, chances were good that the plant would be reopened by a new owner—but without the union. In the early 1980s, bargain seekers always seemed to be waiting in the wings to pick up an ailing plant at a cut-rate price and operate it with low-wage labor (which would not be difficult to find, since jobs were scarce). If the plant were shut down, the new owner could avoid the union by starting with a partially new work force. This, in fact, happened at the old Kaiser Steel Company in Fontana, California. In 1983 the USW district director in California refused to permit the members at Fontana to vote on a concession agreement accepted by local union officers. Kaiser shut the plant and sold it in 1984 to a joint venture, California Steel Company, which reopened it on a nonunion basis. Many former USW members accepted jobs with the new employer. "What could we do," Smith asked, "take machine guns out there to stop people from going in and working? " In most of these cases, however, the union remained intact and accepted lower wage rates.

Therefore, simply letting a plant die was not viewed by McBride as an option in 1982, according to Smith. McBride also had another, more elementary reason for approving concessions at the me-too companies. As president, he could have rejected the concession agreements, since the International— not the district or local union—is the contracting party in each labor agreement. But McBride approved the wage reductions, even if they undercut the master contract, because it was the local members who voted to work for less to keep their jobs. "This is something we have to suffer with," he said in 1983. "It puts us between a rock and a hard place. Do we fight our membership who say, 'Hey, we want to save our jobs'? "

It was a moral issue for McBride. How could he, sitting securely in Pittsburgh, overrule the members' decision and, in effect, wipe out their company? If he did so, he would raise a storm of antiunion criticism. Indeed, there was one such case in the silver-mining industry. The International refused to recognize a vote by members at a silver mine in Idaho to accept a 25 percent wage cut. The union soon found itself in court, answering to charges it had failed in its duty to fairly represent the members. The USW won the case, but not the publicity battle.[9]

In an excess capacity situation, of course, the market eventually would force some company or plant out of business, concessions or no concessions. In 1982, however, McBride could not bring himself to force members of currently troubled companies to lose their jobs in order to save the jobs of unknown members of threatened companies in the future. He must have known that the concessions at the peripheral companies would eventually cause competitive problems for the major firms in the bargaining group. By late 1981 he was already meeting privately with industry negotiators to formulate an approach to their financial concerns. Yet he postponed the hard decision of which members to sacrifice, perhaps hoping—as the steel companies themselves always did—that the demand for steel would come back.

In the prosperous postwar decades, the possibility that a union leader would face such a decision would have seemed as remote as the planet Pluto. It would all be forward, never backward, and certainly a leader should never have to cast a vote for the continued health and prosperity of one member as against another. But the cataclysmic economic events of the 1980s would throw up many such decisions to union leaders. The great irony of 1982 is that McBride came to be viewed by some local union officers—those who were opposed to concessions—as cool, calculating, and lacking in union spirit. This was because he kept demanding that they accept "economic reality" and grant relief to the major steelmakers. The reality of McBride was that his hard head would not let him show his soft heart.

Ed Ayoub's Productivity Concerns

From its first days, the Steelworkers union employed some of the most talented professional staff people in organized labor. Most staffers worked their way up from the ranks, but the union attracted a number of outsiders, generally intellectuals and young college-trained men eager to join "the movement." In the 1930s and early 1940s, among the best known were Lee Pressman, the chief counsel of both the CIO and the USW; the economist Harold J. Ruttenberg; Clinton S. Golden, an organizer and administrator; and Elmer Maloy, a steelworker from Duquesne who became a top-level negotiator. From the mid-1940s on, the USW acquired outside professionals like Ben Fischer; the economist Marvin Miller, who left in 1966 to run the baseball union (Major League Players Association) for seventeen years; Chief Counsel Arthur J. Goldberg, who served briefly on the U.S. Supreme Court and as labor secretary; and a pension expert, John Tomayko.

Ed Ayoub joined the USW staff in 1960 as a young assistant

in the research department. A graduate of Antioch College with a degree in economics, Ayoub had worked as a researcher for the AFL-CIO's Industrial Union Department and a federal agency. Well over six feet tall, lanky and bearded, Ayoub had about him an air of perpetual preoccupation with a problem in higher mathematics. If a hundred union people were standing around in a hotel lobby or drinking beer and eating shrimp at an end-of-negotiations "victory" party, I could usually find Ayoub by standing on tiptoes and gazing across the room. Eventually, I would see his head, with beard and pipe attached, floating ethereally above the sweat and swirl of the crowd.

In 1965, feeling that the union should be opened up to the rank and file, Ayoub risked his career by openly supporting I. W. Abel in the election fight with the incumbent president, David J. McDonald. Most of the Pittsburgh staffers sided with McDonald. After that, however, Ayoub refused to become embroiled in union politics. It was a tough, sometimes dirty business in which the politicians—the elected officials—usually took care of one another, while staff members who endorsed the wrong candidate would be demoted or transferred to a union Siberia. Ayoub stayed on the sidelines in three subsequent battles for the Steelworkers presidency. Yet he remained a key member of the staff, rising to research director and finally to a new position as chief economist and assistant to the president. He was the epitome of the neutral staff man whose brains were valuable to the officers precisely because he refused to auction them off to the highest political bidder. Unlike some union economists (and corporation economists, for that matter), Ayoub refused to bend his interpretation of economic trends to fit the organization's strategy and tactics. In 1982 this would cause his downfall in the union and early retirement (see chapters 10, 13).

Ayoub believed that a union economist should not serve simply as a "numbers man," a sort of human computer who "costs out" bargaining proposals during negotiations. He should also provide broad economic studies to help the union's officers formulate policy. With this goal in mind, Ayoub in the fall of 1980 began preparing a series of reports on the state of the American steel industry. The project would occupy him off and on for many months and place him at the center of a scholarly drama that eventually exposed the union's resistance to change.

Ayoub himself later admitted that the project got started, in a sense, because he was not as good a political analyst as he was an economist. In the fall of 1980, he thought Jimmy Carter would defeat Ronald Reagan for the presidency. In his first term, Carter had shown interest in finding solutions to the problems of the steel industry. Ayoub thought that in his second term Carter might present recommendations to Congress. The USW executive board, therefore, needed "a good background in steel economics" so it could contribute to policymaking.

Although Reagan's election shattered that idea, Ayoub decided to continue the project. Between November 25, 1980, and July 10, 1981, he produced seven memorandums which examined such areas as steel production, employment costs, productivity, production capacity, imports, capital needs, and financial performance of the companies. The seventy-nine mimeographed pages of analysis and fifty-one pages of statistical tables provided a welter of data. There were no explicit conclusions. But if the board members read carefully, they would have detected a sharp focus on wage and productivity trends and an implicit conclusion about the strategy that the USW should pursue in adapting to a radically changed world steel situation. It was not a conclusion that union leaders could easily digest.

Ayoub reported that the average steelworker's real earnings (corrected for inflation) increased by nearly 37 percent from 1967 to 1979, or about 2.6 percent per year. Total employment costs (including wages, benefits, and government mandated costs) rose from $5.677 per hour in 1970 to $15.921 in 1979, a rise of over 180 percent, or an annual rate of 12.1 percent. Productivity, however, was not rising fast enough to offset these costs. In the decade of the 1970s, real earnings rose 3.5 percent annually, compared with a 2.3 percent productivity growth. The reasons for the meager productivity increase included low operating rates of capacity, too many marginal plants, outmoded technology, and "poor labor-management relations at the production level." Ayoub concluded: "This is a subject that even the union cannot ignore because of its inevitable impact on both earnings and employment."[10]

But the union did ignore it. Ayoub had distributed the memos at board meetings in 1980 and 1981. A few directors and Lynn Williams, who was then International secretary, complimented Ayoub on his work. Otherwise, the memos and their disturbing findings dropped out of sight. Ayoub went on to other projects. In his mind, he had clearly stated the dilemma faced by the union. As he later conceded, he did not at that time vigorously press his views on McBride or the board members.

However, this was the beginning of Ayoub's growing concern about the USW's lack of strategy for dealing with the steel industry's economic difficulties. "We've got a very real problem in the labor movement," Ayoub said in 1983. "Do we remain ostriches with our heads in the sand, encouraging our members to believe everything the company says is wrong? Can't we get loyalty to the company without destroying the union? We should understand as much about the industry's side as possible. Talking about management's problems doesn't mean you have sold out to management."

McBride's neglect of Ayoub's memos, however, was all the more curious, because at the time Ayoub submitted his final memo, in July 1981, the USW president was engaged in quiet talks with J. Bruce Johnston, the industry's chief negotiator. Johnston was trying to convince McBride that something should be done to slow down the rise in steel labor costs. What might McBride have done with Ayoub's analysis? It clearly demonstrated that the wage-productivity relationship was moving in the wrong direction. One had to assume that the steel companies would come up with the same data and use it to make a strong case for the proposition that steelworkers' wages were chiefly responsible for the industry's loss of competitiveness. To avoid being pushed into such a defensive posture, the USW leaders had to answer the following question: How could they maintain their members' standard of living while helping the industry address the competitive problems? McBride wanted to aid the industry, but he lacked an offensive strategy for dealing with the problem early and on an industrywide basis.

The time to react would soon be upon the Steelworkers. By late 1981 Bruce Johnston and his negotiating partner, George Moore of Bethlehem Steel, had devised a strategy for approaching the union. The companies were locked into union contracts which would not expire until August 1, 1983, but which called for significant increases in wages and benefits during 1982 and 1983. Johnston and Moore hoped to persuade McBride to "reopen" the contracts at an early date. The initial management strategy was aimed at obtaining cost relief, but it later expanded into a much broader effort to change the labor relations system in the basic steel industry.

It has become increasingly difficult to understand the language of labor contracts. This was not true in March 1937, when John L. Lewis on behalf of SWOC, and Myron C. Taylor, chairman of U.S. Steel, negotiated the first labor agreement in the modern steel industry. The terms they agreed to were written into formal contracts at each of the corporation's subsidiaries. When converted to pocket-size booklets, the pact between Carnegie-Illinois Steel and SWOC filled just fifteen pages of text. Any steelworker with about a sixth grade education could read and understand it.

That was the golden era of clarity in labor relations. The expansion of benefits and ever finer definition of rights and responsibilities later led to obfuscation and legalistic nitpicking. By the 1980s the contracts had become so complicated that each new round of bargaining produced a virtual library of booklets that were jointly published and distributed to steelworkers by the union and the companies.

For example, six separate documents were needed to set forth the 1980 USW agreement with U.S. Steel. The "basic" agreement covering rates of pay, hours of work, seniority rules, work practices, grievance procedure, and various "memoranda of understanding" and "letters of agreement" was contained in a 219-page pocket-sized booklet. Additional booklets explored all the nooks and crannies of the USW's benefit plans. These included: a 99-page booklet spelling out all matters pertaining to pensions, such as eligibility, age, and formulas for calculating monthly stipends under varying conditions; a 102-page description of life, hospital, and medical insurance plans; a separate, 58-page outline of health insurance benefits for pensioners and surviving spouses; a 53-page agreement on supplemental unemployment benefits; and a 16-page booklet describing the industry's endlessly complicated vacation plans. In addition to these joint publishing efforts, the union itself distributed to members a 32-page exposition of steel's "lifetime security" program (see below) and a 68-page booklet summarizing everything that was new in the above explanations.[11]

Even this was not the whole list. There were additional

booklets for salaried employees represented by the union. The Experimental Negotiating Agreement, a document unto itself, was not routinely distributed to members.

By the 1980s, the mass of verbiage required to list all the rules governing life in the plants, including modifying and qualifying clauses and provisions specifying exceptions to the modifiers and qualifiers, had made the agreement hopelessly remote from ordinary workers. There were few individuals, in the company or the union, who understood the relationship in its legal entirety. Both sides had lawyers to explain elements of it to people who needed to know, and increasingly they did so on a specialized basis, thereby emulating the fragmentation of skills in the workplace that had already helped alienate the worker.

Not the least of the problems facing industrial relations today is that the classical terms of economic trade no longer accurately portray the concepts and systems used by the practitioners of industrial relations. The most elementary example is the word "wages." In economic theory, "wages" means the price paid for labor, and that price in the days of Adam Smith, David Ricardo, Karl Marx, and other great labor theorists almost always was a form of monetary remuneration computed on an hourly, daily, weekly, or piecework basis. When other forms of compensation, such as bonuses, commissions, or benefits, were added to labor's price, "wages" was inflated, semantically, to become all-embracing.

Having one term that refers to all remuneration for labor is sensible and rational to everybody but the laborer. To the auto worker and steelworker, "wages" means the paycheck, a practical definition that lingers from the time when "wages" were, in fact, synonymous with take-home pay. From their point of view, it is common sense to distinguish between a form of compensation which they can get their hands on and put to swift use and the payments made in their behalf by employers to pension funds and insurance premiums which are beyond their reach.[12]

The term "fringe benefits" is often used to designate the latter items, but it also has limitations. It joined the family of labor terms under the wage control program of World War II. Unable to negotiate wage increases beyond a limit set by the National War Labor Board, the unions searched for other ways to raise the standard of living. The NWLB members representing employers wanted all compensation frozen at prewar levels. But the public members voted with the labor contingent to allow wage increases "on the fringes." The first fringes were shift premiums (extra pay for the afternoon and night shifts) and pay for holidays and vacations.

In the beginning, the fringe pay added no more than a penny or two per hour worked to employer costs. In 1946, when the average hourly wage (cash payment) of auto workers and steel-workers was slightly above $1, the cost of the fringes averaged 3¢ per hour, or 3 percent of the hourly wage. During the prosperous postwar years, as rising take-home pay satisfied immediate needs for food and shelter, the unions elected to bargain increasing amounts of labor's share in other fringes. With the Steelworkers leading the way in a 1949 U.S. Supreme Court case, unions won the right to negotiate pension benefits. A future pension benefit can be considered a deferred wage. To provide for it, employers pay actuarily determined amounts into a pension fund.

A major victory came in 1955, when both the UAW and the USW converted labor's long-time demand for a "guaranteed annual wage" into a new provision called "supplemental unemployment benefits" (SUB). Although the two unions had managed to increase hourly pay at a fast clip, their efforts were partially adumbrated by the volatility of demand for steel and auto products. Hourly workers bore the brunt of adjusting to a slump in demand, for the companies laid them off without pay as production needs declined. By keeping non-union salaried employees on the job, employers retained their loyalty and clearly signaled that they were the favored class of workers.

This point was not lost on blue-collar workers who, with their increased purchasing power, were buying homes in the suburbs next door to white-collar employees. Here, the double standard was all the more visible. When production declined and production workers were laid off, they could contemplate with rising gorge the sight of their white-collar neighbors driving off to work each morning. SUB merely made the double standard more palatable by supplementing jobless pay which was provided by each state in the form of unemployment compensation (UC) benefits. The combination of UC and SUB benefits totaled, in the 1950s, about 60 percent of gross pay. In the 1970s the UAW claimed its benefits totaled more than 90 percent of take-home pay—a much lower amount than gross pay—and steel paid almost as much.

The SUB principle didn't spread much beyond steel, autos, rubber, and mining. But employment in these industries was more volatile than in many others. SUB, of course, is a form of deferred wage, on the order of pensions. Each employee earns SUB credits for each week worked and collects weekly SUB pay when laid off until his credits are exhausted. The employer maintains a SUB fund by contributing a certain amount of money per hour worked by all union members covered by the fund. In the steel industry, this amount rose from 5¢ per hour worked in 1955 to 19¢ in 1981. When the fund reaches a "maximum funding level," the firm may stop contributing.

Although SUB originated as a benefit for layoffs that could not be prevented, the concept changed somewhat over time. The UAW especially began to view SUB as a penalty which would force employers to find ways of keeping hourly workers on the job instead of laying them off. But by the late 1970s, two decades of experience with SUB indicated that it seldom had this effect. The companies continued laying off hourly workers at the first sign of an economic downturn. Some authorities ascribed this anomaly to a peculiar bifurcation; plant managers, who were responsible for balancing production with demand, thought of SUB as a corporate payment. Because SUB payments came out of a central company fund and were not entered on the expense side of each plant's balance sheet, plant managers felt no pressure to avoid the peaks and valleys of employment levels. In the end, management thought of SUB as a costly benefit, not as a goad to efficiency.[13]

Another problem arose during the huge layoffs of the 1980s. Many steel-company SUB funds were exhausted long before everybody was laid off; some workers were not paid. SUB had not been intended to subsidize massive numbers of jobless employees in a year-long recession. Like other industrial relations principles that had been born in the 1950s era of low competition and profligate spending, the SUB idea was turned on its head thirty years later.

A further complication with the use of "wages" arises from the cost-of-living adjustment (COLA), a fringe which is most certainly a cash payment but which must be distinguished from the regular method of dispensing cash. A COLA raises pay by some formula that takes account of rising consumer prices as expressed by the Consumer Price Index. The steel and auto contracts in 1981 called for a CPI review every quarter. For each rise of 0.3 points in the index over the quarter, steelworkers received an additional 1¢ per hour in their paychecks; in the auto industry, the same pay increase was triggered by a 0.26 point rise.

The penny-per-hour payoff may appear insignificant, but during periods of double-digit inflation in the late 1970s it produced substantial wage boosts each quarter. For example, in 1979, when prices rose 11.5 percent for the year, the CPI increased from 209.3 in March to 216.9 in June. The USW's formula produced a COLA payment of 25¢ per hour for that quarter alone, payable on August 1.

In the 1960s, the fringes began to roll up rapidly: life, hospitalization, major medical, and dental insurance (often fully paid by the employer); a "short work-week" benefit for workers who are not laid off but who work less than thirty-two hours per week; and allowances for jury duty and funeral leave. Time off with pay rose swiftly as employers granted additional holidays and increased vacation time. The USW in 1962 negotiated an "extended vacation" plan calling for thirteen weeks of vacation every five years for the senior half of the work force.

In 1976 the UAW set off on what it hoped would be a long-term strategy to reduce the work week to four, eight-hour days. It established "paid personal holidays," or PPH days, in addition to regular holidays. By the beginning of 1982, workers at General Motors and Ford could look forward to nine PPH days per year, or a four-day week almost once each month. At the same time, auto workers received twelve regular holidays and vacations ranging from one week for the newest employee to five weeks for workers with twenty years of service. As a result, GM and Ford were committed to provide annually an average of forty-one days off with pay (based on average length of service of fifteen years). The average was slightly higher in steel.[14]

At about the same time, the USW began moving toward "lifetime job security" for its steel members, a program that has proved to be both enormously costly to the industry and a lifesaver for workers who qualified. Interestingly, the union demanded and won this program in 1977 not because the leaders perceived that it was a primary concern, then, of steelworkers. When I. W. Abel first proposed it at the USW's 1976 convention in Las Vegas, his primary consideration was ensuring the defeat of Ed Sadlowski in the 1977 election. Abel had asked his advisers to come up with an idea for his keynote convention speech that would take the political momentum away from Sadlowski. Noting that Japan's "lifetime job security" was beginning to stir interest in the United States, Ben Fischer and Bernie Kleiman framed a broad concept of comprehensive security for steelworkers and wrote it into Abel's speech.

The idea attracted widespread support, and the union leaders, of course, had to follow through on it in 1977 negotiations. They conceived a plan for protecting senior workers during an abnormally long layoff or plant shutdown. Normally, steelworkers could retire after thirty years of service, or at age sixty-two. Workers in their forties or fifties with less than thirty years on the job were highly vulnerable in shutdowns, being too young to retire and often too old to find another job. Although the union hoped eventually to cover most workers under the Employment and Income Security Program, it proposed at first to put only workers with twenty or more years of service into the highest category of protection. If laid off, they were guaranteed SUB pay for two years instead of the one year of benefits available to other steelworkers. If the SUB fund was exhausted, the company was obligated nevertheless to continue paying SUB from other sources. (This became known as "guaranteed SUB.")

If the layoff lasted more than two years, or if the plant was declared permanently closed, the company had to offer employees with twenty or more years' service "suitable long-term employment" at another plant in the same region. If no jobs were available, the worker could retire under the Rule of 65. This required that age and years of service totaled sixty-five. In other words, a forty-five-year-old steelworker who had worked just twenty years could receive a regular pension, *plus* $400 a month extra until Social Security kicked in at the age of sixty-two.

Steelworkers could qualify for an early pension under a number of other age-service combinations. Another plan used extensively was known as 70/80. The Rule of 70 covered workers who were at least fifty-five and had fifteen years of service; the Rule of 80 had no age minimum, but the total of age and service had to reach eighty. Recipients of these pensions also received a $400 monthly supplement. Even without a permanent shutdown, an employee could retire on a 70/80 by mutual agreement between him or her and the employer. These so-called mutuals were often given by management as a trade-off for local-union agreement to eliminate jobs in an active plant.

When the plan was negotiated, no one anticipated the convulsions that would sweep through steel beginning in the late 1970s, and so the companies did not offer much resistance to the plan. It was expected to cost very little, less than 5¢ per hour, because a company would not have to grant an early pension if it could offer suitable employment in another plant. In the 1980s, however, the steel industry declined on such a broad front that the companies could not absorb employees idled by plant closures. To live up to their obligations, the steelmakers had to give early pensions to many thousands of workers with more than twenty years of service, putting great strain on steel's pension funds and, ultimately, forcing the federal government to step in and take over underfunded pension plans (see chapter 18). One study found that during 1985 alone, U.S. steel paid out $163.7 million in pension and medical expenditures to recipients of early pensions nationwide, including $39.8 million to those in the Monongahela Valley. Almost 90 percent of the early retirees in the valley received an annual pension of $11,000 to $16,000.[15]

Finally, "fringe benefits" could be construed to include employer payments to the federal government for Social Security taxes and to the state government for workers' compensation for job-related injuries. According to a survey of private-sector industries by the U.S. Chamber of Commerce, all of these extra-cash emoluments and benefits in 1981 constituted 40 percent of the total cost of employing workers. Indeed, the cost of providing benefits could no longer be thought of as a mere "fringe" at the margins of cash pay. In 1982 the steel industry's total hourly employment cost—the cost of providing wages and benefits, including the government-mandated payments—averaged $22 to $23 in the eight companies that bargained as a group. Of that amount, about $13 went to pay and $10 to benefits.[16]

Nevertheless, the benefits were in most cases a deferred payment, the collection and personal consumption of which depended on the individual worker's longevity. So it was understandable that steelworkers heatedly objected to statements that they were "paid" $23 per hour, although this was technically correct.

For purposes of clarity, it seems reasonable to talk of "wages" and "benefits" as two separate costs in a detailed analysis of the labor price. Nevertheless, "wages" still remains the best term to refer to total remuneration if the economic and literary context is clear. While it may have gathered plenty of costly moss at the fringes, the word is still one of the most evocative in the English language. Shakespeare used it in yet another context:

> Fear no more the heat o' th' sun,
>   Nor the furious winter's rages,
> Thou thy worldly task hast done,
>   Home art gone, and ta'en thy wages.
>
> *(Cymbeline, 4.2)*

It might be said that the many thousands of steelworkers who were forced into early retirement in the 1980s had finished their worldly task. Fortunately, their final wages (pensions and insurance benefits) were among the highest furnished to industrial workers anywhere. An extraordinary forty years of collective bargaining had seen to that.

Nevertheless, it could be argued that the bargaining relationship between the Steelworkers and the steel industry never came to grips with a central issue of overriding importance: relating compensation to the quality of work and productiveness of the workplace itself. This is the great failure of the industry and the union, and it is linked to the origins of both. Before continuing the narrative that started in early 1982 in Lloyd McBride's office, I must dip back into history to explore those origins in the Monongahela Valley.

Chapter 4

On Strike in McKeesport

Friday, August 1, 1986. Masses of smooth, gray clouds are closing in on the sun, rather like the halves of a domed roof sliding shut over an empty stadium. It is late afternoon. I am driving south on Lysle Boulevard In McKeesport, passing the mile-long National Works on my right. The three blast furnaces have vanished, dynamited to rubble, leaving a long gap in the mill's jagged skyline. For the first time since 1889, when the first two furnaces were constructed here, a passerby could have a clear view of the mountainous cliff across the river. One tends to think that ninety-seven years of existence indelibly etches the landscape. Not so. Now a big green sign perches there on a pile of rusty junk. "USS Industrial Park of McKeesport, Pa.," it says, and gives a phone number to call, should you want to rent space to build a research laboratory or a high-tech "facility" on the banks of the Monongahela.[1]

I continue south for another half mile and park my car on the boulevard. Right around the corner on Locust Street, four men are sitting on folding chairs in the middle of the sidewalk about seventy-five yards from National's main gate. A makeshift shelter of plastic stretched between two-by-four posts stands against the wall of a parking garage. Inside the shelter are a couple of placards: "USW Local 1408 locked out by USX." Since 12:01 A.M. today the Steelworkers union has been engaged in a nationwide work stoppage at USX, formerly U.S. Steel.

The four men are picketing the plant that (the union claims) locked them out, though not with much enthusiasm. Shutting down this plant seems anticlimactic to them. National only has an active work force of 150, down from over 4,000 a

82

few years before, and the men have worked only three days in the past week.

Two of the pickets are men in their fifties who have worked as millwrights for more than thirty years. They are bitter because, under a local agreement negotiated in 1985, they now must take whatever job the company assigns them, operating cranes (for which they are not qualified), laboring, cleaning latrines, fixing machines. "It's degrading," says Robert Cross. "We used to be proud of our jobs," adds Albert Tancibok. "This new setup has just created bitterness between the guys. What the company is asking all the other locals to do, we already got. We gave them everything."

They are good union men, but they are dispirited—not by the work stoppage. They simply see little left for them, either in their plant or the valley. There isn't even much point in blocking the gate, and so they sit in the shade of the four-story parking garage. Although there are plenty of unemployed people in the valley, the pickets feel that USX isn't likely to open the gates to strikebreakers. Indeed, the question produces a reaction of near indifference, as if to say, "The scabs are welcome to it."

"I think the company wants out of the steel business anyway," says Mike Katchur, a grievance committeeman, who is in his twenties. "I feel sorry for these guys," he adds, nodding at the millwrights. "These guys got brains, but now they're cleaning shithouses. I'm glad I'm not as old as they are. I can start in something new."

"Well, I'd hate to be young today," says the fourth man, a USW staff representative named Ridsy Calderone. "There are no jobs. You love this community, this area, you want to stay here. But you've got to get up and get the heck out of here and make a future. You can't make a future out of your wife working down here as a clerk for four bucks an hour, maybe you picking up a couple of bucks doing odds and ends. That's no future. You don't buy homes that way, you don't buy cars that way, and you don't provide anything to the community."

As I drive away, my memory returns to the 1940s and the picture of steelworker pickets massing at the Locust Street gate. I see workers marching through town with strike placards and hear people talking constantly about John L. Lewis, Phil Murray, and their striking miners and steelworkers. In comparison with those days, the 1986 strike (or lockout) seems like two groggy fighters trying to deliver the knockout blow.

How did ninety-seven years (and more) of steelmaking in the Mon Valley come to this?

From Carnegie Steel to U.S. Steel

The beginning of the 1870s was precisely the right time in the industrial development of the United States for the appearance of a steel industry. Iron had served its purpose in the early stages of the Industrial Revolution, but the infrastructure for the next great leap—railroads, bridges, factories, and office buildings—required a stronger, less malleable material. The Bessemer steelmaking process, invented and improved upon in the 1860s, was ready for large-scale application. Making steel on such a scale, however, required raw materials, investment capital, entrepreneurial genius, a large reserve of labor, and a place where these elements could be brought together in a new form of capitalist organization.

By a mysterious process of geological and historical timing, all except two of these elements happened to be present in Pittsburgh and the Monongahela Valley in the early 1870s. The region had coal, limestone, and some iron ore, navigable rivers for transportation, an accumulation of capital from Pittsburgh's early success as a manufacturing center, the creativity of hustling entrepreneurs and experienced iron-makers, and a growing supply of skilled industrial workers.

The valley lacked large numbers of unskilled workers to supply the hands, arms, and legs needed to shovel, haul, and lift raw materials and great vats of molten metal. But a vast pool of unskilled labor lay only an ocean away, ready to be tapped, in places like Austria-Hungary, where people endured lives of "endless poverty and oppression which were the birthrights of a Slovak peasant in Franz Josef's empire." Assorted other kings, czars, princes, and chancellors presided elsewhere over oppressed peoples who were desperate to escape from rural serfdom. The steel mills of America soon exerted a strong gravitational pull on the European masses. As Thomas Bell, the author of the above quote, depicts in his novel of immigrant life in the Mon Valley mill towns, *Out of This Furnace*, the first generation of immigrants faced a life of industrial serfdom in America. Their sons and daughters eventually made a better life for themselves, and not least because they helped launch the USW in the 1930s.[2]

The second missing element, a new kind of enterprise, was made possible by the Monongahela River itself. Flowing out of the "mountains" of West Virginia and cutting a 128.73-mile, winding gorge northward to Pittsburgh, the river became nature's equivalent of an assembly line. Outfitted with bridges, locks, dams, and a nine-foot channel, its banks lined with railroads, mills, furnaces, coal mines, gravel pits, coal docks, tipples, trestles, pipelines, tanks, cranes, conveyor belts, and practically every other item of industrial equipment known to man, the Monongahela enabled entrepreneurial capital to use and exploit the valley's minerals and men on an awesome scale.

Blast furnaces in Massachusetts and New Jersey were smelting iron long before the first small furnace was built in western Pennsylvania (Fayette County) in 1790. Over the next eighty years, Pittsburgh gradually became an important metallurgical center in the first great surge of industrialization west of the Alleghenies. By the early 1870s, the city's iron industry consisted of many small firms concentrated on the banks of the Allegheny River from Lawrenceville north to Etna. The small blast furnaces of the day turned out pig iron ingots, which were rolled into iron railroad rails and various structural shapes for bridges and buildings, or melted down and cast and forged into tools and agricultural implements.[3]

The Monongahela, Allegheny, and Ohio rivers gave Pittsburgh an obvious advantage in importing raw materials and exporting finished products. Iron ore and limestone, the ingredients of ironmaking, were mined locally, and fuel was abundant in the form of charcoal and coal. When the local supply of ore thinned out, supplies were barged up the Ohio River from the ore fields of Missouri and, later, shipped across the Great Lakes from northern Michigan and Minnesota.

However, the city owed its early dominance in steelmaking "chiefly to the presence in its vicinity of the first large bed of coking coal to be developed in the country." This was the "Connellsville seam" of low-sulfur coal which underlay a large part of Greene County southeast of Pittsburgh. Chicago, also a big producer of iron in the 1860s, lay closer to the Lake Superior ore fields. But Chicago ironmakers had to import coal and coke from the Connellsville region, giving Pittsburgh the first advantage in its long steelmaking rivalry with Chicago. It cost less to transport ore to Pittsburgh than coal and coke to Chicago.[4]

Pittsburgh could not have held its lead as the steel-producing capital of the world well into the twentieth century were it not for the organizational genius of the men who built Andrew Carnegie's steel empire. This was not evident at the beginning. By the early 1870s, Carnegie had amassed a small fortune in various businesses, including iron furnaces and rolling mills. But he was slow to get into the steel business. The new Bessemer furnaces were installed at ten other locations in the east and midwest before Carnegie was persuaded to invest in steel.

When it became apparent that he might lose his railroad customers to producers of steel rails, Carnegie joined with seven other businessmen and built the Edgar Thomson Works about eight miles up the Monongahela from Pittsburgh. A combination of luck, highly skilled labor, and good financial management made the plant a success from the beginning. Carnegie managed to acquire one of the best steel men of the day, Capt. William R. Jones, to run the plant. Two high officials of the Pennsylvania Railroad, including Chairman J. Edgar Thomson (after whom Carnegie thoughtfully named the Braddock plant), invested in Carnegie Steel and saw to it that the Pennsylvania purchased a large portion of the rail-mill output.

Carnegie and his associates introduced new accounting procedures and waged an unending battle to lower production costs. From 1878 to 1898, the plant reduced the cost of producing steel rails from $36.52 to $12.00 a ton. David Brody concludes that this "impulse for economy" shaped the entire steel industry. "It inspired the inventiveness that mechanized the productive operations," he writes. "It selected and hardened the managerial ranks. Its technological and psychological consequences, finally, defined the treatment of the steelworkers. Long hours, low wages, bleak conditions, anti-unionism, flowed alike from the economizing drive that made the American steel industry the wonder of the manufacturing world."[5]

This was only the beginning. Having got into steel, Carnegie increased his market share by expanding internally and buying out competitors. He built two blast furnaces at Edgar Thomson in the late 1870s, thereby integrating iron- and steelmaking facilities in a single plant. Until then, the works had used iron smelted by Carnegie's Lucy Furnaces at Fifty-first street in Pittsburgh. In 1883, Carnegie acquired a two-year-old steel plant that had been built by a competitor at Homestead, a village of six hundred inhabitants two miles down the Mon from Braddock. The Homestead Works eventually became the largest plant in the Mon Valley.

To supply Homestead's steel furnaces with molten iron, Carnegie in 1884 built the first of several blast furnaces across the river in Rankin—the plant became known as Carrie Furnaces—and connected the two plants with a railroad bridge. In 1890 he bought out another competing firm and acquired the newly built Duquesne Works. This plant, constructed in 1889 on farmland adjacent to the Mon just a few miles up the river from Braddock, had steel furnaces and rolling mills that threatened Carnegie Steel's market position.

The Edgar Thomson, Homestead, and Duquesne Works, along with the Carrie and Lucy Furnaces, formed the core of Carnegie's steel chain. It had been put together almost by accident, and individually the operations produced good profits. But the key link was yet to be forged by a young Connellsville bookkeeper named Henry Clay Frick. He saw the immense value in the coal that lay under the rolling hills of southwestern Pennsylvania.

In the early 1870s, Frick began buying huge blocks of coal lands and beehive coke ovens in the vicinity of Connellsville and Uniontown, some fifty miles southeast of Pittsburgh. The beehive was a simple brick chamber in which coal was baked to drive out impurities and produce coke, a nearly pure carbon form of coal, which made an excellent blast furnace fuel. Huge banks of ovens were scattered throughout the coalfields, but the largest concentration existed in the Connellsville area, which gave rise to the term "Connellsville coke," also known as "Frick coke." When Carnegie discerned the importance of the Connellsville coke ovens, he bought an interest in Frick's company and invited Frick into the steel firm. By 1889, when Frick assumed the presidency of Carnegie Steel, he owned or controlled thirty-five thousand acres of coal land and nearly two-thirds of the fifteen thousands beehive ovens in the region. Starting in the 1910s, the beehives were gradually supplanted by more efficient byproduct coke ovens, installed at steel-plant sites.[6]

Coal was hauled out of the mines, baked in nearby beehives, and transported to the Monongahela in wagons, later trains, to be loaded into barges. Although the Mon seemed to bend and twist aimlessly as it flowed north and west toward the Ohio, metaphorically it was as true as a plumb line stretching from the heart of one of the richest coalfields in the world to what would become the largest concentration of steel mills in the world.

In building Carnegie Steel, Frick and Carnegie pulled together many disparate parts into a business monolith that had no corporate peer in the 1890s. Already holding coal mines and limestone quarries, the company bought enormous reserves of high-grade iron ore on the Mesabi Range in northern Minnesota. It acquired Great Lakes freighters to transport the ore to Conneaut, Ohio, and a railroad to haul it south to Pittsburgh.

What Carnegie and Frick did is today called "vertical integration," the combining of separate functions into a whole in order to make the firm immune to outside business pressures. J. H. Bridge described the resulting company as "a solid, compact, harmonious whole, whose every part worked with the ease and silent motion of the perfectly balanced machine." This is an important point, and Bridge elaborates it in an arresting style: "This mammoth body owned its own mines, dug its ore with machines of amazing power, loaded it into its own steamers, landed it at its own ports, transported it on its own railroads, distributed it among its many blast furnaces, and smelted it with coke similarly bought from its own coal mines and ovens, and with limestone brought from its own quarries. From the moment these crude stuffs were dug out of the earth until they flowed in a stream of liquid steel into the ladles, there was never a price, profit, or royalty paid to an outsider." Bridge exaggerated a bit. Carnegie Steel also used other railroads to haul a portion of the raw materials. But the description is substantially correct.

By the time Carnegie sold his steel interests to J. P. Morgan to form the United States Steel Corporation in 1901, the Mon Valley coal-to-steel chain was producing phenomenal profits. With an original capitalization of only $25 million, the firm earned $4 million in 1892, $11.5 million in 1898, $21 million in 1899 and nearly $40 million in 1900. Converted to 1986 dollars, Carnegie's profits in 1900 totaled $528 million on a capitalization of $330 million.[7]

Carnegie and his men created a mighty organization, which made Pittsburgh preeminent as a center of manufacturing,

finance, and inherited wealth (most of Carnegie Steel's partners and executives, as well as their bankers and suppliers—
and their heirs—became millionaires). Like all mighty
machines, it rolled on under its own momentum, bigger than
ever as U.S. Steel. It dominated the steel business, dominated
the corporate world, and dominated the Mon Valley. Carnegie's plants at Homestead, Rankin, Braddock, and Duquesne, of course, became part of U. S. Steel. The corporation
also acquired these other valley plants, either during or after
the merger:

1. The National Tube Works at McKeesport, the headquarters plant of a chain owned by the National Tube Company.
In 1901 this was the largest pipe-producing plant in the world.
McKeesport, known as Tube City, was growing so rapidly at
the turn of the century that McKeesport's inhabitants thought
it would become the Pittsburgh of the universe. Located at the
junction of two rivers, the Monongahela and the Youghiogheny, McKeesport had a history extending back to the 1750s.
National Tube became a subsidiary of U. S. Steel in the great
merger.

2. The Clairton Works, acquired in 1904. This medium-size
steel plant was built just below the town of Clairton, about
five miles upriver from McKeesport. Its importance increased
immeasurably in 1918, when U.S. Steel installed more than
six hundred byproduct coke ovens there, making Clairton the
largest center in the world for this kind of coke production.

3. The Donora Works. Started in 1899, this plant specialized in wire, rods, and nails. It became part of U.S. Steel in
1902. Later, a zinc-producing plant was built here, with disastrous consequences nearly a half-century later. Its fumes,
which had denuded the hills all around, killed some twenty
residents in 1948 during an especially heavy inversion.

4. The Christy Park Works at McKeesport. This small
plant, located a few miles up the Youghiogheny from the
National Works, was built in 1897. It made specialty tube
products and ordnance used in the Spanish-American War,
the two world wars, and the Korean War. In World War II, the
plant employed seventy-five hundred people and became a
prime producer of 250-, 500- and 1,000-ton bombs. U.S. Steel
also owned another small McKeesport plant, the Wood
Works, located next to National Tube.

5. The Irvin Works at West Mifflin. Opened in 1938, this
turned out to be the last major plant built in the valley. It

imported semifinished steel from other plants in the valley and turned out sheet steel for cars and appliances on high-speed hot strip lines. West Mifflin is located between McKeesport and Clairton.[8]

The Making of a Corporate Bureaucracy

From Pittsburgh to Donora, a distance of thirty-seven miles on the winding Monongahela, U.S. Steel operated eight major steel plants by the early 1900s (not counting the Irvin Works, which came later). The scale of operations could be compared to a gigantic assembly line, each plant being a work station along the river serviced by subsidiary lines which distributed raw materials. Trains and barges loaded with iron ore went up the river, dropping off supplies at mills along the way, and coke flowed down the river, first from the Connellsville bee-hive ovens and later from the Clairton byproduct ovens. Gas from the Clairton ovens was piped to each downriver plant to serve as heating fuel for the furnaces. Hundreds of other supplies moved up and down the valley, by road, water, and rail (U.S. Steel's own interplant railroad), connecting all the plants in a finely tuned interlocking system.[9]

The assembly-line simile is not exact, for each plant specialized in one or more final products, with the exception of Carrie Furnaces. There were many instances, however, of a plant adding value to a product passed along by another plant. For example, coke from Clairton was used to make iron at the Carrie Furnaces, and the iron from Carrie was converted into steel at Homestead; the Irvin Works used semifinished slabs from Edgar Thomson, Homestead, and Duquesne; and sometimes National used coils from Irvin. In a production sense, the system was highly integrated.

In terms of administration, however, the organization consisted of individual mills spotted along the Mon, rather like workers staggered along an assembly line, some this side, some that side. Each focused narrowly on what it did best, ignoring its neighbors on either side, and taking commands from the big corporate brain in Pittsburgh, which took its orders from the bigger brain at U.S. Steel's financial headquarters in New York City. The sense that a remote, all-knowing authority held ultimate power over the mills and their workers is reflected in expressions used by employees in the valley.

In *Out of This Furnace*, which is in part a documentary

novel based on real events, all important decisions affecting workers at Edgar Thomson are said to emanate from "City Office," an impersonal *place* whose inhabitants are nameless but all-powerful. When the workers struggle to form a union in the 1930s, mill managers avoid settling their grievances by sending the complaints to "City Office," where they vanish in a "bottomless pit." In the 1970s and 1980s, workers in the valley used the equally impersonal term "Pittsburgh" to refer to U.S. Steel headquarters. Frequently heard comments were, "Pittsburgh says..." or "Pittsburgh wants...," usually accompanied by a shrug of the shoulders. It was common to distinguish between orders given by local management and those that "came down from Pittsburgh."[10]

The steep hills and valleys that separated the towns also helped Carnegie and U.S. Steel keep the working populations segregated in the various communities and thus relatively ignorant of each other's problems. By an accident of geography, they conformed to what has become today's conventional wisdom among manufacturing managers. The smaller the plant, the easier it is for management to foster a family-like atmosphere and maintain control of the work force.

This is a large reason why manufacturing companies, starting in the 1960s, have built new, small plants in the South and Southwest rather than expand their huge old plants in the Northeast. A rule of thumb is that management's "span of effective control" limits the worker population to about five hundred. In larger plants, management bureaucracies bear down on the workers, forcing them to define their own interests separately from management's. The individual psyche is less penetrable, less open to management influence, and worker solidarity grows correspondingly.

The Mon Valley plants far exceeded the modern rule-of-thumb limit, each employing from three thousand to nearly ten thousand workers. But these numbers were controllable in the old days, when managers could intimidate, bully, and fire at will. The division of labor among the many plants impeded the growth of valleywide solidarity among the workers.

The scale of operations in the valley grew to awesome proportions. At its peak in the late 1940s, U.S. Steel employed some forty-eight thousand USW members in production, maintenance, and clerical jobs. A fifth again as many people worked in managerial and supervisory posts at the valley plants (not counting employees in the Pittsburgh headquarters), for a total of nearly sixty thousand. With the addition of employees at Pittsburgh Steel in Monessen and J&L in Pittsburgh, about eighty thousand or more people derived their income directly from steelmaking in this one valley. By mid-1987, the number had declined to less than four thousand.[11]

The corporate organization emphasized separateness in the Mon Valley. For the first sixty years of its existence, U.S. Steel was merely a holding company of wholly-owned, decentralized subsidiaries. Thus, the plants at Homestead, Rankin, Braddock, and Duquesne were run by Carnegie Steel (later Carnegie-Illinois), the National and Christy Park Works at McKeesport by National Tube, the Donora Works and McKeesport's Wood Works by American Steel & Wire, and the Clairton Works by yet another subsidiary which finally was folded into Carnegie Steel.

The corporation was not necessarily villainous because it organized its affairs this way; most corporations of the day eventually did the same. In its early years, U.S. Steel was a paragon of modern, hierarchical organization, so corseted in numerous layers of management that the head couldn't see, much less communicate with, the feet. It was one of the first truly bureaucratic enterprises—that is, a giant organization which exploits economies of scale by its largeness and integration. To cope with the unwieldiness of giantism, it tries to force people at all levels to adhere to rigid rules and regulations that remove the uncertainty of the "human factor." All power resides at the top, none at the bottom, although the bottom level—the shop floor—is where the necessity of change is first felt and where innovation must first occur. However, as Michel Crozier's classic study of bureaucracy points out, "a bureaucratic organization does not allow for such initiative at the lower echelons; decisions must be made where power is located, i.e., on top." But the higher echelons have no advance warning of trouble brewing at the bottom because the rigid rules of behavior do not allow for upward communications: "As a result, a bureaucratic system will resist change as long as it can; it will move only when serious dysfunctions develop and no other alternatives remain," Crozier writes. The need to change can only be met by crisis management, issuing sweeping orders that merely establish another set of regulations. The bureaucratic system actually cannot correct its behavior but only make rules. It is *"too rigid to adjust without crisis to the transformations that the accelerated evolution of industrial society makes more and more imperative."* It cannot change with the market.[12]

In a bureaucracy, management controls employees and gains their commitment—to the extent it does—through reward and punishment rather than through participation. Carnegie, Frick, and later U.S. Steel established a social and political culture of control in the Mon Valley. It was an ideal setting for the "perfectly balanced machine." Each mill town was a work station on the line, complete with all the necessities (if few amenities) needed to sustain a permanent working force. The early mill towns became little more than locker rooms for increasingly dense worker populations.

Carnegie's ingenious system for achieving great economies of scale later was corrupted under U.S. Steel's bureaucratic management. Most companies in the mass-producing industries adopted this form of organization. But its highest flowering occurred in a steel industry comprised of bureaucratic behemoths that could not adapt quickly to changing situations—something that Carnegie was never guilty of. This is why the American steel industry, such a puissant force in world trade for a century, had become a second-rate steel power by the mid-1980s.

Steel's Shortsighted Business Strategies

It ought to be axiomatic in the world of business that one century's victorious strategy is apt to become the next century's strategy for failure.

In the nineteenth century, Carnegie Steel became the wonder of American capitalism when it reached a stage of total integration, from mined ore to finished steel. One of the major steps in reaching this coveted goal was the acquisition of large reserves of high-grade iron ore on the Mesabi Range. In the 1950s, however, the Mesabi ran out of high-grade ore. Was a new strategy called for? Should vertical integration be cut short and modified by the use of low-cost imported ores? No! The prospect of losing control of any link in the ore-to-steel chain, especially to foreign outsiders, was unthinkable. American steelmakers thereupon embarked on two courses of action. First, they began an extremely costly development of ore fields in remote areas of northern Canada. Second, they made major investments in plants to upgrade and pelletize (convert ore dust to pellets that suited the taste of U.S. blast furnaces) the remaining low-grade ores mined on the Mesabi and Upper Michigan ranges, as well as in Canada.

This might have worked, except that U.S. demand for steel shrank, making redundant a large part of this vast new ore capacity, which still had to be paid for. To make better use of their ore processing capacity, the steel companies reduced imports of high-grade ores from Australia and Brazil. These were available in some markets at less cost than the U.S. and Canadian ores. The U.S. producers had become captives of their high-cost mining operations. At the same time, Japanese steelmakers were importing Australian ore at nearly half the price that U.S. firms paid for the Canadian ore. This helped the Japanese underprice U.S. steel on the West Coast.

By the 1980s, many of the American and Canadian ore operations were idle. But Donald F. Barnett and Louis Schorsch, authors of *Steel: Upheaval in a Basic Industry*, point out that the debt incurred in developing them became a heavy millstone around the industry's neck. "This 'strategy,'" they add, "replaced lower cost ores with higher cost ores, largely because of the asset structure of U.S. integrated steel companies."[13]

This is just one example of the many wrong turns that the U.S. steel industry has taken since World War II. At the end of the war, the industry was preeminent in the world. It had set stunning production records during the conflict and accounted for 54.1 percent of the world's raw steel production in 1946.

By 1970, however, the U.S. share of world production was down to 20.1 percent. What was thought of as the *traditional* American steel industry—that is, the big, integrated companies—had already lost its competitive edge over foreign producers and domestic minimills. Production share continued to fall and was only 11.8 percent in 1984.[14]

What happened?

Admittedly, hindsight permits a clear view of events that were enveloped in fog when they occurred. The decisions made by the steel industry in the postwar era probably seemed right when they were made. Yet that was the trouble: The decisions *seemed* right because the industry did such a poor job of looking ahead. While business managers aren't expected to be clairvoyant, they should have better than normal foresight about their own business. The steel industry, unfortunately, compiled a dismal record of forecasting demand and supply, of estimating the speed of development of new technology, and of judging the ability of foreign producers to pose a competitive threat.

In large part, the poor forecasting can be attributed to the philosophical and psychological mindsets that industry executives brought to bear on the problems: market myopia and organizational rigidity.

Market myopia. The best explanation of this syndrome came from George Moore, after he retired in 1984 as vice-president of industrial relations at Bethlehem Steel. He had worked at Bethlehem for twenty-five years as a staff lawyer and executive, in positions that required him to adapt the company's labor and industrial relations policy to its commercial philosophy.

"What the steel industry's management did over the period of the late sixties and into the seventies was rely too much on history," Moore said. "Steel had always been a cyclical business. There were quarters in which you'd lose money or make damn little for what you had invested. But it was always *mañana*. It will turn around. Production will pick up and everything will be fine. Therefore, don't tinker with the toy. And that was exactly what had happened in the thirties, forties, and fifties. But what they overlooked in the sixties was that the downturns were getting deeper and the upswings were getting less high. They couldn't get the price in a downturn because of foreign competition, and in an upswing foreign competition was skimming off the demand. So, this great commandment, that volume will take care of everything, was not longer applicable."

Other observers came to a similar conclusion. Eugene J. Keilin, a partner in the investment banking firm of Lazard Frères, became deeply involved in the steel industry in the 1980s. He designed the financial plan under which Weirton Steel became an employee-owned firm, and served on its board. When Lazard Frères later became the USW's investment banker, Keilin studied the financial structure of many of the larger companies. Even in the early 1980s, he said, steel management believed that "the bad times of the past ten years was a trough. The crest was ahead. Their decisions were based on being around for the next crest. In this frame of mind, they didn't deal with the excess capacity problem and made investment-decision mistakes. They didn't deal effectively with the human problems that would come with handling the excess capacity problem. Their mistake was looking for the good times."[15]

Organizational rigidity. The force of tradition ruled the steel industry. Through thick and thin, regardless of changing circumstances, it clung to two major organizational strategies.

One strategy was vertical integration. Over the decades, the nature of this concept changed from a smart business strategy in certain circumstances to an all-absorbing self-view. No steel company could feel like a *real* steel company unless it operated on an integrated basis. There was a feeling of power tied up in this. When "Pittsburgh" announced an important decision, temblors were felt from the remote iron mines of the Upper Peninsula and the coal pits of Appalachia, to manufacturing centers around the country, and eventually to the political institutions in Washington. In the end, this sense of pervasive control was illusory. There would come a time (there *did* come a time) when Mesabi ore and Connellsville coal would be exhausted. The one facet of the ore-to-steel process that the industry could not control, nature's provision of ore and coal, was the most important. Had the industry developed some humility in the face of this unalterable reality, it might have contemplated modifying the integration strategy while there was still time.

The second age-encrusted strategy involved the control of steel prices. Andrew Carnegie had competed on the basis of cutting costs and, if need be, lowering prices. Within ten years after the formation of U.S. Steel, Carnegie's business philosophy had been replaced by one of cooperating with other producers to maintain price levels. This change was mainly attributible to the strong-willed Judge Elbert H. Gary, who served as U.S. Steel's chairman from 1901 until his death in 1927. Preaching the values of harmony and stability, Gary pulled the largest producers into an informal system (a formal agreement, of course, would have violated antitrust law) of "administered prices." The industry thus became an oligopoly: A limited number of companies exerted controlling influence over prices.

Even in periods of weak demand, the large, integrated producers could manipulate prices to maintain profits. Although Gary himself admitted that such an arrangement existed, steel companies in the postwar years denied it. However, Barnett and Schorsch present a strong conclusion on this point. The American steel industry, they contend, was "a classic example of an oligopolistic industry" from 1901 until the 1960s. "U.S. Steel corporation, as the unchallenged industry leader, in effect determined the industry's overall price structure. Other large firms were generally more efficient, so that it was relatively easy for them to prosper under the umbrella of their unwieldy rival and leader. Price competition was as a rule banished from the world of Big Steel, and with it much of the incentive for reducing costs."[16]

Similarly, the industry modeled its industrial relations structure on the oligopolistic concept. Under the industry-wide bargaining setup, the large companies crowded under U.S. Steel's umbrella to protect themselves from being picked off by the union in single-firm strikes. In return, although with much grumbling, they permitted U.S. Steel to assume command in collective bargaining and to enforce the resulting uniformity of costs. This enabled USS to prevent its less efficient competitors from gaining advantages through lower labor costs and in some cases eliminated cost efficiencies at the smaller firms.

Not that managements of the other companies objected to having autocratic authority; some firms were more bureaucratically rigid than USS. However, when the need for reform began to be perceived by some companies in the 1970s, they faced a dilemma. They could not change under U.S. Steel's leadership, but they were afraid to get out from under the umbrella and dance alone under the rain drops.

These underlying attitudes can aid us in understanding where the steel industry went wrong in the postwar years. Perhaps the key mistake of the entire period, the one that would start an unraveling process that has not ended, came in the 1950s. The domestic industry expanded its production capacity from 100 million to 148 million tons during that decade. This was done partly at the urging of the federal government, which believed the nation faced steel shortages. But the companies themselves were operating in the perpetual cloud of optimism described above. Not only did they expand too much; they expanded in the wrong way, building large, new, open hearth shops. The more efficient basic oxygen furnace (BOF) had already been invented in Germany, but the early applications were in relatively small operations. The domestic industry stuck with open hearths because they provided greater volumes of steel to feed large rolling mills.

American steelmakers increased open hearth capacity by more than 35 million tons, while Japanese and European producers rebuilt their industries with improved BOFs. The U.S. firms also began their costly development of iron mining in Canada, while foreign steelmakers gained a cost advantage by relying on better quality ores. As a result, the large expansion in the U.S. absorbed great amounts of money without maintaining America's technological lead. Furthermore, steel consumption did not grow as anticipated in the 1950s, forcing the industry to run its plants at low operating rates. As unit costs rose, the companies maintained or increased prices to prop up profit margins.

By the end of the 1950s, the U.S. industry was on the brink of losing its lead in technology. However, overcapacity, slow demand growth, and rising costs prevented it from erasing the mistake of the 1950s. In the meantime, the companies also profited on a short-term basis by selling steel technology to industry overseas, further strengthening potential rivals. The U.S. industry, argue Barnett and Schorsch, "acted as though its preeminence was unassailable: Investment was undertaken with little regard for the potential of technological progress, and oligopolistic behavior was sustained in spite of increasing evidence of" the budding competitiveness of foreign steelmakers.

In the 1960s, the domestic industry began a massive shift to the BOF—too late. Sagging profit margins turned investors away from steel, forcing the companies to take on new debt to modernize. Limited in the amount of capital available, the firms could modernize only on a selective basis, leaving them with many obsolescent plants. By then, the cost advantages possessed by Japanese and European firms enabled them to move steel into the United States. For the first time since the turn of the century, steel imposts exceeded exports in the 1959, the year that U.S. mills were closed by a strike for four months. Practically every year thereafter saw a rise in imports. Although domestic consumption of steel grew 1.8 percent per year between 1960 and 1981, steel shipments by domestic companies increased only 1 percent per year. Imports captured the rest of the growth, almost half of it.

In the 1970s, investment in new equipment fell 25 percent from the level of the 1960s. The industry also failed to develop its own new technologies. The BOF was a European development; advances in blast furnace technology were made in

Japan. But the U.S. industry devoted only 0.6 percent of net sales revenue to research and development, compared with Japan's 1.6 percent. By 1980, Japan accounted for 16 percent of world steel output and the United States only 14 percent.[17]

As the industry repeatedly charged in the 1970s and 1980s, some foreign steel was unfairly "dumped" in the U.S. (sold at less than cost). At the same time, governments in European and developing countries subsidized steel to promote industrialization or to maintain employment. Under pressure from the declining steel companies (and the USW), the U.S. government in 1968 began a series of programs to gain voluntary or negotiated restraints on imports. None of these have worked especially well; in each case imports exceeded the limits, however they were established. This is because the foreign advantage in material and labor costs, and the resulting low prices, has inevitably pushed products into the United States in the absence of a policing mechanism at the borders.

By the mid-1980s, perhaps the biggest problem had become the growth of steel industries in developing countries such as Taiwan, South Korea, and Brazil. With low material and labor costs, modern equipment, the support of government policies, and a world market growth to harvest, producers in these nations have enormous advantages. But the market hasn't expanded at the same ratio as new capacity. In 1986, excess steel-producing capacity in the world totaled about 200 million tons. The steel "crisis" of the 1980s, conclude Barnett and Schorsch, is "a competitive struggle over where capacity reductions will occur." Although the American industry can argue that it did not contribute to the problem of global over-expansion, "the competitive pressure provided by imports is likely to ensure that it contributes to the solution."[18]

The domestic industry made another major forecasting error in the mid-1970s. Once again, the industry and its analysts on Wall Street looked ahead and saw a very healthy increase in steel consumption by the 1980s. Robert W. Crandall, an economist at the Brookings Institution, has assembled data showing how far short these forecasts fell. The industry's trade association, the American Iron and Steel Institute (AISI), in 1975 predicted a need for raw steel production by U.S. companies in 1983 of 170 million tons. The steel companies, it must be said, were encouraged to make such optimistic projections by eminent authorities. William T. Hogan, S.J., of Fordham University, a steel economist and historian, estimated in 1972 that production capacity should be increased to 190 million tons by 1980.

As it turned out, the firms in 1983 produced 85 million tons, 50 percent less than the amount predicted by the AISI. Even in 1980, the trade association called for government policies that would enable the industry to expand capacity from 154 million tons to 168 million by 1988. As of 1985, however, demand for steel amounted to less than 100 million tons. In the same year, the Office of Technology Assessment of the U.S. Congress declared that "the current overcapacity in the world steel market is likely to disappear soon" and urged the industry to build more mills.[19]

The steelmakers' view ahead was most cloudy regarding the development of rival materials that would substitute for steel. These include aluminum, plastics, and ceramics. Steel has been ousted especially from the auto market. Although the auto industry in 1986 still consumed more than one-fifth of steel shipped in the United States, the tonnage had dropped from 23 million tons to 12 million over ten years. There will be further shrinkage. A GM executive said that by 1990 his company would produce 1 million cars a year with plastic outer panels, compared with 150,000 in 1986.

The integrated companies also badly underestimated the growth of domestic minimills. These are small steel plants which produce steel from scrap in electric furnaces, rather than from smelted ore, and specialize in products like wire and small bars. Using the highly efficient continuous casting process, minimills usually have higher output per man-hour than the integrated mills. This result also comes from high-yielding incentive pay plans and efficient work practices. The minimills' market share, an insignificant 3 percent in 1960, soared to about 18 percent in 1982, and the chairman of the largest minimill firm predicted that it would expand to 30 percent in the early 1990s.[20]

Although the integrated companies lost the rod and wire business to the minimills, they didn't profit from the lesson. In the 1980s, urged on by the government and the Steelworkers, the big producers kept investing in new equipment. But instead of concentrating capital resources on areas and plants that had most promise, they plowed money into questionable projects. They still behaved as if they could continue to be *integrated* firms, producing a wide variety of products and being all things to their customers. From 1982 through

1985, steel investments actually exceeded depreciation at a time when the industry was losing $6 billion, according to a study by Gene Keilin of Lazard Frères. The companies got the cash by cannibalizing assets such as coal mines and inventory. "They thought the problem was to survive the trough," Keilin said. "They wound up with some marvelous machines scattered over the landscape but not hooked up into an efficient system."

Finally, the steel industry also committed errors of judgment in dealing with the Steelworkers union. It was a curious thing: More than any other U.S. industry, steel attempted to solve its trade and commercial problems by modifying its labor policy. In the 1970s, it might be said, the steel companies overadjusted in the labor area.

To explain this curious behavior, I must pick up where I left off near the beginning of this chapter—with pickets massing at the plant gates during the big strikes of the 1940s and 1950s.

The 1940s and 1950s: Strikes and Strike Threats

In the first fourteen years after World War II, the USW conducted five major strikes against the major steel producers (in 1946, 1949, 1952, 1956, and 1959). The fifth strike, in 1959, involved 519,000 steelworkers and lasted 116 days. An additional 250,000 workers outside steel were idled by the stoppage. In terms of impact on the national economy, it was the largest single strike in U.S. history.

But the actual strikes were only part of the story. The *possibility* of strikes hovered constantly over the industry and steel communities. The entire contract could be bargained only every two or three years, but the USW had the right to reopen on wages in the intervening years. Indeed, the two sides went to the bargaining table in ten of the eleven years between 1946 and 1956. No sooner would one round be over, it seemed, than a new one would start.[21]

All of this negotiating, striking, and threatening to strike exacted a psychological toll on the residents of steel towns in Pennsylvania, Illinois, Indiana, Ohio, and elsewhere. The tension was especially noticeable in western Pennsylvania, because John L. Lewis and his Mine Workers were also striking, or threatening to strike, just about every year. Coal and steel were deeply intertwined, both commercially and in family terms. Many steelworkers had started as coal diggers and still had brothers, uncles, and fathers in the mines. In many Mon Valley communities, coal miners and steelworkers lived side by side, and when the mines went down, the mills were soon to follow, and vice versa.

Nobody could grow up in the Mon Valley in the late forties and early fifties and not be aware of the growing contentiousness over labor issues. I was too young to have many specific memories of the 1930s (though I remember for some reason two young men marching along Sumac Street in our residential McKeesport neighborhood with anti-CIO signs hanging from their necks). But the late forties, when I was in high school, stands out in my memory. There was continuous turmoil, excitement, anger, a buffeting of competing interests. I drove a beer truck for my father and went into taverns crowded with men talking about the mill, the union, the bosses. Union organizers were moving up and down the valley, campaigning for new members in fabricating shops, trucking concerns, retail stores, restaurants. Picket lines here, picket lines there. Smoke and dust descended on us like a flimsy gray shroud, and National Tube throbbed with life.

A few months before a negotiating deadline in steel or coal, the word "strike" would begin floating around, wispylike, just hinting of things to come, but becoming more common day by day. Weeks would pass. Steelworkers could be heard talking of "saving up" for the walkout. Then the announcement came that union and management negotiators were meeting. Union and company statements were broadcast on the radio. Rumors abounded: The union would settle for 15¢, or management wasn't giving in this time. A week or so before the deadline, strike was on everybody's mind. The newspapers carried stories about strike preparations, accompanied by the inevitable photograph of union members—smiling bravely for the camera—painting "ON STRIKE" placards. Picketing assignments were handed out. Tavern owners ordered extra kegs and cases of beer, knowing that their bars would be lined with bored strikers for the first weeks of a strike, until the financial situation began "getting tight." A few days before the deadline, union and company officials exchanged angry charges, and people began to say, "It don't look good."

In the last day or two, blast furnace crews began "banking the furnaces," arranging the ore and coke loads and adjusting the draft to reduce the burning rate so the fire would not have to be doused; re-igniting a cold furnace was a long, expensive process. Arguments broke out in bars between militant and passive steelworkers. In grocery stores and bakeries, women sighed and said, "Well, they're bankin' the furnaces." What a shame it was that the union and company couldn't "get together." Someone would ask, in a disgusted voice, "What's it all about this time? " If a good union member was present, one who kept himself informed, he'd tell you how much the cost of living had risen since "the last contract" (the word "contract" was a shorthand way of referring to contract negotiations). Sometimes a sharp argument would break out between a union hater who would declare that "the goddamned unions are ruinin' the country" and a union sympathizer who would reply, "Well, you're damned lucky we got the union, because nobody'd get a raise without us."

This, in fact, was true in the steel towns after the USW organized the industry. Steelworkers' wage increases tended to raise pay levels for everybody. For this reason, many retail merchants, whatever their political philosophy, were the friends of steelworkers and extended them credit during long strikes. But unions as institutions, and union leaders, were still viewed with suspicion by union member and nonmember alike.

The strike deadlines were always at midnight, and everybody went to bed not knowing if it was strike or no-strike and woke up at 7 A.M. knowing: If the mill siren didn't give out its low, organlike moan at 7, the strike was on. If it did sound, you knew the whole business was over—until the next time. The sigh of relief around town was almost palpable.

When the Steelworkers did strike, it always seemed to drag on interminably. The valley was unnaturally quiet, but it was a pregnant silence, so to speak, nothing like the deadly silence that enveloped the Mon when the mills closed for good in the 1980s. In the 1940s and 1950s, the steel companies did not attempt to make steel during a strike. Aside from some minor scuffling, there was no violence on the picket line. But everybody knew a strike was on. The Steelworker locals were large and powerful and commanded a high degree of member loyalty.

If a study had been done of the psychology of a steel town in those days it would have registered, I believe, widespread feelings of depression in the days and hours preceding a strike deadline. The mere threat of a strike held entire towns, the whole valley, hostage to apprehension. This had nothing to do with the rightness or wrongness of the union's demands and the company's response. The strikes hurt many people. Neither the USW nor the UMW paid regular strike benefits to their members, and the families of strikers suffered most of all. It took some families years to make up the financial losses they bore in the four-month strike of 1959.

Labor and management cannot always resolve crucial differences without wielding economic weapons. Some things are worth striking for, or, on the company side, "taking" a strike for. By striking in 1949, for example, steelworkers forced the companies to negotiate with the union over pensions—a concept that management had resisted. But that decade and a half of strikes and threatened strikes in the steel industry turned out largely to be a misspent time both for the union and the companies.

Government Involvement in Wages and Prices

The prevalence of steel strikes in the immediate postwar years is attributable to several factors. Steel was only one industry swept along in the massive strike wave of 1946. Preoccupied with organizing the mass production industries before the war, the industrial unions had accomplished little in raising workers' living standards. By 1946 a pent-up demand for wage increases gave them a reason to display their economic muscle. The 1946 round of strikes seems to have been inevitable, given management's resolve to contain unionism after the war.

Another factor is that during the war, steel management and the union had not been free to construct by trial and error a relationship that suited their separate and mutual interests. After the war, the two sides made progress on a number of fronts, including the USW's vital grievance-arbitration and seniority systems. A monumental job evaluation and classification study, carried out jointly by union and management in the mid-1940s, enabled the industry to rationalize its chaotic (and unfair) wage-rate system.

These were important provisions that laid a permanent foundation for fair and equitable treatment of workers. But the USW and the steel companies had much less success coping with the complex relationships between the major issues that are at the heart of every union-management bargain: wages, productivity, competitiveness, and the sharing of power. Dealing with these issues required a strategy for the future, and neither side developed anything more than short-term goals. Management's goal was to keep the union out of managerial affairs and maintain high profit margins (by raising prices, if need be, to offset cost increases). The union's short-term goal was to push to the limit for economic benefits, regardless of whether they would be generated solely by the growth of the steel business, or whether the consumers of steel would have to chip in by paying higher prices.

Collective bargaining became a kind of three-party chase round the maypole: the union pursuing the companies for wage boosts, the companies hounding the government for approval to raise prices to cover the wage hikes, and the government (especially Harry Truman's administration) dashing after the union to obtain strike restraint (as well as labor votes).

Maintaining the industry's competitiveness was not a major consideration during those heady years of U.S. world dominance in steel. Phil Murray concentrated on increasing wages, eliminating wage differentials among steel-producing regions, and establishing pension and insurance benefits. To some extent, he used tactics aimed at producing government intervention, expecting at the very least fair treatment from a Democratic president. This was a realistic recognition of the fact that the government *would* try to avoid the economic damage caused by steel shutdowns. In addition, as the historian Ronald Schatz has pointed out, Murray believed the American labor movement could succeed only in alliance with, and under the protection of, the federal government.

The government time and again intervened in steel bargaining with mediation, fact-finding boards, and President Truman's seizure of the mills to prevent a strike in 1952 (the U.S. Supreme Court ruled the seizure unconstitutional and the strike occurred anyway). In 1959 the Eisenhower administration obtained a court injunction under the "emergency disputes" provision of the Taft-Hartley Act to halt the steel strike. In most cases, the timing and nature of the interventions influenced negotiations; the steel companies were unwilling to settle without an indication from the government that they would be able to raise prices to cover any wage increase.

In practically every bargaining year, the industry raised prices immediately after a labor settlement (sometimes before, too), and in many cases the size of the price hike was specifically approved by government officials. There was always the suspicion, dating from the very first agreement between SWOC and U.S. Steel in 1937, that the companies raised prices more than was necessary to cover the wage boost. That price increase in 1937, in fact, was estimated by government economists to be double the outlay in wages.

Because of the complexity of steel pricing, an immediate calculation of the net gain or loss to the company was seldom possible. Very likely, the industry took a beating on wage increases in some years, particularly from 1956 to 1959, when labor costs increased by 30 percent. Moreover, steel was not an overly profitable industry. From 1940 to 1959, steel's rate of return on stockholders' equity was lower than the all-manufacturing average in all but three years.

The USW, it is true, usually maintained an official silence about the industry's pricing policy, contending that it bargained on behalf of the workers and let the companies worry about prices. The USW's actions behind the scenes, however, suggest that its public position was hypocritical. In the aftermath of the 1946 strike, according to Ben Fischer, the union interceded with the government to get permission for small steel producers to raise prices. Another instance occurred in 1952, when the union actually bargained with the government for an increase in iron ore prices to offset a wage boost for ore miners. Not publicized at the time, this maneuvering was revealed many years later by former USW President David J. McDonald after he retired from the union.[22]

No simple conclusion can be drawn about the wage-price relationship in steel, except to say that both the union and the companies looked to price increases as the source of wage boosts rather than productivity. And so, the dollar—the wage dollar and the price dollar—became the focus of steel labor and management, not relations in the workplace where the coming competitive battles would be won or lost and where authoritarian bosses stifled worker initiative.

Finally, the mode of bargaining in steel made strikes more likely than in other industries. Starting in 1946, the USW usually negotiated the pattern-setting agreement with the industry leader, U.S. Steel. Dozens of other steel-producing and fabricating companies followed the leader, either in settling or in taking a strike. Essentially, this was "industrywide bargaining," although only one company participated. The reality of this relationship was recognized in the 1950s, when several large companies began bargaining as a group (see chapter 9). Since a strike was against the entire industry instead of a single company, no firm suffered more than another. "There is," says the most authoritative history of steel bargaining, "less pressure to avoid a strike in steel...because steel strikes are industrywide and virtually all companies are shut down."[23]

Whatever advantages industrywide bargaining had for both sides, it had a major disadvantage. Just as the people who lived in the steel towns became apprehensive about strikes, so did steel's customers. To avoid having to shut down for lack of steel, they would buy extra metal and stockpile it, starting months before a strike deadline. This "hedge-buying" had little adverse impact on the industry until the 1959 strike, which lasted so long that customers turned to foreign sources for steel. Imports more than doubled over the previous year, to 4.4 million tons, making the United States a net importer of steel for the first time since the 1800s.[24]

Normally, union leaders like to be taken seriously when they raise the strike-warning flag; it gives them leverage in negotiations. But USW negotiators found themselves in the strange position of hurting their own members by talking strike. In each bargaining year after 1959, the mere threat of a walkout caused a boom in steel-buying before the strike deadline, followed by a big bust after. To meet the artificially high demand, the steelmakers had to reactivate standby equipment (old, inefficient iron and steel furnaces), put work forces on costly overtime, and hire new workers. When demand returned to normal, the market would be glutted with steel, forcing the producers to lay off workers and deactivate the old furnaces.

Even when U.S. mills ran flat out, they could not satisfy the hedge-buying demand for steel products. Steel users increasingly turned to foreign steelmakers, who frequently demanded long-term supply contracts. Imports surged to 10.4 million tons in 1965, to 18 million tons in 1968, and to 18.3 million tons in 1971, and stayed high in intervening years. The 1971 imports constituted 18 percent of the domestic market, an intolerable intrusion in the view of both the steel companies and the USW.

The only possible way to discourage hedge-buying and reliance on foreign sources was to give customers an ironclad assurance there would be no strike. Was it possible? Could the industry and union leaders find a way of eliminating the threat of strike or lockout and still retain their rights as independent entities? The American industrial relations system was based on the idea that the conflicting interests of management and labor could be compromised only by allowing the two sides to fight it out in the marketplace, guided by a few rules to prevent unfair play—and may the strongest side win. It was an imperfect system, but whatever replaced it might carry even more risks. Above all, taking the drastic step of eliminating the "right" to strike called for a union leader of uncommon credentials. That man was I. W. Abel.

Chapter 5

The No-Strike Agreement

In 1973, I. W. Abel, a former foundryman, was serving his third term as USW president. At the age of sixty-four, white-haired, bass-voiced, his face a tribulation of bumps and wrinkles, Abel had a working-class stolidity about him that fit the public stereotype of union presidents. When he sat beside George Meany at an AFL-CIO executive council meeting, it appeared that he had been dozing there for years amidst the old heads and swirling cigar smoke. This was not a true image, for Abel was a fairly progressive leader and a competent administrator. His major achievement—and it was no small thing—was that he had opened up the USW to rank-and-file concerns and put an end to the head-bashing of dissidents that occurred at USW conventions under his predecessor, David J. McDonald.

Abel, however, was not a militant's delight. He was a man of glacial patience who, like McDonald, had developed some sympathy for management's efforts to solve the growing profit and import problems. He had been involved in every major steel strike since the founding days of the union and believed that walkouts often were more damaging than useful. As a district director in Ohio, he conducted thirty-four strikes in one year. "You soon learn from your membership," he told the USW's 1966 convention, "that this is a costly procedure and an unpopular procedure."

In 1967 R. Conrad Cooper, the steel industry's chief bargainer since 1959, conceived the idea of negotiating under a no-strike agreement and proposed it to Abel. The USW executive board heatedly rejected the plan, and the 1968 steel talks almost ended in a strike. The settlement, however, proved unpopular with steelworkers because of compromises on

109

vacation scheduling. When Abel ran for election to his second term in 1969, the rank and file stunned him by giving 40 percent of their vote to a little-known staff lawyer, Emil Narick.[1]

Thus, the 1971 steel negotiations loomed as critically important to Abel's aspirations to a third term. Meanwhile, the steel industry had developed very severe problems, despite a huge capital spending program in the 1960s largely aimed at installing the basic oxygen furnaces that had not been built in the late 1950s. With steel imports taking an increasing share of the U.S. market, domestic producers could not digest a $3 billion annual investment. Long-term debt had more than doubled since 1959, to $5.3 billion. But the modernization program had yielded little; productivity rose, but not as fast as wages. Earning only 2.3 percent on sales in 1970, the companies refused to cut prices to beat back Japanese imports. For the first time, government economists expressed concern about the very "survival" of the industry.[2]

At this critical juncture, the companies still believed that new technology plus import protection would solve their problems. They could bounce back if only Washington would help, and the Nixon administration did help by renegotiating a voluntary program under which Japanese producers limited steel exports to the United States. The industry's new chief bargainer, R. Heath Larry of U.S. Steel, argued for wage restraint in 1971 negotiations. However, Abel faced a rank-and-file revolt. The USW had given up its cost-of-living (COLA) protection in the early 1960s, and inflation since 1968 reduced steelworkers' average weekly earnings by 3.5 percent.

Although Abel still wanted to find an alternative to crisis bargaining, the gap between management and labor interests was too wide. He used the strike threat to good effect in July 1971 and, aided by pressure brought to bear on steel executives by the Nixon administration, emerged from last-minute bargaining on August 1, 1971, with an enormous settlement. The wage increases alone amounted to 31 percent over three years, and the union regained its COLA provision. When Abel presented the agreement to local union presidents, they literally whooped with joy and quickly ratified it. But Abel lectured them sternly, saying the union could not continue dipping into the industry's well for huge contract gains if the well ran dry. He appealed to the unionists to help "save" the industry by cooperating in a program to raise productivity.

This was the beginning of an acknowledgment by labor and industry leaders that workers *could* make a contribution to raise output. Unfortunately, the "productivity committees" mandated by the 1971 settlement turned out to be feeble creatures, dominated by the existing management philosophy that "participation" by workers should consist of a willingness to eliminate their own jobs. Nor did the USW demand that the companies elicit workers' ideas to increase output *in return for a share of the monetary gains and power in the companies* (see chapter 11).

The steelworkers' joy about the 1971 wage increases evaporated in the months ahead. Steel customers had stockpiled about 17 million tons of steel. After the settlement, layoffs, begun in June, accelerated. About one hundred thousand steelworkers were furloughed in the next few months, and ninety thousand were out of work for six months. "Sure, we negotiated a helluva agreement," Joe Odorcich told me at the time, "but not everybody was around to enjoy it." For the companies, the triennial buildup and letdown was a waste of resources and effort which cost a total of $80 million in 1971.[3]

The 1971 bust period made a deep impression on Abel. If the price of threatening to strike was that one hundred thousand members would be laid off while the remaining two hundred thousand received the immediate 10 percent pay hike, would the jobless members think the price was worth the principle? This experience added to his growing conviction that the industrywide strike might be outmoded. As USW secretary-treasurer during the long 1959 strike, he received hundreds of letters from workers and their wives asking for financial help. But the union did not then have a strike fund, and Abel could approve aid only for "hardship cases." "We had no money, and I had to tell our people to go on welfare," he recalled. "I know what it is to be in the midst of a strike and have the ranks desert. So when you talk of a strike involving five hundred thousand people, you have to do some serious thinking, some avoidance perhaps."[4]

In late 1972, Abel and Larry began a series of secret talks to develop a method of avoiding hedge-buying. Abel was in a more favorable political position by then. He had no opposition for reelection to his third term in early 1973, and after that he would be a lame duck president (he would be too old under USW rules to run in 1977). The two sides finally reached a tentative agreement in March 1973, and

Abel called a meeting of the Basic Steel Industry Conference in Pittsburgh.

Most of the local presidents who convened at the Pittsburgh Hilton on March 28 were shocked when Abel laid before them a proposal to sign a no-strike pledge with the steel industry. The more militant unionists condemned the plan and tried to organize a battle against it. But Abel had the support of his executive board, and in meetings that lasted a day and a half he also won over a large majority of the BSIC delegates. They ratified the agreement on March 29.

The arrangement was contained in an eleven-page document entitled Experimental Negotiating Agreement, which became widely known as the ENA. It was a set of rules under which steel bargaining would be conducted in 1974. The union pledged not to strike the companies, and the companies pledged not to lock out the union. If negotiating issues remained unresolved on April 15, they would be submitted to a panel of arbitrators whose decision would be binding. This meant that steel customers had an ironclad assurance that steel production would not be interrupted when existing labor contracts expired on July 31, 1974.

The agreement was unprecedented in U.S. labor history. In a few instances, a single company and a union had submitted bargaining issues to arbitration. During railroad labor disputes in the 1960s, Congress had imposed *compulsory* arbitration on that industry and its unions. But this was the first time that a major union and an entire industry had agreed *voluntarily* to settle bargaining disputes by arbitration rather than strike or lockout.

In some ways, the ENA was an ingenious piece of work. Abel had insisted on a crucial provision that probably persuaded a large number of BSIC delegates to vote for it. While there would be no industrywide strike, the ENA expressly granted local unions the right to strike individual plants over strictly local matters. These included such issues as washup and relief time on the job, the condition of locker rooms and parking lots, and other items that had long been festering in the plants. The locals had never had authority to strike over these matters. Over the next several years, many locals would discover that this new-found strike ability was far more limited than it first appeared. But it helped win approval of the ENA.

ENA critics contended that the union should not have given up the strike weapon without a vote of the membership.

But the USW constitution did not require rank-and-file ratification, and a federal district court dismissed a suit filed by ENA opponents. This became a central issue in the Sadlowski-McBride election battle in 1977 and continued to rankle for many years. But the majority of steelworkers were willing to live with the ENA, and increasingly so after they began to collect the benefits that it conferred upon them.[5]

To the steel companies' eternal regret, the ENA turned out to be the most incredible money-making machine ever invented in collective bargaining. This came about because of the two sides' reluctance to expose themselves to the decisions of arbitrators on provisions in the contract that each side considered inviolable. To prevent the arbitrators from tampering with "sacred cow" issues, the steel negotiators devised a system of guarantees and exclusions, and the union got by far the better deal.

The companies, for example, insisted that a contract clause declaring that management had the sole right to run the business be excluded from arbitration. Abel extracted from the industry a guarantee that wages would rise 3 percent annually. The USW could bargain upward from the 3 percent floor, but the industry couldn't bargain the figure down. Furthermore, the USW's recently regained COLA provision was excluded from arbitration. Even if COLA payouts damaged the companies financially (as they did in the late 1970s), the industry couldn't ask the arbitration panel to revise the existing formula. As long as the industry wanted to be protected from a strike threat, it would be stuck with one of the more generous COLA formulas in American industry.[6]

Perhaps because both sides feared arbitration, it was never invoked. Although the ENA governed three rounds of steel talks (in 1974, 1977 and 1980), the parties managed to settle all issues on their own. This was an impressive display of negotiating ability—as far as it went. The problem was that the most critical economic issue—the rate of rise in wages—was neither bargainable (downward) nor arbitrable. From 1974 through 1982, steelworkers received automatic, annual pay increases of at least 3 percent (in most years it was more), while COLA payments rose with inflation.

Once this machine started operating, its generative ability was awesome. The average wage rate at companies covered by the industrywide agreement, including the 3 percent increases and COLA payments, jumped from $4.27 per hour on August 1, 1972, to $11.91 on August 1, 1982. COLA alone contributed $5.11 of the increase. In that ten-year period, the steel wage rate climbed 179 percent, while inflation rose 132 percent. Compounding was the key to this phenomenal rise. Each year, the COLA hikes were added to the base rates, thereby raising the value of the next 3 percent raise and boosting vacation, overtime, and holiday pay (which are calculated on the base rate).

The ENA did, indeed, eliminate hedge-buying in the years that it was used. It did not halt the growth of steel imports, and in fact exacerbated the problem because it increased production costs. "In effect," conclude Barnett and Schorsch, "the industry believed its own propaganda; namely, that the periodic threat of a steel strike provided the foundation for ratchetlike surges in the import share."[7]

There was little economic rationale for the steep rise of steel wages in the 1970s. Output per man-hour had increased by only 1.5 percent per year in the decade before the ENA was signed, and the negotiators were well aware of this. The 3 percent guarantee, Abel told me at the time, "wasn't an easy thing to sell to an industry that hasn't enjoyed a 3 percent productivity improvement over the years." Why, then, did the industry buy it?

It was not because USW negotiators submitted overwhelming evidence to support 3 percent. The union cited two influencing factors. First, the UAW received automatic 3 percent pay raises in the auto industry, and the USW, among other unions, felt its wage rises should be in the same range. Second, Kennedy administration economists had suggested in the early 1960s that if an industry met many qualifying conditions, including the way it priced its products, annual compensation increases of 3.2 percent (reflecting the national productivity growth rate) would not be inflationary.

Neither of these arguments could be economically persuasive, given steel's productivity problem. Steel management, however, being perennially optimistic, speculated that a high-volume crest lay in the future. Very likely the persistence of the union negotiators, rather than their arguments, finally wore down the industry negotiators. After all, they may have thought, 3 percent of the 1972 average wage rate of $4.27 an hour was only 14¢. The oil shocks and high inflation rates of the mid and late 1970s were not foreseen in 1973.[8]

Looking back on the ENA negotiation, it is not overly harsh

to conclude that the industry's eyes were bigger than its stomach. It coveted no-strike for its presumed benefits but couldn't digest its costs. In terms of carrying out traditional bargaining roles, the Steelworker leaders represented the interests of their constituents with greater shrewdness than the management bargainers did theirs. The union leaders pushed, as they were supposed to do in that form of traditional bargaining. Management gave up too much to solve a problem that was only a manifestation of the industry's central problem: It had lost its competitive-cost advantage.

When the ENA was first announced, many commentators hailed it as a great feat of "labor diplomacy." Ten years later, the discovery that steel's hourly labor costs ranged from $22 to $25, drew hostile criticism from many of the same sources, especially editorial writers who blamed "high-wage" steelworkers for the industry's decline. As Barnett and Schorsch point out, however, the union "was only one of the two signatories to the ENA."[9]

Bruce Johnston: Labor Took Too Much

By 1982, when the steel industry was falling apart, the suspicion and anger engendered by the no-strike pact among militant local leaders remained a barrier to coordinated union action. Some company executives also had disliked the ENA from its inception, though for much different reasons. Among them was the man who later became the industry's chief labor negotiator, J. Bruce Johnston of U.S. Steel.

No one could claim that Johnston was an absentee landlord. Born in Donora, he had strong ties to the Mon Valley, the steel industry, and even the Steelworkers. His father emigrated from Scotland, having been recruited by agents for U.S. Steel to work for the company in Donora and play on the plant soccer team. The sport was popular among immigrants who settled in the middle regions of the Mon Valley, and the elder Johnston earned $100 per game as a player and later coached the team. He went to night school, advanced to foreman and became superintendent of the Donora zinc works.

Bruce Johnston has good memories of growing up in "the wonderful melting pot" of Donora, where people of different nationalities stood in separate lines in the butcher shop so they would be waited on by a butcher who spoke their language. In high school, he played football and baseball. He earned both his undergraduate and law degrees at the University of Pittsburgh, although he also attended Harvard Law School. During this period, he worked six summers in the Donora plant. Johnston's official company biography indicates the importance to him of this background. The third sentence, the point at which most executive biographies begin listing major accomplishments, notes that Johnston "started as a laborer with the company and worked regularly in steel mill jobs while attending college and law school." The Mon Valley heritage was like a chestful of combat ribbons, and Johnston was not shy about displaying them to union negotiators. Indeed, he could be contemptuous of union lawyers who had never lived or worked in a steel community yet presumed to speak for "our members in the Mon Valley."[10]

Johnston joined U.S. Steel in 1957 as a labor attorney and in 1966 was put in charge of labor relations for nonsteel divisions and subsidiaries. He rose rather quickly through U.S. Steel's thick layers of assistant managers, managers, and general managers. In 1973, at the age of forty-three, he became vice-president of labor relations for the corporation, only two steps below the highest position in labor and personnel. I first met him that year, at the time the ENA was signed, and later remembered that he seemed skeptical about the agreement. A lean man of medium height and black hair, Johnston had a sharp face with deep-set, restless eyes. In a face-to-face conversation, those gray eyes would scroll up and down, as though reflecting the brain's activity as it registered and categorized incoming information and prepared a rejoinder. There always was a rejoinder, usually a pointed one, for Johnston was a debater.

During his senior year in high school, 1948, he had won various debating championships. Johnston possessed the characteristics of a good debater—a quick mind, a facility for pithy expressions, intellectual aggressiveness—and he developed a hatful of forensic tricks. A superb storyteller, he had a large fund of anecdotes and often opened a bargaining session with an off-color joke. Johnston also was a gifted mimic who enjoyed acting out parables of economic behavior. He often recounted episodes from the business affairs of Andrew Carnegie, who he admired, quoting Carnegie in a Scottish brogue. If the occasion or issue called for it, Johnston could descend to ridicule. Once, during an interview, he stood in front of me with rounded shoulders and swinging, apelike arms, pretending to speak the mind of a selfish steelworker who wanted only "more."

According to company colleagues and union officials, Johnston was a formidable opponent at the bargaining table. He disarmed people with an astonishing ability to recall facts and statistical data. He could take command of a meeting by engaging in long, extemporaneous disquisitions on the state of the steel industry or labor relations. With these tactics, he tended to control the agenda of a negotiating round.

Brooding and moody, Johnston was known in the company as a "real loner," says a former U.S. Steel executive, Alvin L. Hillegass, who retired in 1982 as vice-president of steel operations. "He doesn't like to get involved at any level, except with Dave Roderick [U.S. Steel chairman]. Johnston keeps himself sequestered on the sixty-first floor. He talks to Roderick, calls people in, makes speeches. He's a spellbinding speaker." Johnston often turned staff meetings, as well as interviews with reporters, into scolding lectures on labor and economic issues. When I mentioned to a retired member of Johnston's staff that I'd had such lectures, he laughed and said, "Haven't you ever walked out of there and said, 'What did he say?'" To inform people he considered ill-educated about labor relations, Johnston frequently sent them clippings of newspaper and magazine articles on which he had scrawled critical comments in red ink. Among the recipients were union officials, professors, consultants, and journalists.

For all his contentiousness, Johnston was by no means a union buster of the far right. He had absorbed the "steelworker ethos," as he liked to put it, while growing up and had familial links to the Steelworkers and Mine Workers. An uncle on his mother's side of the family, H. Charles Ford, served as director of the USW's District 7, Philadelphia, in the 1940s. Various other uncles and cousins worked in the Donora mill and nearby coal mines, and some served as officers of local unions. Johnston left law school as "a certified, Brandeis-Holmes, card-carrying liberal," he told me, and favored John F. Kennedy in 1960. But he came to view the valley, and unionism, in a far different light than his uncles and cousins.

In later years, when he drove through the Mon Valley and contemplated the mills and furnaces stretching along the river, Johnston could only think, with deep admiration, of Carnegie and other steel magnates who made it possible. "What a colossal corporate structure that was!" he said. It existed only because the early entrepreneurs and stockholders had risked their capital to put it in place. Organized labor and government, however, inhibited capital investment. The USW had sprung to life for good reason, but in later years it had done nothing but "consume," Johnston came to believe. "Only the union took a premium out of the business," he said, leaving practically nothing for the shareholder or for reinvestment.

By the time the ENA was negotiated in 1973, Johnston had joined the ranks of hardliners who wanted to restrain wage growth. He disliked the automatic wage and COLA escalation embodied in the agreement and criticized the plan in a written analysis. However, he was not in a policymaking position, and the memo had little effect. Heath Larry, U.S. Steel's chief labor executive in 1973, could not remember such a memo when I asked him about it in 1986.[11]

In 1974, however, Johnston found an important potential ally in his opposition to the ENA, another rising executive in U.S. Steel named David M. Roderick. An accountant and former marine, Roderick had been based for many years in Europe as vice-president of the company's international operations. He returned to the United States and was elected a director and chairman of the company's powerful finance committee in the very month that the ENA was signed. Johnston chatted with Roderick and discovered that the latter "had a serious concern about the ENA." It locked the companies into long-term contracts with a rich cost-of-living provision. Inflation would rise in a growing economy, but there was no guarantee that steel's profitability would. "I don't know how you can guarantee water out of a spigot when back at the house there is no water," Johnston said later.

Roderick and Johnston developed a strong association. The former became chairman and chief executive officer in 1979. By 1986 Johnston was executive vice-president in charge of all employee relations and a member of Roderick's top-level corporate policy committee, a position of unusual influence for executives in the human resources area. "Dave thinks there's nobody in the world better than Bruce," said Alvin Hillegass, who was involved in high-level policymaking in 1980 and 1981.

Neither Roderick nor Johnston, however, had enough power in the mid-to-late-1970s to cancel or modify the no-strike agreement. Negotiating under the ENA in 1974, the industry granted large pay and benefit increases and improved the COLA formula so that it would yield higher payments. The size of the package didn't bother the steel companies at the time, because they shipped record amounts of steel in 1973–74. For the same reason, the industry committed itself to using the ENA procedure again in 1977. Total employment costs rose by 36 percent between 1974 and 1976.[12]

Johnston seized an opportunity in late 1976 to speak out publicly against the ENA. At the time, Lloyd McBride and Ed Sadlowski were campaigning for the USW presidency. Sadlowski attacked the union establishment, in which he included McBride, for refusing to submit the ENA to the rank and file for ratification. McBride was more circumspect but didn't issue a ringing endorsement of the agreement. Johnston wrote a stinging speech and, with Larry's approval (even though it implied criticism of the industry), delivered it on December 16, 1976, at a meeting of the Pittsburgh Personnel Association. He attacked the USW presidential candidates for using the ENA as a political punching bag, without acknowledging that it guaranteed steelworkers high wages. Pointing out that 1977 bargaining would take place under the ENA, Johnston said the industry had obligated itself to a $3.20 per hour increase in employment costs "before we open our mouths in negotiations." This was more than "the Steelworkers ever got in any pre-ENA settlement when they were flexing their so-called strike muscle every bargaining year." Although Johnston pretended not to take sides, comments like this probably were intended to help McBride's candidacy as well at attack the ENA. As the establishment candidate, McBride would stand to gain votes from a management criticism of the rich settlements produced by the ENA.[13]

Larry retired in early 1977, and Johnston was named to lead the industry in the 1977 talks. In his initial performance as chief bargainer, he not only failed to restrain the USW but allowed a larger wage increase than the ENA required. The package also contained the new lifetime job security plan with its unprecedented income guarantees to long-service workers affected by plant shutdowns.[14]

The last issue to be settled in the negotiations was what to do about the ENA. Johnston had recommended to the industry's chief executive officers that they not renew it and return to negotiating against a strike deadline. The big runup of oil prices in the mid-1970s had swollen the inflation rate, producing COLA increases of $1.22 per hour for steelworkers from 1974 to 1976. Industry leaders, however, indulged in wishful thinking. They believed the high inflation rate an aberration, and they still had hopes that the long-predicted shortage of steel would finally arrive. They did not want strikes to interrupt a period of high-volume production. "Most of the CEOs, backed by their commercial departments, still were scared to lose that insurance policy," said George Moore, who in 1977 was Bethlehem's second-ranking labor executive. Although Roderick supported Johnston's recommendation, U.S. Steel Chairman Edgar Speer rejected it in favor of using the ENA again in 1980. "I told them the insurance premium was more costly than the indemnification," Johnston said later. "But I couldn't get a consensus of the CEOs."

It was a different matter in 1980, however, because Roderick had assumed command of U.S. Steel in 1979 when Speer retired. A hard-headed numbers man, Roderick did not have Speer's romantic attachment to the "Big Steel" of the past. Speer had dreamed of recapturing U.S. Steel's world dominance by building a new integrated steel plant at Conneaut, Ohio. Roderick cancelled the project when he took over. As for the ENA, he noted that it had not halted a steady rise in steel imports by eliminating hedge-buying. But it had cost a pretty penny. Inflation spurted upward again in the late 1970s, increasing the industry's cost-of-living bill by $1.71 per hour between 1977 and 1979.

When the steel chief executive met in early 1980 to discuss the coming steel talks, several of them argued for keeping the ENA protection for three more years, despite the acceleration of labor costs since the early 1970s. According to Johnston, who attended the CEO meetings, only Roderick took a firm position against retaining the procedure.

In addition to Johnston, labor relations officials in other companies were trying to wean their chief executives away from the no-strike agreement. At Bethlehem, for example, Anthony St. John, who served under Moore, advocated pulling out of the coordinated bargaining group, contending that the ENA and the oligopolistic approach to labor relations were both outmoded by the late 1970s. "Moore and I had some substantial arguments about it," St. John recalled in 1986, after he had left Bethlehem to become head of labor relations at Chrysler. Moore was willing to jettison the ENA but wanted to keep coordinated bargaining. Bethlehem's top management followed his advice. At Jones & Laughlin, John H. Kirkwood, who later became chief negotiator, began recommending a withdrawal from coordinated bargaining in 1977—to no avail.[15]

During several weeks of steel bargaining in the spring of 1980, the debate over the ENA continued in internal company sessions and at the CEO level. By the end of the talks, the anti-ENA people had prevailed, and Johnston was authorized to reject the union's offer to renew the no-strike pact for use in 1983. The two sides announced settlement on a new, three-year labor agreement on April 14, 1980, but said the ENA issue would be deferred. They could sign a new ENA any time in the next two years and still halt stockpiling before the 1983 talks.

Although Johnston was not bashful about his part in killing what he felt was a bad deal from the beginning, he carefully refrained from publicly criticizing his predecessors. "The ENA was a noble experiment," he said in a 1986 interview. "It looked like the answer. But it was an answer in a market that no longer existed."

In one sense, the ENA had been the first attempt by the union and the industry to forge a wage-productivity relationship for the long term. Unfortunately, the formula had been faulty. But now, belatedly, it was gone, and the two parties had to decide what to put in its place.

Quiet Talks on the ENA

For the remainder of 1980, however, that decision was low on the priority list. The USW focused its attention on trying to reelect President Jimmy Carter in the fall of 1980. The companies were preoccupied with a business slump that reduced shipments of steel in 1980 to 83.9 million tons, down from 100.3 million in 1979. The federal government's "trigger-price" mechanism for restraining imports had come to an ineffectual end, and U.S. Steel pressed ahead with major court suits aimed at forcing the government to take action against foreign producers that dumped steel in the United States.[16]

The USW and the companies also continued their involvement in the Carter administration's Steel Tripartite Committee (STC). Created on the recommendation of a government task force on steel, the STC was cochaired by the secretaries of labor and commerce. This was a promising approach to

solving some of the industry's problems, based on the concept
of gaining a consensus of management, labor, and government
on what each could contribute to the process. The recommendations of the STC, issued in late 1980, included depreciation
allowances to encourage investment (eventually provided by
Congress), amendments to the Clean Air Act to allow the
industry more time for compliance (also provided by Congress), measures to promote development of steel technology,
revival of the trigger-price process, and improved assistance to
workers and communities affected by shutdowns.

Arriving at a consensus on these issues might have served
as a base for resolving more important matters. But the committee was disbanded when Carter left office. The Reagan
administration, with its ideological antipathy to government
intervention in the marketplace, had little use for tripartite
committees. The administration did conduct tripartite discussions in 1985, but the government's half-hearted commitment to the process barred any real accomplishments. It is
true that establishment of an industrial policy based on tripar-
tism faces hazards, chiefly the tendency of labor and management to seek government help without sacrifices on the part
of the company and the union. Ray Marshall, secretary of
labor under Carter, is convinced, however, that the government could have extracted concessions from the USW and the
companies if the Carter committee had been kept alive.
"Clearly, the union and companies should have been required
to improve their productivity and reduce their costs in
exchange for tax credits, trade protection, and regulatory
relief." It is unclear whether either side would have given up
anything, had the process reached that point. But it did not.[17]

With these activities under way, and with McBride recover-
ing from a serious illness, the USW and the steel bargaining
group, Coordinating Committee Steel Companies (CCSC),
ignored the ENA issue in 1980. It was revived in 1981, in
conversations between McBride and Bruce Johnston. The
position of the industry had not changed since negotiations
were concluded in 1980: The companies *might* sign another
no-strike pact but only if the wage and COLA guarantees were
modified substantially. McBride wanted as much as he could
get but was willing to negotiate. The idea of mounting a
national steel strike in 1983, given the government's reluctance to hold imports at a lower level, was out of the question.
He also understood what a bonanza the ENA had been for

steelworkers. Under its protection, the USW had obtained wage and benefit increases at least equal to, and usually better than, unions in other major industries throughout the 1970s— and without even having to think about going on strike.

It was traditional in the steel industry for the companies' chief negotiator and the USW president to remain in constant contact. They discussed problems that arose in administering the industrywide agreement. So it was by no means unusual that Johnston and McBride talked frequently in the summer and fall of 1981, sometimes by phone, sometimes in meetings. They would meet in Johnston's or McBride's office, or occasionally have lunch together. The two had built a special relationship. Johnston respected McBride as an honest, direct man, and the USW president felt that he could trust Johnston. They usually met alone. Occasionally, Joe Odorcich and Bernie Kleiman, the USW's general counsel, accompanied McBride, although infrequently before 1982.

In these meetings, Johnston represented the industry, not just U.S. Steel, and to broaden company participation he sometimes invited George Moore, the fifty-four-year-old lawyer who in 1978 had become Bethlehem's chief negotiator. He was named in 1980 to the industry's two-man bargaining committee, along with Johnston.

Tall and sharp-featured, Moore wore owlish spectacles, pinstripe suits and vests, and could have passed for an old-time banker. In reality, he was a genial man with a wry sense of humor. A graduate of the University of Pennsylvania Law School, Moore was known as "a good concept man," one who can frame labor issues in their legal and political context. In one of his most important assignments, he chaired an industry committee in the early 1970s which negotiated with the USW and government agencies to produce a consent decree under which the black and female victims of past discrimination were compensated and rules were established to halt racial and sexual bias. Unlike his predecessor at Bethlehem, Moore did not mind playing second fiddle to Johnston and, in fact, formed a close partnership with him. The two thought alike on many issues.

Johnston and McBride had been meeting on various problems for some time in 1981 before McBride began suggesting that the time was ripe to negotiate a new ENA. A number of informal, exploratory discussions ensued. Neither side could have struck a deal without appropriate approvals, Johnston's

from the CEOs and McBride's from his executive board and
the Basic Steel Industry Conference (BSIC). The USW president kept pressing for negotiations to renew the ENA. Johnston and Moore said they doubted that the companies could
live with the automatic commitments for wage and COLA
increases.

Moore sensed a change in McBride's position at a meeting
in late September or early October 1981. "When I left that
meeting," he said, "I had no doubt in my mind we could have
the ENA without the 3 percent wage increases and with some
modification of COLA." Moore admitted that this was a sub-
jective impression gained from an informal dialogue, and a
number of USW officials have expressed doubts that the USW
president would have made such a statement without authority. Nevertheless, McBride many months later affirmed that
he would have recommended such an agreement to the BSIC.
"I think if the industry had been willing to enter into an ENA
that only had the bare bones of a COLA with no continuation
of the 3 percent increases, but leaving the subject open for
negotiations as opposed to a guarantee, probably we could
have worked it out. But even the COLA cost, they said, was
too much for them."[18]

Johnston and Moore reported this discussion at a meeting of
the eight chief executive officers and were instructed to
explore the modified ENA approach. But they decided to go
beyond that charter. In their own strategy sessions in the late
summer of 1981, the company men had come to the conclusion that the ENA must go. Modification, they thought,
would result in political complications for the union. If
McBride agreed to reduce the wage guarantees, he would have
a hard time selling the proposal to the local union presidents.
It would be regarded as a step backward. Even if the BSIC
ratified a scaled-back ENA pact in 1982, the problem would be
only partially solved, Johnston and Moore thought. In early
1983, less than a year later, the two sides would have to start
bargaining on a new wage agreement. Because of continuing
financial problems in steel, the industry negotiators were
already certain they would have to ask for a wage freeze in
1983 and some reductions in benefits. For the second time in a
period of months, McBride would have to present a "negativ-
e" proposal to the local presidents. Getting two yes votes on
concessions out of such a cantankerous body would be virtually impossible, Johnston and Moore speculated.

Johnston suggested a way around this problem: Persuade McBride to forget about the no-strike agreement and, instead, reopen the 1980 contracts long before they expired on August 1, 1983. Without setting a strike deadline, the two sides could negotiate lower wage and benefit increases to replace those called for in the existing pacts and extend the expiration date to 1985 or beyond. In this scenario, McBride would have to go to the local presidents only once to plead the case for concessions. The modified wage terms would hold down the rise in labor costs, thereby helping the companies ride out the current economic slump.[19]

McBride probably would not have welcomed the suggestion that management should protect him from the political processes of his own union. However, Johnston and Moore did not reveal their strategy to him. They approached their goal in a more roundabout way, by confronting McBride with the overwhelming evidence of the domestic industry's competitive problems.

Gathering this evidence had begun many months before. During the 1980 round of ENA negotiations, the industry assembled reams of financial and other data to support its case if arbitration were necessary. In the spring of 1981, Johnston decided to use the same process in a continuing effort to keep the union apprised of competitive trends. He commissioned Putnam, Hayes & Bartlett, an economic research firm in Boston, to make a study of employment, energy, and raw material costs in the Canadian steel industry, which then was taking business away from U.S. steelmakers. Canada at that time had a currency advantage over the United States, a much faster depreciation rate for writing off new equipment (four years vs. twelve years), and lower employment costs in steel. Johnston presented this information to McBride in June 1981.

Later, Johnston told Putnam, Hayes to develop a statistical methodology for comparing production costs in Canada, Japan, and West Germany with those in the United States. Included also were the labor costs of four domestic unionized industries: chemical, automotive, electrical, and petroleum refining. In late September, Johnston began showing this data to McBride. The USW president was "taken aback" when he first saw the raw numbers, Moore said, and the union president ordered his chief economist, Ed Ayoub, to review the data.[20]

Ayoub collaborated with Howard W. Pifer III, chairman and

managing director of Putnam, Hayes, for several weeks in late 1981 and early 1982, in meetings, phone calls, and correspondence. The USW economist suggested a number of technical changes and a few substantive ones. He noted that the industry's study ignored employment-cost trends in the 1960s, when productivity increased at twice the rate of labor costs in the U.S. steel industry. Ayoub also contended that the industry study dwelled on employment costs as the cause of the industry's problems, to the exclusion of management mistakes, such as the failure to modernize.

For this reason, Ayoub produced a parallel study, which came to be known as the "Blue Book" (because it had a blue cover). "I wanted to broaden the perspective of the problems of steel," Ayoub later explained. "I pointed out that employment costs weren't the only problem. You guys [the companies] missed the boat in a lot of respects. You can go back to the early sixties and see the technology lag between us and foreign producers. In 1980, continuous casting was used in only 20 percent of our production. Even the Koreans had more than that. You guys were not keeping up. I pointed out that in the seventies real capital investment went down in the steel industry."

Meanwhile, steel usage in late 1981 dipped still lower as high interest rates depressed the auto, appliance, construction, and other important steel markets. Only a continuing demand for steel pipe in the petroleum industry's exploration boom prevented the steelmakers from experiencing huge losses on steel operations. The companies had strongly supported the Reagan administration's economic program, but it began to appear that the 1981 recession—which had been caused in part by Reagonomics—was no ordinary slump. By January 1982, the steel companies' cash pools were drying up as the cost of borrowing soared. Except for oil country goods, forecasts for steel shipments in 1982 were dismal.

As they always did when a financial crisis loomed, the steel companies looked to labor for relief. The fastest way to stanch the cash drainage was to dam the outflow caused by rising labor costs. The question was how to approach the union. When the issue had last been discussed by the eight CEOs, in the fall of 1981, some were still grasping at the ENA as a solution. The situation, however, was considerably different when they met in January 1982 in San Francisco, where the

CEOs were attending a meeting of the International Iron & Steel Institute.

Johnston and Moore flew to San Francisco and presented the results of the Putnam, Hayes study to the chief executives. The data had been charted and graphed and entered on slides. In a darkened conference room, Johnston provided the running commentary as the slides were projected on a screen. He painted a very bleak picture.

As recently as 1960, the U.S. steel industry had had lower costs and better productivity than Japanese, Canadian, and German producers. By 1980 these advantages had been wiped out. It took 7.4 man-hours to produce a ton of steel in Japan and 8.5 in Canada, compared with 9.0 in the United States. It cost U.S. producers $307 to produce a ton of steel, compared with $218 in Canada and $183 in Japan. The study contained other comparisons showing American producers at a disadvantage. The most startling, though the least meaningful, figure was a projection of labor costs. If the wage and inflation trend of the 1970s continued, it would cost $66.34 per hour in 1990 to employ one steelworker.

The CEOs were shocked by the full extent of their competitive problems. Their predecessors had fallen asleep in the 1950s, content with their worldwide leadership in the steel business, and the present group of chief executives had awakened to a nightmare. (The group included David Roderick, U.S. Steel; Donald H. Trautlein, Bethlehem Steel; Thomas C. Graham, Jones & Laughlin; E. Bradley Jones, Republic Steel; Howard M. Love, National Steel; Harry Holiday, Armco Steel; Frank W. Luerssen, Inland Steel; and Richard Simmons, Allegheny-Ludlum.) Some of these men—Roderick, Love, and Holiday—had been aggressively diversifying their companies, and now the other leaders might well have wondered why *they* were still in the steel business.

Johnston outlined the labor policy alternatives. The union still wanted to renew the ENA, he said. This would prevent hedge-buying in 1983, but it would not solve the present problem of cash drainage. Significant labor cost increases were scheduled for the remaining eighteen months of the existing labor pact, including six quarterly cost-of-living adjustments and a 3 percent wage boost on August 1, 1982. Johnston proposed that the industry abandon the ENA and seek an early contract reopening. There was no need for debate. Johnston and Moore left the meeting with a mandate to put their strategy in place.[21]

Back in Pittsburgh, the industry negotiators continued meeting with McBride. One meeting, probably on February 8 in McBride's office, was particularly somber. "We told him we didn't think it was feasible from his standpoint or ours to talk about renewing the ENA, when what we really ought to be talking about was addressing the core problem, and that was the escalating employment costs," Moore said. McBride was disturbed and irritated. "I'm tired of being the only one [in the union] worrying about all this bad news," he growled. He thereupon invited Johnston and Moore to address a meeting of the USW's executive board in April.[22]

McBride previously had asked the board members for approval to invite the industry men. Even though he was used to shouldering burdens, this problem overwhelmed McBride. Other union officers, including Joe Odorcich and Lynn Williams, had seen versions of Johnston's slide presentation. However, because McBride usually talked alone to Johnston, without the presence of aides, the president was the sole elected union official who realized the extent of the industry's competitive problems and knew the full story of his own discussions with Johnston. If he felt overburdened, it was largely his fault. Now it was time to share his burden with other USW leaders.

McBride, however, still hadn't formulated an overall strategy for dealing with concession demands. He later insisted that he had no inkling that the industry bargaining group wanted to renegotiate the master economic agreement at an early date. It was not until after the UAW and General Motors reached early agreement on a slimmed-down contract that McBride became aware that the steel firms were pressing for early talks (the GM agreement was ratified on April 9). "Before that time, it was simply a question of research [on] things we could establish as fact with respect to the negotiations that would take place in 1983," McBride told me in September 1982. "We were continuing to talk about the possibility of an ENA."

This question of timing is important. During this period, McBride continued to approve concessions at me-too firms which would undercut the union's position with the large companies. One would think that the sooner he realized that the industry group would demand wage concessions, the sooner he would formulate a strategy for all concession bargaining. In fact, though, he never did devise such a strategy.[23]

A Mission to Linden Hall

The executive board meeting was held at Linden Hall, the USW's Labor Education Center, located about sixty miles southeast of Pittsburgh near the town of Dawson. The site of an old mansion, this was an ironic choice for the kind of mission that Johnston and Moore had taken on.

Linden Hall was a relic of nineteenth century industrial entrepreneurialism in the coalfields of Western Pennsylvania. It was near Jacobs Creek in Fayette County, a center of the Whiskey Rebellion of 1794 and the source of iron ore and coal used in the region's early iron industry. Sarah Cochran, the widow of a man whose family had made a fortune in the coal and coke business, moved into the thirty-five-room Tudor mansion in 1913. After she died in 1936, the 785-acre estate was converted to a country club with a golf course, swimming pool, and tennis courts. In 1976 the USW bought the estate and added a seventy-five-room motel and classroom facilities, where unionists could learn their trade—how to handle grievances, how to run an in-plant health and safety program, and so forth—and have fun doing it. So comfortable and expensive-looking was the layout, that union opponents of I. W. Abel branded it "the officers' playground." And, indeed, it was a powerful symbol of the Steelworkers' rise in status and power.[24]

That the USW owned a country club represented a curious turn of events. Golf had always been "the rich man's game," inaccessible before the 1950s to most steelworkers in the Mon Valley and elsewhere because of cost and location. Golf was played in shallow ravines and gently sloping meadows many miles from the mill towns, far out of the path of the trolley cars that jerked and swayed along the main streets. When steelworkers took up sports, it usually was bowling, or soccer in the middle Mon towns, or—the most popular—fast-pitch softball. In the 1930s and 1940s, the mill towns were dotted with softball fields, rutted diamonds that lay tentatively between a gorge on one side and a hill on the other. Teams sponsored by churches, city wards, and mill departments competed for league championships and drew large audiences.

So it was into the early 1950s, when a younger generation of steelworkers, driving Chevies and Dodges, began finding their way out to the new public golf courses, which charged only $1.50 for eighteen holes. The trend grew rapidly and by the 1970s golf had become a sort of national steelworkers' sport. Every year the USW conducted a tournament, matching district against district, with the best golfers advancing to the championship match, first at famous courses and later at Linden Hall. The winners, fashionably dressed in monogrammed polo shirts and Arnold Palmer slacks, posed for photographs to be used in *Steel Labor*, the union's monthly publication. Gone were the pictures of softball players kneeling in the dirt in smudged uniforms. The softball fields fell into disuse and, eroded by rain, eventually washed into the gullies.

It was a fascinating cultural change. As recently as the Abel-McDonald election in 1965, the Abel forces had criticized McDonald's membership at the South Hills Country Club in a Pittsburgh suburb and pictured him as hobnobbing with "management pals" on the golf course and in the bars. In 1982, however, some of the Steelworker board members who had been most outraged by McDonald's "tuxedo unionism" were habitual users of the Linden Hall golf course. There was nothing wrong with this, of course. Why shouldn't steelworkers play golf? The union had won them the means to take up the game, even to have their own golf course.

But on April 6, 1982, the steel industry came to Linden Hall and told the assembled officers and district directors that the entire structure—the high wages, middle-class culture, and hundreds of thousands of jobs—was in danger of collapsing. In the coalfields, greedy operators of yore had often extracted every last ounce of coal from the mines by "pulling the pillars," or digging out the pillars of coal that should have been left to support the mine roof. This caused a general subsidence in the area and sometimes worse. Now, it was as if Linden Hall, the rustic symbol of everything the union had accomplished, was in danger of plunging into the abyss.

Ed Ayoub addressed the executive board in the morning. He distributed copies of his Blue Book, a sixty-four-page document which started with this introduction: "Steelworkers are the highest paid industrial workers in the United States. There is not another domestic industry whose combined earnings and benefits exceed those which have been negotiated by the United Steelworkers of America." As he had done in his earlier series of studies, Ayoub outlined the problems caused by rising imports, lagging technology, falling demand for steel, and increasing production capacity around the world. He had undertaken this project on his own, without orders from McBride, because he felt that the industry's study "came down too heavily on employment costs as the key problem." The economist felt that the industry's data was canted against wages because it concentrated on the rise in unit labor costs in the 1970s. The Blue Book pointed out that the industry's problems started in the 1960s when unit labor costs were declining but the technological lag was growing.

But Ayoub didn't slight the fact that steel's "real" employment costs (corrected for inflation) rose at a 4.3 percent annual rate through the 1970s, compared with a 2.4 percent growth in productivity. This meant that real unit employment costs rose 1.9 percent per year, choking off profits and investment. In all other major manufacturing industries, unit labor costs declined over the same period but began to rise again in the early 1980s.

Ayoub noted that David Roderick had recently been quoted as saying that it was "futile to increase real wages faster than productivity growth." Ayoub said he interpreted Roderick's statement to mean the industry's leading company had adopted a new policy of slowing wage growth. In what would turn out to be a highly significant comment as 1982 wore on, the economist added: "Where does this leave us? One, you can moderate employment costs to bring them into line with productivity growth, or two, you can look at what needs to be done to increase productivity. Or you can do a little of both."[25]

Although couched in the economist's seemingly neutral language, this was a suggested strategy for bargaining. Once again, however, McBride and the other board members seemed to pay little attention, with one exception. Bruce Thrasher, director of District 35, Atlanta, and the union's chief negotiator in the aluminum industry, approached Ayoub at the end of the morning session. "I like the concept of moderation," he said. "That's what I want to do when we get into aluminum negotiations." Thrasher followed through on this idea in 1983 by agreeing to a wage freeze in aluminum.

In the afternoon, Johnston went before the executive board and narrated his slide presentation. Probably influenced by Ayoub's Blue Book, Johnston acknowledged that wages weren't the only problem. After the hour-long program, Johnston and Moore responded to questions for two more hours. It was a cordial session, although the questions were sharp. Why did the industry export technical knowhow? How could the Japanese afford "lifetime job security" and still be at a cost advantage over the Americans? If things were so bad, an attentive director asked Moore, why did Donald Trautlein (Bethlehem's chairman) receive a huge salary increase in 1981? In that year, his first full year as the Bethlehem CEO, Trautlein received $555,985, nearly double the $280,880 he earned in 1980 as executive vice-president. Moore tried to answer, but the board members didn't buy his explanation.

Particular attention was devoted to U.S. Steel's purchase of Marathon Oil. This $6.4 billion deal had been consummated in March 1982. The steel corporation had sunk $1.4 billion in cash into the merger and had borrowed an additional $4.7 billion. It was another sign that U.S. Steel had set out on a determined course to reduce its exposure in the steel industry. The company had declared that Marathon would give it an "energy hedge" for its steel operations and raw materials for its chemical business, as well as make the company less vulnerable to a downturn in steel and chemicals. But U.S. Steel had bought Marathon on speculation, expecting that the high oil prices of the late 1970s, 1980, and 1981 would last for years. Almost at the very moment that U.S. Steel concluded the deal, oil prices began to fall, precipitously. Indeed, to digest the Marathon acquisition, USS in 1982–83 had to sell over $1 billion in assets, including coal mines, its Pittsburgh headquarters building, and other real estate. When oil prices turned back up in 1986, Marathon produced large profits for the corporation.

The Marathon deal produced a fury of criticism practically everyplace that U.S. Steel had a steel plant. Monongahela steelworkers had long believed, rightly or wrongly, that the corporation was not making major investments in new equipment for valley plants. No matter how strongly Roderick defended the purchase, it was still a slap in the face. One would have expected U.S. Steel management to anticipate a negative reaction from its work force and to recognize that this would damage its relationship with the union and the rank and file. Certainly, it was a curious action to take when the company hoped to get concessions from its employees. Some large corporations by the 1980s were beginning to understand that major business decisions should not be made without considering their impact on labor relations policies. U.S. Steel was not one of these. "I doubt that they gave it a thought," said Alvin Hillegass, who was vice-president of steel operations when the Marathon purchase was made—and who disagreed with the decision.

At Linden Hall, the USW board members asked Johnston how it was that U.S. Steel had enough money to buy Marathon Oil but not enough to modernize the plants. Essentially, he said that the company couldn't afford to borrow money to modernize steel plants at the interest rates that banks would charge for investing in a sick industry. On the other hand, the banks were willing to loan money at market rates for investment in petroleum. Johnston's reply seemed to satisfy the board. The members didn't like U.S. Steel's action but found it difficult to contest the decision in strictly economic terms.

The meeting ended amicably. "At the end, we were all bemoaning the plight that we were in," Moore recalled. "We went to the bar and had a drink before we left. And a number of the board members came up and said it was the best presentation they had ever seen."[26]

If Johnston and Moore hadn't secured the undivided commitment of the union leaders to the concession cause, they had gained some sympathy for, and understanding of, the industry's problems. In the mill towns, however, there was much less understanding. Bruce Johnston could not put on one of his "spellbinding" performances before an audience of three hundred thousand workers and supervisors. What was the mood of the rank and file?

Chapter 6

Something of Importance in the Mon Valley

We New York journalists who specialize in economic reporting can take the economic pulse of the nation without leaving our offices. We read the Dow-Jones ticker for the news on mergers, acquisitions, and bankruptcies, punch up the latest stock-market prices on a Bunker-Ramo, and scan the reports issued daily, weekly, and monthly by the U.S. Treasury, Federal Reserve, Commerce Department, and Bureau of Labor Statistics. A formidable array of data practically leaps at us from all sources—figures showing money supply, housing starts, ten-day auto sales, retail sales, consumer and producer price indexes, new claims for unemployment insurance, gold and commodity prices, raw steel production, crude-oil refinery runs, number of people employed and unemployed, bond yields, and so on. Some of my colleagues can immerse themselves in these numbers and produce images portraying the state of the American economy at any given moment. As I understand the process, the images unreel in their minds like a 16-millimeter negative, displaying shadowy integers cavorting in various patterns which are converted to hard print as economic forecasts. This approximation of macroeconomic reality sometimes even proves to be almost correct.

Economic imaging is beyond the capacity of people like me who can form judgments only by seeing, touching, hearing, and smelling, down at the micro-level of economic activity. By early 1982, the recession had been under way, as recessions are defined by economists, since the previous July. The steel industry was reported to be suffering, since unemployed people tend not to buy products made of steel, and I needed to get a sense of what was happening. In late April 1982, I returned to Pittsburgh for the first time in nearly a year.

134

I had quit McKeesport in the mid-1950s for good, I thought. With the arrogance and bravado of the young, I talked of leaving the old home town in the dust, a metaphor that was especially apt for McKeesport. A few years later, drawn back by a family gathering, I was driving on River Road, a winding, two-lane highway that runs up the west bank of the Monongahela from Duquesne to Clairton. Across the Mon the furnaces and shops of the National Works stretched from the McKeesport-Duquesne Bridge all the way to the mouth of the Youghiogheny. I parked on the road shoulder to get an unhurried look at the place.

As if staging it for my personal viewing, the Bessemer crew began a blow. Having poured molten iron and steel scrap into a reclining furnace, they tipped the vessel's mouth skyward and blew air up through her innards to purge the iron of carbon and other impurities. At first I could only hear the blast of air. Suddenly, a great stream of flame shot out of the furnace, lengthening and widening, shading from golden yellow to a brilliant red, and carving a bloody wound in the sky. It went on for more than ten minutes, and I felt lonesome when it subsided and died out. I had seen this display thousands of times without thinking much of it. It was part of your life if you lived in a steel town. The open hearths poured out rusty smoke, the blast furnaces coughed up metallic dust, and other stacks issued steam and varying thicknesses of particulate matter. But the orange glow that lit the night sky came only from a Bessemer blow.

On this day, however, I was mesmerized by the sight. I hadn't been close to a steel mill for a few years and realized that I missed the environmental drama. Somehow, this grand gesture—created by a strong blast of cold air hitting a molten bath of 3000° F.—convinced me that something of importance had happened, and was happening, in my home town, in the Monongahela Valley. From then on, I never stopped going back. The Bessemer steel formed the pipe that inched its way across America from the oil and gas wells of the southwest to factories and storage agreas all over the country. National Tube had contributed to my upkeep (and I in my small way to its) and kept McKeesport working. Perhaps I should be embarrassed for allowing another dirtying of the valley's sky to spur in me such an intellectualism. To this day, I carry the remembrance of a Bessemer blow as a very personal image of economic activity—and pollution.

In the 1950s the Bessemer steelmaking process was a century old and already obsolete for most purposes. The National Works still used Bessemers, combined with open hearths, in a "duplex process" that turned out a high grade of steel, especially for use in seamless pipe. The postwar Japanese and Germans, however, were not installing either Bessemers or open hearths. They were erecting the much more efficient basic oxygen furnaces. McKeesport's Bessemer shop, one of the last to operate in the United States, was shut down in 1967.

On my trip in April 1982, I sensed a disaster in the making as I traveled through the Mon Valley. I know that steel is a cyclical business. Mill towns always live on a desperate margin. If they aren't in the middle of a business slump, strike, or corporate move-out, they fear that one or the other is around the corner. One year's boom is the next year's bust in one-industry towns.

But 1982 was different. I had never seen the region in such bad shape, except for what I saw through a boy's eyes in the 1930s. People seemed to be stumbling around in shock. Doors and gates were shut, and window blinds were drawn. Trash piled up in the entrances of abandoned stores. But I was struck most of all by the lack of sound. As I drove from Homestead to McKeesport, I felt that I was floating in a cavernous silence. Nobody was hammering on the "anvil of democracy," as someone once called the steel valley.

On Fifth Avenue in McKeesport I drove past a half-mile stretch of unused parking lots which had displaced much of the city's business district in an abortive redevelopment effort of the early 1970s. At the corner of Fifth and Locust, where shoppers, mill workers, cars, streetcars, and B&O trains (which until 1970 ran through the middle of town) used to intermingle in dense traffic jams, the intersection was empty. A vendor was selling fruit off a wagon in the middle of the street. I parked the car and walked to the state unemployment office, located in a one-story brick building on Lysle Boulevard. Inside, hundreds of men and women were standing in lines that snaked back and forth across a large room, It was the biggest assemblage of people I would see during this trip. There was little talking and much staring at the floor and ceiling.

The reason for the glut of people in this particular office was that National Tube and its sister plant in Duquesne had laid off thousands of people. (In 1969, after blast furnaces and open hearths at the National Works were shut down, the plant was combined with the Duquesne plant in the National-Duquesne Works. Duquesne supplied steel ingots which were rolled into blooms and converted into pipe at National.) Just two months before my trip, the National-Duquesne Works was going flat out, employing more than ten thousand steel-workers and clerical and managerial employees. U.S. Steel had stated publicly—and repeatedly—that the strong market for oil-country tubular goods would last another ten years. In late February 1982 the market collapsed. It would never return for these plants.

A few years later, the corporation would permanently close the McKeesport and Duquesne plants, by implication blaming the workers and their union. The workers' reluctance to accept wage cuts and change outmoded work practices did contribute to the growing inability of these plants to compete with imported steel. However, this reluctance stemmed largely from U.S. Steel's own decision making and the consequences that flowed from it. The chain of events extending from that illogically bullish forecast caused personal anguish not just for employees but also for many thousands of residents in the affected communities.

Pouring Oil on Troubled Steel

In 1971 the price of oil on the world market was $2.10 per barrel. A little-known Developing World group called the Organization of Petroleum Exporting Countries (OPEC) formed a cartel which quadrupled oil prices in 1973—and brought on massive recessions in the United States and Europe. By 1981, after major price hikes in 1979-80, petroleum cost $34 a barrel. Under President Jimmy Carter, the United States began a national conservation program, and per capita energy use fell by 20 percent between 1978 and 1981. At the same time, the high price of oil devastated the automobile, chemical, heavy machinery, and textile industries, which had prospered on cheap energy. Their decline not only reduced the need for oil but also for steel.[1]

However, one steel market remained strong. To escape dependence on foreign oil, Washington decontrolled prices for domestic oil and gas, resulting in a surge of domestic oil drilling. Through most of the 1970s, American pipe producers enjoyed a booming demand for pipe used in drilling—large diameter casing, smaller diameter tubing, and the actual drill pipe. The National-Duquesne Works made enormous profits for U.S. Steel, exceeding $100 million a month in some periods. In 1981, when the recession began to shrink steel's automotive, appliance, housing, and capital goods markets, tubular operations alone supported the steelmakers. U.S. Steel at that time did not report profits by product segments. However, Alvin Hillegass, who was vice-president of steel operations in 1980-81, later told me that the company's pipe mills— located at McKeesport, Loraine (Ohio), Gary, and Baytown (Texas)—produced $850 million in operating profits in 1981. Since the entire steel segment reported profits of only $386 million that year, the pipe operations offset losses of about $465 million in other parts of the steel business.

Hillegass, a native of McKeesport and graduate of the University of Pittsburgh, had started his career at National Tube in the late 1940s. As U.S. Steel's line manager in charge of steel operations in 1981, he recommended that the company build a new, technologically advanced seamless pipe mill at the Fairfield Works, near Birmingham, Alabama. He decided against an expansion at McKeesport, his hometown, for several reasons, including a lack of space along the Monongahela for a 600,000-ton capacity pipe mill. More importantly, Fairfield was closer to the pipe market and had a new blast furnace and other units that were underutilized. "I knew the decision would be the handwriting on the wall for McKeesport and Duquesne," Hillegass said in 1987. "But I never dreamed it would happen as fast as it did. It looked like another ten years, based on market predictions."

The Iranian revolution of 1980, which took 6 million barrels of oil a day out of the market, had strengthened the perceived need for increased domestic production. Meanwhile, Exxon and other big oil companies had published long-range forecasts showing that crude oil prices would remain high. The U.S. Department of Energy did the same. In October 1981, Hughes Tool Company, which manufactures drill bits and keeps a semiofficial count of the number of drilling rigs in operation, forecast for 1982 a 15 percent increase in rigs over the record number of 1981. The steelmakers' forecasters, in turn, calculated from the "rig count" how much pipe they could sell in 1982. Based on these forecasts, most American steel companies made plans to expand pipemaking capacity.

At U.S. Steel, the board of directors approved the Fairfield project, and the company's 1981 annual report said that tubular goods was the company's most profitable product line, adding that *"strong demand [was] projected for these products for at least a decade"* (emphasis added).[2]

At $34 to $40 a barrel, oil looked like the best business going, and everybody wanted a piece of it. In 1981, $8 billion to $9 billion of outside venture capital flowed into Texas and Oklahoma, the banks practically pursued wildcatters to give them loans without collateral, and 91,600 new wells, an all-time high, were drilled. In short, the giant oil producers, the U.S. Energy Department, wellhead suppliers, steel companies, banks, institutional investors such as pension funds—all were caught up in something like market hysteria. "There was a big panic," Hillegass said. "Everybody figured there would be a shortage of pipe, and the customers were willing to pay anything to get product. It was a superheated market, and we were paying more attention to shipping the product than to the depth of the problem."

In late 1981 and early 1982, Albert L. Voss, the general superintendent of the National-Duquesne Works from 1977 to 1984, stated publicly and to various employee groups that National-Duquesne would have enough pipe business to continue normal operations for ten years. The forecast, he said later, came from U.S. Steel's marketing department, which based it on predictions made by the oil industry. Because of the demand for pipe, National-Duquesne had worked seven days and twenty turns a week (out of a possible twenty-one) and hired some fifteen hundred new workers in 1980-81. And no end was in sight. In December 1981, 4,530 rigs, the most ever, were pumping oil in the southwest.

Unknown at the time to the experts and their computers, some 6 million tons of pipe sat unused in storage yards "downstream"—that is, in the Texas and Oklahoma area—at the end of 1981. The rig count suddenly began dropping in January 1982 and continued falling week after week. Such a decrease in drilling hadn't occurred in 1980 and 1981, but the experts passed it off as a return to a traditional seasonal drop-off in drilling activity. McKeesport and other pipe mills continued shipping great loads of pipe into the $6 billion worth of inventory that already existed.[3]

"We were still getting orders in January, February, and March of 1982," Voss told me. "But one day the cancellations started. It was amazing! Any orders not in process were cancelled. I spent thirty-six years in the pipe business and never saw it happen like that." On March 20, 1982, the company shut major departments in the National plant and laid off twenty-five hundred of its forty-two hundred hourly workers. Layoffs continued and by fall only two hundred were still working. When I visited McKeesport in late April, the National plant was working at only 15 percent of capacity and Duquesne (which had other customers for its steel ingots and bars) at 40 to 50 percent. For all practical purposes, this was the end of major pipemaking operations in McKeesport and raw steel production in Duquesne.[4]

Admittedly, it is easy to say that the petroleum producers, the U.S. government, the banks, and the steel industry should have known that the bubble would burst. After all, published figures on oil imports had shown a dramatic decline year by year, starting in 1978. OPEC accounted for 70 percent of oil imported into the United States in 1977 but only 40 percent in 1982. Other industrialized countries had reduced their dependence on oil. Big new oil reserves in the North Sea and Mexico were being exploited. How could OPEC maintain the $34 per barrel price? In fact, the cartel was losing control as producers began discounting prices. By March 1982, although the official price remained at $34, crude oil was selling for $26 a barrel on the spot market in Amsterdam. The price later fell as low as $15.

Many changes in the marketplace are unforeseeable. Nobody can predict with certainty a sudden change in consumer preference or when a frost will damage a citrus crop. However, the relationship between price and supply in a major international industry like oil, constantly monitored and analyzed, should not be a mystery to corporations that pride themselves on their belief in the free market. Yet in 1981, U.S. Steel and most other steel producers—an industry which frequently boasts of its sophistication in analyzing and forecasting future steel needs—had let themselves be gulled by speculators in the oil trade.

Many people and institutions were taken on a ride in the oil spree. But U.S. Steel suffered more than most. The high selling price of oil in 1981 was a major factor in the company's decision to acquire Marathon Oil. It made sense to diversify away from the cyclical—and declining—steel business. Looking at projections that oil prices would rise to at least $50 a barrel, U.S. Steel management thought it was buying "a money machine," as Al Hillegass put it. He, and other steel-oriented executives at U.S. Steel, were disconcerted by the Marathon purchase. They believed that the company should have split its resources, investing some in diversifications and some in steel modernization. Spending $6 billion for Marathon, Hillegass thought, was "electing to swallow a whale," taking on too much debt. Diversification could have been accomplished by buying two $1 billion companies.

The Marathon acquisition alienated a very large percentage of U.S. Steel's production and supervisory employees in the Mon Valley and other regions. But that transaction was not the end of the chain of events forged by a faulty reading of the demand for pipe, which turned workers against the corporation.

To finance construction of the new Fairfield pipe mill, Hillegass used an innovative "taker-pay" scheme. He persuaded fourteen oil producers to invest $1 billion in the project under contracts that earmarked an equivalent amount of pipe production for each producer when the mill began operating, in the late 1980s. In the meantime, these companies demanded assurances of a steady supply of pipe. Since U.S. Steel's existing pipe mills were operating at full capacity, Hillegass negotiated a $600,000 contract with an Italian steelmaker, Dalmine, to deliver 120,000 tons of pipe a year for five years.

Dalmine began shipping pipe to the United States in early 1982. Some of it had to be processed further at U.S. Steel plants. In April 1982, when I visited the valley, bargeloads of this pipe sat on the Monongahela, awaiting processing at National. At the very time that U.S. Steel stopped making pipe, and while thousands of steelworkers were being laid off at McKeesport and Duquesne, the company brought truckloads of Italian pipe, with Dalmine's name stamped on each pipe, into National Tube. This enraged the workers. "When the domestic business gets to the point that thousands of our steelworkers are laid off, nothing should come in to the docks," declared Local 1408 President Dick Grace. The union, however, could do nothing about it. Word of this spread up and down the Mon Valley, further angering people already annoyed about the Marathon purchase.

What could it do? U.S. Steel said, it had a five-year contract with Dalmine. However, the contract had a "hardship" clause which said that if the pact worked a hardship on either party, the two sides could review and adjust the terms. Hillegass had

left U.S. Steel in early 1982, and nobody spotted the hardship
clause in his absence. Later, talking to acquaintances at U.S.
Steel, he called attention to the provision. Only then, in 1983,
did the company move to terminate the contract. "They took
pipe for a year they didn't need to take," Hillegass said.

In 1980 U.S. Steel had spent more than $100 million to
renovate old equipment and install new equipment at National. This included a $65 million "quench and temper" line,
used to strengthen pipe for deep-well drilling, and an auto-
mated pipe-coupling shop. Some of the new coupling machines were never used.[5]

The abrupt turndown in the pipe business surprised everybody in U.S. Steel management, including Voss, Hillegass,
Chairman David Roderick, the board of directors—and somewhere a lawyer who should have known about the hardship
clause in the Dalmine contract. If these people were surprised,
the hourly workers at McKeesport and Duquesne were thoroughly confused. They had heard announcements of the Marathon acquisition, the building of the new pipe mill at Fairfield,
and the ten-year prediction for their own plants. They knew
that foreign-made pipe continued to enter the country, even
after a large part of National was shut down. They saw the new
machines sitting unused in the coupling shop and recalled the
hiring of new workers in 1980-81. Finally, the employees knew
that their plants were producing at a high rate of capacity
practically until the day they went down.

Workers in the Mon Valley did not know what or whom to
believe. Many became convinced that the turndown was
either temporary or contrived. In 1981 some ten thousand
hourly and salaried workers had been employed at the
National-Duquesne Works. After the layoffs in 1982, Al Voss
asked the local unions for concessions to increase productivity—and got nowhere. When he called a community meeting
to try to explain what had happened to the pipe business, only
two hundred people showed up. "We really got no response,"
he said. "They wouldn't listen. Most people didn't believe
what we were saying. There was an inherent mistrust of management. They had seen turndowns before, and the business
had always come back. They didn't realize this was a different
scene."

Management itself, however, had only belatedly come to
the recognition that the scene had shifted. It made little effort
beyond the routine articles in house organs to communicate

this to the rank and file. In making major business decisions, U.S. Steel paid too little attention to the implications for employees and labor relations. "If there's a right way and a wrong to do labor relations, they'll do it the wrong way," Hillegass said.

Hillegass conceded that "we were all fooled by those fore- casts of high oil prices"—perhaps he more than anybody. Cap- tivated by the thought of cashing in on the booming pipe business, he planned a $500 million minimill in Arkansas. He said good-bye to U.S. Steel at the end of January 1982, about a month and a half before oil prices collapsed. When they did, his financial supporters backed out of the deal. Hillegass and his wife went to Florida to live.

The Mon Valley Unemployed Committee

In late April 1982 unemployment was higher in the Mon Val- ley than at any time since the Depression. Eleven percent of the labor force in the four-county Pittsburgh metropolitan area was out of work. People at the International headquarters in Pittsburgh felt sympathy for their laid-off members but hadn't implemented programs to help them. The union had never seen itself as a social welfare agency but rather as a political organization. In periods of rising unemployment, it lobbied in Congress for economic stimulation, retraining pro- grams, and import restraints. But union leaders had no idea how to cope with mortgage foreclosures, psychological prob- lems, and the need for developing new jobs.[6]

However, a new group called the Mon Valley Unemployed Committee (MVUC) had been organized to offer aid of various kinds to laid-off employees. I also had heard vague talk of "radicals" traveling the valley, exploiting the unemployment issue in an effort to turn steelworkers against union and com- pany alike. To old-style conservative unionists, however, any member who acts on political and social issues outside of the union structure is a radical. This attitude still lingers from the 1940s, when Phil Murray of the USW and other CIO leaders drove Communist-led unions out of the organization. The fossilization of that attitude prevented the USW and many other unions from using the energies and ideas of some of their best and brightest members.

Since the 1960s, I had met many USW "dissidents"—the term applied to any member who bucked the union establishment—but few of them were political radicals. In the 1980s, the best-known dissident in the Mon Valley was Ronald Weisen, president of Local 1397 at U.S. Steel's Homestead Works. He had attracted wide publicity by engaging in unrelenting attacks on USW leaders and U.S. Steel. He seemed to take a uniformly negative view of all actions and statements by USW leaders, a position that distinguished him in my mind from constructive dissidents. However, Weisen apparently had a genuine concern for the jobless. His local was the first in the valley, perhaps in the union, to set up a food bank for the unemployed.

I was curious about what the MVUC did and where it stood in the political spectrum. On the first Saturday in May, I met two of the committee's organizers, Steffi Domike and Robert Toy. Over coffee in the Sheraton Motel restaurant on Lysle Boulevard, they described the organization as a rank-and-file "self-help" group of the kind that had sprung up in industrial communities during the Depression. The fact is that neither government welfare agencies nor organized charities, such as the United Way, are equipped to handle many of the psychological, social, and economic problems that accompany large-scale unemployment because they call for cutting through government red tape or putting pressure on private industry.

Domike and Toy themselves had been laid off, respectively, from U.S. Steel's Clairton and Duquesne plants. Both were bright, well-educated people in their late twenties. Domike, an earnest-looking woman in bobbed hair and sandals, had been one of the founders and the first chair of the committee. (The other two founders were Barney Oursler and Robert Anderson, of the Irvin Works.) Domike had worked at Clairton since 1976 and was now an apprentice "wireman," or electrician, and would become a journeyman (the steel contracts had not yet embraced the term "journeyperson") in eight more months—if she was called back to work.

Some fourteen thousand people were laid off from all U.S. Steel plants in the valley, Domike told me. "The local unions weren't used to dealing with so much unemployment," she said. "Some have layoff committees, but they weren't doing much. The International wasn't doing anything. Nobody was helping the jobless. I don't blame the union. Dues money is way down, and the local officers have to work part time in the mills. They can't work full time for the union. Somebody had to do something. People need help."

The committee was staffed by volunteers like Domike and Toy from twelve USW locals in the valley. Some were local officers, but most were rank and filers. With contributions from churches and individuals, the group set up food banks and established a hot line. Many workers needed advice on collecting unemployment compensation, or applying for food stamps, welfare, and government retraining classes. Some of the unemployed would soon exhaust their UC payments and wanted to know what to do next. They could go on welfare. But if they owned a home, the state would put a lien on it. Furthermore, they would have to sell their car, and without a car finding a job would be nearly impossible.

How did the valley get into this fix? Toy, a slim man with a gentle manner, leaned over his coffee. "Some people say it's imports, 'Stop the imports,' they say. We say we ought to look at why there are imports. U.S. Steel has pursued a disinvestment policy over the last twenty years. The mills in this valley are run down." Domike picked up this point. "If they had applied the $6 billion [used to buy Marathon] in this valley, it could have been totally modernized. Up at Clairton, they were still running a rolling mill with a rope-driven engine. Roderick is a money man. He's in steel to make money and not steel. If they're going to ask us for concessions, they're going to have to cut top-heavy, low-efficiency management. The foremen are tripping over one another. What they mean when they talk about cutting costs is, they're cutting us. If they'd ever listen to us, they could be a lot more efficient."

The MVUC also lobbied for legislation to help the jobless. (In 1983, the committee was instrumental in winning passage of a law under which the state would loan money for mortgage payments to people who became delinquent "through no fault of their own.") It organized rallies and demonstrations of various kinds, particularly at banks that foreclosed on mortgages. To some degree, the committee was inciting activism on the part of rank and filers who had never taken part in union activities, Domike and Toy said, Monthly meetings were now drawing one hundred or more members, in contrast to the normal twenty to thirty.

The formation of a self-help group was an unusual event in the Mon Valley, where decades of political domination by the companies and—all two frequently—low-grade politicians, had bred cynicism about government and political action. The fact that people like Domike, Toy, Oursler, and Anderson even worked in the Mon Valley mills was itself a matter of note. Domike, for example, had earned a B.A. degree in economics at Reed College in Oregon but wanted to work with her hands. She saw an opportunity when the major steelmakers and the USW in 1974 signed a consent decree with the federal government, pledging to end discrimination against blacks and women in hiring and promotion. From then on, U.S. Steel and other firms were required to concentrate on hiring minority and female workers to bring their percentage in the steel work force up to their proportion of the regional labor force. "As a young feminist, I wanted to get into the man's world," Domike later told me.

Barney Oursler, whom I met later the same day, went to work at the Irvin plant in 1975 after getting degrees in philosophy and history from Penn State and the University of Buffalo. An anti-Vietnam War demonstrator and community organizer, he wanted to get into union work. Another early MVUC leader, Paul Lodico, was a well-educated former staff representative of the United Electrical Workers. He, Oursler, Domike, and Toy obviously would be activists in any line of work. Articulate and smart, they leaned politically to the left—Domike said she was a socialist—but they were by no means doctrinaire radicals. They were angry at U.S. Steel, angry at the Reagan administration, and disdainful of the way government worked. While they were disappointed with the union, they didn't attack it and, indeed, got along well with USW officials in Pittsburgh. They undoubtedly shared some of Ron Weisen's opinions about corporate behavior. But they didn't join Weisen and a small group of steelworkers and ministers who later expressed their outrage against "the system" by stuffing dead fish in bank vaults and interrupting services at churches attended by company executives.

Obviously, however, the MVUC leaders were atypical steelworkers. Having learned about their work, I wanted to find out how ordinary rank-and-file workers—the kind who never ran for union office and seldom went to meetings—felt about the company, the union, and the economic circumstances. Union activists and local officers tend to criticize the company, but this can be attributed partly to politics. In the political atmosphere of the Mon Valley, incumbent union officers had to be wary of expressing cooperative attitudes. On the other hand, perhaps the rank and file really was angry and militant.

At this point, I came up against the reporter's age-old dilemma: How do you find, and recognize, the "average person" on the street?

Views of the Rank and File

In 1907, when John Fitch collected material for *The Steel Worker*, his classic study of life and work in the Mon Valley mill towns, he could easily identify steelworkers. They were male, wore rough clothing, usually carried lunch buckets, and looked out of place in America. In succeeding decades, that stamp of immigrant "mill worker" gradually faded into an undifferentiated look of a broad working-class, or lower middle-class. When I was growing up in McKeesport in the 1940s, working-class neighborhoods still existed where a knock on any door would produce a mill man or his wife. Taverns near the plant gates were filled with steelworkers shortly before and after a shift change. And every mill town had at least one well-known place, usually near the main plant gate, that by long custom and tradition was a gathering place for steelworkers.[7]

In McKeesport that place was located at the main Baltimore & Ohio Railroad crossing. By an accident of history, the B&O tracks ran through the center of the city, and the passenger depot stood at the corner of Fifth and Locust. Directly across from the depot, a chest-high iron railing stretched some thirty yards along the Locust Street curb in front of two taverns, dividing the sidewalk from the B&O tracks. If a person came staggering out of one of the bars, either under his own power or propelled by a bartender, he would encounter the railing before falling into the path of a train. Located only a block from the main gate of National Tube, the railing was a social center for steelworkers before and after shift change.

Up to the 1950s most workers traveled to work by trolley or bus and strolled down Locust to the plant. The older men, especially, liked to socialize before going into the mill and came into town an hour or more before the start of their turn. They would line up at the railing (it accommodated fifteen to twenty leaners), forearms resting on the top bar, lunch buckets dangling from their hands. They smoked cigarettes and pipes, chewed tobacco, gazed at the crowds of shoppers on Fifth Avenue, and talked. Small talk usually—the weather, doings in the plant, the dismal Pirate baseball season. Occasionally heads would bob in the heat of debate and fingers would shake in grim faces.

When a southbound passenger train came into the station, the locomotive would stop just short of Fifth Avenue, snorting clouds of steam which enveloped the men at the railing. There would be coughing and spitting. But the train stayed in the station only a few minutes. When it pulled out, the leaners would reappear, all heads turned to the right, watching the last coach sway around the bend. If an early comer left the railing to get a quick beer or go to work, someone would take his place. Eventually, one by one or in pairs, the men would flip their butts away, knock out their pipes, and saunter toward the plant gate.

The B&O trains were finally rerouted around McKeesport in 1970. By then, however, the Locust Street railing was not the socializing center it had once been. Many steelworkers had moved out of the city and into nearby suburbs, like White Oak, or Elizabeth, or dozens of other communities within forty or fifty miles. They drove to work and parked on the street or in parking lots near the plant gate. Only retired people were left in the old working-class neighborhoods. Streetcars no longer ran through town, and bus traffic had diminished because outlying shopping centers had sucked most of the retail trade out of the city. The Steelworkers union and Henry Ford had succeeded in dispersing the steelworker population, the first by raising the standard of living and the second by putting a means of private conveyance within reach of the ordinary worker.

By the 1980s, one could no longer venture into a mill town and say with any confidence, as if pointing out a wren in a group of sparrows, "There's a steelworker!" Even saloons lost their appeal to steelworkers in the era of traffic jams and steaming radiators. Most workers would dash out of the gate and go directly to their cars, trying to beat the jam. Many mill-town taverns went out of business.

It was true enough that steelworkers and saloons had always occurred together in the popular imagination, like baseball and hot dogs. And the mill towns apparently were hard-drinking places in the early days, when the only amusement that men had time to indulge in, after working twelve hours a day, six days a week, was sloshing down beer and whiskey in the corner bar. It seemed that nobody could describe the steel communities of the early 1900s without specifying how many taverns existed in each town, as if a tavern-counting agency provided such data on a current basis. "In 1906, McKeesport had 40,000 people and 69 saloons," Fitch wrote. At about the same time Braddock had 65 saloons and Duquesne had 30.[8]

Habitual saloon-going, I believe, was not a pastime of the typical steelworker by the time I came of age. In the late 1940s my father operated a hotel, the McKeesporter, which had a bar and grill on Ringgold Street, across from the main bus station. I occasionally tended bar and came to know many steelworkers who stopped on their way home from work after the day turn. Most drank a glass or two of draft, sometimes with a shot of whiskey (a combination known locally as "a puddler and a helper" but elsewhere as a "boilermaker"), and dashed across the street to catch a bus. This is not to say that steelworkers stopped drinking. They simply did it outside the mill towns, in gaudy restaurants on Routes 30 and 51, at backyard barbecues, and certainly in front of the television set.[9]

In any event, it was becoming increasingly difficult to find the species steelworker close to his working habitat. The Unemployed Committee offered me an opportunity to talk to some rank and filers that John Fitch would not have recognized in any costume. They were women. By the 1980s women were not new in the steel industry. Thousands had worked in the mills during World War II, only to be displaced by returning veterans after the war. But the consent decree, which had lured Steffi Domike to Pittsburgh, also made opportunities for many less privileged women in the Mon Valley.

On that same Saturday afternoon in early May, I talked to three female steelworkers at the Local 2227 hall near U.S. Steel's Irvin Works in West Mifflin, about two miles up the valley from McKeesport. The women had volunteered to work for the MVUC, and Barney Oursler had just spent a couple of hours giving them tips on public speaking.

It was a sunny day, and the five of us sat on a low retaining wall outside Philip Murray Hall. We were near the top of a steep hill on the west side of the river and could look far across the valley and down on the Irvin Works, which sat on a plateau directly below us, some two hundred to three hundred feet above the river. At the bottom of the bluff, a narrow strip of riverbank afforded space only for railroad tracks and the two-lane River Road. West Mifflin was neither a river town nor a mill town but a huge bedroom suburb that sprawled for miles along the top of the ridge and dipped down the slope to pick up taxes from the Irvin Works.

Below the plant, the cliff bulged outward, obscuring the river and the industrial mess on its banks. The hills across the valley seemed warm and green on this spring afternoon. On our bluff the mill—five or six metal-sided buildings, each more than a half mile long—was an island surrounded by trees and shrubs tangled in the dense, lush growth that is characteristic of western Pennsylvania. I could only see the tops of Irvin's shallow, triangular roofs glinting in the sun, forming a washboard pattern. When layers of mill soot covered the surrounding patches of foliage, you could fold them over the top of the plant and give them a good sudsing.

Each of my average, rank-and-file female steelworkers had been laid off for at least a month. René Garcin, thirty-one weeks on layoff, had worked as a blue-collar clerk in the Irvin Works. Her husband had skipped out on her some time ago. She had four children but had to get along on UC payments of $175 per week. She couldn't collect welfare because she owned a car. Ellen Mason, also laid off from Irvin, was in a better situation because her husband was still working at a Wheeling-Pittsburgh Steel plant in Monessen. Lynn Stovall, a single woman in her mid-twenties and the most militant of the three, had been a crane operator at the Duquesne plant.

Collecting unemployment pay was a new experience for all three women, and they were angry about the "crazy" rules they had to follow. Garcin and Mason were now facing new regulations, because they had exhausted their first thirty weeks of UC and were starting on a new program for thirteen additional weeks. Under the old regulations, they went to a state unemployment office once a week, declared that they had made "an active search" for work, and picked up a check. The new rules required them to submit proof of having contacted five different employers each week—four personal visits and one phone call. Garcin and Mason lived in Monessen, several miles up the valley in a semirural area. Except for steel companies, there were few employers of any size, and no jobs to be had. Yet every week, hundreds of jobless steelworkers in the Monessen area converged by foot or phone on the same employers to obtain a "no jobs available" note.

Another rule prohibited contacting the same employer more than once a month. Garcin had exhausted her list of employers within easy distance of Monessen. Next week, she would have to drive to Uniontown, and the cost of the gas would reduce the amount of money she could spend on rent and food for her children. "Because I have to have transportation to find a job, I'm going to starve," she said. Mason added, "I've been driving up and down the valley so much, it's like playing Pac-man."

Satisfying the bureaucratic rules of the state employment agency had produced absurd situations in the Mon Valley. Steffi Domike had told me that to comply with the five-contacts per week regulation, she organized her search according to the alphabet. In her first week, she visited or phoned only companies starting with an "A." But a clerk at the employment office told her this wasn't allowed. "What they told me was, I had to make a random search for a job," Domike told me. "You're not allowed to be organized."

Garcin, Mason, and Stovall said they had not been activists in the past but had come to feel that they ought to "take a stand for ourselves," as Garcin said. They had the feeling that other rank and filers felt the same way. "A hundred people showed up for a layoff meeting at Duquesne recently," Stovall said. "They stayed for hours, and the questions were political for the first time. There was a real feeling that this recession was consciously done." The more she talked, the angrier she became. "If U.S. Steel made 135 percent higher profits last year, why should they come to us for concessions? The annual shareholder reports are glowing."

But Garcin demurred on the subject of concessions. "I'm not proud," she said. "I'd take a cut to get back to work."

Stovall frowned. "They'll hope you will say what you just said. It makes me mad. They're riding the wave of concessions."

A few days later, on May 6, I went to USW Local 1408 in McKeesport and interviewed, at random, workers who showed up at the union hall to apply for UC (under a special procedure that saved them the time of standing in line at the state office.) There were a number of "no comments" and "I'm in a hurry." Three workers, however, had reasonably coherent views. All had just been laid off from the National plant.

Martin Guy, thirty-one, a stocker with nine years of service, said he earned $32,000 in 1981. "My old lady was angry because I was working a lot of overtime," he said. "All of a sudden our mill doesn't have any orders. U.S. Steel got involved with Mobil [sic], and now they say they want to cut costs. My buddies tell me we can't give up anything. But if you don't have a job, you definitely have to give up something. I'd give up a wage increase, a $1 an hour or so, but not the cost of living. We need that."

"They told me I'd be lucky to get three weeks' work the rest of the year," said John Liscinsky, thirty-four, an inspector with seven years in the plant. "I don't know what's gone wrong. Other guys say they'd let the company shut down the plant rather than give up anything. But I don't mind giving up something. I could see a wage freeze, but if it weren't for the cost of living, we wouldn't be making anything."

Ernest Humenik, forty-one, a test carrier with four years of service, had read the news more closely. "U.S. Steel went out and acquired Marathon Oil instead of putting the money into steel," he said. "All the mills in this valley—the company sucked them dry and got out all the money they could. I've worked in a couple of plants, and some parts of them are held together with bailing wire. But last year only 38 percent of U.S. Steel's profits were derived from steel. That scares me. Everybody feels there should be some concessions, but it can't be all for them and nothing for us. We need job security and prenotification of shutdowns."

My random interviewing didn't produce anyone who was firmly set against wage concessions, although such workers certainly existed. On the other hand, no one was contemplating deep wage cuts. There was no conception in May 1982 that National might be a candidate for permanent shutdown.

Dick Grace, the president of 1408, was sitting in his tiny office. I mentioned the UAW's low-cost labor settlements at Ford and GM. How would his members feel about deferring COLA and freezing wages?

"Concessions!" Grace exclaimed. "You can't say to the workers that U.S. Steel can acquire Marathon and come to us and ask the workers to give up cost of living and wages." His members were annoyed about the Marathon deal. Moreover, younger workers, those hired in 1980 and 1981, complained about their treatment by foremen. The young workers, Grace thought, had a different attitude about supervision than older employees. The bosses didn't know how to deal with them. (Yet another "new generation" of workers, I thought, remembering the stories I had written a decade ago about the "new breed" workers of the "baby boom" generation and how they had challenged union and management alike.)

"These kids are smarter," Grace went on. "They read the contract. They question authority. Before, the boss was Messiah. He always said, 'You do what I say or go home.' These kids say, 'Screw you' and go home. A couple of weeks ago, the company brought in a new foreman on a grinding job. He called the guys 'bastards.' There was a slowdown. That wouldn't have happened before. What did management do? They threatened to send these jobs out of the plant, special processing jobs that could be done somewhere else. But I complained, 'Hey, who wants to be called a bastard?' Now, the foreman is lying low. But management didn't want to back down.

"The U.S. Steel philosophy is, they put a white hat on a foreman's head, and he gets the idea he's better than you," Grace continued. "I went to some of those prayer breakfasts in Pittsburgh [an annual 'labor-management prayer breakfast' started in 1979 by a personnel executive, Wayne Alderman]. I sat next to management guys and said prayers and listened to speeches about getting along with each other. And the next day, back in the plant, it's the same old thing."

I left Grace as he prepared to go jogging. His comments were remarkably like those I had heard from union officials in the early 1970s. One of the eternal verities, I thought, must be that people keep getting more resistant to "authority"—a verity that many managements, including U.S. Steel's, had not discovered.

U.S. Steel management rarely replied to statements made by employees and union officers. If asked to comment, a public relations person would usually say that the press should not rely on the opinions of uninformed people. Whether the employees were right or wrong, however, U.S. Steel's problem was that its employees had these anticompany attitudes. The corporate management never had communicated well with people in the plants, relying on the "voice-from-above" technique rather than more intimate contact. The rank-and-file anger about dictatorial bosses, management mistakes, and the

Marathon deal could not be dispelled with harsh words by David Roderick or Bruce Johnston in the company's house organs about the need to face economic reality. U.S. Steel faced a substantial credibility gap.

The company had failed to justify the Marathon purchase even to its salaried employees. A few days before my interview with Grace, I visited a U.S. Steel supervisor, Larry Delo, at his home in White Oak. He and I have been close friends for many years. We traveled in the same crowd in grade school and high school and had in common one trait that bonded us a peculiar way: On Friday nights we sat together at stag dances, morosely watching other guys dancing with girls that we were too shy to ask. A good fast-ball pitcher in high school, Delo played a couple of years at Talahassee in the Pirates' farm system before his arm gave out. He graduated from Purdue and joined U.S. Steel in 1956. He spent some twenty years as a first-line foreman in a number of Mon Valley plants and in 1982 became general foreman of the shipping yards at Duquesne. Tall and slightly stoop-shouldered, Delo looked and acted less like my stereotyped notion of a foreman than anybody I know. He was exceptionally mild-mannered, even deferential, and I have rarely heard him curse or swear.

I usually went to see Larry and his wife Maureen, a gifted organist who played at church services and taught piano, when I visited the Mon Valley. He provided a sort of running commentary about life in the mills and the state of the steel business. Unfortunately, the news only got worse in the 1980s. In May 1982, rumors were spreading that USS would lay off up to five hundred managerial employees in the valley and freeze salaries. Delo did not foresee much improvement in the near future. USS was not investing money in the Mon Valley plants. "As a member of management," he said, "I feel U.S. Steel has let this area down. Why should I give up an increase in salary unless they say they'll modernize the plants? Marathon doesn't do us any good."

Delo was not an antiunion man, but he was irritated by the attitude of some USW grievance committeemen. They refused to negotiate over management proposals to reduce idle crew time and allow the combining of duties from different job classifications. In Delo's view making the plant more productive would lengthen its life. "A few people are willing to pitch in and help when it's outside their classification, but for the most part they'll only do what is in their function," he said. "That is so ingrained in our system, it would be almost impossible to change."[10]

The Mon Valley was an old industrial region in which union and management had made some fundamental compromises fifty years ago in order to work with one another. They had developed procedures for dealing with their legal relationship—the grievance procedure and arbitration—but they had become mired in a deep operational rut. When I mentioned this thought to my friend Larry, he became agitated. "Oh, no!" he said. "You make us sound old and tired. We've got plenty of life left in us." But he didn't appear to have convinced himself.

A Mechanism for Reform: LMPTs

I was wrong, however. Union and management could have changed, though it would have taken an almost superhuman effort. On May 5, two days after talking to Larry Delo, I saw such an effort under way at Jones & Laughlin's Aliquippa Works. At the company's invitation, I visited this plant, down the Ohio from Pittsburgh, to interview workers and supervisors involved in a Labor-Management Participation Team (LMPT) program. J&L and USW Local 1211 had set up about a dozen of these teams at the departmental level in the plant. Composed of from eight to ten hourly workers and a foreman, the teams met once a week for an hour or so, on company time. They discussed ways of reducing costs, eliminating production waste, raising product quality, and improving health and safety conditions, among other things.

The USW and the eight companies in the steel bargaining group had established the LMPT concept in the 1980 contract. The provision stated that "both sides recognize that a cooperative approach between employees and supervision at the work site in a department or similar unit is essential to the solution of problems affecting them." The LMPTs were similar to Quality of Worklife (QWL) committees established in the early 1970s by the United Auto Workers and General Motors. Both forms borrowed from worker participation efforts of the 1920s and World Wars I and II. The USW itself had experimented with participation in its early days but abruptly dropped the program after the second war (see chapter 11). LMPTs could not solve all of steel's problems, but they had a large potential for making the companies more competitive.

By 1982, however, only four of the large steel firms, Jones & Laughlin, Bethlehem, Republic, and National, had mounted strong efforts to get an LMPT process started in their plants.

The LMPT concept was built on the idea that workers knew as much, and frequently more, about running their own operations in a steel plant than the supervisors. Steel plants are large, complicated places, filled with enormous machines and furnaces that do batch processing and must be linked by efficient logistical systems. Unlike an assembly line, every steel furnace, every rolling mill, has its own mechanical personality and requires long periods of trial and error before it can be run at anything approaching the peak of efficiency. The hourly worker who works in intimate contact with the mill or furnace hour after hour, day after day, thinks of shortcuts that can be taken to speed production or ways to make the shop layout more efficient. But if the employee distrusts management, or feels that his shortcut will eliminate his own job, or thinks that the boss will holler at him for making a suggestion, he is likely to withhold the idea.

Most companies use suggestion boxes to solicit ideas from employees. The suggestion box, however, does nothing to dispel alienation or actively involve employees in running the business; nor does it ask questions. Problem solving by committee sets up a dialogue between team members that constantly refines and improves an idea. This process also carries a large psychological reward: It enables workers to speak out, to criticize poor management techniques, and to have a voice in matters that vitally affect them. In the end, it can also save jobs by making a plant more competitive.

I had visited many factories and offices to report on and write about the participatory process (it is a "process" in the sense of never reaching a definable end, while "program" implies a more limited undertaking which halts within a specified time). The enthusiasm and energy that workers and managers invested in the team process impressed me everywhere I went. Where both union and management had made a commitment to participation and persisted with it despite failures, it had added a significant new dimension to working life and carried a powerful potential for reforming union-management relations and increasing productivity. People involved in the successful projects had learned, however, that participation should not be aimed merely at making people "feel good" about themselves, or at improving the work environment (for example, by installing fans in hot, smelly places). If a team addresses only "comfort" items, the program usually collapses. A second and more important goal is to improve economic performance. This sounds like a management statement, and union opponents of participation—and there are many—usually view the advocates as talking the company line.

At the Aliquippa plant, I sat as an observer in an LMPT meeting in the coke department. This team had seven hourly workers, including two black men, and a turn foreman. The foreman and one of the hourly employees were co-leaders of the group. In this session the foreman led a discussion on improving the maintenance of coke oven doors. The hourly workers did most of the talking and admitted that some of the oven-door patchers had a "bad" attitude. However, this could be traced in part to poor supervision, a lack of work materials, and inadequate communications, they said. The foreman agreed. Identifying the possible causes of faulty maintenance took up the entire meeting, and the team decided to discuss solutions at the next session.

It appeared to me that the discussion was open, honest, and frank. The team members, all volunteers, had been given a full week of training in problem-solving techniques and group dynamics before the meetings began in January. By consensus the team itself, not management, decided what problems to tackle. Contrary to what management skeptics might guess, the first issue did not involve something the workers wanted for themselves, such as a new vending machine. The team launched an investigation of energy loss from the ovens and, after weeks of study, proposed a method of halting steam leaks which would save the plant $449,000 a year. The plant's LMPT steering committee, consisting of local union officers and plant management, approved the plan, as well as expenditures for necessary equipment, and saw that it was carried out.

One of the black workers, Ford Newsome, told me he had seen a significant change in management attitudes toward the work force since the LMPT process had been inaugurated. In the past, it had been difficult for hourly workers to talk to upper-level managers such as the plant superintendent. "Some doors have been opened for us which haven't been opened before," he said.

Tom Porter, another team member, said the hourly workers had noted with particular interest that management responded quickly to the team's proposals, implementing all that were approved by the steering committee. Nothing can kill a participation process more quickly than management refusal to act on the proposals. "If all ten thousand people in this plant went through this training," Porter said, "we'd be the lowest-cost steel producer in the United States."

The hourly team members readily conceded, however, that many employees were still suspicious of LMPTs. Some charged that it was just another management method to break the union and reduce jobs. LMPT, the opponents said, really meant "Less Men Per Turn." Although J&L had required all foremen to attend orientation sessions on participation—an absolute must for any company that wants to change its management style—many supervisors still resisted the idea of taking suggestions from hourly workers. This kind of opposition is to be expected. Managing a factory or an office in the participatory style requires breaking away from tradition, changing customs, and undermining old habits.

Problem-solving committees like those at Aliquippa represent an early phase of participation, one which must evolve into higher forms if the process is to have a lasting impact on management style, labor relations, and the organization of work. Nevertheless, the very existence of LMPTs at Aliquippa put J&L light-years ahead of most of the steel industry. Eventually, the USW and J&L set up LMPTs throughout the Aliquippa plant and achieved major cost-savings. J&L's corporate parent, LTV Corporation, later acquired Republic Steel and merged the two steel firms in a wholly-owned subsidiary, LTV Steel. By 1985 many hundreds of LMPTs were at work in most LTV plants. But in that same year most of the Aliquippa plant was closed, and LTV later went into bankruptcy. Some USW critics of participation cite the Aliquippa shutdown as evidence that employees involved in shop-floor cooperation are only working themselves out of a job. It is more likely, however, that the LMPT process prolonged Aliquippa's life by some period of time—a year, a month, or perhaps only days.

J&L had embraced the LMPT concept largely because of two men. One is Thomas C. Graham, who is regarded by many observers both inside and outside the industry as one of the best steel managers of the 1970s and 1980s. In 1975 he became president and chief executive officer of J&L, with headquarters in Pittsburgh's Gateway Center. Eight years later Graham moved to the other side of Pittsburgh's Golden Triangle and took command of U.S. Steel's faltering steel operations. It was at J&L, however, that he first made his mark.

A native of Greensburg, Pennsylvania, Graham joined the firm in 1947 and spent many years as a plant-level engineer, becoming an expert in steel technology and production practices. As chief executive, Graham launched a series of moves to improve productivity throughout the company by lowering the man-hours required to produce a ton of steel. He was not averse to aggressively slashing jobs and people, as his later career at U.S. Steel demonstrated, if he deemed that necessary. But he preferred to handle "people problems" through quiet negotiation rather than combat. His first step was to change J&L's management style. In the mid-1970s, Graham later said, J&L operated under "an authoritarian style of management that was more appropriate to the Prussian army than to a modern business enterprise."[11]

Like most chief executives, Graham could out-Prussian a Prussian. He didn't ask lower level managers to adopt the participative style; he ordered them to do so. As contrary as this may seem, experience at many companies has shown that the participative approach must be spread down through a firm with the active and visible interest of the chief executive. If it starts at the bottom, it will stay at the bottom. Graham himself was by no means an unheeding, dictatorial executive. Although impatient and inclined to plunge ahead, he invited the ideas of subordinates, listened to their advice, and even admitted mistakes, his J&L associates have told me.

In the late 1970s Graham's efforts began to show results and, partly because of an aggressive public relations campaign, he became known as the industry's productivity expert. By introducing new technology, upgrading existing equipment, and reducing manpower, J&L was cutting cost per ton. Among the people that Graham most relied on was John H. Kirkwood, vice-president of industrial relations and one of the more innovative labor executives in the industry.

Kirkwood was another of those sports alumni of the Monongahela Valley who seem to dominate this story. He grew up in Clairton, spent some of his youthful summers working on the coal docks of U.S. Steel's Clairton Works, and played halfback for Clairton High in the early 1950s. He attended Brandeis University on a football scholarship, and later earned a law degree at Duquesne University. He worked as an industrial engineer and personnel director at the small Jessop Steel Company in Washington, Pennsylvania, and then as an industrial relations staffer at Crucible Steel.

In 1967 Kirkwood joined J&L's legal department and was assigned to dealing with the USW on grievance matters. In the grievance procedure under USW contracts, worker complaints that cannot be settled by the local union and management go to the company and the International union at the fourth step. At this point Kirkwood would meet with an International representative and decide whether to settle the grievance, drop it, or push it into arbitration.

In a 1983 interview, Kirkwood told me that this experience was the beginning of his disaffection with the labor system in the steel industry. "We [the industry and the union] got ourselves enamored of this adversary procedure, where we have intricate precedents for resolving problems at the fourth step. If we can't settle it, we both of us become great lawyers and go try arbitration cases. You get into a macho game. I've seen grievance committeemen admit it's just a game with them. How much can they get from the company as a trade-off by filing a grievance, whether there's merit or not? I've seen company guys, who could have resolved problems on the plant floor, tell the worker, 'Go file a grievance!' It's a tremendously bureaucratic process. And in industrywide bargaining, we have more goddamned top-level committees than you can imagine, supposedly studying problems. What we should be doing is working with the guy on the floor in a participative sense and letting him use his own resources."

Kirkwood was a compact five foot, eight inches, with a square face, squinty eyes, and black hair. Both in his conversation and interests, he seldom stayed long in one place. He was constantly moving, stirring up the dust. He left J&L in late 1983, formed a management consultancy, and later bought a small steel company in Pittsburgh. In the early 1980s he would call me every now and then when he visited New York, inviting me to meet him for a quick drink at a certain bar in a large hotel. After a breezy greeting, he would pump me for news, plant a few ideas of his own, and dash off to a meeting. The breeziness disguised the fact that Kirkwood believed deeply in the participation concept and was willing to take chances to establish it at J&L.

The general manager of labor relations under Kirkwood

was a tall, blonde, imperturbably calm man named A. Cole Tremain. To implement Graham's productivity strategies, Kirkwood and Tremain established good relations with International and local union officials and managed to negotiate local agreements under which J&L could achieve higher output with fewer people. The company called this "problem solving," and decided to apply it to resolve production bottlenecks. In 1979, with the cooperation of a USW local at Louisville, Ohio, J&L set up task forces consisting of hourly workers and supervisors. Just as in the later LMPTs, the production and supervisory employees developed ways of cutting costs and increasing output.

I doubt that Tom Graham practiced the participatory style from instinct. I particularly remember that in a 1980 interview I asked him whether the hourly members of the Louisville task forces could suggest for discussion issues that concerned them, such as safety matters. "Oh, no," he said, "if you let this degenerate into a generalized bitch session, you destroy the whole thing."

Under Kirkwood's influence, Graham apparently changed his mind and accepted the idea that participation should not be all "give" on the workers' part. But if he was not inherently a "democratic" manager, Graham did appreciate the contribution that rank-and-file workers could make to improve the firm's competitive ability. What happened under him at U.S. Steel is another story, told elsewhere in this book.

The LMPT concept was introduced at only two U.S. Steel plants, Braddock and the Irvin Works, despite requests by USW locals at other plants to join in a cooperative program. The idea that labor should participate in any significant way in decision making on the shop floor, much less at higher levels in the steel plants and the corporation itself, did not fit in with U.S. Steel's management style and philosophy.

Chapter 7

The Mon Valley mill towns did not spring full grown from the iron breast of Andrew Carnegie. Most of them had a rural antecedent, though it might not have been more than a country crossroads with a few houses. Upon this pre-Revolutionary civilization, with its century-old political divisions and rural characteristics, was imposed a harsh factory economy. Out of the inevitable clash grew a new way of life with social and political patterns that would have a profound effect on relations between workmen and bosses in the steel mills.

Less than three hundred years ago, Monongahela country was an unexplored wilderness. A vast hardwood forest stretched across rolling hills and plunging valleys. Ancient Indian tribes lived here periodically, leaving large earthern mounds for white men to puzzle over, and the Delaware and Shawnee fought rearguard battles along the Monongahela, Allegheny, and Ohio as they migrated westward. One Indian legacy was the name "Monongahela," the white man's corrupted form of "Menaungehilla," which meant "high banks breaking off and falling down at places." The river's shale and limestone banks were underlaid by large deposits of red and yellow clay. Moving water gradually washed away the clay, which gave the river a muddy look and caused the banks to collapse. Rather than simply naming the river their equivalent of "muddy," the Indians observed cause and effect and captured a timeless process in five lilting syllables.

Scotch-Irish settlers began trickling into the region in the middle of the eighteenth century. McKeesport was one of the earliest communities, and its history was fairly typical of the region. George Washington made the first recorded visit of a white man to the spit of land where the Youghiogheny flows

162

into the Monongahela. While scouting the French presence in and around Pittsburgh in 1753, Washington paid homage to the old Indian Queen Aliquippa at the mouth of the Youghiogheny. In his journal he wrote of presenting her with "a matchcoat and a bottle of rum, which latter was thought much the best present of the two." In 1768, after the British had driven the French out of the territory, David McKee—a North Country Irishman—settled at the mouth of the Youghiogheny with his family and acquired title to it. He farmed the land and operated ferries over both the Mon and Yough (pronounced by residents with a hard "gh" as in "Yok"). John McKee, one of David's sons, laid out a village in 1795 and solicited buyers for lots in McKee's Port, listing among its attractions the fact that the village was "at least twelve miles nearer to Philadelphia than Pittsburgh is."

Other families established themselves along the river and in the hills and hollows for miles around. They cleared and farmed the land, fought off the last of the Indians who made sporadic raids on white settlements until the 1780s, and became adept at operating stills. A brisk whiskey trade grew up in Monongahela country, and in 1794 farmers rose up in the ill-fated Whiskey Rebellion to protest a federal tax imposed on shipments of the product.

Coal was abundant throughout the region, often appearing as thick outcroppings on the river bluffs. The settlers at first used it to heat their cabins. By the early 1800s, Pittsburgh residents already burned so much coal in homes and forges that black smoke hung over the city and penetrated homes. "Even snow can scarcely be called white in Pittsburgh," noted an observer in 1811. The demand for coal here and, later in other downriver communities as far south as New Orleans, created the first large-scale industry in the Mon Valley. Up and down the Mon and Yough, farmers dug coal from the hillsides and floated it down the river on flatboats. In the 1850s, the flatboats began disappearing, replaced by steamboats which pushed strings of coal barges down the river.[1]

Meanwhile, a large boatbuilding industry sprang up in river towns such as Brownsville, Elizabeth, Belle Vernon, and McKeesport. At first they provided boats for homesteaders and traders bound for the regions west of Pittsburgh: flatboats, pirogues, skiffs, bateaux, keelboats, arks, broadhorns, barges, packet boats, and Kentucky boats. Many of the most famous paddle-wheelers used on the Mon, Ohio, and Mississippi in the late nineteenth century were built in the valley. Elizabeth even turned out ocean-going steamers which sailed to New Orleans and thence into the Atlantic.

Nourished by mining and boatbuilding, smaller industries came to life: saw mills, planing mills, grist mills, tanneries, liveries, and brick yards. Blacksmith shops and forges were scattered about. The first iron rolling mill in this part of the valley was established in McKeesport in 1851. Built by W. Dewees Wood, it turned out high quality iron sheets and became known as the Wood Works. Later a part of U.S. Steel, this works lasted for nearly a century before it was shut down in the 1940s. Railroad lines crept up the valley in the 1860s, led by the B&O.

By the 1860s, on the brink of the steel era, Pittsburgh was already a metropolis, with a population of more than one hundred thirty thousand. It had a large concentration of iron works, foundries, glass factories, railroads, boat yards, and banks, as well as mansions, theaters, and a growing cultural life. Industry had spread northward along the Allegheny and southward along the Monongahela. The industrial crawl in the Mon Valley, however, apparently had reached its limit three to four miles from the mouth of the Ohio, in the vicinity of what is now Southside. Beyond this point was little but farmland and patches of small industry. Standing at the Southside's far boundary, one could look upstream and see the river vanish as it veered northward around a high, tree-covered cliff. Rounding the bend would have been like going backward in time. The inhabitants of the little villages up the valley had discarded their buckskins and long rifles, but they were still considered "country people" by Pittsburghers, living on the other side of a cultural and economic gap that would never be entirely closed.

McKeesport was a hamlet of only twenty-five hundred people in 1872 when Boston industrialist John H. Flagler came to town and established the National Tube Works. Most of the other prospective mill towns at this time were small farming and coal-mining communities. Homestead was "a quaint country seat," according to J. H. Bridge. In 1885 what would become the city of Duquesne consisted largely of "fields of waving grain," as a local newspaper described it. Woodlands still blanketed hills and ridges behind the river communities, and for brief stretches the Mon ran between blunt, uninhabitable cliffs. Boat traffic was heavy, however, and traveling up and down the river must have been a bit like driving today through patches of semiwilderness and farmland on Interstate 80. While some eight thousand men sweated in the iron works of Pittsburgh, less than a dozen miles to the southeast people still hoed, sawed, hammered on anvils, rode horseback, and fished in the rivers—waiting in a sort of rural expectancy for something to happen.

And then came the steel industry. Mill buildings were erected, blast furnaces were raised, railroad spurs built. Smoke poured into the skies, and flames leaped from the furnaces. All sorts of mill suppliers moved into the towns, and related industries sprouted along the slices of riverbank not yet gobbled up by the first iron and steel interests. A brief rundown of the industries that settled in McKeesport indicates the remarkable pace of development. In 1872 a plant producing stainless iron was established on the bank of the Monongahela north of the Tube Works. This plant, which eventually became the headquarters of Firth-Sterling Steel, made the shells that sank the Spanish fleet at Manila Harbor in 1898 and later introduced the first stainless steel in the United States.[2]

National Tube, meanwhile, expanded rapidly, adding several mills, a foundry, and pipe furnaces in the 1870s. As the use of natural gas increased in the 1880s, the firm met the market by finding ways to produce larger diameter pipe in longer lengths. By 1890 the Tube Works employed three thousand men and soon blew in the first two blast furnaces to supply iron for the pipe plant. A changeover to steel pipe called for a unit of Bessemer furnaces, which were built in 1892–93 and employed one thousand workers.

Iron- and steel-making establishments now extended four to five miles in an unbroken line along the east bank of the Monongahela—the Wood Works, National Tube, and Firth-Sterling. The Duquesne Works was built directly across the river from Firth-Sterling. On the other side of McKeesport, mill buildings housing the McKeesport Tin Plate Company covered thirteen acres along the south bank of the Youghiogheny. Started in 1903, this firm quickly became the largest tin plate manufacturer in the world. The Christy Park Works of National Tube, built in 1897, occupied part of the north bank of the Youghiogheny. Fort Pitt Steel Casting Company, a large foundry, was built nearby in 1906.

The mill towns grew with amazing speed. The population of Braddock increased from 1,290 in 1870 to 19,357 in 1910, or

1,400 percent. From 1870 to 1890, the number of residents in McKeesport soared from 2,500 to 20,751. Said to be the fastest growing community in the nation, McKeesport in 1891 became a third-class city, having far exceeded the 10,000 population required under Pennsylvania law. Other mill towns expanded almost as fast.

Typically, more attention was paid to the machines of production than to the needs of the people who would operate them. The immigrants were packed into row houses and tenements which crammed the narrow strips of flat land adjacent to the mills. The homes of skilled workers, bosses, and middle-class merchants lined the hilltops in the rear of town. As the immigrant populations grew, quick-buck developers marched their ramshackle housing up hollows and down ravines and across forty-five-degree slopes. Streets switchbacked up the hills, and pedestrians climbed from level to level on wooden staircases of fifty to one hundred steps. One of my sharpest memories is of old women with slipper like shoes and muslin stockings climbing the steps, carrying shopping bags with twine handles and pausing every tenth step or so to catch a breath.

In most of the towns, homes eventually struggled over the brink of the hill and blanketed the ridge tops. But in towns like Braddock a sheer cliff halted the growth, and one can still see the high-water mark, a ragged line of white, gray, and blue frame houses that cling to the hillside some two-thirds of the way up. The space available for low-cost housing simply ran out in many of the towns. As late as 1930, Braddock still had 131.5 persons per acre and Homestead 116.2.

"Everything that has made this area ideal for the building up of our steel civilization is balanced by factors that make it unfit for urban living," concluded a social research team in the 1930s after studying the mill towns. "Only the most ingenious city planner and the most far-seeing civic pride, combined with the capacity to forego immediate profit and to build for the future, could have made this scenic marvel into a city of homes."[3]

Industry Equals Progress

That the entrepreneurs of the 1870s could jam an industrial revolution down the throat of a beautiful river valley without provoking a human revolution says much about the character of the region and the people, and the aura of the times. This was the "century of progress," the fruits of which were abundantly evident in Pittsburgh, and the old farmer-tradesmen stock of the upper valley—village leaders and workers alike— were eager to share those fruits. At the same time, these descendants of the original settlers still bore the imprint of a rural background which bred respect for private property and home ownership. A sense of community sharing was strong, and the family was the focus of social life. There was a strong distrust of foreigners, outsiders, and "radical ideas."[4]

When a new factory moves into town today, the normal reaction is to welcome the addition of new jobs and the expansion of economic activity, but also to worry about the changes that will be wrought in the community. Corporations are required to present evidence of the social and environmental impacts of industrialization, even on creatures as remote in the chain of life from humans as snail darters. It can be safely assumed that village leaders on the Mon demanded no such information from Andrew Carnegie, John Flagler, or Henry Clay Frick. The resources were there to be used for profit and expansion, not to be conserved. It is important to note that this was the attitude not only of the robber barons but also of people in their prospective baronies.

In those days of environmental innocence (or indifference), the sight of flames leaping into the sky, smoke pouring out of stacks, and waste water gushing into the rivers evoked a great sense of community pride. Up and down the Mon the heightened civic feeling that came with each plant opening, combined with the first awed look at the wonderful new industrial machines, was reflected in local newspapers. Each town seems to have had a poet laureate who sought to paint a word picture of what the coming of steel meant for the community.

In Duquesne, a local writer compared Carnegie's steel works to "the meteor that has darted out of space and cut a brilliant path across the sky." This plant, the writer went on, was "the acknowledged young giant and the mastodon of the unconquered and the unconquerable Monongahela Valley." Across the river, two writers who collaborated on a history of McKeesport for the city's centennial celebration in 1894 declared its residents were "strong in the conviction that the great Ruler of the Universe regards the community with a special favor."

A Dr. Frank Cowan in Braddock, writing before 1903, con-

trasted the slaughter of General Braddock and his troops in
1755 with the activities taking place in Braddock's field sometime around 1900. "Where the cannon of Braddock were
wheeled into line," he wrote,

> There the turning converter, while roaring with flame,
> Pours out cascades of comets and showers of stars,
> While the pulpit-boy, goggled, looks into the same—
> Thinking little of Braddock and nothing of Mars.

The growth of the community also was important, and
these poet-prophets were not shy about claiming big things
for their towns. The first political milestone was to convert
from borough to third-class city. This prospect in 1882 called
forth a rhyme from a resident of Steelton, an early mill town
located on the Susquehanna River south of Harrisburg:

> To a city shall this borough run
> Its rapid changes through
> Before the child is twenty-one
> This saying will come true.

At about the same time, an observer noted that Steelton had
the largest steel plant in the country, two railroads, a river,
and a canal and asked, "Why shouldn't Steelton boom?"

The toting up of natural and man-made assets, like counting the rooms in one's house for a real estate appraisal, seems
to have been a common thing in steel towns. As late as 1934,
McKeesport proudly claimed for itself 54,632 inhabitants, 73
churches, 61.50 miles of paved streets and 61 miles of sewers,
3 railroads, 2 streetcar companies, 5 bus lines, 4 banks, and
"rivers, two, Youghiogheny and Monongahela."

One of the more ambitious of the mill-town odes was a
piece entitled "McKeesport" by a William Bingham Kay.
"Where Aliquippa's swarthy tribe roamed river's brink and
hill's steep side," it starts, and, after four stanzas of history,
describes the present this way:

A mighty host here dwells and toils. Black, flame-shot smoke
  blue heaven soils,
Uprising from Inferno-fires where dazzling, blue-white metal
  boils
    And heavy thunders quake....
Men everywhere upon the sphere use products manufactured
  here.

Our tin bears food to Arctic lands, our pipe is laid o'er tropic
sands.
And foreign foeman well may fear projectiles fashioned by our
hands
    Should war its fetters break.

While describing smoke belching from mill stacks as "King
Steel's grim gonfalons," Kay had large hopes:

This city, then, will be the heart and wealth-enjoying central
mart
Of that great district that to fame will better advertise its
name.—
The greatest district on the chart of manufacture. Who can
claim
    Too much for what's to be?

The bard of Tube City seemed to be prophesying here no
less than a rude overthrowing of Pittsburgh as the steel capital. But this apparently was common in the days when anything could yet happen. Down the Ohio River at Aliquippa,
according to the *Post-Gazette,* "citizens were confident that
the area along that stretch of the river would be mightier than
Pittsburgh."[5]
Buoyed by an unquestioning belief in economic expansion,
the town fathers were eager to cooperate with the companies.
What was good for the company was good for the community,
and vice versa. Labor unions were anathema. Immigrants
from southern Europe could be treated like peasants without
rights, as they were in the steel plant.

The Immigrants Divided

The demand for labor was inexhaustible. The first workers
came from the original stock of Scotch-Irish and English settlers, but this supply quickly ran out. In the 1870s, word
quickly spread through Europe that the steel mills of Pittsburgh and Chicago needed men. And they came, in an ava-
lanche, practically burying the ill-prepared mill towns.
Carnegie and other steelmakers sent agents to Ellis Island to
recruit workers and placed ads in European newspapers, offering "good opportunity" in the mills of America. Although
some skilled workers such as iron puddlers, moulders and

machinists came from England and Germany, the great bulk of the immigrants were unskilled.[6]

The first wave of immigration, dating from the mid-1800s, generally included Germans, English, and Irish. Along with the old-stock families, they became the supervisors and skilled workers. The unskilled came in succeeding waves, starting in the 1870s and lasting for the next forty to fifty years. Included in these groups were Swedes, Italians, Slovaks, Croatians, Hungarians, Poles, Romanians, Serbs, Russians, and Greeks. Mexicans and blacks later were imported by the companies to act as strikebreakers.

Each ethnic group tended to work in the same mill department and live in the same part of town. Earlier immigrants helped newcomers get jobs, a process that frequently included a payoff to the foreman. A series of ethnic enclaves formed in the plants, with Anglo-Saxons dominating the better jobs in the rolling and finishing mills. The southern European immigrants and blacks were confined to the dirtiest, most hazardous departments—for example, Slavs at the blast furnaces and blacks on the coke ovens. Rampant discrimination against all "hunkies" enforced the segregated working and living patterns. A "hunky" was anybody who looked remotely Slavic or spoke a difficult tongue—Slovaks, Magyars, Croats, Serbs, Poles, and others.

Segregation in the plant suited management's interest because it kept the men divided. The old Amalgamated Association of Iron, Steel & Tin Workers abetted the pattern by refusing to admit unskilled workers, forcing the immigrants to define their interests as sharply different from the union's. The national steel strike of 1919 failed, in part, because the Amalgamated had aroused such animosity among immigrants and black migrants. Refused admittance to the union, black steelworkers felt no compunction about working during the strike. The steel companies imported an additional thirty thousand blacks to Illinois, Indiana, and Pennsylvania to break the strike.[7]

Lacking democratic traditions and accustomed to authoritarian leaders, the European immigrants accepted American life as they found it. Along with the blacks, they were denied promotion and confined to low-paid laboring jobs, a fact that is exceptionally well documented by John Bodnar's excellent study of ethnicity in a steel town. Without economic advancement, there could be no social mobility. Revolt was impossible without an institution to guide it, and management's smashing of the Amalgamated in the early 1900s snuffed out any possibility that it might evolve into an all-embracing union.

With all other doors closed, the immigrants turned "inward," as Bodnar describes what happened. "Confined in lower-level occupations in the steel plant, housed in separate row homes, unable to rise occupationally, subject to economic vicissitudes, and lacking positions of power in the steel town, the newcomers turned inward. Croats, Serbs, Slovenes, Bulgarians, and blacks displayed almost no regard for Anglo-Saxon concerns such as civic reform or local politics. Immigrants, especially, faced problems which concerned their own congregations, homelands, and ultimately their own identities. Unsure of their status in a new land and faced with rejection and criticism from the old stock, they debated issues that were peculiar to their own ethnic communities. And in the process, they acquired a new ethnic consciousness which surpassed anything they had known in Europe."

The story of the immigrants' fight for independence in the mills and communities is largely obscured by the overall labor struggle. But the two movements, the immigrants' push for fair and equal treatment and labor's drive for independent unions, finally merged in the 1930s. Reliance on the ethnic community, Bodnar says, taught immigrants "the value of confronting social and economic problems with large, formalized institutions." While ethnicity was a divisive force before 1930s, it helped lay the basis for "the type of cooperation that would be necessary for the eventual triumph of the CIO."

Before the 1930s, however, community life was fragmented by the large number of nationality groups. According to the 1930 census, 58 percent of McKeesport's 54,632 population was either foreign born or native born of foreign or mixed parentage. A survey in the early 1930s showed that McKeesport residents had been born in thirty different countries, (twenty-nine for both Duquesne and Clairton). In 1926 the McKeesport Chamber of Commerce boasted that the city had seventy-two churches, representing twenty-three denominations.[8]

The fact that these groups maintained their national pride and traditions, in the end, gave the valley a rich diversity. But their early isolation helped block the development of a wider civic consciousness and abetted political corruption. Each group had leaders who controlled the votes of the group and dealt with the city administrations through a patronage system.

In practically every steel center—the Mon Valley, Bethlehem, Steelton, southeast Chicago—the companies dominated town councils, school boards, and political affairs. If the elected officials were not, in fact, also the company officials, the former were greatly influenced by the latter. John A. Fitch, a writer who lived in Mon valley steel communities for six months in 1907–08, described how U.S. Steel foremen rounded up the steelworker vote for Republican candidates. Even native-born mill workers voted as told, Fitch says, because they "have a sort of superstitious feeling that somehow the boss will know if they vote wrong."[9]

The GOP held most municipal offices in the Mon Valley from the late 1800s until the late 1930s. In McKeesport, Republican George H. Lysle served seven consecutive terms as mayor, from 1913 to 1941. Like most elected officials, Lysle was rabidly antiunion. In 1919, when William Z. Foster led a national AFL campaign to organize the steel industry, the valley towns refused to permit organizing meetings. Lysle issued a proclamation prohibiting public assemblages of more than three persons and enforced it by stationing three thousand "special policemen" around the city. He contended that "Reds" and other "outside agitators" were stirring up the foreign-born populace. The *Daily News* editorialized that the real question was not whether a union should be formed but whether "the foreigners are going to run the affairs or the Americans."

In Duquesne, Foster and Mother Jones—the famous coalfields organizer—were among forty people arrested when they tried to hold a meeting in a vacant lot. The forty were convicted of disorderly conduct and fined $100 each by Mayor James Crawford (not only did he, as mayor, prohibit the meeting; he also, apparently as magistrate, enforced the order). When a New York rabbi asked to speak for the union cause, Crawford replied: "Jesus Christ himself could not speak in Duquesne for the A.F. of L.!" Another long-term Republican mayor, Crawford had intimate connections with the steel interests: His brother was the president of McKeesport Tin Plate.[10]

The steel companies were also possessive about the pool of labor in the mill towns and surrounding countryside. They discouraged other large industries from entering the area. As a result, the valley remained a one-industry region. While the companies encouraged the growth of good educational systems, it was recognized that they would make the first claims on the output. Consequently, the high schools developed curricula enabling students to specialize in industrial skills. In the late 1930s, McKeesport opened a separate Vocational High School that was recognized as one of the finest in the state, with well-equipped electrical and mechanical shops.

Because the city had a relatively large middle- to upper-middle class, the main high school, McKeesport Technical High, offered a strong college preparatory curriculum, but even as late as the 1940s, when I attended high school, the so-called college counselors overtly discouraged students from applying to first-class universities, unless they were of the "right" parentage—management people, doctors, lawyers, and so forth. You were "counseled" to educate yourself for a mediocre place in life, not a large one involving intellectual ability. It was a mildly humiliating experience for me, and I wasn't even from a steelworker family. A friend who *was* the son of a steelworker recalls that a counselor asked him, "Why do you want to go to college, you're just the son of a steelworker?"[11]

Gradually, the towns opened up to broader ideas, but their parochialism was bolstered by another fact of life in the Mon Valley. The original river towns and farming villages had grown independently of one another, separated by impassable cliffs, deep ravines, and the Monongahela's elbow bends. Roads connecting the towns had never been good. And when the mills and railroads settled into the riverside flats, pedestrians and vehicles were diverted to the ridge tops. Even today, in the mid-1980s, some Mon Valley communities remain landlocked and river-locked, practically inaccessible from major expressways that crisscross the region.

Studded with many small communities, each with its own political structure and traditions, the valley never developed a regional political framework for dealing with problems common to all of its towns. The growth of the steel industry on a piecemeal basis, from site to site, helped institutionalize the divided political structure, and the decentralized organization of U.S. Steel further abetted it.

Steel communities followed the corporation's hierarchical organization: There was no reason to do otherwise. Until the

1960s, the valley's entire history was one of decentralization, subdivision, and separation. This legacy became a barrier to regional development programs that might have helped the mill towns adjust better to the long-term decline of steelmaking.

Life in the Mill Towns

According to the accounts of Thomas Bell, John Fitch, and other observers, life for the great mass of unskilled workers in the valley was grim for the first forty to fifty years. They endured unsanitary, substandard housing conditions, alarming death and accident rates in the mills, and discrimination and political suppression in the community. Discrimination on the job eased somewhat for the second and third generation of white immigrants but not for blacks. In a recent book on black mill workers in western Pennsylvania, the historian Dennis C. Dickerson—himself the son of a Duquesne steelworker—records in comprehensive detail how blacks were denied promotion in steel plants. Steelmakers hired blacks during business expansions but only in certain departments, such as the coke ovens. With the lowest seniority, blacks were laid off first. In 1944, 11,500 or 14 percent of all mill employees in western Pennsylvania were black, but the proportion fell to 6.8 percent by 1966 as jobs declined.

Although black workers supported and helped organize the USW, many locals with white leaders, abetted by management, practiced blatant discrimination against their black members. Rarely were blacks admitted to apprenticeship programs for skilled craft work. In production units, many locals arranged seniority schemes so that blacks could not rise above unskilled and semiskilled jobs. Clairton's coke batteries were attended almost exclusively by black workers, except for a few whites who occupied choice positions as foremen, heaters, and gang leaders. More than six hundred blacks went on strike at Clairton in 1944 over discriminatory treatment. In 1966 blacks held only 3.1 percent of all skilled jobs in Pittsburgh-area mills.[12]

Discrimination against blacks on the job mirrored racial prejudice in community life. During World War II public housing projects were located near black neighborhoods, so that segregation continued. Harrison Village in McKeesport and Blair Heights in Clairton housed only blacks. A municipal swimming pool constructed in Clairton excluded blacks, causing public protests. Although the Pennsylvania legislature in 1935 outlawed racial segregation in public places, blacks were denied service in many restaurants and taverns. The Ku Klux Klan was active in the Mon valley well into the 1940s. A chilling sight that has remained with me from those years was the burning of a huge cross on top of a hill in White Oak.

Working and living conditions in the valley improved at a very slow pace. The twelve-hour day came to an end in 1924, and some companies sponsored welfare programs to raise community health standards and provide recreation. However, mill slowdowns in the 1920s caused sporadic employment, and the Depression brought long periods of unemployment and shortened work weeks for unskilled and skilled worker alike.

The contemporary picture of the overpaid steelworker implies that this was always the case. In fact, by many standards mill employees were underpaid up to World War II. On the eve of Pearl Harbor, the average hourly worker at U.S. Steel grossed $35.92 a week, or $1,868 per year, less than 80 percent of the amount needed to sustain a family of four on a minimum budget.[13]

Skilled workers, managers, and merchants fared better, although all were hurt by the Depression. The business boom that started with the war, combined with the USW's collective bargaining efforts, finally created the tide that carried all steelworkers to higher-income ground and led to a significant increase in living standards.

There was much more to life in the mill towns than discrimination and poverty. In the early days of industrial growth, an almost joyful exuberance was in the air, displayed albeit in episodes relating to drink. The first issue of the McKeesport *Daily News*, dated July 1, 1884, reported on the front page of "a small-sized riot" that took place the previous night when Jeremiah Fitzgerald invited friends to share a keg of beer. "The Burgess [magistrate] thought they had about five dollars worth of fun and they all paid it except Jeremiah, who is in the borough coop listening to the sweet music of the waves as they beat against the rock-bounded shores of the placid Youghiogheny."

Across the river in Duquesne the steel plant built in 1885 attracted "a rough class," as a newspaper described the newcomers. "Everybody seemed to have money and everybody seemed willing to spend it," the account continues. "Gambling was very common and it was no unusual spectacle to see a crowd of men and boys playing poker over a keg of beer along Duquesne Avenue Crap shooting on the street in the daytime was an everyday happening and the use of intoxicants was very common, and it was a cold night when a scrap was not precipitated at the corner of First and Grant. Law and order were cast to the winds and a state bordering on anarchy prevailed." Older citizens formed a "vigilance committee" and patrolled the streets at night.

The exuberance was displayed in other ways. Quite early in its history, McKeesport became something of a theatrical center. In 1883 White's Opera House opened in the city, spawning a tradition of serious drama, opera, and vaudeville that lasted until the 1920s. The lead story in that first 1884 edition of the *Daily News* told of the current offering at White's, a three-act burlesque entitled "Prince Chow Chow, King of Pickeldom." Over the next several decades, White's and other McKeesport theaters imported many of the leading repertory companies and actors of the day. This tradition was continued by John P. Harris, who owned theaters in both McKeesport and Pittsburgh. In 1905 he opened the nation's first movie house— "nickelodeon," he called it—in Pittsburgh and followed with the second one in McKeesport.

One of McKeesport's most famous sons, playwright Marc Connelly, says he got his inspiration from the musical shows at White's and the many carnivals and pitchmen that visited McKeesport in those years. Born in 1890 of show-business parents who operated a small hotel in McKeesport, Connelly left town in 1908 to work on newspapers in Pittsburgh. He later collaborated on several plays with George S. Kaufman, another Pittsburgher. Connelly's drama *The Green Pastures* was awarded a Pulitzer Prize in 1930.

Other McKeesporters who achieved some degree of fame included Helen Richie, the 1930s aviator and first woman airline pilot, and Henrietta Leaver, the Miss America of 1935. Lt. Gen. George D. Miller, a high school classmate of mine and football co-captain, went to Annapolis and later flew combat missions in Vietnam. He retired from the U.S. Air Force in 1986 and was named secretary general of the U.S. Olympic Committee.

The entire Mon Valley was noted for the many first-rate football, basketball, and baseball players who went on to stardom in college and, some of them, in professional sports. To try to name them all would be to offend many whom I don't know. One of the most unusual, however, was Betty Dingledein, a McKeesport High graduate of the 1940s, who became the first American woman matador in Mexico under the name Betty Ford.[14]

Because it was a trade center, McKeesport had a large upper-income population in comparison with most mill towns. The city's fast growth made millionaires of an extraordinary number of McKeesport businessmen. Numbered among these were owners of real estate, meatpacking and wholesale food businesses, as well as large clothing, furniture, and general merchandise stores. G. C. Murphy Company, the large five-and-ten-cent-store chain, originated in McKeesport in 1906 and expanded rapidly after a new management took over in 1911. The chain grew to more than four hundred stores but maintained its home office in an ugly tan building on Fifth Avenue.[15]

People who invested early in locally owned foundries and mills, such as McKeesport Tin Plate, also became millionaires, as did some bankers, lawyers, and doctors, and the owners of the *Daily News.* Steel-company executives and mill superintendents may not have been millionaires, but they lived quite comfortably. The upper-income group formed the power elite of the city whose social center was the Youghiogheny Country Club, located on a hill behind the Yough and far from the urban hoi polloi. Few major decisions were made in the city without the tacit approval of the leaders of this group. In upper crust and middle-class social circles, steelworkers were looked down on as "mill men."

Not only did the McKeesport rich derive their income from the inner-city businesses; most of them also lived in the city in huge brick and stone houses that lined tree-shaded streets on the hill above downtown. In the 1940s and 1950s improved transportation enabled them to move to suburban retreats, followed first by the more affluent middle-class and later by many working-class families.

This process of suburbanization was largely the same in Duquesne, Braddock, Homestead, and Clairton. John Bodnar, the closest observer of social and economic mobility in mill towns, says that in Steelton it was largely the old stock Anglo-Saxons who migrated to suburbia, leaving behind "white ethnics" and an increasing number of blacks.

In the Mon Valley, the white ethnics also moved out of the mill towns in large numbers. As a general proposition, however, Bodnar's conclusion about Steelton seems broadly applicable to the Mon Valley towns: "Abandoned by the affluent old stock, these two groups [working-class ethnics and blacks] were left alone in the urban arena to face its awesome problems. The legacy of enforced isolation which characterized their historical experiences, however, left them unprepared to deal not only with these problems but with each other."[16]

McKeesport prided itself on its business leadership of the valley and its tradition of independence and self-sufficiency. It was big enough to support the only brewery (Tube City Beer) and the largest daily newspaper in the valley outside Pittsburgh. Led by the *Daily News*, McKeesport stayed aloof from all moves toward regional economic planning or valleywide tie-ins such as water and sewage systems (an aloofness it would have cause to regret in the 1980s).

McKeesport also developed a reputation for getting itself involved in bizarre political conundrums. The most enduring was the decades-long fight to rid itself of the B&O Railroad tracks that ran through the center of town. In 1855, McKeesport leaders paid $150,000 to a predecessor of the B&O to lay its tracks along McKeesport's northern perimeter. The town grew so fast in the ensuing steel boom that within a few decades the B&O tracks ran through the heart of the business district. In the 1960s, thirty-five freights and several passenger trains rumbled through the city every day, shaking buildings and halting car traffic on every major north-south street. By the 1930s, everybody wanted to get rid of the trains. Not until 1970, however, could the city and the B&O agree on a rerouting scheme and the financing. Federal and state agencies paid $7.5 million and the city of McKeesport $400,000 to remove a problem that was created at a cost of $150,000 ($623,000 in 1970 dollars).[17]

Politics McKeesport Style

Until the late 1930s, local government in Mon Valley communities unabashedly supported corporate interests. Social stability, reasonable tax rates on mill property and equipment, a ready pool of labor dedicated to basic steel—these were the primary considerations in governing the mill towns. In return, steel management backed Republican officeholders like Lysle and Crawford in elections and advised them on financial and planning matters. Needless to say, none of the planning involved a diversification of the industrial base.

The depression, combined with the popularity of President Roosevelt and the unionization of steel, dramatically changed this picture. In 1937 Duquesne and Clairton elected their first Democratic mayors—Elmer Maloy in Duquesne and John Mullen in Clairton—both of whom were union activists and later served on the USW staff. McKeesport, with its strong, local business interests, continued to elect Republicans locally, although in 1936 its citizens switched to the national Democratic ticket to vote for Roosevelt. With labor gaining a voice on city councils, overt control by the companies came to an end. But many other sociopolitical patterns were too deeply embedded to be changed.

An intimate connection between the numbers racket and local government remained unchanged, for example. "Playing the numbers" flourished for decades in many American cities before the legalization of state-run lotteries. But the racket was a particularly basic fact of life in steel towns and contributed to a festering corruption in political affairs. Betting a dime, quarter, or half-dollar every day on a three-digit number was practically as common as buying a loaf of bread. You would bet the number "straight" or "boxed"—that is, any combination of the three digits.

Everybody knew where and how to bet. Nearly every large workplace had a "numbers man" who would make the rounds in the morning to collect bets. You could phone in a bet to a telephone booth during certain hours. Most betting, however, was done in taverns, or in confectionery stores. The proprietor would write the number on a tiny pad of white paper and give the bettor a carbon copy. An agent would pick up the numbers slips and bets, or the proprietor might read off the numbers on the phone. I remember, as a boy, getting impatient with clerks who ignored my urgent need for ice cream while they droned on at the phone, reciting meaningless numbers.

In late afternoon, a specified three digits of the daily handle (revenues) at a race track would determine the winning number, and it would spread through town by word of mouth. This was grumbling time for losers who just missed a big payoff by one digit or forgot to box their number. But there was always tomorrow. The mill towns wouldn't run out of numbers—or numbers writers.

The numbers racket was controlled by shadowy gangs who

now and then engaged in bombings and shootings to keep competitors out of their territory. Revenues were undoubtedly used in loan-sharking, prostitution (although the latter flourished quite nicely on an independent basis), and other criminal activities. The names of the top men in the business were well known to police, politicians, newspapers, and other leading citizens. Every cop on the beat knew where numbers could be played and even frequented these places on coffee and cigarette breaks, although they were conspicuously absent at collection and payoff time. Banks would open their doors after hours to allow the chief bagman to deposit the day's collections.

In fact the numbers racket was regarded as a quasi-legitimate business, the benefits of which offset its quasi-ness. It is tempting to speculate that allowing people to gamble was one method of keeping the laboring class satisfied with its lot, even if this meant tolerating a certain amount of racketeering. This much is certain: The numbers racket propped up economic activity, allowed people to indulge their passion for gambling, and contributed significantly to the political stability of the mill towns—all of which accrued to the benefit of the steel companies.

The racket chiefs assured stability by bankrolling election campaigns, usually of incumbent officeholders. There may have been some penny-ante payoffs on a daily basis to cops and city officials. But this was discouraged, because too much of it would have led to anarchic greed. Instead, the entire town administration, if it had cooperated by not crusading against the numbers men, was rewarded at campaign time by a donation sufficiently large to buy the support of ward bosses and precinct workers. In McKeesport, the amount was $30,000 to $40,000. If the incumbents were reelected, which was a practical certainty with such a large campaign treasury, rewards would filter down in the form of patronage. The chief of police would be reappointed. Beat policemen might get relatives on the payroll.

Although most people suspected that somebody was being paid off, documentation of the link between numbers and politics rarely appeared in print. In late 1965, I interviewed dozens of people in the Mon Valley for a story on its decline as a steel center. One of these was Joseph Sabol, Jr., who was finishing his one and only term as mayor of Duquesne. He had become disillusioned by corruption and the lack of civic cohesion to solve the valley's growing problems. "Without the numbers you wouldn't have city government in this valley," he said. "They're the big boys. That's an industry around here. They make the biggest campaign contribution, and they put the people in office." He paused and added vehemently, "They're a cancer!" Sabol has long since retired.[18]

I received further and more specific corroboration of the numbers-politics nexus from Andrew J. "Greeky" Jakomas, who served as mayor of McKeesport from 1954 to 1966. One of the more colorful public officials in the valley, Jakomas was regarded by some residents as a showman who did "crazy things," but by others as a leader who made an effort to halt the city's economic slide. However he was viewed, his long mayoralty illuminated much about mill-town political life.

The son of Greek immigrants, Jakomas first made a name for himself as a standout tailback on McKeesport High teams of the early 1920s. He attended college and worked in the foundries and mills of McKeesport as both hourly employee and supervisor. After the war, he bounced around town as a beer distributor before being appointed director of public works. This job enabled him to promote himself politically. During snowy weather, he personally drove bulldozers through residential neighborhoods to clear the streets. The secret of politics, Jakomas told me, is to "fill the people with service."

Although he had got his job as a Republican, Jakomas switched to the Democratic party and ran for mayor against his boss. He cashed in on his popularity—and the fact that he was a Mason and a Shriner—to become the first non-Anglo-Saxon mayor of McKeesport. The second generation of immigrants was taking over.

Greeky was a barrel-chested stump of a man with a quick temper and a flamboyant personality. He had a free-wheeling attitude about parliamentary procedure at council meetings. One councilman was fond of rising to expound on almost any subject, at length. An observer who attended all the meetings remembers how Jakomas would let the councilman ramble on for a while. "Finally, Greeky would push down his glasses on his nose, and he'd say, 'Harry, all I want out of you is a yes vote. So shut up and sit down.' Harry would sit down."

Political calculations were Greeky's forte. Two residents showed up at a council meeting to complain about kids playing ball on their street. At the next meeting, Jakomas said he had personally investigated and counted fourteen boys play-ing whiffle ball. They didn't destroy anything and were gone by 8 P.M. "Each of those kids," he said to the complainants, "has two parents. That's twenty-eight votes. There are only two of you. On election day, I win, twenty-eight to two."

Jakomas was an activist mayor who promoted McKeesport every way he could. He got President John F. Kennedy to campaign in McKeesport during the congressional elections of 1962, and the city later built John F. Kennedy Memorial Park where the president stood. Actually, it was not Kennedy's first visit. In 1947, as a young congressman, Kennedy and Richard M. Nixon came to McKeesport for their first debate. The topic was the Taft-Hartley bill, which was passed later that year.[19]

When McKeesport's retail business began to sag in the 1960s with the growth of suburban shopping centers, Greeky borrowed a European idea to lure shoppers. The Mayor halted car traffic on a two-block stretch of Fifth Avenue, the city's main business street, and turned it into a shoppers' mall with benches and potted trees and plants. The street itself was painted from curb to curb in brightly colored geometrical designs of no particular pattern. Flags were strung overhead. Greeky had envisioned sidewalk cafes—but that was for later.

It was like importing Munich's Marienplatz to McKeesport, minus the fifteenth-century town hall. Bewilderment seems to have been the overwhelming reaction of McKeesporters. An overhead photograph appearing in the *Daily News* on opening day, October 22, 1963, shows businessmen and pedes-trians standing in the middle of the street, hands on hips, looking as if they were about to engage in a mass scratching of heads. Before long, the merchants demanded that the car-jammed old street that they knew and loved be returned to them. The pavement was repainted in mill-town gray, the trees and plants were carried off, parking meters were re-installed, and Fifth Avenue reacquired its exhausted look of yore. Business continued to decline.

I had met Greeky in the 1940s when I was a teenager and both he and my father were in the beer business. I had left McKeesport by the time he became mayor, but years later he remembered me as a kid from the old days. In early 1986, I found him one lunchtime in the bar of the Elk's Club in McKeesport, the last refuge of a gang of retired cops and other city employees. They meet there daily at noon to drink coffee, play bridge, and—if the occasion presents itself—reminisce about McKeesport's salad days. Now in his eighties and bald, Greeky had expanded laterally, his barrel chest having thickened into great rolls of fat. Still flamboyant and loud, he was the central figure of the group, a sort of arbiter of clashing memories about the "old McKeesport."[20]

He told me stories about my father, dead for thirty-seven years, and talked about his own mayoralty. Although he had been something of a populist, Greeky was never opposed to the company. "We had a fine relationship with U.S. Steel," he said. "Before we passed the city budget, we'd call the superintendent of National who would call Pittsburgh [U.S. Steel], and they'd send up a comptroller to help us work out the taxes. He'd say, 'If you go above this figure, you'll take us out of competition.' It was a very affable arrangement."

Greeky still spoke with some scorn of the city of Pittsburgh, reflecting the old fear of McKeesporters that their city would be gobbled up in a huge, regional annexation. He recalled, laughing, the year that McKeesport's sewage disposal plant broke down. He told the plant employees to discharge the sewage into the Youghiogheny. "We dumped it in the Yough, which took it to the Mon, and the Mon took it to Pittsburgh," he said. "What the hell, we had two fast-flowing rivers. Send it to Pittsburgh and let them take care of it." When the state insisted on a halt to this practice, the city refused to join the Allegheny County Sanitary Authority (ALCOSAN) which was building a regional system and instead built its own plant—a display of independence that condemned the next generation of McKeesporters to higher sewage bills than residents in ALCOSAN.

Finally, I told Greeky that after all these years—twenty since he had been mayor—he could speak candidly. Had he, as the rumors always had it, taken payoffs from the numbers people?

"Nope," he said without hesitating. "Personally, I never took a dime. The way it happened was, every four years a campaign contribution would show up. To this day, I have no idea where it came from. I'd tell the chief of police what we needed to run, and it'd just show up, $30,000 to $40,000 usually, in several payments. Once it came in a watermelon." He laughed. "Yeah. The melon had been cored. I took the top off and inside was $5,000 wrapped up in a newspaper." The practice had been going on for years, he said. "All the mayors before me did it this way except one guy. He tried to shake them [the numbers men] down. He wasn't reelected."

The Effect of Social Environment

There is a theory that the social environment of a community has a deep and lasting effect on the climate and practice of labor relations in the community. In the early 1940s, two high officials of the young steel union applied this concept to the Mon Valley in *The Dynamics of Industrial Democracy*, which laid out their vision of how industrial peace might be achieved through labor-management cooperation. Their insights, though long neglected, are critical to an understanding of what has happened in steel labor relations—and what might have happened.

The authors, Clinton S. Golden and Harold J. Ruttenberg, argued that the "dominant social influences" in a community "have a far greater effect on the state of industrial peace" than collective bargaining. They illustrated their point by comparing the labor climate in the steel and auto industries. Organizing the steel union in the Monongahela Valley in the 1930s had been a peaceful process, they noted, while at the same time sitdown strikes and other turbulent events accompanied the unionization of the auto companies. The social environment made the difference, declared Golden and Ruttenberg.

By the time the Steel Workers Organizing Committee (SWOC) began organizing in the mid-1930s, two generations of steelworkers had worked in the valley mills. They had been suppressed and beaten down by company and town officials. The towns "had become stabilized over several decades," and "community disciplines had been established." The authors continued: "One of these disciplines learned through long, bitter experience and passed on from generation to generation, was not to join unions rashly, as that meant strikes, and hardships, and losing out in the mill."[21]

The first task of SWOC organizers was "to break down the fear that joining the union meant strikes." Only after it was clear that SWOC's momentum would carry it to victory did its membership soar. After joining the union, workers in the Mon Valley "assisted, or at least did not impede, the orderly establishment of union-management relations." They did this "because they were dominated by a social environment that had cast them into accustomed routines which they did not lightly upset."

The social environment in Detroit and Flint was just the opposite. The auto industry expanded fastest in the 1920s, when immigration was restricted, and almost half of Detroit's auto workers migrated from other parts of the United States. This first generation still dominated the auto plants in the 1930s, though its members had suffered long spells of unemployment during the Depression. Detroit, Flint, and the other auto towns "never had a chance to settle down as Pittsburgh and the older steel communities had done long since," and auto workers did not develop a sense of belonging.

Golden and Ruttenberg carry their argument too far in attributing the wave of auto strikes in 1937–38 to "social instability." They neglect the fact that the auto companies' refusal to recognize the UAW spurred militant local union leaders to organize sitdown strikes. Local unionists had much less influence in SWOC, which was controlled from the top by Murray. In any case, the UAW's success in forcing General Motors to negotiate played a strong role in convincing U.S. Steel to voluntarily recognize SWOC in early 1937. If USS had taken a hard position against unionism, labor strife might have exploded in the Mon Valley. The Golden-Ruttenberg argument also is partly self-serving, since Golden and Ruttenberg were trying to convince employers that SWOC was not strike-prone.

Nevertheless, their point that fear and repression had produced social stability in the Mon Valley seems well taken. From the vantage point of the 1980s, we may add discrimination, parochialism, an authoritarian leadership in hock to racketeers, and a top-heavy economic organization that resisted reform.

The work ethic of labor in the Mon Valley was never in question. The men and women who went into the mills worked hard—that is, as hard as the management bureaucracy, jealous of its prerogatives, would allow. (This concept of "hard work" assumes that people want to use their heads as well as their hands.) Forced to accept the authoritarian tradition, the union confined itself to fighting over grievances and protecting a concept of narrowly defined jobs which impeded productivity. In time, a low-grade factionalism poisoned the political life within many of the local steel unions. A light glaze of moral decay settled in the valley, characterized by a willingness to "go along with" a mildly corrupt political system and a less-than-best production system in the mills, imposed by the companies in the name of scientific management.

New ideas found little foothold in the mill towns, or the mills themselves. Above all, there was no world view, or even a valleywide view, of forces that would inevitably require change. It is extremely difficult to introduce wide-ranging changes in any social environment, but doubly difficult in one where the patterns of life, politics, education, and business have forced people into a narrow mold of behavior. That, unfortunately, is what happened in the Monongahela Valley.

Growing Up in McKeesport: The Forties

A certain unreality is involved in compressing real life into abstract concepts like "social trends" and "patterns of behavior." When I was growing up in the Mon Valley, my senses didn't supply me with trends and patterns but with sights, sounds, and the feeling of things that I shall never forget. I knew that calling anybody a "hunky" was wrong—my parents taught me that—but I didn't think of patterns of discrimination when my friends and I went to a Polish picnic at Renziehausen Park on a Sunday afternoon in summer. Beer would be flowing liberally from kegs on one side of an open-air pavilion, while on the other side a hundred people would be whirling and stomping to the "beer barrel Polka," or one of dozens of obscure polkas known by true polka lovers, but not me. All I knew was that the brassy polka band created a wonderful sense of wanting to whirl and whirl until the senses were numb. And then stagger, with sweat dripping from every pore, to the great tables of food, laden with kielbasa, golumpkes, potato salad, and cakes and cookies. Was I Polish? No. Did I look Polish? Whatever that look is, I probably didn't have much of it. But they welcomed us.

The 1940s were my impressionable years and, coincidentally, perhaps the best for the Mon Valley. For me and many of my friends, McKeesport was a good town to grow up in. It was an unusual decade of war boom and postwar prosperity, and a gritty vitality and energy coursed through the mill towns. The plants were working at near capacity for much of the time. My father's beer distributorship fronted on Fifth Avenue, with a warehouse and garage entrance in the rear on Lysle Bouleyard. A few hundred yards to the west, across the four-lane road and beyond the railroad tracks, stood National Tube's four blast furnaces, feeding on enormous mounds of iron ore. On rainy days the ore piles made me think of sticky red licorice. Skip cars, moving on tramways, carried the ore up the furnace face and dumped it in the top. They always seemed to pause with their rear ends in the air, as if peering down into the boiling mass, before backing down the tramway.

When the ore piles dwindled rapidly, as they did in those days, you knew that business was booming. As finished pipe came out of the seamless mills to the south, it was piled in triangular loads on gondola cars, and made-up trains often stretched the length of the plant before they pulled out. You could see them crawling down the east bank of the Monongahela, car after car of pipe, headed south and west, where men were drilling for oil and laying pipelines across vast territories.

The mill towns still had a turn-of-the-century physical appearance. Old brick row houses, built for an early wave of immigrants, crowded much of the flat land next to the steel plants. They were filthy and crumbling, with cement lintels sagging over windows and door frames cocked askew. Most of the mill workers had moved uptown, and the row houses had largely become squalid tenements housing the elderly poor, blacks and Mexicans (who were kept out of better neighborhoods), and widowed families—all constituting one class of "social problem" that grows in the wake of industrialization.

The main business street was lined with two- and three-story brick buildings, built in the early 1900s, with ground-floor shops and apartments above. Retail shops ran in an unbroken string for nearly two miles on Fifth Avenue in McKeesport. In the prime district on upper Fifth Avenue, a few large merchants had remodeled old buildings, but shabbier structures dominated for the next mile and a half going northward. They housed butchers, bakers, clothiers, confectioners, five-and-ten-cent-stores, furniture emporiums, groceries, movie houses, restaurants, and two or three saloons on every block. The line was broken here and there by a self-conscious bank edifice with stone columns supporting pretentious porticos, or the modernistic tile-and-glass front of the first of the fast foods, the White Tower. Red and green neon signs glared in the store-front windows or stuck out vertically from the facade.

Each town had a similar street: Braddock Avenue in Braddock, Eighth Avenue in Homestead, Grant Avenue and Duquesne Avenue in Duquesne. Clairton had two business districts, black and white. To work the coke ovens, U.S. Steel imported blacks from the south, and they settled in the "bottoms" near the plant, in the Blair district. State Street, which ran parallel to the long batteries of coke ovens that lined the riverbank, was the blacks' shopping area. Sulfurous mists from the quenching of hot coke settled on State Street, ruining the paint on cars and homes. The whites were not so vulnerable, living up on the hill and down the other side. Their retail shops lined St. Clair and Miller Streets. A long cultural and social distance separated the hilltop from the bottoms.

McKeesport was the retail hub in the days before shopping centers. People from up and down the valley poured into town on Saturday, and by midafternoon the pedestrian traffic was so thick that the current could carry you past the store you wanted to shop in. The B&O tracks added to the congestion, with people massing at the crossing gates, awaiting the passage of a lumbering freight or a shorter passenger train. Cars inched along Fifth Avenue, slipping into and out of the trolley tracks. Ungainly yellow streetcars moved with spasmodic jerks and impatiently dinging bells down the center of the street, their trolleys sputtering and sparking on the overhead power line. Meanwhile, men with lunch buckets would be trickling through the crowds, moving toward the Locust Street gate for the 3 P.M. shift change. When the siren sounded, hundreds of day workers poured through the gate and hurried up Locust to catch a bus or streetcar. Where the mill workers and shoppers intersected, at Fifth and Locust, it was like one army passing through another.

In 1948, 718 retail establishments in McKeesport grossed $73 million, a large sum in the pre–Korean war period of low inflation. Braddock's 319 businesses brought in $32 million that year. Homestead ranked third with $21 million. A large portion of these revenues came from one of the chief sources of entertainment, the movies. McKeesport had four movie houses, offering productions that ranged from Busby Berkeley musicals and other first-run films, which were shown in the elaborately ornate Memorial theater, to Saturday morning serials featuring the evil Fu Man Chu, which were viewed by foot-pounding kids in the shabby Capital. Braddock had two or three movie houses, and Stahl's Theater in Homestead was said to be the first $1 million movie theater in the nation.

All kinds of sports were popular, but high school football was the focus of attention. The Saturday afternoon or Friday night game was the big event of the week for the entire town. Huge crowds packed McKeesport's football field on a hill overlooking downtown, and the roar of the fans resounded for miles around. After the game, we walked down the hill, exhilarated by the cool night air, as cars filled with chanting kids went skidding down the brick streets. There were soda fountains and ice cream parlors with juke boxes in those days, and we would pile into the booths and talk about the game as we sipped cokes and milk shakes and kept an eye peeled for girls.

Exhilaration and fun aside, high school football played a crucial role in the mill towns. All the sameness in the valley created a mass need for a sense of community distinctiveness, and the hard, young bodies of high school athletes provided the means to that end. Each town had a booster club which raised money to help the team, and they demanded a winning team. The players were to win not for themselves but for the community. A defeat by a long-time rival was a blotch on the town's pride. And so the mill-town teams played rock-'em, sock-'em football, and rivalries in the valley were intense, bitter affairs. When McKeesport played Duquesne, it was like Brazil versus Argentina in soccer, and the police turned out in force to prevent fights or riots. Not so incidentally, the intra-valley football rivalries intensified the separateness between communities.

Apart from movies and sports, the pleasures and amenities of mill-town life were sparse. The greatest natural resource of the valley, its rivers and streams, had long ago been made unfit for recreation. Before the coming of large-scale industry, the rivers apparently were widely used for fishing, boating, swimming, and ice skating. Probably by the turn of the century, however, they were befouled by mine acids, steel wastes, chemical effluents, and sewage.

On a rash impulse one day in about 1947 a friend and I swam across the Monongahela near the mouth of the Youghiogheny. The current was swifter than it looked from shore, and it pelted me the whole way with a variety of slimy objects that, fortunately, I couldn't identify in the muddy water. There being nothing on the other side but a weed-filled bank and a railroad track, we swam back and carried from the river a stench of sewage that didn't endear us to our mothers.

The river was cleaner upstream of the biggest population and industrial concentrations. I was fortunate in having relatives who rented a summer cabin at Bunola, an old mining village between Clairton and Donora. The Monongahela was muddy at this point, but I didn't feel as though I was being assaulted by garbage. The 1940s were the last days of the stern-wheelers, and we would swim to midstream to breast their waves. As we swam close, we would wave and shout, and sometimes the captain would give a couple of short blasts on his air horn.

As for riverside parks, where one might sit under a tree and contemplate Old Man River, the very thought of such a thing in a Monongahela mill town is laughable. It was virtually impossible even to get near the river. One day in 1966 or so, my wife and I packed our young sons and a picnic basket into our car and set off on a trip up the west side of the Mon, thinking we surely would come across a tiny patch of shady riverbank *somewhere* south of Clairton. We drove nearly to the West Virginia line without finding so much as a square yard that we could claim for an idyllic lunch. Where there weren't tracks or mills, we encountered fences with Private Property—Keep Out signs, or impenetrable undergrowth leading up to the water, and—finally—a muddy road in which I mired the car up to the hubcaps and had to be towed out. My sons were amused by this episode but remained uneducated about the ways of rivers.

Today, although barge traffic has declined dramatically and the rivers are reputed to be much cleaner (pure enough to support fish life), they are rarely used for recreation. People drive across them every day but, out of generations of habit, ignore them.

Any one who writes about steel is tempted to quote William Blake's "dark, satanic mills" in describing the mill towns. Blake used his poetic license in another time for his own reasons, and I have no intention of borrowing it. But it reminds me faintly of a personal feeling about the Mon Valley that results more from weather and dirt than a deathly battle between heaven and hell. Nothing was more characteristic of the mill towns during their heyday in the 1940s than the damp, dark, chilly days when inversions, which were common in the valley, sealed up all the escape routes of the mills' spewings and embalmed us in dust and ferrous crystals. Winter, spring, and fall these days descended on the valley, slowing time, fraying tempers, exacerbating emphysema and its symptoms, and—who knows?—perhaps causing widespread depression. When the dark sky hung practically low enough to touch, still the columns of black smoke rose from the mill stacks and coal vapors seeped from tens of thousands of home chimneys, and we seemed to be smothering under the belly of the world.

I didn't think ill of anybody at the time for the pollution, not knowing that precipitators had already been developed that could arrest large particles of dust as they went up the stack, if not the invisible sulfur dioxide. But the normal day's pollution was nothing compared to those times when a blast furnace "slip" occurred. Sometimes the coke, ore, and limestone, the "stock," which was poured in at the top, would adhere to the furnace walls, creating a gap between the "burden" at the top and the molten iron at the bottom. Eventually the burden dropped in a mass and met the upward blast of air, which blew tons of stock out the top. Instead of a steady drizzle of soot, a dark blizzard would descend on everything in the immediate vicinity. You could actually see white changing to black, as in a litmus-paper lab test.

Usually, however, we forgot about the dirt and lived life as it came. In those days many mill workers still lived within walking distance of the part of the plant they worked in. Quite a number of blast furnace and open hearth employees lived on the steep streets leading up from Fifth Avenue near our business. My father was a friendly man who enjoyed having a shot and a beer with the men next door at Musulin's Cafe, and sometimes he'd take me along. Men would come in for lunch, standing at the bar and sitting at small tables. Talk of the day's work floated around—so many casts of iron, so many heats of steel, a breakdown here, a soona-m'beetch of a foreman there. At the age of ten or so, I knew nothing about steelmaking, but I learned to identify men by their jobs. The jobs carried social status. At the bottom of the heap were common laborers, often grizzled old fellows with gnarled hands who spoke broken English. Men of the cast-house crew were big and tough. Millwrights, riggers, and bricklayers were independent characters. Open hearth workers tended to swagger and show their strength, especially "second helpers" who performed most of the hard work on the furnace. The boss of each furnace crew was a first helper, an hourly man who worked up to that position over many years. He didn't need to show off.

One second helper, a Frenchman named Phil Lejeune—an oddity in this neighborhood of Slovaks and Croatians—often helped us load trucks on his days off from the mill. A big, hearty man with a moustache and bulging biceps, he could lift a full barrel of beer from the loading platform to the truck bed—a task that usually took two men. He'd wave everybody else away and hike the barrel to his bent thighs and thrust it up and into the truck. We'd shake our heads, and he'd go off with a smile.

One man I came to know only by sight at Musulin's was a quiet, older man who always sat by himself at a small table next to the wall. He would put his hat on the empty chair and sip one shot of whiskey—no more—while reading a newspaper. I remember someone telling my father, in a low voice that impressed me with a sense of importance, that the lone man was a first helper. He seldom talked to anybody, and the other man kept a respectful distance. Old John Musulin, the white-moustached proprietor, addressed him as "Mr." In terms of social status, the first helper was rivaled only by rollers on the blooming and bar mills.

Musulin's, I believe, must have been one of the last boarding hotels for immigrant mill workers. Five or six single men lived in tiny rooms upstairs. At lunch and dinner time, they would eat at a big round table just off the kitchen, talking in their own language, sopping up stews and soups with hard bread. Mrs. Musulin and one or two daughters prepared the stew or ham and cabbage in large iron pots, and an overpowering, warm pungency filled the place.

Musulin's has been closed for years, but I would guess that a trace of the odor lingers there. Most of the old buildings on that section of Fifth avenue are now abandoned. People no longer come down the hill to go to work.

My impressionable years did not really last very long, for me or for McKeesport. The city's population peaked during World War II at about 55,400. Global conflict seems to have a broadening effect on such people as it doesn't kill. When the war was over, McKeesport and the Mon Valley looked a lot smaller, dirtier, and less offering of opportunity to the men and women who came back, as well as those growing up. People began leaving, especially young people. An estimated fifty-eight hundred people between the ages of twenty and forty-four left the city for good in the 1950s. I was one of them.[22]

The United States has space to spare. People often have created whole towns to indulge in the entrepreneurialism of the moment, searching for a rumored vein of silver, or gambling that a railroad will lay tracks through town and bring prosperity. Many such dreams have died and their towns with them. The ghost town is as natural to America as the restless individualism of its people. And so a mythology has grown up around ghost towns and played-out places, that they were dedicated to foolish adventures and therefore doomed from the start.

In a sense the Mon Valley mill communities have become ghost towns, not abandoned to be sure, but condemned to wither and fade for a generation or two before a new economic activity takes root. These towns, however, did not arise as the result of a dice toss, or the defective judgment of a citizenry bent on finding a shortcut to riches. They performed a very basic, critical service in the industrialization that swept the country—and that *needed* a Monongahela Valley. The mill towns did not look pretty in the process. They contributed their share of corrupting influences and failed to make the most of their opportunities. Still, one cannot help but wonder about a social and economic system that sentences the thousands of people who are left—who cannot move—to lives of despair.

Chapter 8

The "Downside" Cycle: The UAW

Hard times always seem to produce pithy expressions that describe, if not explain, the nation's predicament. In the early 1980s it became voguish to contrast the long prosperous years of economic growth after World War II with the sudden, deep plunge starting in the late 1970s in gravitational terms. What had once been "on the way up" was now "on the way down"— and would continue falling for some time to come. The issue for the labor movement—and the American system of industrial relations—was how to convert labor's "upside" gains to "downside" compromises with economic reality. For institutions that had no experience in reversing directions, however, this posed immense difficulties.

By April 1982, the auto and steel industries were clearly riding the same downside cycle and holding on for dear life. The United Auto Workers had swallowed its pride and granted cost relief to Ford and General Motors in February and March. The Steelworkers, having engaged in decades of upside wage competition with the UAW, could hardly avoid a similar downside retreat. Nonetheless, after steel management officials presented the industry's case for wage deceleration to the USW executive board in early April, Lloyd McBride continued to resist a contract reopening. He personally supported the need for concessions but worried that local union officers would put up a stiff fight to prevent early negotiations.

In the auto industry, many GM locals had reacted with hostility when UAW President Doug Fraser negotiated concessions. Union critics suggested that he had violated a long-standing principle, supposedly enunciated by Walter Reuther. "Reuther must be spinning in his grave," commented Bob

White, the outspoken UAW director in Canada, when Fraser announced the GM agreement in March 1982.

Fraser was stung by this criticism. He was a "Reutherite," as the late president's closest followers were known, and believed that he carried on the Reuther tradition of bargaining militantly with a social vision. After he retired in 1983, Fraser continued to urge a change in union behavior, in speeches to UAW groups. There were always critics present who challenged him, invoking Reuther's name as the final authority.

Fraser told me about one such meeting when I visited him in 1985 at his office in the Walter Reuther Library at Wayne State University. "I went into one lion's den last week, a union meeting," Fraser said, "and there was this old Commie there who I knew would raise that precise issue, 'Reuther spinning in his grave.'" Fraser chuckled and rummaged in a desk drawer. He pulled out a mimeographed text of a speech.

> So I brought this along and read it: "All industries and all companies within an industry do not enjoy the same economic advantages and profit ratios. We cannot blind ourselves to this fact at the bargaining table. As an employer prospers, we expect a fair share, and if he faces hard times, we expect to cooperate.... Our basic philosophy toward the employers we meet at the bargaining table is that we have a great deal more in common than we have in conflict, and that instead of waging a struggle to divide up scarcity, we have to find ways of cooperating to create abundance and then intelligently find a way of sharing that abundance."

Fraser showed me the first page of the text. It was an address delivered by Walter P. Reuther in 1964 at the University of Virginia.[1]

Only rarely, however, was the UAW called upon to share an employer's bad times during Reuther's presidency, which extended from 1947 until he died in a plane crash in 1970. This was the most prolonged upside period in U.S. history. For some thirty-five years after World War II, collective bargaining was geared to the great drive-wheel of American life: rapid economic growth. By 1982 the wheel was decelerating, and economic forecasts indicated that it would continue to turn at a slower speed for years to come. The U.S. economy expanded at the rate of 2.8 percent per year during the 1970s, but the rate decreased to 2.4 percent between 1980 and 1986. Although less than a percentage point, this difference represented a 14 percent decline in the annual growth of goods and services (gross national product) produced by the economy. Because employment rose annually by 1.6 percent during the 1980s, the GNP grew only 0.8 percent for each employed person.

A crimping of the national lifestyle was unavoidable. Some individuals and sectors, including the low-income poor and middle-income workers in basic industries which were shrinking because of foreign competition and technological changes, would suffer more than others. No longer could unions in these industries extract 10 percent annual wage gains as they had in the late 1970s. [2]

In January 1982, the floodgates of wage concessions suddenly burst open. The Independent Brotherhood of Teamsters reached agreement January 15 on a new contract covering some three hundred thousand drivers and warehouse workers. Teamsters were used to getting wage increases of close to 10 percent per year, but the new three-year pact actually rolled back wage increases scheduled for 1982 by 3.7 percent. Hundreds of small, independent operators (many of them self-employed) were taking advantage of deregulation to start up a trucking business with wages that undercut the union pay rate by $5 an hour and more.

While trucking-industry negotiations normally did not set a wage pattern for manufacturing, the Teamsters' action could not be ignored. Industrial workers were acutely aware that the Teamsters union, despite its reputation of invincibility since the days of Jimmy Hoffa, had bowed to economic pressures without even much talk of striking.

Much closer to steel, the UAW on February 13 granted Ford moderate wage relief. The more militant UAW locals at General Motors balked at the Ford pattern. But the closure of several plants forced them to relent, and UAW negotiators agreed to a Ford-type deal at GM on March 21. It was important to note that the UAW—the ideological leader of the labor movement—had flashed a message to all unions: Labor *could* take a backward step without losing its purpose in life. (The UAW, of course, previously had reduced wages at Chrysler, but that action was taken to pull a failing company back from bankruptcy or liquidation. Ford and GM had serious competitive difficulties, but neither was close to failure.) Radical unionists would disagree with the message. But it had been given nonetheless, and it carried considerable authority for historical reasons.[3]

Wherever the auto industry went in terms of slowing wage acceleration, steel eventually would have to follow. This was true because of determinative links that connected the two industries and their unions. The auto industry constituted the largest steel market—about 15 percent of total mill output went directly into the manufacture of cars—and General Motors was steel's single largest customer. GM, in particular, used its market power to keep supplier prices in line. Since labor costs help determine prices, the steel companies looked to the auto industry for a very rough guide on increases (or decreases) in labor costs.

The UAW-USW Rivalry

The UAW and USW had been linked politically since birth as the premier unions in the CIO. At the end of World War II, they and other CIO unions attempted to coordinate an inter-industry bargaining assault in the manufacturing sector. This failed partly because of ideological differences between the unions. However, the USW, UAW, UE (United Electrical Workers), and URW (United Rubber Workers) all settled for an 18.5¢ hourly wage boost in 1946 (they had demanded much more) in their respective industries. Average hourly earnings in steel and autos in 1947 were only 2¢ apart, $1.45 and $1.47 respectively.[4]

In 1948 the UAW and GM made what most expert observers refer to as *the* key labor bargain of the postwar years. Up to then, there was little standardization in wage bargaining. The UAW and most unions threw every possible factor into their justifications for higher wages: They demanded a "catchup" wage hike for past inflation, a prospective boost for future inflation, an increase to match productivity gains, and sometimes a further hike to relieve management of "exorbitant" profits. In GM's view, this led to a chaotic bargaining process which absorbed too much time, tended to cause costly work stoppages, and produced unpredictable results in terms of labor costs. To avoid these problems, GM president C. E. "Engine Charlie" Wilson sought to eliminate wage bargaining and replace it with an automatic wage-setting procedure.

GM proposed a two-pronged wage concept. First, it would pay a fixed annual wage increase roughly equivalent to the long-term national rise in productivity, and, second, it would protect workers' purchasing power (though not up to 100 percent) through quarterly cost-of-living adjustments (COLA). At the outset, the productivity raise, or "annual improvement factor" (AIF), amounted to 2 percent of base wages, but the UAW later bargained this up to 2.5 percent and then to 3 percent. For the next three decades, the AIF-COLA concept served as the basis for auto's wage settlements. GM achieved cost-predictability, but—like the steel companies with their ENA wage guarantees—the auto maker practically gave up negotiating the size of wage boosts.

Ford and Chrysler accepted the GM bargain, though with no large enthusiasm. Outside the auto industry, few companies wrote the AIF-COLA formulation into their labor contracts. But many others might just as well have done so, because the size of the AIF increase, in dollars and cents, strongly influenced wage negotiations in many industries. Many unions also adopted the COLA concept. It is impossible to quantify the degree to which the wage bargains in one industry are patterned after those in another, economists say. However, the UAW has occupied such a pivotal position in the labor movement, and the auto industry is linked commercially with so many other industries, it is likely that the UAW's wage deals over the decades have had a far-reaching impact. Indeed, Michael J. Piore and Charles F. Sabel, authors of the provocative book, *The Second Industrial Divide*, contend that the GM-UAW wage-setting formula became "the keystone of the whole system of macroeconomic stabilization" in the United States.[5]

Even if Piore and Sabel overstate their case, the UAW's wage-bargaining was immensely influential. After 1953, steel bargaining usually occurred about a year after auto negotiations. With the exception of only a few bargaining rounds, the UAW's AIF gain of the previous year became the USW's starting point in wage discussions. A rivalry developed between the unions. To some extent, economic circumstances in the two industries determined bargaining outcomes. As a result, auto workers' pay spurted ahead in the early 1950s and again in the 1960s but fell behind as steelworkers' wages sprinted ahead in the late 1950s and throughout the 1970s. In every negotiating year, a desire on the part of both unions to keep up with, or go ahead of, the other played some part in the final settlement. But productivity rose significantly less in steel than in autos from 1950 to 1980.

The rivalry was more intensely felt in the Steelworkers because of personal and institutional animosities. Dave McDonald, the USW president from 1952 to 1965, intensely disliked Reuther—who received better press notices than McDonald—and tried to show him up at the bargaining table. Moreover, the UAW always had the better public image as a "democratic" and "socially progressive" union, while the USW was pictured as stodgy, conservative, and undemocratic. The UAW's image largely reflected the articulate leadership styles of the UAW's extraordinary series of presidents: Reuther, Leonard Woodcock, and Doug Fraser. Whereas Reuther, for example, was always in the front ranks of civil rights marchers during the 1960s, McDonald refused to participate in demonstrations and parades—including Martin Luther King's 1967 march on Washington—although the USW had a large black membership and supported civil rights legislation.

As to which union was more "democratic," it depends on the test used. Auto workers had the right to vote on their contracts, while basic steel pacts were ratified in committee—a method that USW leaders, rightly or wrongly, believed most appropriate to industrywide bargaining. On the other hand, the steel rank and file had the right to elect top officers and district directors every four years, while UAW leaders were elected by the more easily controlled convention-vote procedure. Many union officers would prefer that a contract rather than their personalities and leadership qualities be voted on.

In the 1970s, said Ben Fischer, who was a high-level staff man in the steel union before retiring in 1979, USW leaders felt especially "harassed, criticized, and demeaned" by liberal critics, who held up the UAW as a model of democracy. As a result, Steelworker officers went into bargaining with "the attitude that we had to show up these guys [UAW leaders] in negotiations." This may seem an especially wrongheaded basis on which to establish wages, but there was in addition a very human justification for wage comparability.[6]

Until very recently, the United States had a producer mentality: The early entrepreneurs who learned how to extract minerals from the ground and convert them to consumer goods were considered the princes of industry and commerce. The nation thought of itself as primarily a producer and only secondarily a consumer. It was natural for the men and women who dug the coal and iron ore that go into steelmaking to believe their labor was worth as much (or more in the case of the underground coal miner) as that of the men who dumped it into blast furnaces. In turn, steelworkers who spent eight hours a day in the vicinity of a hot iron or steel furnace demanded a higher wage than auto workers who applied screws and bolts to pieces of steel and plastic on an assembly line. In keeping with the producer mentality, all three groups—miners, steelworkers, and auto workers—felt that they deserved higher pay than retail sales people and government clerks.

With such a strong feeling in the ranks, not even the most saintly union leader could *always* avoid the temptation to become a hero by winning more than another union. Comparing wages, however, is by no means the exclusive preoccupation of blue-collar workers. The same desire to outstrip the other fellow helps explain why the salaries of corporate chieftains in the United States are far higher than those in Japan and Europe.

After nearly four decades of this one-on-one competition, where did the USW and UAW stand in relation to one another? By 1980, hourly earnings no longer told the tale; the cost of benefits also had to be included in the overall "wage" paid to steelworkers and auto workers. For all steel companies in 1980, average hourly employment costs averaged $18.45, compared to $15.82 for the auto industry.[7]

Auto Bargaining in 1982

By 1982 the U.S. auto industry largely had been abandoned by American consumers in favor of better-built Japanese cars. GM and Ford needed relief from accelerating labor costs so they could invest in new products and climb out of a two-year slump. Ford had lost $2.5 billion over two years but needed to invest $4 billion per year in a new product program for 1984 so that it would remain competitive.

In late January, the UAW agreed to defer three quarterly COLA payments that would fall due in 1982 for eighteen months each. This would mean a big cash saving for Ford in the first year, when the company most needed the money.

The UAW also gave up two future AIF increases of 3 percent each, as well as the paid personal holidays first negotiated by Doug Fraser in 1976 (see chapter 3). In the midst of the 1981–82 recession, which had already idled more than three hundred thousand auto workers, Ford employees preferred to give up their nine PPH days rather than pay. The Ford agreement, therefore, was a "cost avoidance" pact, one that slowed down the growth of wages but did not reduce existing labor costs. Actually, hourly employment costs would rise by 17 percent under this contract by the time it expired in September 1984 for two reasons: COLA payments, though deferred, would still be paid, and the cost of merely maintaining existing medical and hospitalization benefits would rise by about 15 percent because of the national inflation in health care costs.

Almost immediately, management critics in other industries questioned whether the modifications agreed to at Ford went far enough. They had hoped, somewhat unrealistically, that the UAW would renounce the AIF-COLA formula. Instead, it was merely set aside. Steel executives said nothing publicly, but Bruce Johnston later told Lloyd McBride what he thought of the auto pattern. In terms of cutting costs, he said, the auto companies had bellowed like an elephant—and produced "a mouse-fart agreement."

Yet the Ford agreement, in particular, suited the company's needs. Peter J. Pestillo, Ford's vice-president of labor relations, wanted to negotiate a new contract long before the strike deadline in September to avoid any chance of a work stoppage. "We had the most extensive product changeover in history under way, and it was critical to have production," he said. The amount of the concessions, he said, indicated "how much we were willing to take and still retain the good will of the employees as opposed to how much we might have extorted and lost their good will."[8]

The difference of opinion over the 1982 Ford contract reflected a wider debate taking place at the time. Was the industrial relations system undergoing fundamental alterations in response to the changing economic environment? Practitioners and observers generally agreed that it should be reformed but disagreed sharply over whether permanent changes were actually taking place.

The auto settlement convinced many academic and government observers that nothing "new" was happening. The

UAW merely had pulled back a little from its aggressive bargaining of earlier years. It would always be thus: A union that
won excessive wage increases in good times would be forced
into a position of restraint during the next downside cycle.
Although the wage-concession trend was growing, this in
itself did not mean that basic changes in compensation methods, for example, could be expected. John T. Dunlop, the
nation's foremost wage theorist, a former secretary of labor, and an economics professor at Harvard, was the leading exponent of this view. Those who speculated that
wage concessions would sweep through the economy were
"creating unrealistic expectations and bargaining problems,
which somebody will have to settle," Dunlop told me in 1982.
"It is not a major new era, and that I'm willing to sign my
name to."

On the other hand, the insistence that nothing is new under
the sun could discourage management and labor from searching for ways to make their firms more competitive. Audrey
Freedman, a labor economist at the Conference Board, repeatedly challenged Dunlop and other academic experts in 1982,
contending that the elaborate systems of pattern bargaining
that existed within and between industries were breaking
down. She noted where old wage links had been severed and
predicted that competitive pressures would force labor and
management to negotiate on the basis of individual company
performance. "When the recession abates," Freedman wrote,
"there will be no going back to the 'model' of the 1970s. We
are returning to the individual conditions of the enterprise, for
good."[9]

"For good" is a long time, but in 1982 I thought Freedman
was correct in believing that a fundamental change was in
process. It would be a few years before some of the more
important patterns of old could be declared dead. For example,
in 1982 the USW still aimed at reproducing the auto wage
pattern, and it would not be until 1986 that the steel companies were able to bargain on their own circumstances. The
patterning of wage terms between the steel, aluminum, copper, and container industries also had mainly disappeared
by then. In addition, more flexible work practices and the
involvement of workers in decision making on shop-floor production issues and actual business decisions at the plant and
company level (product planning and design, work scheduling, the introduction of technology, job training, equipment

purchases) would eventually begin to reform the system from the ground up.

That Ford and the UAW had begun this ground-up type of reform in 1981 seemed to me the deeper significance of their 1982 agreement.

The 1982 Ford Contract

Two farsighted men entered the Ford scene in 1980. One was Peter Pestillo, a somewhat brash and outspoken labor executive who had worked at General Electric and B. F. Goodrich. A wiry man in his early forties, Pestillo was a fast talker with an irreverent wit. His thin lips were usually cocked in a half-grin, and the eyes that peered out of black-rimmed glasses always hinted that mischievous plots or shrewd calculations were going on.

The second protagonist was Donald F. Ephlin, a UAW vice-president and chief negotiator at Ford. He was tall and heavy-set, a formidable man with a direct and forceful way of talking in measured sentences that smacked both of traditional union oratory and Walter Reuther's social uplift rhetoric. A former local union president at GM's big plant in Framingham, Massachusetts, Ephlin had served as UAW regional director in New England and, before that, as a technician on the UAW's central staff in Detroit. He was at one and the same time a practical negotiator and a visionary, very much in the Reuther and Fraser mold. Ephlin's outspoken interest in innovation did not sit well with the more conservative UAW leaders, which is one reason why he was passed over for the presidency when Doug Fraser retired.

Pestillo and Ephlin formed an alliance to change the union-management relationship at Ford, which had lagged far behind GM during the 1970s in democratizing its plants. Both were strong advocates of worker participation. They traveled together to Japan to study the participation and "lifetime employment" systems in effect at large Japanese concerns.

Ford and the UAW called their program of worker participation Employee Involvement (EI). Whatever resources were needed to establish EI at individual plants—money, technical help, training—Pestillo and Ephlin provided. They tried to ensure that it was not perceived as simply another "program" forced upon the plants by corporate edict; it was a new process actively promoted by—and to a large extent personally implemented by—the chief labor spokesman of each side. This visible commitment by top-level people was critical. In early 1982, some form of Employee Involvement was under way at fifty-eight of ninety-four plants and offices.

With the broad-based participation in the plants serving as an underpinning, the two negotiators thought they could frame a new collective bargaining policy for the company. Most importantly, they had the support of Doug Fraser and Ford Chairman Philip Caldwell.

As early as 1980 Pestillo and his labor relations staff began fashioning a negotiations strategy. They knew that the UAW was concerned about a shrinkage of Ford jobs, caused by rising imports and by Ford's decision to farm out the production of many auto components ("outsourcing" was the term used in Detroit) to lower-cost domestic and foreign suppliers. More than fifty thousand Ford workers were on indefinite layoff. Like most American manufacturers, Ford had always responded to technological and market changes by laying people off, rather than retraining them for other jobs.

In late 1981 Pestillo proposed to Ford's policymaking committee that the company replace that decades-old practice. "We made the corporate policy decision that a trained work force is an asset to the company and that labor would no longer be treated as our most variable cost," Pestillo said. Ford would try to reduce fixed labor costs, such as annual wage increases, and costs which vary with economic conditions, such as COLA. In return, the company would take on new obligations to protect workers from market and technological changes by strengthening income and job security programs.

The confluence of UAW and Ford strategies produced significant new provisions in the February 1982 contract. Under a Guaranteed Income Security (GIS) program, laid-off workers with fifteen years' seniority would receive fifty percent of their last pay until retirement if they were no longer eligible for unemployment benefits. The company and the union established a jointly operated center to train the unemployed for jobs that opened up either in Ford or in other industries. It was the first formal training program ever set up in the auto industry. The company also agreed to a two-year moratorium on plant closings that would have resulted from the farming out of production work. These provisions were aimed at boosting morale and convincing workers that they could increase productivity without endangering their jobs.

To bind the workers closer to the company, Pestillo and Ephlin negotiated a profit-sharing plan. In addition, Ephlin would have the right to address Ford's board of directors twice a year. The two sides set up another information-sharing mechanism called Mutual Growth Forums in which plant managers and local union officials would meet quarterly to discuss the business and employment outlook. Despite the UAW's retreat on holidays and wages, Ford workers ratified the contract by a plurality of 70 percent.

None of the new provisions in the Ford pact were revolutionary. All of the collaborative elements—employee involvement, information-sharing, job security, worker training—had been tried in the United States at some time, in some place. The combination of these provisions, however, provided the beginnings of a new industrial relations system. It still had to prove itself, of course, and the company's labor costs were still very high compared with some foreign car makers. But Ford and the UAW had made a concerted attempt to create new mechanisms to solve problems and to install a democratic system of plant management that would enable the company to adjust quickly to economic and technological change.

I made a reportorial tour of four Ford plants in the summer of 1984 and found a highly participative atmosphere. In each case the amount of collaboration between the local union and managers came close to joint management of the plant. The most important result had been a substantial improvement in product quality. An independent research firm reported that a survey of sixty-five hundred buyers of 1984 Ford cars showed that "things gone wrong" had declined 55 percent, compared with a 1980 survey. By 1984 thousands of employee involvement teams were functioning in eighty-six of ninety-one plants and depots. I reported at the time that Ford and the UAW "have established what may be the most extensive and successful worker participation process in a major, unionized company."

Ford not only bounced back from its huge losses of 1980–81. In 1986 it earned higher profits than giant General Motors for the first time since 1924. Earnings totaled $3.3 billion, enabling Ford to make profit-sharing payments to its UAW workers averaging more than $2,100. GM, meanwhile, announced its 1986 profits were too slim to require profit-sharing with union employees under the UAW contract.

There is no way to quantify the degree to which Ford's labor initiatives helped produce this turnaround. Technological and financial factors were obviously important. But the company's new model cars had caught the attention of critical American car-buyers, and this could not happen without a major improvement in quality. Having seen the spirit of cooperation in Ford plants (and comparing it with what I saw twenty years before), I believe that Ford's new industrial relations system contributed enormously to the company's success. Other observers agreed.[10]

Failure at GM

Within a month after the Ford settlement in 1982, General Motors took action to *force* the issue of concessions with its own employees. The corporation announced it would close seven plants and threatened to shut more. In effect, GM found it necessary to frighten its employees to obtain their agreement to enter into the same kind of contract that Ford had gained with a 70 percent vote of employees. This should have alerted the steel industry, especially U.S. Steel, to the problems that it would confront because of employee attitudes.

Both U.S. Steel and GM had failed to establish an absolutely crucial principle of industrial relations: By words and actions, management must win and retain the trust of employees. This was demonstrated repeatedly during the efforts to reform collective bargaining in the 1980s. In no case was it "easy" to convince workers that they should give up hard-won gains to improve the company's competitive ability. But firms that had worked hard to apply consistent employee policies, follow through on promises, avoid unnecessary threats, and share information with employees—these companies were able to make fundamental changes with relatively little turmoil.

General Motors was not among the latter. For decades its relationships with union employees ranged from bad to worse. It managed the work force from the top down, in a gigantic, many-layered bureaucracy that tried to regulate human behavior as if the employees were robots. U.S. Steel also suffered from bureaucratization, especially from the 1960s on. I have known only two major corporations that actually engendered feelings of hatred among their employees, GM and U.S. Steel.

After the UAW won recognition at GM in 1937 by engaging in sitdown strikes, corporate management decided it would live with the union. But it would fight in every way, at every level of the company, to corral the UAW and prevent it from breaking out of a legally prescribed role of merely reacting to management decisions on work-related issues. Carrying out this "tough but fair" policy became the duty of GM's labor relations professionals, who were honest, smart, highly knowledgeable, and very, *very* combative. They operated under the principle that a dog will keep its place if you hit it frequently on the snout. A GM labor relations man who started his career under this tradition but later adopted a cooperative style told me that the older approach was aimed at keeping the union off balance. "There were often two ways of achieving the same result in dealing with the union," he said. "You could talk to them in a quiet, reasonable way, or you could poke them in the eye with a stick. If we had a chance, we'd poke them in the eye." [11]

Obviously, resentment flourished in the work force, producing high absenteeism, poor quality, and frequent work slowdowns. GM finally decided it must change its management style or begin to suffer competitively. In 1973 the corporation and the UAW began a Quality of Worklife (QWL) program, which sought to replace adversary relations with cooperation through various forms of worker participation. It was a pioneering effort and it spread only slowly, but by the early 1980s it had made important headway at many plants, especially in Flint, Michigan, Tarrytown, New York, and Warren, Ohio. Many plants, however, had barely been touched by the new philosophy. A large pool of rank-and-file anger lingered from the past.

In 1981 even mighty GM was losing money because of the invasion of Japanese small cars. Industry officials urged the UAW to reopen its contracts a year early and ease the rise in labor costs. In the summer of 1981, Doug Fraser later told me, UAW leaders seriously considered doing this. Just then, however, an employee survey at General Motor revealed a high level of distrust of management. The union officials concluded they could not win a ratification vote at GM. [12]

Fraser met frequently with GM Chairman Roger B. Smith to discuss the growing emergency, however. During one session, Fraser told Smith he could think of only one way to convince GM's workers that the company needed help. If GM would assure the employees that labor-cost savings would benefit consumers rather than management and shareholders, ratification of a concession pact might be possible. The idea interested Smith, and the two sides secretly negotiated a proposal for reopening the labor pact. It was a novel plan. GM would "pass through" to consumers in price cuts any cost savings generated by UAW give-backs. An auditing firm selected jointly by the company and the union would have access to GM financial data to verify that the savings were going directly to consumers.

This procedure, in effect, would give the UAW a voice in pricing decisions—and produce an ironic twist in labor history leading back to 1946, when Walter Reuther espoused a similar bargaining goal. When the wartime wage freeze came to an end, Reuther—who was a brilliant conceptualist—declared that General Motors could afford a 30 percent wage increase and still hold the line on car prices. If this was not so, the corporation should open its books to prove its inability to pay. The UAW, Reuther said, should not "operate as a narrow economic pressure group which says 'we are going to get ours and the public be damned.'" Therefore, it was willing to reduce the pay demand to keep car prices level. GM, however, declared that such a concept would lead to "a socialistic nation" and refused to open the books.

Reuther's proposal also encountered hostility among other union leaders, including Phil Murray, the USW chief and also president of the CIO since Lewis's resignation from that post in 1940. Many unionists rejected Reuther's contention that unions should consider the general welfare in bargaining for their members. Reuther led the UAW in a 113-day strike at GM. It ended in April 1946 only after Murray, without Reuther's acquiescence or participation, negotiated an agreement with GM's C. E. Wilson. As John Bernard writes in his biography of Reuther, the bold experiment failed "because neither government nor business—nor even most of labor's leadership and rank and file—had the vision or the will to transcend traditional roles. The strike demonstrated the limits of the union's power to create a new distribution of social and economic responsibilities."[13]

In 1982, however, businessmen no longer raised the specter of socialism, as they constantly did in the 1940s and 1950s. The real threat to American industry, it turned out, came less from socialist politics than capitalist competition. Roger

Smith desperately wanted to stimulate car sales. The goal of the "pass-through" agreement was to reduce average GM car prices by $1,000. The concept appealed to Fraser because laid-off auto workers would return to work.

In addition, as a committed Reutherite, Fraser still searched for a breakthrough leading to a rational wage-setting process that would bind labor to the public interest. Once again, however, the idea of transcending labor's traditional role aroused skepticism among many unionists. When Fraser asked local officials who made up the UAW's GM Council to approve early negotiations on the basis of the pass-through proposal, they did so by only a 57 percent majority. Negotiators subsequently failed to reach a compromise on the amount of labor-cost reduction necessary to cut car prices by $1,000. In late January 1982, the effort collapsed.[14]

The UAW then negotiated the pattern agreement at Ford. GM, meanwhile, closed seven plants in March and threatened more shutdowns. The fear of job loss among the rank and file enabled the UAW to resume negotiations and reach agreement with GM on March 21 (GM agreed to reopen four of the closed plants). But the pact was ratified by only 52 to 47 percent.

On the very day that GM and the UAW signed their agreement, the company made what a Detroit newspaper headlined as the "Dumbest Move of the Year." In a proxy statement issued that day, GM disclosed that top executives would be awarded a new, richer bonus plan and other improved benefits. Hourly workers, who had just accepted a wage freeze, raised an uproar. Talk of striking made the rounds. So furious was the turmoil that GM announced a week later that it wouldn't implement the new benefit plans. In terms of creating employee distrust, the GM action ranked with U.S. Steel's purchase of Marathon Oil.[15]

Three years later, when I interviewed Fraser in Detroit, he was still dismayed about the failure of the pass-through agreement at GM. Retired from the UAW, he taught labor courses at Wayne State University and Harvard and served on many boards and committees. Fraser had risen from the Chrysler shops and never lost a down-to-earth openness and humor that made him a favorite of rank and filers and reporters. In 1985 he had thinning white hair and a notably craggy face, but he still pulsed with the energy and ebullience of the man I had known twenty years before.

Fraser and I talked about the political problems that unions must solve to be competitive with nonunion and cheap foreign labor. In addition to the rank-and-file distrust of General Motors, Fraser said, union discipline broke down at GM. In the UAW, once an agreement is approved by the executive board and the GM Council, local officers are obliged to recommend approval in their locals, But many delegates took the opposite course. "I think what happened, and what continues to happen these days," Fraser said, "is that the local guys go back home, and they run into a little opposition. They think, 'Jesus Christ, I'm going to put my job on the line here, I'm unwilling to do that.' They don't work for the agreement. It's sad. One of the big jobs for the union is to see if we can turn that attitude around among the secondary leadership."

Another problem, he added, is that workers measure the depth of a company's competitive problems by what is happening in their own plant. If their plant produces a successful product, they tend to disbelieve management declarations that the company is ill. "Back in '82," Fraser said, "the Tarrytown plant was doing well and had just added a second shift—all young people who didn't have a great stake in carrying the contract. I remember, the local union president called me and apologized because they voted the contract down, by 65 or 70 percent." GM's large plants in Flint, however, had laid off thousands of employees, and local UAW leaders strongly supported the agreement. "They carried Flint by 11,000 votes. We would have been whipped if it hadn't been for Flint, of all places, the historic hotbed of the rebels. But our guys worked at it and stuck with it, and when they do that, you got it."

Moving Toward Negotiations in Steel

The UAW's contrary experiences at Ford and General Motors in 1982 proved the value of communicating with the rank and file long in advance of dramatic changes in the employment relationship. In the steel industry, a one-shot effort, made practically on the eve of negotiations in July 1982, was "too little, too late," as John Kirkwood put it. The union and the industry waited too long to begin the bargaining process.

Although the UAW concessions in Detroit gave the steel negotiators a strong trump card to play in Pittsburgh, they did not play it immediately. An element of self-delusion seems to have temporarily diverted the management men. They believed their case to be so strong that McBride himself would demand a contract reopening to "save the industry." They waited in vain for this to happen. McBride was not the kind of leader who would seize the moment and make a Fraser-type foray into the managerial domain of price-setting. A reactive leader, McBride wanted to create the impression that he was being forced into negotiations.

In the absence of strategy-making by McBride, hard-line attitudes against concessions were forming among USW board members. At a February meeting, a number of directors noted that the UAW had deferred COLA payments in the Ford agreement. They were "very adamant," Joe Odorcich said, that COLA should not be tampered with in steel.

But consumer prices were rising at a 6 percent annual rate in early 1982, and quarterly COLA increases were piling up. Since the previous November, COLA had raised wages 32¢ per hour, or about 3 percent of hourly earnings. Steelmakers had not been able to raise prices by 3 percent and in fact were offering price discounts to compete with foreign steel. The problem would be compounded on August 1, when the four quarterly COLA increases over the past year would be "rolled into" the base wage rate—that is, become a part of the base wage for purposes of computing vacation, holiday, and over-time pay. "It's crazy to keep paying COLA when we're losing money," John Kirkwood complained to me. Realizing that the USW would not give up the COLA provision, he tinkered with the idea of paying cost-of-living as a bonus linked to company performance.[16]

Meanwhile, the USW negotiated a concession agreement with Wheeling-Pittsburgh Steel, which employed over ten thousand steelworkers in West Virginia and western Pennsylvania. The company had undertaken an aggressive moderni-zation program but was having trouble making interest payments on $358 million in long-term debt. Once again, union officials were forced to decide whether to provide relief to a troubled company or allow it to slide into bankruptcy. As in the other cases, McBride decided to rescue the firm and save as many jobs as possible.

In early April, Jim Smith and Paul D. Rusen, the USW's director of district 23, reached agreement with the chief company negotiator, Joseph L. Scalise. To reduce scheduled increases in labor costs by $35 million, the union gave up two weeks of vacation and thirteen paid holidays over a nineteen-month period, as well as a 23¢ per hour wage increase scheduled to take effect August 1. But Smith and Rusen refused to modify the COLA provision. While this position was taken purely on the basis of W-P's situation, it signaled that the union was not of a mind to limit COLA in the rest of the industry. Since Wheeling-Pittsburgh's labor costs had been about $1 per hour higher than other steel companies, these reductions—which totaled about $1—did not give the firm a competitive advantage. However, they extended the life of what had become a marginal company in an industry with a large amount of excess capacity.

In this agreement, the USW introduced a new concept, proposed by Smith. To represent the wage cuts as "investments" on the part of the workers, the company agreed to give each worker preferred stock equal in value to the wage and benefit give-backs. This meant that each employee would receive about $4,000 in preferred stock with voting power. The ownership idea was backed up by a new worker-participation program which would give the fledgling worker-owners more of a say in production decisions.[17]

While the steel industry chiefs hesitated to play their Detroit card, Roger Smith played it for them. He and McBride were members of the Labor-Management Group, a voluntary organization of corporate and union executives coordinated by John Dunlop of Harvard. The committee met periodically to discuss broad political and economic issues. On at least two and possibly three occasions when Dunlop's committee met in the spring of 1982, Smith asked McBride to step aside for a private conversation. According to McBride's recollection, Smith addressed the issue of wage concessions from an oblique angle, never bluntly urging McBride to start bargaining with the steel industry, but making the point in a way that could not be misunderstood.

In June 1982, McBride related one such conversation to a union meeting that was recorded on tape. Smith reminded the USW leader that up until then GM had bought only American steel but that the corporation might change this policy for reasons of price and quality. GM had found through quality testing that it could save $40 to $100 per car by using Japanese steel instead of domestic steel. " 'I don't want to do it,' " McBride quoted Smith as saying. "He said, however, that General Motors is in trouble. Here is what he said to me.

'McBride, General Motors is not going down the drain. If we have to use foreign steel to stay afloat, we're going to use it.'" This would mean the elimination of thousands of steel-workers' jobs, McBride added.[18]

Smith's not-so-veiled warning worried McBride, but it did not embolden him to volunteer to reopen the industrywide labor agreement. Finally, in mid-April, Bruce Johnston and George Moore forced the issue. They told McBride that the industry would make a formal reopening request by letter. Then occurred another delay, occasioned by McBride's insistence that the industry send a long, explanatory document, rather than a one-paragraph notice. Johnston drafted an eight-page letter and presented it to McBride for review. The union objected to only one word, "fungible." Johnston had written that "steel, in a sense, is a fungible product now available in our market from all over the world."

"What the hell does 'fungible' mean?" Odorcich asked when he read the letter.

Johnston explained that whether produced in the United States or abroad, steel is the same and thus "fungible" or "generic." Then why not use "generic?" Johnston agreed to the change.

The letter episode is important only in that it illustrates the posturing that can occur when a union insists on maintaining the appearance of an adversary relationship. McBride realized that the industry needed cost relief, but he couldn't discard the idea that the union must only react to a management initiative. Under Murray, McDonald, and Abel, the union often introduced solutions to joint problems or saved management from making a mistake in labor relations.[19]

Yet another problem delayed the 1982 reopening request. It so happened that the triennial elections for local union offices were scheduled to be held throughout May at the USW's five thousand or so locals. These elections typically produced about a 40 percent turnover of local officers even during normal times. If a request for reopening became public during May, it would introduce an inflammatory issue into the election process that would result in even more turmoil and possibly kill any possibility of renegotiating the contract.

It could be argued that union elections should force candidates to declare their positions on critical issues. On the other hand, a question of political responsibility was involved. The presidents of the basic steel locals made up the USW's ratification committee. It would not be rational political behavior for a candidate to campaign on a proconcession platform. Many candidates would be forced to adopt an anticoncession stance and stick to it when they voted in the Basic Steel Industry Conference. If so, the election process would bias the bargaining process. In any case, McBride asked Johnston to send the letter after the elections.[20]

The industry letter, dated May 28, 1982, requested "that the United Steelworkers of America, effective June 14, 1982, commence negotiating with us for successor labor agreements, which will replace those now in effect." It had taken the industry nearly a year to reach this point from the time Johnston first began his lobbying effort with McBride. It would take another nine months for management and labor to reach an agreement.

Chapter 9

The 1982 Recession Worsens

In the week ending June 11, 1982, 532 companies in the United States declared bankruptcy, the highest weekly total since the Depression. More than 50,000 businesses had already failed during fiscal 1982, compared with 11,432 in 1975, the peak year of the last big recession. When President Ronald Reagan attended an economic summit meeting at Versailles in June, 30 million people were out of work in the industrialized countries. Unhappy European leaders urged him to reduce interest rates in the United States—they ranged up to 25 percent for short-term loans—and lead the world out of the economic morass. The president assured the Europeans that a business recovery was on the way.

In June, however, the U.S. economy was still slowing down. Reagan's 25 percent tax cut had created enormous budget deficits without producing the promised "supply-side revolution." This was to have been manifested by a great surge of capital spending on new plants and equipment which would boost steel production. But the mills operated in June at less than 50 percent of capacity, far below the 65 percent break-even point, and consequently lost as much as $60 on every ton of steel shipped. In this climate, economists predicted, the industry would perforce shrink by 8 million to 17 million tons of capacity, representing some fifty thousand jobs, by the end of the recession.

The supporters of Reaganomics and the tight money policy of the Federal Reserve Board marveled at how the combination had wrung double-digit inflation out of the economy. Less commented upon, but no less true, was the intended squeezing effect on wage earners' incomes. Reaganomics

215

really amounted to a wage control program without the need for bureaucrats to administer it.[1]

The BSIC Debates Reopening

Bruce Johnston's letter asking for a contract reopening arrived at Lloyd McBride's office on Tuesday, June 1. Shortly thereafter, the union sent telegrams to the presidents of all 633 locals involved in the steel industry, summoning them to a meeting of the Basic Steel Industry Conference (BSIC) in Pittsburgh on June 18. This first phase of the renegotiation process encountered trouble from the beginning.

The first problem was timing. The USW did not publicize the contents of Johnston's letter until June 18, when copies were distributed to the local presidents. However, news that the industry had asked for immediate negotiations leaked to the press without explanation. In this void, rumors about the content of the letter sped from local to local and steel town to steel town. According to some rumors, the companies had already made specific demands for concessions. The proponents of conspiracy theories were quick to suggest that McBride had secretly agreed to a wage cut and would attempt to "ram it down our throats" at the BSIC meeting.

Nothing of the sort was true. Nor was the letter aimed at angering the union. Johnston, the principal author, had tried to avoid a belligerent approach. The most inflammatory sentence in the eight-page document was a typical staple of management rhetoric: "Our pay for steel employment has passed all reasonable bounds." The letter described the "crisis" of the steel industry and asked that negotiations begin in mid-June to amend the terms of the existing labor agreement.

Interestingly, the letter declared that labor, management, and the government bore responsibility for the decline of the industry, adding that each of the parties must share "the opportunity and the obligation" to help the industry regain strength. The USW had negotiated excessive wages. The government was faulted for its trade, depreciation, and antipollution policies, as well as for holding down steel price increases through formal and de facto price controls. As a result, the letter said, the industry "has suffered from capital undernourishment for over two decades." However, beyond the one brief, early nod in the direction of management culpability, no mention was made of management mistakes.

   Steel unionists probably should not have expected to read a litany of management errors in a letter *from* management. Still, the omission lent strength to local unionists who were skeptical about the claims of distress. They could not quarrel with the industry's description of job loss, however. To drive home this point, Johnston appended to the letter an eight-page list of every steel plant or portion of a plant that had been closed since 1974. There were 205 listings. Since 1965, two hundred thousand jobs had disappeared from the basic steel industry.[2]

   If the economists were correct, a large number of the currently unemployed steelworkers would be added to that list. In June 1982, steel layoffs had risen to 111,500, more than a third of the 1981 work force. For the first time in this recession, salaried employees were being pruned by the steel companies. U.S. Steel laid off 100 people at its research laboratories in Monroeville and ordered a 10 percent reduction of salaried staffs in the Mon Valley plants. The company continued to close mill departments, leaving each plant with fewer operations. The mill buildings remained standing—large clusters of them in some plants—but the abandoned structures took on a dark, haunted look and began to sag and show the effects of weathering. It reminded me of how unused army barracks quickly develop a worn-down, lifeless appearance when the soldiers move out.

   The same was happening in other big steelmaking centers like Gary, South Chicago, Johnstown, Wheeling, Cleveland, and Baltimore. Some 5,000 workers were laid off at Bethlehem Steel's huge complex at Sparrows Point near Baltimore. Nevertheless, a *New York Times* reporter marveled at the "bravado" of rank-and-file workers when they were asked about wage concessions. "I would rather have the mill shut than give concessions," said a welder at Sparrows Point.[3]

   In addition to the timing problem, the chances of the companies gaining concessions suffered from rising conflict in the plants. In every sharp downturn, plant-floor supervisors tried to reduce production costs commensurate with the drop in output. This meant getting along with the smallest possible hourly work force. To do this, foremen would try to pare the size of mill crews in individual operations by combining duties, eliminating what the foremen decided were unneeded jobs, or assigning maintenance work outside the craft that traditionally performed certain kinds of repair.

Under certain circumstances, these kinds of changes could violate the contract unless they were negotiated with the union. If management made the changes unilaterally, the affected union members could file grievances and possibly win back-pay awards. Until the grievance was settled, however, workers had to accept whatever job assignment they were given.

It was a system ready-made for conflict. Plant managers accepted the risk of losing a grievance and doling out back-pay awards later (if the case went to arbitration, a ruling could be up to two years in the future) in order to get more production at lower cost in the short run. Protecting crew size, on the other hand, became one of the main functions of a union when jobs were scarce and declining. In every prolonged downturn, the locals would complain that management was violating the contract by cutting crews and combining jobs. Particularly annoying to local unionists was that the grievance threat didn't deter the company from unilaterally cutting crews. "Go ahead, file a grievance," the foremen would say.

And so it was in 1982. By mid-June, USW District 15, which covered the plants in the Mon Valley, had thirty-six crew-size grievance cases pending at the third-step grievance level (two steps before arbitration) at U.S. Steel, about six times the normal number. I checked with my friend Larry Delo, who was a general foreman at U.S. Steel's Duquesne Works. "Sure, we're cutting crews," he said. "We've done things in the past few months that we'd never tried before. Of course, we have a lot of grievances. But the company is at the point of just ignoring the union so we can operate the place as efficiently as possible."[4]

Delo did not like to manage this way, but he felt there was no choice. Plant managers ordered foremen to ignore the union. Local union officers realized that cutting costs in a slump might enable the company to conserve jobs in the long run. But a local officer could hardly negotiate a temporary job-cutting plan unless he or she believed that management would return to the pre-emergency work rules and reinstate laid-off workers in the upturn. Only a long history of trust between a management and a union could justify such a belief, and in June 1982 few steel plants in the country had that kind of history.

In the weeks before the June 18 BSIC meeting, management

and labor interests became more polarized just when they should have been coalescing. The combination of the increasing layoffs, spreading rumors, deteriorating relations in the plants, the union's failure to communicate more fully with the locals, and the willingness of some local unionists to politicize union-management issues—all of these had a deleterious effect on local attitudes. Many locals hastily called membership meetings and adopted a "no-concession" policy. As a result, a substantial number of local presidents (perhaps as many as a few dozen, though nobody knows the number) showed up at the BSIC meeting under orders from their locals to oppose reopening the agreement.

At about noon on June 18, some seventy-five or so demonstrators gathered on the plaza in front of the U.S. Steel Building in Pittsburgh, across the street from the William Penn Hotel where the BSIC meeting would be held. Carrying signs with anticompany statements and the no-concessions slogan, they milled around on the plaza, posing for TV cameras and handing out literature. The demonstrators included a sprinkling of local union officers, such as Ron Weisen of Homestead and Mike Bonn of the Irvin Works. It was by no means an overwhelming show of anti-McBride sentiment, but I had long ago learned never to measure the strength of opposition to USW leadership by counting demonstrators. These gatherings always attracted political activists from leftist groups who were not USW members. On the other side of the ledger, some steelworkers who opposed union leaders' policies disliked demonstrating against their own union.

For all the publicity attending the BSIC meeting, less than half of the 633 presidents who had been summoned actually showed up. This can be explained only by looking at the structure of the steel conference, which would come to be a divisive issue before the end of 1982.

The BSIC membership theoretically represented the entire basic steel industry, which included every unionized plant, mine, and warehouse that produced or sold iron and steel products. These included large integrated steel plants (many of which had two local unions, one representing production and maintenance workers and the other, office and clerical employees), iron ore mines, fabricating units, and distribution centers owned by the major companies, and dozens of plants operated by me-too companies which traditionally accepted the wage pattern set by the majors.

Despite this large diversity in size and type of local, each president had one vote in formulating wage policy and ratifying the "industrywide" agreement made with the eight major companies. Presidents who had nothing to do with the eight companies nonetheless had a voice in establishing wages and benefits for members whom they did not represent. The theory was that the industrywide agreement set the pattern for all employers in steel-related businesses, though in fact it did not.

Structures like this had persisted from the earliest days of the union, when it copied the Mine Workers' industrywide bargaining format. The USW's ultimate aim had been to force all metals producers to negotiate with the union as one body on a national basis. In practical terms, the USW had long since given up any hope of achieving that goal. Yet, like most unions, its organizations and political customs were governed by the past.

McBride convened the BSIC meeting at 1:30 P.M. in the Grand Ballroom. He briefly recapitulated what had happened in the industry since the 1980 negotiations, including the reduction in demand, layoffs, and rising imports. He reported that some companies had cash flow problems and that one firm could face bankruptcy before the contract expired at the end of July 1983. He criticized U.S. Steel's purchase of Marathon, saying that it had "created a breakdown in the relationship between the employees...and U.S. Steel." But McBride also counseled against allowing "our venom [to get] the better of our good judgment." After his opening statement, Johnston's letter was distributed and read aloud to the delegates.[5]

That morning the union's executive board had approved a resolution that the BSIC "should authorize discussions [with the industry] for the purpose of exploring possible solutions to the problems of our members and the industry." The motion carefully avoided mention of reopening the contract, or negotiating new terms. "Reopening" the agreement was a formal legal step which, if taken, might enable the companies to declare that no contract existed and therefore that they could establish new conditions of work. The safer course was to negotiate new terms and simply substitute them for the provisions in the existing contract. If the negotiations failed, the companies would still be obligated to live with that pact. (Although the union actually had eight separate contracts with the eight companies, I shall use the singular "contract" or "pact" to refer to the agreement that was bargained with the industry committee and translated into the individual contracts.)

Many delegates probably assumed that there could be no negotiations without reopening the contract (after all, the entire process was new to them), and McBride did not set the matter straight. In response to one delegate who said he was against reopening, McBride replied flatly, without explanation, that reopening was not at issue. "Why are we voting then?" the delegate asked. "Because McBride isn't the only guy who will be discussing," the president said. "If there's going to be any discussions, you guys are going to be in it, and you're going to be discussing. And if you say no, there will be no discussions. Believe me, that will end the discussions until next year."

This piece of semantic obfuscation was not worthy of McBride or a democratic union. From his talks with Johnston and Moore, he knew that the companies wanted to renegotiate the wage and benefit terms. He had already formulated a negotiating plan: He would be willing to eliminate wage and COLA increases scheduled to take effect on August 1 if part of the $100 million savings were used to raise unemployment pay. McBride referred vaguely to this idea at the BSIC meeting and mentioned it more specifically at a news conference.

Yet he could not bring himself to ask the BSIC directly for authority to negotiate new terms, apparently believing that such a request would be voted down. Many, if not most, of the delegates undoubtedly understood that the resolution gave McBride authority to negotiate, regardless of the words used. They could also see that the USW president deliberately tried to circumvent the issue, and this would add fuel to the rumor that McBride was conspiring with management to bargain a concessionary agreement.

Ed Ayoub was called upon to give the same slide presentation that he made to the executive board in April. As before, it showed labor costs running far ahead of productivity. But Ayoub emphasized that he was not merely parroting industry figures. "I personally calculated every single number," he told the delegates.

In the debate that followed, eight presidents spoke against the resolution and five in favor. Most of the opponents said they had a "mandate" from their locals to vote against any-

thing smacking of concessions. Mike Bilcsik, president of
Local 1256 at the Duquesne Works, said that while two-thirds
or more of his local were laid off, "every steelworker I know is
against reopening." Bilcsik would later change his mind, but
his position at this meeting was reinforced by another president who said, "Even people laid off say, 'Don't give them
a goddamned thing.'" Al Forney, president of Local 1157,
thought that increasing unemployment pay merely "prolongs
the agony" of those already laid off. Many would never get
their jobs back. In any case, the union should not go "crawling" to the companies.

Charles Grese, who headed Clairton Local 1557, made the
opposite plea. The union ought to consider how to bring the
one hundred thousand jobless steelworkers back to work. "If
it's going to take 10 or 15 cents an hour to get these people
back to work, I say let's do it." Jim Brown, of Local 1066 at
U.S. Steel's finishing plant in Gary, said it was the union's
duty to "help all concerned." The sister local of 1066 was
1014, one of the USW's largest locals, which represented
twelve thousand members who operated the iron and steel
furnaces and rolling mills at Gary. Its president, Phillip
Cyprian, scolded the opponents for "kidding" themselves. "If
we try to tell ourselves the company doesn't have financial
problems, we are wrong," he said.

By this time, the meeting had lasted most of the afternoon.
Many delegates began calling for a vote. McBride made a final,
somewhat impassioned statement. He recounted his conversation with Roger Smith (see chapter 8) to suggest that a no
vote could have unpleasant consequences, including the loss
of steel's automobile market to the Japanese. If the delegates
rejected the motion, McBride concluded, "I can carry my
share of it [the consequences], and I will do it. But I can't carry
your share of it. You've got to do that."

It was a theme that McBride would bring up repeatedly in
the months ahead. Driven by a hard, Puritan-like morality,
McBride had become convinced, as one top aide later said,
that steelworkers (including union officers) made too much
money and that the union over the years had failed to instill a
sense of responsibility in local officers and the rank and file.
The result of the balloting that followed McBride's comments
probably reflected less an acceptance of guilt by the BSIC delegates than a realization that the steel industry was in sufficient trouble to warrant some discussion. The vote was 263 in

favor of entering "discussions" with the industry, and 79 opposed.

In the next few days, McBride and Johnston laid out a format for their discussions. If the wage and COLA increases due on August 1 were to be avoided, the two sides had to reach agreement before that date. The delays in reaching this stage of the renegotiation process now began to assume importance. The eight producers that belonged to the Coordinating Committee Steel Companies (CCSC) had barely one month to prepare their employees for a major change in compensation policy. If a new agreement with a wage freeze and a limit on COLA payments was to be ratified by the politicized BSIC, company negotiators told their chief executives, the work force should be "conditioned" to help their representatives see the light. The eight CEOs thought this a splendid idea and ordered their labor and employee communications specialists to carry it out.

The ensuing "education" campaign was a perfect illustration of one of the fundamental problems in American industrial relations.

A Communications Fizzle

For all their vaunted expertise in persuading people to buy their products, American corporations have been exceptionally poor at another form of communications. They have failed miserably at telling their employees what the business is all about and why they should be committed to its success. The main reason for this is that employers by and large, however much they may proclaim their respect for the common man, essentially believe, as Henry Ford I put it so eloquently, "All men want is to be told what to do and get paid for doing it."[6]

This "economic man" view fit quite well with the organization of work for which Ford's mechanical assemblyline was the paradigm. Each worker on the line was trained to perform a few rudimentary tasks such as fitting fenders and tightening bolts. Management did not want him to think, if thinking would distract him from tightening bolts. Management also thought that it deserved the worker's full commitment by paying him a competitive wage, or—if that wasn't enough—an extra bonus or incentive that appealed to the average person's greed. Nothing else was required, certainly not knowledge about the business, about profit and loss, capital investment plans, sales projections, new product development, marketing strategies, and plans for technological change.

Henry Ford's attitude about workers best typifies a leftover assumption of nineteenth-century economic theory upon which business policy was largely based. The economist David Ricardo conceptualized it this way: Society consists of a horde of unorganized individuals, each of whom acts in a manner calculated to secure his self-preservation or self-interest. This "rabble hypothesis," as it was termed by Elton Mayo—a Harvard sociologist who pioneered the study of work—in the 1930s, was eagerly embraced by early laissez faire theorists to justify a Darwinian approach to business. "And the general public, business leaders, and politicians are left with the implication that mankind is an unorganized rabble upon which order must be imposed," Mayo wrote.[7]

One of the most important American theorists of work innovations was Douglas McGregor of the Massachusetts Institute of Technology. (Contrary to popular notion, the theories and research supporting worker participation did not originate in Japan. Practically all of the critical research on motivational theory, alienation, and the impact of different management styles was done in Britain, Norway, Sweden, and the United States.) In the 1950s, McGregor developed the famous Theory X versus Theory Y concept of management. Although thirty years old, this concept still provides the best description of the assumptions about human behavior that the two contrasting styles of management are based on.

Theory X postulated the traditional, authoritarian concept which holds that managers must control and direct the work force as if they were military commanders. This view is based on the following assumptions: "The aveage human being has an inherent dislike of work and will avoid it if he can. [Therefore], most people must be coerced, controlled, directed, threatened with punishment to get them to put forth adequate effort toward the achievement of organizational objectives." Theory Y posed the participatory style, in which management assumes that workers want to be committed to their jobs.[8]

Steel management by the 1980s had shifted away from the extremes of authoritarianism described above, and not every manager in the mills of 1982 carried out his or her role on the basis of these assumptions. Most probably would have scoffed at them. But the steel management *system* still rested on that old foundation of beliefs about the worker as "economic man." The companies' communications efforts reflected this philosophy.

"If you're going to treat a guy as merely a 'contract' employee, who works eight to four, nine to five, you don't communicate with him," said John Kirkwood of Jones & Laughlin in a 1983 interview. "You try to coerce him. You pay him an hourly wage and try to get more work out of him by also paying an incentive, but that isn't really communicating with him. That's been a fact at most of the steel companies for years. If you have a big plant, with thousands of employees, most of the people are never touched by any kind of dialogue. They come in and raise hell with the company, and there's no reaction, or the company reacts poorly. There's no real communication. The company tends to communicate with the union representatives, and everybody else is kind of floundering in a big plant."[9]

Organizational systems do not change by themselves; management must proclaim a new philosophy of work and root out all the procedures that are based on the old beliefs. In the early 1980s, the communications procedures in most steel firms grew out of the old need-to-know test: Since workers didn't care about the business, management concluded that they should tell workers only what they needed to know to perform their narrow jobs. If each worker did what he was told to do—that is, if each cog rotated in the precise manner for which it was set in place—all gears would mesh when management pressed the button.

Such little information as workers need, or can understand (according to the Theory X view), can be supplied by an annual report, a monthly or quarterly company newspaper, and perhaps a periodic plant newsletter. In most corporations, and steel was no exception, the "news" in these publications had little interest for employees. To most, it was as plain as the nose on your face: The company newspaper never told you what management didn't want you to know. And so the descriptions of wondrous employee benefits and virtuous management policies were read with skepticism. Managers who consistently extolled their stewardship in glowing accounts of quarterly profit results to shareholders could not expect employees to believe the opposite—that the company was in trouble and needed an infusion of cash from wage cuts.

Confronted with this argument, some managers said that employees ought to look beyond the short term and consider the long-term interests company. But as one veteran observer of labor-management practices has said, "Companies that have always encouraged workers to take the short-term view of their jobs, which are paid by the hour, should not expect them to turn around and take the long-term view of the company."[10]

To be fair, I must say that the major steel companies made a bigger effort than many corporations to discuss the industry's commercial problems *with union officials*. One company, Allegheny-Ludlum, had a long tradition of meeting annually with local officers and so did several others. U.S. Steel was the exception. At the beginning of each set of industrywide negotiations, hundreds of managers and local and International union officials gathered in "sound-off" meetings in which each side aired problems and complaints. In some years, particularly in the 1970s and 1980s, industry executives gave elaborate presentations on cost, productivity, and steel consumption at the sound-off meetings.

However, this data for the most part showed very broad, industrywide trends. It was difficult for local union officers to apply this information to their specific situations and almost impossible to make it tangible to rank and filers back in the mills. Furthermore, the steel companies only rarely opened the books on what they considered to be proprietary information, such as competitive cost data for individual plants, even to small groups of union officials who could be expected to keep it confidential.

The tradition of discussing industry problems was limited to the representative level in the union. When the companies set out in early July 1982 to inform the rank and file that the industry was in a crisis, one that had to be resolved within a month, they immediately violated one basic principle: If worker education has never been company policy, don't begin with a last-minute plea for help.

Most of the companies sent copies of Johnston's May 28 letter to the home of each employee. U.S. Steel, among others, devoted a special issue of its company magazine to the industry's problems, based on the study it had commissioned from Putnam, Hayes & Bartlett. In addition, most or all of the companies hastily made up video tapes of slide presentations using the same information and showed them to all employees, department by department. U.S. Steel, for example, put on the "slide show," as it came to be called (much to management's annoyance), before about one hundred thousand employees, and even some unemployed workers, at twenty-one steel plants.

Since this crash PR campaign was carried out industrywide, it may well have been the largest single attempt to influence workers' attitudes on a labor-management issue in industrial history. The only effort that came close to it occurred in 1973, just before the CCSC bargaining group and the Steelworkers announced agreement on the first ENA. To "soften up" union members for the no-strike pledge, the USW and the industry jointly commissioned a twenty-five minute propaganda film which was shown in every steel plant and on television in seventeen steel towns. Entitled *Where's Joe?*, the movie was a blatant piece of racial and political chauvinism. Joe, of course, was the American steelworker who had lost his job to a foreigner, and the film depicted 130,000 Joes being eliminated as more and more steel came into this country from several nations. The Japanese, however, were selected to be the villains. "Their ambitions were bigger than we once knew," the narrator said ominously over shots of Japanese technicians taking notes on a tour of American mills. A montage of Japanese mill scenes flashed closeups of intensely competitive, Oriental faces. Here are these little slanty-eyed guys, the film suggested, the same guys who came at you in World War II and are now trying to get your jobs. The import issue was real enough, but the presentation of the problem hardly comported with the union view of international labor brotherhood.

Between *Where's Joe?* in 1973 and the slide shows of 1982, steel management had made little effort to get in touch with the attitudes of rank-and-file workers. Thus the first mistake of the campaign was made before it started. There were others. One was U.S. Steel's decision not to allow dissemination of the data until July. The company gave the findings of the Putnam, Hayes study to the chief negotiators of all CCSC companies in April 1982 but instructed them not to release it to employees at that time.

U.S. Steel could issue such an order under its authority as the designated bargaining leader. U.S. Steel management reasoned that if the information were distributed to employees it would leak out and expose the industry's financial plight to customers, who would begin deserting the sinking ship. A high-level executive at one company, who supported with-holding the data, pointed out an even more basic reason for the decision. "In this study," he said, "management for the first time was disclosing what happened to labor costs under the ENA. If it got out to the stockholders how lousy this industry was, who would get hung? "

Kirkwood and others, however, believed that the overriding concern should have been gaining the understanding of em-ployees about the crisis. He wanted to mount a major educa-tion program with the rank and file immediately after Bruce Johnston addressed the USW executive board in April. "Sending a couple of letters and giving a great big slide presen-tation on a one-shot basis isn't really educating people," Kirk-wood contended. "You've got to be working at it all the time."

However, only after the USW agreed in mid-June to open negotiations were the companies allowed to make up their own slide shows (U.S. Steel would not permit duplication of the original presentation). Under a plan adopted by the com-panies' coordinating committee, each firm displayed the data to their local union officers who were summoned to Pitts-burgh in early July. At the same time, videotape and slide presentations were made in the plants—only two to three weeks before the industry hoped to reach agreement on a new contract.

It is impossible to say what kind of impact the tape and slide shows had on ordinary workers in the plants. But the effort backfired in some cases. A session for supervisors at U.S. Steel's National plant in McKeesport occurred shortly after the corporation imposed a 5 percent salary cut and reduc-tions in medical benefits for managerial employees. The fore-men sullenly watched the presentation, which seemed to blame the industry's competitive problems on the high wages of hourly workers. In the question-and-answer session, said one foreman who was there, "Our guys wanted to know, if the wages were so high now, who the hell gave it all away back in the 1970s? The answer was top management. We were push-ing to get the production out, and it was their duty to hold the line. But they didn't."

In mid-July, negotiators for Bethlehem Steel and USW local officers gathered at an Allentown hotel for a bargaining meet-ing. Four times in one day, company labor relations officials tried to present the slide show, but on each occasion the dozens of local union officers walked out. Charles Z. Molnar, president of Local 2635 in Johnstown and chairman of a coordinating committee of Bethlehem locals, told me what happened. "We didn't want to see slides and graphs. We've had enough of that Madison Avenue stuff, and now we want to know what their bottom line is."[11]

To some degree, the boycott was a tactical and public relations move by the Bethlehem local officials to show the members back home that they were heroically resisting concessions. But the presidents also expressed real annoyance about being subjected repeatedly to slide shows on statistics. They had seen Ed Ayoub's presentation at the June 18 USW meeting, and now they had to view again much of the same material. If carried on too long, a softening-up process can produce the opposite result, a hardening of attitudes.

The entire campaign was wrong, Kirkwood said, because it was oriented to the union representatives. "It didn't have an impact on the hourly guys out in the plant," he said. "This industry has always failed to see that. We have tended to communicate only with the union and not on down the line." In his view, a communications effort should start with line-operating management people. "When managers understand the issues facing us, they can communicate to their people face-to-face. The really effective thing is to have supervisors talk to the key opinion leaders in the hourly group. Of course, this gives the dissidents a chance to shoot at you, but that's part of the communications effort."

How Industrywide Bargaining Started

The dissatisfaction expressed by some steel companies about the education program typified the strains and tensions inside the industry's bargaining group. The CCSC was as imperfect for the conduct of a rational collective bargaining policy in the steel industry as the BSIC and, like the BSIC, it owed its survival to the paralyzing fear of change.

Bargaining on an industrywide basis began in 1956, although in reality something approaching industrywide negotiations had occurred for some years. Before 1956 the USW usually set the wage pattern for the industry at U.S. Steel and extended it more or less intact to other large, integrated steelmakers and in varying degrees to dozens of smaller companies. One rationale was that wage increases in steel tended to be paid out of price hikes. If a smaller company had settled first with the union and attempted to raise prices, U.S. Steel—the price leader—conceivably could have refused to validate the price rise. Actually, USS consulted to some degree with other major companies on labor policies during the 1940s and early 1950s. But the corporation had to pay the cost of negotiating for the entire industry, and it bore the brunt of public obloquy when a strike occurred. These were among the many factors that eventually led to industrywide bargaining. Pressures for centralization also had come from government boards which called for consolidated presentations by the steel companies during World War II and the Korean War.[12]

The USW had pushed for this negotiating format so that it could fulfill its goal of eliminating wage differentials in the industry. Although negotiating and striking the companies individually may have enabled the union to whip-saw one firm against another, this tactic really didn't appeal to USW leaders. Forcing its members in one company to strike to raise wages for all members in the industry would create political problems. Furthermore, as Bruce Johnston would say, steel was a "fungible" product. A ton of steel was a ton of steel no matter who made it. Therefore, a strike at a steel company, unlike a work stoppage at an auto maker, would not penalize the firm by damaging customer loyalty.

In 1955–56 Dave McDonald moved to tighten the union's control over contract terms throughout the industry by assuming the chairmanship of all negotiating committees at the six biggest companies. In response, twelve firms named a committee of four negotiators to bargain major issues. In 1959 the twelve formalized the bargaining group and gave it power to negotiate on all issues. As mentioned earlier, both USS and its competitors could benefit from centralization, though in different ways. Although the union didn't want to strike companies individually, that threat still existed. The smaller firms feared that if they were struck, U.S. Steel—which at one time covered the entire steel market—could take their customers away. By grouping together under the protection of the largest company, the smaller firms eliminated the threat of individual company strikes. The main impetus to industrywide negotiations was the smaller companies' hope of gaining a voice in bargaining decisions.[13]

Furthermore, the idea of competing on the basis of labor costs had receded as the USW moved closer to industrywide wage uniformity in the 1940s and early 1950s. With some 25 percent of the steel market, U. S. Steel might have been strong enough to stand alone against the union. But USS had another reason to welcome group bargaining. "It made sense for the dominant producer to control the rate of wage rise and spread the cost results evenly across the competitors," Johnston explained. By exercising control over the negotiations, the corporation could prevent other companies from gaining a competitive advantage in a labor area. At the same time it could eliminate advantages other firms already possessed.

In 1968, for example, USS led the industry before a panel of arbitrators on an incentive pay issue that remained unresolved after bargaining. The USW had demanded that incentive coverage be expanded to many workers who did not have it at that time. Silently, USS must have welcomed this demand, because it already had one of the highest percentages of incentive coverage in the industry, about 85 percent. In other companies, coverage ranged as low as 50 percent. The arbitrators' award ordered the coverage to be at least 85 percent for all companies. "U.S. Steel structured the award to impose high incentive costs on the rest of the industry," said the labor relations vice-president of another company. "By controlling coordinated bargaining, U.S. Steel can prevent the rest of the industry from achieving competitive inroads in bargaining. They've been very effective in doing that."[14]

Internal strife broke out in the CCSC early in its existence. A number of companies disagreed with a decision by U.S. Steel's Conrad Cooper in 1959 to raise the inflammatory issue of local work rules in industrywide bargaining. Cooper, then the industry's chief negotiator, wanted to make sweeping changes in working conditions to prevent local unions from protecting established crew sizes as a "past practice." The injection of this demand enabled McDonald to unite a previously split union and resulted in the 116-day strike. Near the end of the walkout, Kaiser Steel withdrew from the bargaining group and negotiated its own settlement. The industry didn't succeed in changing the work rules, but it did win a low-cost settlement.

The CCSC companies had dwindled to eight by 1982, and many of them were chafing under U.S. Steel's leadership. Negotiators such as John Kirkwood of J&L, Tony St. John of Bethlehem, and Stanley Ellspermann of National Steel questioned the relevance of industrywide bargaining at a time when the oligopolistic approach no longer suited the nature of the steel market. Complaints about U.S. Steel pursuing its own goals to the detriment of other companies and refusing to consider other than its own bargaining ideas had been voiced by a number of companies. Ellspermann put it this way: "What happens, typically, is that because U.S. Steel and Bethlehem did the talking in negotiations, guess whose problems got addressed?"[15]

Bruce Johnston, who became the industry's chief negotiator in 1977, told the companies that he welcomed their ideas. In late 1979, he asked them for recommendations for bargaining in 1980. Kirkwood and his colleagues at J&L accepted the invitation and proposed that the industry make a big issue of productivity in the negotiations. In a letter to Johnston, Kirkwood suggested a plan under which some COLA payments would be made only if output per man-hour had reached a certain level—a level that would vary from company to company. This would force local plant and union leaders to focus on productivity and, eventually, result in different wage levels in the industry. "I knew it was going to be unpalatable to Bruce and the union and everybody else," Kirkwood said, "but we had to get the dialogue going. We had to start relating productivity to wage increases."

But the dialogue did not get going. Kirkwood had been right. Johnston wrote a letter rejecting the idea, saying it was unworkable and would be rejected by the union. (Indeed, Johnston ridiculed the proposal in a 1980 interview with me.) U.S. Steel's disapproval was enough to kill an idea without debate among the companies. "That was the way bargaining worked in those days," says a current industry executive. "Bruce had an intolerance for things that weren't his ideas. And if it wasn't something U.S. Steel liked, they just stiffed you." Ironically, a proposal somewhat like the J&L scheme was introduced in late 1982 bargaining.[16]

Many companies complained that they had no voice in coordinated bargaining decisions and that, once negotiations started, they received too little information. The actual coordinating committee, which consisted of the industrial relations vice-presidents of the eight companies, met periodically to discuss issues but did not make important decisions. The smaller companies on this committee—Inland, Armco, Allegheny-Ludlum, and National—had very little influence on the conduct of bargaining.

The real power resided in a smaller group originally designated as the bargaining team and dominated by the larger companies. From the 1950s to the 1980s, the team numbered four people, two from U.S. Steel, and one each from the next two largest companies, Bethlehem and Republic. As the industry contracted, U.S. Steel agreed to give up one of its positions on the four-man team, and J&L joined the group. Johnston, however, insisted that, to prevent unwieldiness, only two executives sit at the main table, he and George Moore, Bethlehem's chief negotiator. In 1982 Kirkwood and John Wall, a Republic vice-president, also belonged to the four-man team. Supposedly, they would be briefed thoroughly after most sessions, although it was understood that some issues might be too sensitive to talk about outside the bargaining room. Kirkwood, however, was dissatisfied with the amount of information he received from the top negotiators, especially in 1982.

The vice-presidents of smaller CCSC firms like National Steel felt shut out completely. Stanley Ellspermann, National's vice-president of labor relations, likened the communications system to "receiving a call from a telephone booth, someplace, and getting a cryptic description of what went on someplace else." In early 1983, when the industry was trying for the third time to reach agreement with the USW, National worked hard to settle all plant-level issues with its local unions and was the first company to do so. Ellspermann, however, had no control over the main economic issues. "So we [he and his staff] sat over there in the Pittsburgh Hilton, waiting, and all of a sudden a light bulb went on and we thought, 'Jesus Christ, we did all that we set out to do, and now whether we go on strike or not is a function of how well our biggest competitor does our talking for us.' I promised myself that I never wanted to be in that position again." National withdrew from the bargaining group in 1984.

It could not be said, however, that U.S. Steel acted arbitrarily toward the other companies all the time. And the chief negotiator had a difficult job in gaining a consensus of eight industrial relations vice-presidents, eight chief financial officers—who had a voice in setting the bottom line in bargaining—and "eight CEOs with big egos," as one steel executive put it. Johnston acknowledged that some of the companies complained about coordinated bargaining, but he pointed out that "nobody dragged them into it." If U.S. Steel sometimes hoarded information and acted without consulting the other firms, it also provided most of the staff expertise, computer time, and background studies needed in the bargaining process.[17]

In the 1950s and 1960s, the coordinated approach generally satisfied all the companies because all faced more or less common problems, or perceived—perhaps wrongly—that they did. Contrary to the situation in the 1980s, however, the USW viewed U.S. Steel as the "liberal" member of the bargaining group. In the 1960s, top-level union negotiators such as Ben Fischer and John Tomayko found that they could form alliances with influential USS bargainers to deal with complicated problems. One such executive was Warren G. Shaver, a vice-president who had to retire in the mid-1970s because of ill health. Shaver and USW aides resolved many grievance, arbitration, incentive, and seniority problems causing unrest in plants across the industry. USS had "gutsy" leadership in those days, Fischer said, while other firms resented USS's accommodating ways and were more hostile toward the union. But they could not bring themselves to leave the bargaining group and lose U.S. Steel's protection and negotiating expertise.

As the industry deteriorated in the 1970s, the companies' interests diverged. In 1980 the CCSC ejected Wheeling-Pittsburgh Steel from the group because it negotiated an allegedly substandard agreement with the USW, the first departure for this reason. According to former USW and Wheeling-Pittsburgh officials, the charge was false. After the 1980 industrywide settlement, the company renegotiated "runaway" incentives at the firm's Allenport plant to reduce the yield. W-P claimed that companies had the right to make such changes at the plant level, but other CCSC firms disagreed and tossed W-P out of the group.[18]

Some firms, especially U.S. Steel and Armco, diversified so they wouldn't be so dependent on steel, while others—Bethlehem, J&L, Republic, and Inland—were buffeted harder by the noncompetitive factors arising in steel. Inland Steel had the most reason to remain committed to coordinated bargaining. It had only one plant, no other businesses, and would be most affected by a strike. Inland complained more than other firms about the conduct of bargaining but did little to improve its own labor relations before the mid-1980s.

By the 1980s several companies were expressing dissatisfaction with coordinated bargaining. But they had no leverage to force changes. U.S. Steel remained the dominant force. It replied to complaints, one negotiator said, by saying, " 'If you don't like the way it is, don't let the door hit you in the ass.' That was a typical U.S. Steel response."

Despite the increasing divergence of interests, the eight steel producers that still remained in the Coordinating Committee Steel Companies were united in 1982 on one thing: They wanted concessions from the United Steelworkers.

I. W. Abel (right), retiring as the Steelworkers' third president, swears in his successor, Lloyd McBride, in June 1977. *James Klingensmith/Pittsburgh Post-Gazette*

The McBride team at the beginning of a second term, pictured outside the USW headquarters in Pittsburgh on Sept. 1, 1981. They are (left to right) Frank McKee, treasurer; Lynn Williams, secretary; Lloyd McBride; Leon Lynch, vice-president; Joseph Odorcich, vice-president. *Pittsburgh Press*

USW President Lynn Williams, Oct. 13, 1986, in the third month of the lock-out/strike at USX. *Harry Coughanour/Pittsburgh Post-Gazette*

Bernard Kleiman, the USW's general counsel, in 1983. *Harry Coughanour/ Pittsburgh Post-Gazette*

(Below) Edmund Ayoub in 1985, shortly after he retired as the Steelworkers' chief economist. *Pittsburgh Press*

USW officials in 1985 being interviewed on their attempt to save the Dorothy Six blast furnace at U.S. Steel's Duquesne Works (plant in background but furnace not visible). Mike Stout (foreground, speaking into microphones), grievance committee chairman of Homestead Local 1397; Andrew "Lefty" Palm (behind Stout), director of USW District 15; Mike Bilcsik (far right), president of Duquesne Local 1256. Listening are congressmen Bob Edgar of Pennsylvania (left foreground) and James Oberstar of Minnesota (in dark suit, behind Edgar). *Tony Tye/Pittsburgh Post-Gazette*

Mike Bonn (center, with child), president of Irvin Works Local 2227, and Ron Weisen (right), president of Homestead Local 1397, at demonstration against Mellon Bank, 1985. *James Klingensmith/Pittsburgh Post-Gazette*

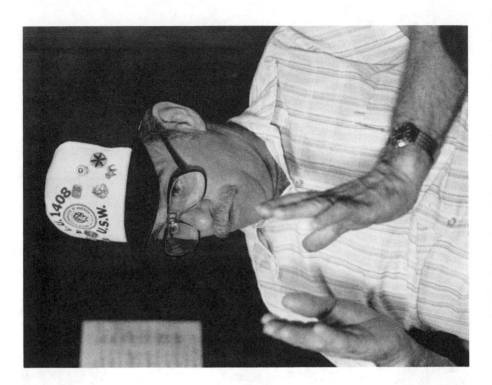

Dick Grace, president of Local 1408 at U.S. Steel's National plant, shown in his McKeesport office, 1982. *Jim Judkis*

(Below) Delegates of the USW's Basic Steel Industry Conference, meeting in the Grand Ballroom of the William Penn Hotel, Nov. 18, 1982, the day before they repudiated Lloyd McBride by rejecting his tentative agreement with the steel industry. *Bill Levis/Pittsburgh Post-Gazette*

David M. Roderick, chairman and chief executive officer of USX Corporation, appearing at a news conference in early February 1987. *Mark Murphy/ Pittsburgh Post-Gazette*

Thomas Graham, appearing at a news conference on Dec. 28, 1983, when he was vice-chairman of U.S. Steel and chief operating officer of steel operations. In 1986, he became president of the steel division of the renamed USX Corporation. *Joyce Mendelsohn/Pittsburgh Post-Gazette*

In April 1980, Joseph Odorcich (left), USW vice-president, and J. Bruce Johnston, senior vice-president of U.S. Steel and the steel industry's chief negotiator, announced agreement on a new contract. Odorcich had taken over as the union's chief bargainer when Lloyd McBride became ill. On the right is George A. Moore, vice-president of industrial relations at Bethlehem Steel, who, together with Johnston, constituted the industry's bargaining committee. *Bethlehem Steel*

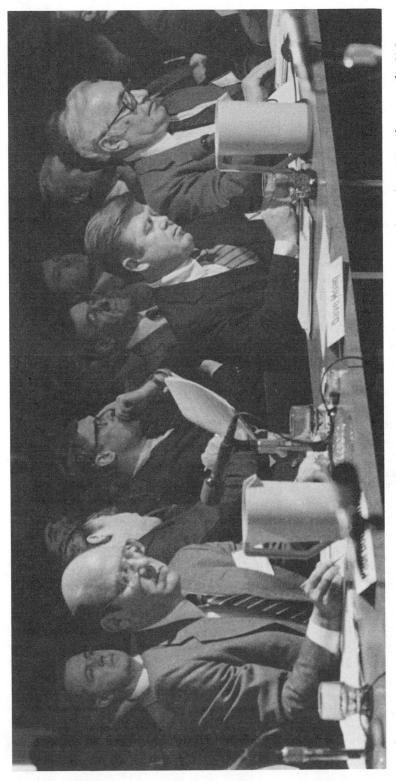

Jones & Laughlin Steel and the USW installed many Labor-Management Participation Teams in the early 1980s. Shown at the J&L-USW first annual LMPT conference in 1983 are David Hoag (left), president and chief executive officer of J&L after Thomas Graham moved to U.S. Steel; John H. Kirkwood (center), J&L's vice-president of industrial relations; and Sam Camens (right), a special assistant to the USW president and director of the union's LMPT activities. *LTV Corporation*

Chapter 10

The 1982 Talks Begin

The steel negotiations of 1982 began on July 5 in a sixty-first-floor conference room in the U.S. Steel Building, about twenty-five yards from Bruce Johnston's office. This was the elite floor at U.S. Steel—some called it "ulcer heaven"—where the corporation's top executives, including Chairman Roderick, worked in large, well-appointed offices sealed off from outsiders by a thick glass door operated by electronic locks. The conference room, which was long and narrow, contained a rectangular table and a few prints on the walls. The five negotiators would see much of this room over the next several months.[1]

Generally, union and management negotiators scowl across the table at one another in equal numbers. In this case, however, only two men represented the industry, Johnston and George Moore, against three for the Steelworkers, McBride, Joe Odorich, and Bernard Kleiman, the union's general counsel. This curious inequality resulted partly from tradition. Since the 1930s, most companies had given lawyers prominent roles in negotiations, reflecting the American tendency to legalize labor relations. The union had followed suit. Kleiman was the latest in a series of exceptional USW chief counsels that included Lee Pressman, Arthur Goldberg, David Feller, and Eliot Bredhoff.

In 1982 Kleiman was fifty-two years old, a lean man of medium height who kept in shape by jogging. Wearing gray business suits and smartly styled hair, Kleiman could easily have passed as a corporate lawyer for a high-tech, forward-looking "growth" firm. He graduated from the Northwestern University School of Law and worked in a Chicago law firm until Abel appointed him chief counsel in 1965. As a key

236

adviser, strategist, and negotiator, he could arguably be characterized as second only to the USW president in his influence on union policy-setting. In bargaining, he typically assumed a deceptively soft-spoken manner, as if to say, "Let us all reason together as honorable men." But his negotiating style of pushing relentlessly, even on side issues, and seldom conceding a point, annoyed management bargainers. Kleiman, however, was adept as a "closer," meaning that he knew when it was time to draw all the issues together and make an agreement. A personal tragedy struck him in late 1982 when his wife Lee contracted cancer. She died in 1984.

Kleiman had been involved in every major steel negotiation since 1965, and McBride could not do without him. Joe Odorcich became the third member of the union team by dint of a combative stubbornness that characterized his union career (see chapter 14). During the 1980 talks, he had substituted for McBride when the latter suffered a heart aneurysm only 10 days before a settlement deadline. In the USW tradition, only an elected officer could give the final yes or no in major negotiations. Odorcich hadn't been involved in the bargaining up to that point and had to lean heavily on Kleiman's knowledge and advice. The experience convinced Odorcich that at least two elected officials should be involved in all discussions bearing on negotiations.

In late 1981, when it became apparent to him that McBride would continue the two-man team approach, Odorcich confronted him in his office. "I chewed his ass out because in the last negotiations he gave me the goddamned ball in the middle of the ball field," Odorcich told me. "I told him he ought to have another officer in those meetings, and I didn't care who it was. Then I drove the point home by reaching over and feeling his pulse. Lloyd said he'd have to talk to Johnston. I said, 'I don't think you ought to talk to Johnston, you ought to *tell* him.'" The idea of consulting Johnston on a union decision reflected McBride's belief that management authority must be respected. Some months later, he relented and put Odorcich on the committee.

In addition to the top committee sessions, two subcommittees met at other offices in the U.S. Steel headquarters and the William Penn Hotel across the street. A benefits group discussed matters such as health insurance and supplemental unemployment benefits, and a problem-solving subcommittee considered "noneconomic" issues such as seniority, job classifications, and "contracting out," the company practice of hiring contractors with non-USW employees to perform construction work and equipment repair both inside and outside the plant. The subcommittees were to negotiate their issues within parameters set by the top teams. The top teams, however, couldn't agree on the parameters.

The industry had two goals at the outset of bargaining: to get immediate relief in terms of cash flow and to slow down the rate of rise in labor costs. Johnston and Moore referred to this as "leveling out the trend line" or the "moderation" approach. They asked for a new, four-year contract but, because of union opposition, agreed to negotiate on a three-year basis.

For purposes of calculation, the two sides had agreed to use the figure of $22.05 as the average hourly employment cost (wages plus benefits) for the eight companies, as of July 1982. (Actually, the figure was much higher for some companies because they had to continue paying medical benefits and funding pensions for the thousands of workers who were laid off during the recession. With fewer people working, each hour of work had to support a higher proportion of the total employment cost. But $22.05 was accepted as the cost at a "normal" level of operations.) If the industry had its way, the "terminal" employment cost on July 31, 1985, would be about $22, instead of the $26 to $27 that would have been generated under ENA bargaining with the 3 percent guarantee and unlimited COLA. The companies at this stage of the game looked for a cost-avoidance rather than a cost-reduction pact.

Not all company officials agreed with this strategy. When the chief executive officers had discussed bargaining goals in the spring of 1982, some expressed hopes of cutting the employment cost immediately to about $18, roughly the price of labor at nonunion minimills. This would have made U.S. mills competitive on a cost basis with Japanese steelmakers who at that time paid their workers only about $10 or $11 per hour but who had to bear significant freight costs to get into the U.S. market. The CEOs also offered specific ideas for slashing labor costs. There was talk of cutting pay, reducing the number of paid holidays, eliminating some health-care benefits and shrinking others, wiping out COLA, and reducing eligibility for unemployment pay, among others. "Adding it all up," said a labor relations official who attended the meeting, "you'd probably have done away with the contract and have everybody working for the minimum wage."

Johnston and Moore reduced the CEOs' "wish list" to a workable number of items. In the benefit area, they wanted to raise major medical insurance deductibles from $100 per family to $200, reduce dental benefits, cancel eye care benefits, and eliminate an extended vacation plan—or "EV" as it was called—under which older workers received a thirteen-week vacation once every five years. In the wage area, the industry's priorities were to eliminate a 23¢ wage hike and a 25¢ COLA boost that were due to be paid on August 1, freeze the wage rate at its current level, substantially cut future COLA payments, and stop the practice of adding COLA to the base rate once a year (to escape the compounding effect).

When the dickering began, however, the USW negotiators made it clear that they would not bend on health insurance and the EV. McBride was willing to eliminate the August 1 wage and COLA boosts and divert 50¢ of the savings to rejuvenate the companies' exhausted SUB funds. This would benefit thousands of the unemployed who currently received no jobless pay. The two sides agreed that workers with five or more years of service would receive $100 to $220 a week. The idea that active workers would give up a pay increase to help their suffering brothers was a noble but unpopular concept, as McBride later discovered.

Not much happened in the first two-and-a-half weeks of talks. McBride indicated that the industry should withhold its specific demands for the time being. He had scheduled a meeting of the local presidents on July 21 to assure them that he was keeping them involved. But the appearance of involvement overshadowed the reality. "Lloyd wanted no proposals before that meeting, because he wanted to be able to stand up and say, 'All we've been doing is exploring. I have no proposals to report,'" Moore said. A premature disclosure of company demands would give the no-concession forces specific issues around which to rally opposition.

When the BSIC assembled in Pittsburgh on July 21, McBride reported truthfully that the companies hadn't presented any demands. It was as if he threw his hands in the air and said, "What can I do if they won't tell me what they want?" The delegates grumbled and fussed, and many of them suspected what McBride's ploy was, but having nothing specific to vote against, they agreed that he should continue the "discussions." Once again, though, McBride's dissembling added fuel to the conspiracy theory.

The "real" negotiations occurred in the last week and a half of July. Neither side, however, seemed to have a sure grasp of how to conduct "give-back" bargaining. The industry refused to consider any plan but its own for arriving at its goal, and McBride—although convinced that the companies needed financial help—couldn't decide how much help the workers ought to give, or how much the BSIC would approve. He received little aid from staff members on his advisory committee, because essentially they opposed concessions. In normal bargaining, union staffers usually are bursting with ideas for getting more. In talks involving getting less, a scarcity of ideas prevailed.

The advisory committee, or "think tank," as McBride called it, met after each bargaining session. Its members included Kleiman, Jim Smith, and Ed Ayoub; Sam Camens, a special assistant to McBride and the USW's expert on "noneconomic" issues and worker participation; and Thomas Duzak, the chief negotiator on benefits, and one of his assistants. Carl Frankel, an associate general counsel in charge of litigation, usually attended these meetings, while James English, an associate general counsel in charge of internal legal matters, sometimes did. Of the USW's other four principal officers, only Odorich attended the think-tank sessions. Even Secretary Lynn Williams, who was presumed to be McBride's choice to succeed him as president, was not invited.

Of all these people, only McBride and Ayoub sympathized with the wage-moderation approach. "Most of us," Carl Frankel said, "thought the steel industry was in a cyclical downturn. We thought it would come back. It always had before." Yet none of the staffers openly criticized McBride's negotiating policy. They realized that he, as president, had decided that the industry needed help. In the Steelworkers, loyalty—not intellectual stimulation—was the first duty that staffers and subordinate elected officials owed the president. So McBride's brain trust fell in behind him but without enthusiasm. "I think a lot of us, really, without even necessarily analyzing it, were dragging our feet—and I think Lloyd knew we were dragging our feet—because we weren't convinced that the case [for concessions] had been made," Kleiman said.

The advisers felt that the union should insist on *quid pro quos* as its price for economic concessions and had presented a list of such items to company negotiators. It included restrictions on contracting out, a moratorium on plant closings, a "neutrality clause" under which union organizers would have a free hand in the unlikely event a company opened a new plant, a change in the disciplinary procedure to prevent workers from being fired or suspended until they were proved guilty of a wrongful act, and a provision requiring that companies invest labor-cost savings in plant modernization. Achieving some of these improvements, the staffers believed, would help the union "put a good face on" what essentially would be a concession agreement.

McBride, however, ignored the *quid pro quos*, making no attempt to trade them for economic concessions. Kleiman summarized McBride's approach this way: "Lloyd certainly didn't attack the problem of leveraging relief against the industry as strongly as I would have. I would have said, 'Look, we are going to talk about your problems, but only if you talk about our problems and only when it's clear that you're going to deal with those significantly.' That did not occur. I think Lloyd believed those issues were important. But I think he didn't believe they could be accomplished." In Johnston's and Moore's view, these issues faded to insignificance when compared with the companies' financial problems.

The Failure of Round One

The last four days of July were dry but not too hot. Anxious to forget their economic troubles, Pittsburghers invaded Point State Park in huge numbers—more than twenty-five thousand on Saturday, July 31—to attend the annual Three Rivers Regatta. The events included most kinds of aquatic and aerial competition known to man, including races by power boats, sternwheelers, canoes, tourist clippers, rafts, swimmers, and hot-air balloons. Whistles, screeches, sirens, and the angry buzzing of speed boats filled the air. Sky divers floated down from feathery skies, water skiers zipped up and down the rivers, and the Civic Light Opera gave evening concerts.

In the U.S. Steel conference room, the negotiations "just sort of frittered out," as Kleiman put it. McBride and Johnston had done most of the talking. Kleiman and Moore took notes and occasionally commented. Odorcich, sitting at the end of the table, listened and doodled. Johnston had argued, pleaded, and cajoled. He invented a parable to represent the steel industry's competitive problem. Here was the American steelmaker pushing a wheelbarrow up a hill (he mimed a man staggering behind a loaded wheelbarrow) with 2,200 pounds of rocks ($22 per hour), while the Japanese wheelbarrow contained only 1,100 pounds and the Korean, 200 pounds. "The Japanese and Koreans are already ahead of us, but every three months we have to stop and pile on another COLA." He groaned under the load. The union men smiled. Johnston could be diverting.

Diversions aside, the industry negotiators proposed cancelling all COLA payments in the first year of a three-year contract. In each of the last two years, COLA would be limited to 50¢ per hour regardless of the rise in consumer prices. How much this would reduce future COLA payments depended, of course, on the inflation rate. If the CPI rose 8 percent each year, as the two sides assumed for calculation purposes, unlimited COLA adjustments from 1982 to 1985 would total $2.50 per hour, compared with $1 under the new industry plan. The USW spurned the idea of capping COLA in the last two years and offered only to cancel the August 1, 1982, payment and defer the next three quarterly hikes for eighteen months each. It would freeze wages for the duration of the contract and reduce Sunday premium pay from time-and-a-half to time-and-a-quarter.

The union plan would give the companies substantial cash savings in the first year of the pact. The union bargainers defended this form of wage deceleration, as they had justified wage acceleration for thirty years, by citing what the UAW had given up in the auto industry—the deal that Johnston had described as a "mouse-fart settlement." The USW went further than the UAW by giving up three 3 percent wage boosts, instead of two. Kleiman, in particular, insisted on the auto-steel linkage, contending that the Steelworkers should not be expected to sacrifice much more than the Auto Workers. Johnston and Moore argued that the two industries faced different situations. GM and Ford, they said, acted unwisely in granting a 17 percent settlement in a recession year. "Whatever they elect to do is up to them," Moore said, "but we shouldn't add to our problem by following them blindly down the path." The USW men dismissed this argument.

On Wednesday, July 28, Johnston produced the industry's final offer. He asked the union to study it carefully, and the two sides recessed until the next afternoon. The industry had backed away from some demands, but the proposal still contained elimination of the extended vacation plan, which would have saved companies an average of 48¢ per hour, and the $1 cap on COLA. After diversion of 50¢ to SUB, the gap between the industry's demanded cuts and the union's offer was about 65¢ per hour, $1 versus 35¢.[2]

Although inflation had diminished, McBride and his advisers worried about what would happen to steelworkers' wages if prices soared by 10 to 13 percent as they had in the late 1970s. They also remembered that the USW had lost an earlier COLA clause in settling the 1959 strike. This surrender had caused political problems in the union when inflation picked up in the late 1960s. Not until 1971 did the union win a new COLA provision. To lose it again, or limit the yield, would represent, in the perception of many unionists, a humiliating defeat.

A few days earlier, McBride had suggested a possible compromise on COLA. Using an idea introduced by Ayoub, McBride told Johnston that additional COLA increases might be postponed if capacity utilization remained below a certain level. But Johnston and Moore showed little interest and declined to explore the idea. "We could not get them off of their proposal," McBride later told me. "They said the only way they could survive was to have their concept put in place. If it wasn't, they would have to take a strike next year [when the contract expired]."

McBride decided to reject Johnston's "final" proposal. On Thursday, July 29, as he and Kleiman walked to the U.S. Steel Building, they tried to surmise the industry's reaction. Kleiman, who many times had negotiated down to the wire with steel bargainers, said he thought the industry would pull back from its position and either accept the union's last offer or continue bargaining. There was time to put together an agreement before McBride met on Friday with the local presidents. "No," McBride said. "Johnston meant what he said." He would even bet a dollar or two that he was right. Kleiman accepted the bet.

The meeting was brief and to the point. After a short recapitulation of the last proposals, the two parties caucused to review their positions. When they resumed the meeting, McBride said, "We can't recommend your proposal to our people. In fact, we'll have to recommend against it."

Johnston shrugged. "We've gone as far as we're going to go."

He said he would inform the chief executives and call McBride that evening if the industry changed its position. The two teams shook hands and parted amicably enough. There were no recriminations or venting of emotions. They had all been through this before. It was only 5 P.M. The union leaders went to the Pittsburgh Hilton, where the executive board was waiting, and reported that the talks had failed. The board voted to recommend that the BSIC reject the company proposal. As McBride had predicted, Johnston did not call that evening.

McBride, of course, won his bet with Kleiman. The USW president later told Johnston and Moore about the wager. Moore's interpretation of the bet is interesting. "For many reasons we had an unbroken history, sadly enough, of never sticking with a no. Bernie was so conditioned to us never walking away, that he was sure we wouldn't walk away this time." Four years later, Kleiman recalled the bet and admitted as much. "I thought they were going to come up higher than they came," he said.

There is a second ironic footnote to this story, and it raises the question of whether the bargaining ended prematurely, with potential compromises left unexplored.

Ed Ayoub had expressed sympathy for Johnston's moderation approach in the USW's think-tank meetings. In no sense could he have been labeled "procompany" (later in 1982, he would be the most outspoken opponent of actual wage cuts). However, he believed that productivity increases in the industry did not justify the compensation rise over the past decade. He had warned union leaders on several occasions about the problem and could not understand how the union negotiators failed to see the plight of the industry as he saw it. He did not find it surprising—in fact, thought it reasonable—that the industry should want to cut future COLA increases and wind up with terminal costs about the same as the beginning costs. Later, he recalled Kleiman's reaction after Johnston first rejected the union's proposal to defer COLAs. "Bernie was rather pale," Ayoub said. "The first comment he made was, 'Bruce is irrational.' What he meant was that Bruce had rejected an enormous amount of money that we offered him."

"Bruce is not irrational," Ayoub told Kleiman. "If you were approaching this from the point of view of the industry, knowing the prospects ahead and the terminal costs that were going to be put into this agreement, Bruce was saying that he would rather take his chances on a picket line. His position, from his perspective, was entirely rational."

Ayoub, however, agreed with Kleiman and other advisers that COLA should not be capped. Limiting cost-of-living yields at high inflation rates, when workers most needed protection against income erosion, was unwise. On the other hand, Ayoub felt, COLA shouldn't be paid if the companies were consistently unprofitable. This is why he suggested the COLA-capacity linkage. Under this plan, cost of living would not be paid until the industry operated at least at the break-even point. Ayoub also had worked out another form of compromise, the "free corridor" approach, in which the companies would be exempted from paying COLA when the inflation rate was low.

When the USW negotiators set off for their meeting with Johnston and Moore on July 29, Ayoub did not realize this would be the last bargaining session. He prepared sets of figures showing how the utilization and corridor concepts might work at various levels of volume and inflation. Attending the executive board meeting, he expected that McBride would brief the board and return to U.S. Steel for more bargaining. Instead, McBride declared that the negotiations had ended.

As McBride walked out of the room, Ayoub confronted him. "Lloyd, this is terrible," he said. "I thought you had another meeting coming up." He explained that he could recommend a COLA provision that might be acceptable to the industry. McBride said nothing but shrugged and walked on.

The next day, July 30, the Basic Steel Industry Conference approved McBride's rejection of the industry proposal with the peculiar elation reported in chapter 1.

Johnston reported the bargaining failure to the chief executive officers, and they unanimously supported the negotiating decision made by him and Moore. The CEOs, however, were not told about the union's compromise proposal based on linking COLA increases to capacity utilization. There was no rule that the negotiators had to tell the CEOs about every trial balloon lofted in negotiations, but this was a particularly important one.

When I interviewed McBride four days after negotiations came to a halt, he disclosed the COLA-capacity proposal and faulted Johnston and Moore for refusing to explore it. As a dutiful reporter, I tried to check this with industry sources. U.S. Steel declined to comment. I reached John Kirkwood on the phone and learned unexpectedly that he hadn't known about Ayoub's idea. Irritated that Johnston had not briefed him, he spoke with an edge on his voice. "This indicates to me there was a failure of coordinated bargaining," he said. "We should have known about this and had a chance to discuss it."

Why did Johnston and Moore ignore the COLA-capacity idea? In Moore's view, the proposal did not change the fundamental problem of COLA escalation but merely postponed payments until output rose. Under a version of the plan on the bargaining table, all postponed COLA would have to be paid at the end of the contract. This meant that the terminal hourly labor cost would be $4 to $5 higher than in July 1982. "The union said, 'We'll give you some relief until the customers start appearing at your door, but then we want to be right back up there where we were before,'" Moore said. "Once again, we'd get a cash-flow saving but wind up picking it up again. That's why we did not pursue it."

This explanation, however, glosses over a vital point. The very fact that the union had suggested a compromise on the COLA issue, rather than simply saying no to any limitations in the last two years, suggests that further bargaining should have occurred. Ayoub's concepts could have resulted in retention of the COLA principle without raising the yield much beyond the $1 limit demanded by the companies. If the USW had insisted on restoration at the end of the contract, Moore's reasoning would be justified. This may have been the likely outcome, but no one can know with certainty because the industry prematurely terminated the talks.

Ayoub remembered this episode with a sense of personal failure. He had argued for the moderation approach since the April 1982 board meeting, but no one in the union seemed to listen. Other staff members later recalled that Ayoub took this position but said the economist did not argue it strongly enough to make an impression. Ayoub agreed in part with this view. "What it amounts to is, I was not aggressive enough," Ayoub said three years later. "It haunts me in the sense that I don't feel that I sufficiently argued to persuade anybody." More likely, no one in the union *wanted* to be persuaded.

An analysis of Round One bargaining must take account of what actually happened to inflation in subsequent years. Instead of rising 8 percent annually, consumer prices went up 3 percent in 1983, 3.4 percent in 1984, and 3.5 percent in 1985.

Even under the USW's modestly restrictive plan, the total COLA payment from August 1, 1982, to July 31, 1985, would have been 83¢, well under the $1 maximum proposed by the companies. As sometimes happens in collective bargaining, the entire fight in Round One was largely mooted by subsequent events.[3]

The Propaganda War

In the immediate wake of the bargaining collapse, critics attacked both sides. The union, typically, was blamed for not heeding the industry's problems. However, executives of some steel companies outside the bargaining group found it difficult to understand why the industry negotiators turned down an offer that would have meant immediate cash savings. The USW had declared publicly that its final proposal would have saved the producers $2 billion over three years, compared with the $6 billion demanded by the industry. These numbers were widely reported in the press. "I just don't understand walking away from a savings of $2 billion," said a vice-president of a steel company not involved in the negotiations. As a me-too firm, this company would have received the same deal.[4]

Johnston claimed to have had a good reason for rejecting the union offer. "If we took that," he later said, "it would tell our employees that this was the total size of the problem. It wasn't. We'd have had to come back for more, and there would have been a massive problem in communications to get the employees to believe otherwise."

While Johnston's explanation makes a good point, it is not totally convincing. According to Johnston and Moore, the chief executive officers unanimously endorsed the negotiators' decision. But for Republic Steel, which was in precarious financial condition, the rejection of immediate cash-flow relief must have verged on the absurd. It is possible that another consideration guided the industry negotiators, at least in part. U.S. Steel could better withstand the lack of immediate financial relief than most other firms in the bargaining group. Steelmaking capacity had to be cut, and the corporation undoubtedly preferred that other companies do the cutting. Since U.S. Steel controlled the bargaining, the decision to turn down the union offer could have reflected this reality.

In any case, the union's $2 billion figure vastly overstated the case. To put the union's offer in the best light, Ayoub and Jim Smith used a methodology which involved the entire steel industry and which counted the suspension of 3 percent wage increases in the future as a cash saving. The amount of actual savings at the eight companies would have been much less than $1 billion. Using statistics to put the best blush on a bargaining outcome is done all the time in American labor relations. Immediately after a settlement, the company normally plays down the size of the package in talking to Wall Street analysts, while the union inflates it in describing it to members. The truth lies somewhere in between. What was different in the steel case of 1982 was that the union inflated *how much it had offered to give up.*[5]

The industry also publicized some questionable figures and statements. Shortly after the July 30 BSIC meeting, Johnston issued a statement saying it was "regrettable that the Steelworkers chose to increase wages for a few at the expense of many [the unemployed]." He contended that the USW proposal would have increased hourly employment costs by $4.54. But this figure included the significant compounding impact that occurs when accrued COLA is rolled into the base wage once a year, a feature that was not part of the final union proposal.

A few days later, the eight companies sent letters to all USW-represented employees, criticizing the union for its unwillingness "to accept just a moderation on the size of future wage increases." U.S. Steel's letter, signed by Johnston, characterized the company proposal as involving "no wage cuts, no benefit cuts, and no 'concessions' currently received by any employee." The proposed elimination of the sabbatical leave plan was called by Johnston a "nonrenewal" of a benefit.

Contrary to the management statement, McBride wrote in a union letter that went to all members in the eight firms, he was prepared to offer "significant relief" to the companies. However, the producers "demanded concessions which they knew no decent union representative could accept."[6]

The propaganda war produced no victories. The fact was that both sides had committed mistakes of judgment and timing. The companies lacked credibility with the rank and file, as well as with local union officers, and did a poor job of explaining the industry's problems. Indeed, later evidence showed that the companies did not fully *understand* their own problems, especially that of a management fixated on the past and determined to avoid a meaningful sharing of power with the employees.

The union also had its mind and soul in the past. Although management must bear the final responsibility for communicating the company's financial situation to employees, the union can be faulted for a lack of imagination in translating that situation to its members in terms of the employees' interest. Workers must be able to see the *possibility* that something positive will result from taking a backward step. Single-minded in his attempt to help the industry, McBride failed to perceive the need to "get" as well as "give."

The final factor to be considered in the failure of what became known as Round One of 1982 bargaining was the impact of rank-and-file thinking. Some industry officials commented that McBride let himself be led by the "no-concessions" element in the union.[7]

A month and a half later, when I interviewed McBride at length about Round One, he acknowledged that a "movement" in the steel locals against concessions had had some effect. "These kinds of movements," he said, "are always very loud and very vocal, and yet it's hard to tell how much sentiment there is for them, because they speak out in advance of the debate and in advance of whatever logic would be used in making a decision. So people get hysterical, and they make up their minds and say, 'Don't confuse me with the facts. I've got my mind made up.'" McBride continued: "There was another current, not as loud, but it was there and very, very evident, that the industry's in trouble. 'If they're in trouble, we're in trouble, and let's see what we can do.'"[8]

McBride rejected the industry's proposal both because he himself disliked it and because he thought it would encounter difficulty in the BSIC. A different kind of leader might have insisted that the industry accept the cost relief that he *did* offer in order to keep the weaker companies viable. But McBride apparently believed that the management negotiators knew what they were doing. Inasmuch as he did not try to capitalize on Ayoub's ideas or demand that the industry continue bargaining, it seems likely that McBride simply decided that the time wasn't right to conclude a concession agreement. He had every intention of eventually giving the companies cost relief but wanted the local officers to feel more pressure from the rank and file as the economy worsened.

The "dissidents" in the BSIC didn't force McBride's decision in July—as many would claim—but they obviously influenced it. In a union as large as the Steelworkers, it would have been surprising if a considerable body of opinion against wage concessions did *not* exist. Such a movement sprang up in every union. However, a special virulence characterized relations between union leaders and dissenters in the USW. In Round Two, the dissidents would repudiate McBride and delay financial relief to the steelmakers for a few more months. Why did this sharp division exist? The politics of dissent in the USW grew out of special circumstances that reflected the history of the steel industry itself as a mass-production bureaucracy.

"Democracy" and Dissent in the USW

During the Steelworkers' first thirty years, it maintained the *form* but not the *reality* of democracy beyond the local union level. This paradox came about because of the unique origins of the Steel Workers Organizing Committee (SWOC) in 1936. When John L. Lewis set out to organize steel, he recognized it would be a difficult task in light of the industry's resistance to unionization since the 1890s and the fear of company reprisals in steel communities (see chapter 7). The steel firms possessed great social and political power. The CIO, therefore, had to create an equally massive organization, a highly centralized vehicle designed to organize large numbers of workers in the shortest possible time.

SWOC was not a union in the ordinary sense, and yet it functioned as one for six years. It did not have a constitution spelling out the rights and duties of officers and members and requiring elections. Lewis planned it, financed it, and staffed it as a CIO organization and appointed Philip Murray to lead it. Murray, in turn, appointed his top aides, including Dave McDonald, who became secretary-treasurer, and regional organizing directors. When Murray called a founding convention of the USW in 1942, he and the other leaders were already in place. Although elected in a *pro forma* vote, the first officers of the union actually had been appointed from above rather than elected from below.

The new USW constitution contained a provision for rank-and-file elections of all officers. But Murray had learned form Lewis how to use his political leverage as president to beat down political enemies and naturally felt that adoption of direct election by referendum would carry no risk for him. (In all other major unions except the Mine Workers, elections of top union officers are largely carried out in conventions through elected delegates.)

The steel union, therefore, had a long period of political stability, from the creation of SWOC in 1936 until Murray's death in 1952. His extraordinary power also enabled him to crack down on wildcat strikes and other forms of insurgency. A "responsible" union would gain easier acceptance by employers. In contrast, the Auto Workers, Rubber Workers, and other industrial unions, organized under the leadership of rank-and-file activists, seethed with political fights and factionalism from their earliest days. Rank-and-file activism and leadership was very strong in those unions, but worker activists in steel lacked an "institutional base," as David Brody has put it. This point is important in understanding the deep divisions between USW leaders and local activists that later developed and adversely influenced the union's ability to deal with problems in the steel industry.

In steel, according to Brody's analysis, the old Amalgamated Association of Iron, Steel & Tin Workers was still a chartered AFL national union in the 1930s. Although Lewis had broken from the AFL, he was enough of an old-line unionist not to flout the traditional rule against "dual unionism." Therefore, SWOC simply assigned everybody it organized into lodges of the Amalgamated, which later were converted into USW locals. By and large, SWOC organized the steel industry from the outside, in a top-down fashion. The Auto Workers, Rubber Workers, and Electrical Workers organized in large part from the bottom up. While this argument gives short shrift to many rank-and-file organizing efforts inside the steel plants, it holds as an overall generalization about steel.[9]

Fortunately for the union and its members, Murray was a good man. His special quality, wrote Murray Kempton, was "to touch the love and not the fears of men." Murray demanded absolute loyalty, however, to build a disciplined organization that was responsive to his will. Although the USW had a democratic constitution, Murray's personality and the top-down organization of the USW stifled the democratic impulse—at least in the top tiers of the union. Not until 1955, thirteen years after the USW's founding, was a top officer even opposed in a referendum, and only one challenger up to that time had managed to unseat a district director. Of course, top officers in most other unions, including the UAW, have *never* been opposed in a referendum.[10]

The political atmosphere had begun to change in 1952, however, when Murray died and McDonald took command. Unlike all other USW officers and board members, he hadn't worked in a mine or mill (except for a brief stint as a clerk) or risen from the ranks. McDonald started his career as a personal secretary to Murray and followed the latter into SWOC. Nevertheless, he schemed for years to succeed Murray, and was thoroughly disliked by many executive board members. When Murray died, McDonald by default became the highest officer, and the board members couldn't bring themselves to rebel against the union's top man. They named him president in an interim election. Thus, McDonald became the leader of a major union without ever having received a vote of a single union member in a local, district, or national election.

McDonald was a pompous, self-regarding man who had little interest in steel mills and the people who worked in them. He had one partially redeeming quality: He allowed his best staffers considerable latitude for innovation. But their contributions lay mainly in the field of collective bargaining, not union politics. And McDonald was not a Phil Murray who commanded the respect and awe of rank and filers.

The first explosion occurred at the USW convention of 1956 over proposals for a raise in officers' salaries and a boost in dues from $3 to $5 per month. The union administration hadn't done enough to prepare the membership for the necessity of increasing dues, and the delegates boiled over with anger. The motion passed, but within weeks rank and filers formed the Dues Protest Committee (DPC). They signed up thousands of steelworkers and decided to run candidates against top officers in the USW's 1957 referendum. "What they had going was a rank-and-file movement, a concept that for many of the top leadership bordered on obscenity and treason," writes John Herling in his pioneering book on USW politics. The insurgency was particularly strong in the Monongahela Valley. Donald C. Rarick, a grievance committeeman at the Irvin Works, was named chairman. Other DPC leaders were Frank O'Brien of J&L's North Side Works and Nicholas Mamula of Aliquippa.

Rarick was the first USW rank and filer who qualified in a nominations procedure to be on the ballot for president. After a wild campaign in 1957, McDonald defeated Rarick by only 404, 172 votes to 223,516, according to USW tellers (official vote counters). Almost certainly, this count overstated McDonald's margin, for ballot cheating had been widespread. Years later, one staff representative told me how he simply switched the results at a local in the Mon Valley to make McDonald come out on top. Charles H. Allard, the *Pittsburgh Post-Gazette's* crack labor reporter at the time, always believed that Rarick had won, as did many USW staffers.

At the next convention, in 1958, McDonald and other officers heaped vilifications on Rarick, Mamula, and O'Brien, accusing them of "dual unionism" and treating them as if they were foreign agents. Indeed, the administration tried to have them expelled from the union, a tactic that failed when the trio's own locals tried and acquitted them. At the 1960 convention, however, pro-McDonald unionists gave Rarick a severe beating, and another Dues Protester, Anthony Tomko, president of McKeesport Local 1408, was pummeled by four USW ushers because he dared to pass out anti-McDonald literature.[11]

When I came to know Rarick in the 1960s, I found it hard to understand why the USW leadership considered him such a threat. Rarick's physical bulk—he was six-foot-four with the girth of a small home furnace—impressed me far more than his mental capacity. He made confused, rambling speeches at conventions, advanced few, if any, ideas, and could not properly run a local union, as he demonstrated after becoming president of the Irvin Works local, let alone an international union. Politically, Rarick was a conservative. He personified the discontent that began to emerge in the labor movement in the 1950s: He disliked paying dues, disliked unionism with a "social conscience," and disliked being told what to do. These dislikes, combined with ambition to be something other than an ordinary steelworker, gave Rarick his sole qualification for leadership, a willingness to stand up to intimidation.[12]

Instead of demolishing Rarick in debate, the union officers heaped political and physical abuse on him and, according to Rarick, tried to buy him off. Apparently, they looked upon Rarick and his followers in the same way that Louis XVI and his court must have regarded the hungry populace of Paris in 1789. The rumblings of discontent were getting louder by the hour, and the leaders could think of nothing more imaginative than holding the people down. Although the Dues Protest movement peaked in 1957 and thereafter declined, it had long-lasting consequences. The officers and executive board displayed an intolerance of dissent that was not forgotten by militants who had sincere disagreements with International policy. The officers' attitude helped foster a strong tradition of rebellion among these activists, some of whom came to enjoy the practice of dissent more than constructive action.

From the late 1950s on, dissidence thrived in the USW. The Landrum-Griffin Act, passed in 1959, protected union rebels from arbitrary actions by leaders. The law also authorized the Labor Department to investigate charges of fraudulent voting practices and seek court-ordered reruns of dishonest elections. Although this appeals mechanism had faults, its very existence encouraged members to run for regional and national offices in their unions. This was especially true in the USW with its rank-and-file elections, which provide a forum for dissidents to express their ideas. In other major unions, except the Mine Workers, dissidents have much less exposure to the great mass of ordinary workers and almost no chance of being elected in a convention largely controlled by union leaders.[13]

The rank-and-file balloting procedure is, of course, a purer form of democracy that theoretically opens the way for rank and filers to rise politically. Not unnaturally, it also encouraged political opportunists, as well as principled unionists, to seek office in the USW. For ambitious steelworkers, the political process offered a route out of the daily job in the mill, first into a local union office. This might serve as a base to run for district director, or to win appointment to the district staff and advance from there. With twenty-five to thirty district directorships up for election every four years, the USW's political life took on aspects of a continuous election campaign. From 1965 to the mid-1980s, it became more common than not for directors to face opposition, and several incumbents were defeated despite the support of the USW's "official family" (see chapter 3).

USW dissidents ranged in a political spectrum from far left to far right. The largest number of antiestablishment unionists were nonideological in the sense of not following a particular political line. This group included perennial rebels who opposed the USW leadership for the sake of opposition and who often formed alliances. It also included independents who based their challenges on specific issues, forsook coalitions, and sought union office for reasons of principle and ambition.

Since its early days, the USW also contained a scattering of ideological rump groups, small cadres of activists who pursued a political agenda. The leftists, as usual, were split into various sects ranging from democratic socialists, to communists, to radical Marxists. The core membership of these groups was tiny, but their influence was widely felt in District 31, the Chicago-Gary area. These dissidents were the spiritual descendants of the International Workers of the World (Wobblies) and other radical labor organizations that for decades made Chicago their base. In the 1980s, leftist militants served on grievance committees in a number of Chicago-area USW locals and even controlled a few locals.

The broad "progressive" movement, as it was called by its adherents, reached its zenith of power and influence in the 1970s when Ed Sadlowski became director of District 31 following a long fight in which the International's favored candidate stole the first election. Elected in a government-conducted rerun, Sadlowski (who had socialist leanings but was not an idealogue) became the spokesman for all leftists and radicals, as well as many independents. Sadlowski had the ability to move people. Had he reined in his impatience—and that of his followers—and gathered more experience as a leader, Sadlowski may have made a good president. He was several cuts above many other dissidents who had run for top office. Sadlowski's abilities, along with the liberal and leftist groups that bankrolled his campaign, alarmed the USW administration, which set its powerful political machine to work and got McBride elected over Sadlowski in 1977. James Balanoff, a longtime radical, succeeded Sadlowski as director of District 31. But at the end of his four-year term, the International supported the present director, Jack Parton, and regained control of the district. (Sadlowski stayed in the union as a staff representative and became a subdistrict director after Lynn Williams was elected president.)

There was also a conservative strain of thinking in the USW, embodied in periodic demands for a kind of isolationist policy. In this view, the union should focus inward on shop-floor problems and bread-and-butter issues; it should also sever ties with international labor organizations and stop lobbying on broad social issues such as income redistribution, help for the poor, and civil rights. This kind of thinking cropped up only rarely, but it surfaced in the 1969 presidential election, when conservative rank-and-file groups supported Emil Narick in his failed bid to unseat I. W. Abel.

At times, the agendas of leftists, rightists, and in-betweeners were mashed into an undifferentiated porridge of against-ism. This was manifested particularly at conventions, when rebels of many persuasions would band together to try to defeat the union administration on one or more issues. The issue varied from convention to convention. Proposals for dues increases, salary boosts for officers, and district mergers always produced fierce debates, as did agitation for the right to strike (and an end to the ENA) and rank-and-file ratification of steel contracts. For some years, a demand by black members for an earmarked seat on the executive board was a biennial issue. In this case, however, conservative dissidents joined the majority in opposition, while progressives sided with the blacks.[14]

Political maneuvering on these issues produced interesting conventions. The dissidents spent the evenings trying to organize a concerted floor attack on the administration position. Trudging from motel to motel, they sought out supporters and held strategy sessions in smoky rooms where the combination bureau-TV stand was strewn with empty beer cans, fifths of Seagram's Seven, and platters of bread and cold cuts with the baloney turning up at the edges. At the same time, proadministration district directors would be feting friendly delegates at receptions or in hotel suites equipped with bars that offered Jack Daniels, Canadian Club, and mixed nuts. Here, staff representatives had the duty of lining up backers to speak on the issue in question, or merely to queue up at the floor microphones to prevent the "antis" from monopolizing the debate.

On the convention floor, voice votes were not good barometers of convention feeling, because the dissidents usually shouted louder than the loyalists. Standing votes provided better evidence of the true sentiment, although close votes among three thousand or more standing delegates could be difficult to judge. Not infrequently, the officer chairing the session employed imaginative estimates of the size of the aye vote to avoid an unwanted outcome—and ordered the delegates to sit down before they could be counted individually. The dissidents almost invariably lost these votes but immediately set up a cry of "Roll call! Roll call!" But a roll-call vote required approval of 30 percent of the convention, and the dissidents never garnered that many backers for a procedure that would take many hours. When they appeared to come close to 30 percent; the gavel would bang down to enforce one of those imaginative estimates.

Rarely did the dissidents manage to win a fight at a convention. In the Steelworkers, as in most unions, it was practically impossible to outmaneuver a union administration that controlled the parliamentary rule-setting procedure, as well as appointments to convention committees. Not least important, the officers and the staff could turn the floor mikes on and off from the speaker's platform and thus shut off a delegate who protested too much. In addition, more than 10 percent of the weighted vote of the convention likely was held by loyal staff representatives who could be elected as delegates because they themselves belonged to Steelworker locals.[15]

At their best, the dissidents forced debate on issues that needed debating, a not insignificant achievement, and likely prevented the establishment from becoming more autocratic. Although the administration won most, if not all, convention votes, the officers on occasion bowed to the depth of opposition on a particular issue and amended a resolution before bringing it to a vote. At their worst, the dissidents merely attempted to embarrass USW leaders for political reasons, or they fomented divisive battles over issues that did not really merit so much smoke and fire; for example, proposed dues increases that were clearly warranted by the circumstances.

This discussion may leave the impression that the USW was a totally dictatorial and undemocratic union. This is not true. In collective bargaining, the union administrations were highly sensitive to rank-and-file needs and demands. Even in Murray's day, the union changed priorities to conform to clearly expressed sentiment from below. When pressed to the wall by the industry, the union did negotiate some unpopular agreements. Generally, however, it performed its major function in society by getting what the members wanted and needed.

There existed a certain ethos of leadership in the USW, however, that enabled the top officers and directors to consider themselves a privileged class once they graduated to the board level. This was made all the more possible by the union's unique political history and the structure of the industry. Gradually, an influx of new leaders and new ideas—

along with the crumbling of the monolithic steel industry—brought change to the USW. In 1986, after nearly two decades of intraunion fighting over the issue, the Steelworkers suddenly adopted rank-and-file ratification of steel contracts with practically no debate. Two forces converged to produce this change. One was the deterioration of the steel industry and the companies' decision to dissolve their industrywide bargaining group.

The second factor was the accession of Lynn Williams to the USW presidency in 1984. He had come out of the socialist-oriented Canadian labor movement and believed, unlike his presidential predecessors, that union democracy meant giving responsibility to union members. It had taken a long time for the USW to learn that lesson.

The Dissidents of 1982

In 1982 the old enmity between the USW administration and the dissidents still existed, though with many new faces. It seemed to be an inherited condition, passed from one generation to another both in the official family and the family of dissidents.

This was particularly true of the militant leftists in District 31. Sadlowski had bowed out as a spokesman and leader of the progressives, concentrating on his staff job. But a number of progressives served as local presidents and followed a hardline, anticoncession policy. They included Bill Andrews of Inland Local 1010, Alice Peurala of U.S. Steel's South Works (Chicago) Local 65, and David J. Sullivan of Local 6787, at Bethlehem's Burns Harbor (Indiana) Works—all influential locals with many thousands of members.

In the Monongahela Valley, Ron Weisen, president of the Homestead local, received the most press. He didn't seem to be a programmatic radical, but the phrases "sell-out artist" and "strike the bastards" always sprang to Weisen's lips when he described union leaders and union-management relations. First elected president of Local 1397 in 1979, Weisen attracted attention by attacking the International and U.S. Steel at every opportunity. A group of young, very militant unionists gathered about him as local officers and grievance representatives. This "1397 Rank and File Team" gave both the company and the International a hard time on grievances and other matters.

A onetime amateur boxer and a welder in the mill, Weisen had a fighting Irish face (he was part Irish, part German) with a bulldoggish intensity. He was a street-corner orator with a voice that rang with rebelliousness, like someone crying out in anger from the inside of a barrel. If Weisen had a philosophy, it was that the USW should tear a page from the 1892 Homestead strike and treat mill bosses like Pinkerton goons. Only then would the Steelworkers be a *real* union. In fact, some of the top USW officers and staff *had* become remote from rank-and-file concerns, and certainly U.S. Steel's management style encouraged supervisors to act like little dictators. There is an old saying in labor relations that union members will elect officers who are the mirror image of the managers they must deal with.

The Homestead local twice reelected Weisen, but he lost a 1982 bid for election as director of District 15, indicating that many steelworkers disliked his politics of protest. To try to force U.S. Steel and the banks to invest in the Mon Valley, Weisen and a few other steelworkers—along with a group of militant ministers—stuffed dead fish in Mellon Bank deposit vaults, interrupted Easter Sunday services at a Shadyside church in Pittsburgh, and burned a cross at the home of David Roderick. These tactics alienated people who shared Weisen's anger.

"Ronnie's idea," said Barney Oursler, the militant leader of the Mon Valley Unemployed Committee, "is to get the guy who's doing this to him, dare him out of the bar and beat his ass off. Then he feels better about it. That's an understandable worker attitude, but it doesn't accomplish much." Mike Stout, the grievance chairman of Local 1397 and once a member of the Rank and File Team, came to dislike Weisen's negativism. "If you're going to call someone a sell-out artist," Stout told me in 1986, "you'd better have an alternative, or you'll merely feed cynicism and apathy. Weisen never had an alternative."[16]

Another well-known dissident was Mike Bonn, president of the Irvin Works local and a professional opponent much like Weisen. A shaggy bear of a man, Bonn had a manner about him of crashing through dense thickets in pursuit of he was not quite sure what. Unlike Weisen, Bonn was not totally negative, however. He cooperated with U.S. Steel management in establishing Labor Management Participation Teams at the Irvin Works in 1981 and enthusiastically supported the participation process until things went wrong at his plant (see chapter 16).

Many of the dissidents in 1982, however, were not at all like Weisen. Oursler, Stout, and Mike Bilcsik of Duquesne were examples of militants who, though often at odds with union policies, tried to pursue constructive programs. Some of these efforts, it is true, were not constructive by management standards, particularly the attempt by Stout and Bilcsik in 1985 to establish worker ownership in the Mon Valley by operating a blast furnace shut down by U.S. Steel. But there is no reason to believe that the company's definition of constructive was superior to all others.

The primary lesson of Round One, however, was that this troubled industry—its structural and financial problems ran deeper than even the protagonists realized in July 1982—could not amass enough trust on the part of its employees after a forty-five-year bargaining relationship to make modest changes in compensation. This argued that while the employees worked in the industry and depended on it for a livelihood, they lacked commitment to it in real terms.

The inability of the steel industry and the union to help each other stave off disaster in 1982 arose partly because the *forms* of democracy, both in the mills and in the union, had not been converted to a political *reality* which could comprehend economic reality. Long ago union and management leaders had shunned an approach that might have produced a vastly different result in 1982—and that still could improve the future relationship.

Chapter 11

Mike Bilcsik, Idealist

Mike Bilcsik wore a black brush moustache and had large, ruminative eyes under sooty brows, a combination that somehow made him appear both sinister and passionately idealistic. It was a sort of dark, anarchical look that I associated with those mythical "foreign labor agitators" that the mill town mayors of the Monongahela Valley condemned in absentia during the 1919 steel strike. Bilcsik *was* a Slovak but by no means sinister. He was also an idealist, but a reformer, rather than an agitator. Unfortunately, the one reform that he had wanted to introduce for most of his working life in the end eluded him.

In 1964, after serving in the army and graduating from an electronics school, Bilcsik took a job at U.S. Steel's Duquesne Works, where his grandfather and father had worked. Cocky and sure of his knowledge about electricity and electronics, he expected to be a foreman in six months. His first day in the plant set Bilcsik on a different course. He was shown about the blast furnace department by a veteran "motor inspector," an electrician who specialized in repairing electric motors. Bilcsik spotted a defective wire on the electrically charged third rail of a track used to transport ore to the furnaces. "We ought to fix that," he told his guide.

"No, sir!" the motor inspector said, moving on. "You wait 'til the boss tells you to do it. Then you do it."

It was a classic example of an artifical division of roles that was endemic in the American steel industry. Managers were order-givers and workers were order-takers, and an hourly employee who crossed the line would be slapped down. "After six months in the plant, I hated my job. No matter what input

261

I wanted to make, the foremen didn't want to hear it," Bilcsik said.

He served six years as a grievanceman in Duquesne Local 1256. Running for local president in 1982, he promised to install Labor-Management Participation Teams (LMPTs) in the Duquesne plant. "I thought that would be my contribution. I didn't see labor-management cooperation as something for the company. I saw it as something for the workers." He won the presidency and in early 1983 proposed that the two sides start the LMPT process. "The plant superintendent turned me down flatly," Bilcsik said. "He said he didn't trust me. I thought that's what LMPTs were all about, to build trust."

Bilcsik's relations with management worsened, and the union made its share of mistakes. Negotiations to save some operations failed, and U.S. Steel closed the plant in 1984 (see chapter 16).[1]

For nearly forty years, labor and management had been frozen in the artificial division of roles. And yet industrial unionism in the steel and auto industries started down a far different path in the early days, one that might have led to worker participation as a way of life in the United States. Almost lost in the mists of industrial history is a remarkable story about the origin, growth, and premature death of a very different kind of unionism than that which prevailed in the 1980s.

SWOC and Labor-Management Cooperation

The story starts at the birth of the Steelworkers. On June 21, 1936, Phil Murray addressed a mass rally on the north bank of the Youghiogheny River in McKeesport to open his steel organizing campaign. It was a "beautiful, sunny, Sunday afternoon." The spectators, probably numbering in the hundreds (a crowd estimate is not recorded), stood in a vacant field and flowed out onto Water Street, surrounding a coal delivery truck which served as the speaker's platform. The rally had been restricted to this area, next to the city dump, by George Lysle, the autocratic old mayor who seventeen years earlier had prohibited union gatherings during the 1919 strike.

In 1936, however, the Wagner Act protected union activity, and anger about unemployment and the Depression subordinated workers' fear of authority. George Powers, a rank-and-file organizer who attended the McKeesport rally, remembers that workers brought their wives and children. A group of 350 coal miners even marched to the rally through city streets from Versailles, a community adjacent to McKeesport. Murray gave a "lengthy" speech, appealing to workers to "arise and defend your rights as American citizens." Later, "hundreds" of steelworkers signed membership authorization cards, Powers says.[2]

Despite the apparent receptiveness of workers at that first rally, it took five long years for the Steel Workers Organizing Committee (SWOC) to organize most of the steel companies. To everyone's surprise, U.S. Steel capitulated first. On March 2, 1937, after several meetings with CIO President John L. Lewis, USS Chairman Myron C. Taylor agreed to recognize SWOC as the bargaining agent for U.S. Steel employees who joined the union. But shortly after the victory at Big Steel, SWOC's campaign bogged down. The Little Steel companies (Bethlehem, Republic, Youngstown Sheet & Tube, and Inland) fought organizers with violence, intimidation, and mass firings. A strike called by Murray collapsed in the summer of 1937, and Little Steel remained nonunion until 1941. Meanwhile, SWOC organizers had trouble collecting dues, and the union's treasury dwindled.[3]

Despite these setbacks, Murray and his staff organized hundreds of smaller steel and fabricating firms and negotiated contracts with them. With the help of favorable decisions from the courts and the National Labor Relations Board, SWOC and the other CIO unions began laying the foundation of a unionized industrial relations system in basic industries. The excitement of the times is captured in *Steel Labor*, SWOC's official newspaper, which played up union victories and turned defeats into occasions for attacking avaricious management. From the first issue in 1936, the paper's editorials took an uncompromising "them versus us" position on every issue.

A different tone began to creep into *Steel Labor*'s tabloid pages during the severe business slump of 1938. Phil Murray began talking like a labor diplomat. "We can continue lambasting each other," he was quoted as telling a thousand business leaders in a speech at the Waldorf-Astoria, in which case "human suffering and threatening bloodshed will be the order of the day. The other course we can take is a cooperative one." No cooperative endeavors resulted from this speech, but it indicated the temper of Murray's thinking which enabled SWOC at that early date to begin promoting a form of worker participation.[4]

Two other SWOC officials entered the picture at this point. They were Clinton S. Golden, SWOC's organizing director in the Northeast, who later became a USW vice-president, and Harold J. Ruttenberg, a young intellectual who headed SWOC's research department. Golden, a journeyman machinist and a former railroader, had served the labor movement since the 1910s as a union official, labor educator, and a regional NLRB director. He, too, was a "labor intellectual" but a self-educated one. Ruttenberg had studied economics at the University of Pittsburgh and made an early name for himself as an adviser to rank and filers who rebelled in 1934–35 against the regressive leaders of the old Amalgamated Association of Iron, Steel & Tin Workers. Joining the SWOC staff in 1936 at the age of twenty-two, Ruttenberg worked under Golden.

The two men advocated that the union, in addition to bargaining for a share of the profits, should help the company raise productivity to provide for rising wages. This meant involving the members in systematic efforts to improve efficiency and production methods. Ruttenberg, whom I interviewed in 1986, had learned from his early contact with rank and filers that "they were full of all kinds of ideas as to how you could do things better and more economically." Undoubtedly, he also fell under the influence of the much older Golden.

Once a follower of Eugene Debs, Golden believed in the all-embracing socialist concept of "industrial democracy": that workers should "participate in making the decisions that vitally affect them in their work and community life." In the 1920s, he came in contact with a small circle of work-reform advocates, including Elton Mayo of Harvard, and began to see practical means of implementing industrial democracy. Golden enthusiastically endorsed a program known as the B&O Plan, begun in 1923 by the Machinists Union and the B&O Railroad. In B&O repair shops, worker-supervisor committees were set up to discuss ways of improving operations, equipment utilization, and working conditions.[5]

The B&O Plan was the best known of several "union-management cooperation" programs started in the 1920s. I put the term in quotes because it came to have an unsavory reputation in militant labor circles. Actually, labor-management committees of this type had first been used during World War I with government encouragement, and with some success. After the war, many companies converted such committees into company unions. Aided by a postwar business slump, trade associations and corporations also mounted a broad assault against unionism through an open-shop movement called the American Plan. At the same time, large corporations shifted to the welfare capitalism concept of humane but paternalistic employee relations. They installed safety programs and introduced insurance, pension, and profit-sharing plans. As a result of these developments, organized labor was gradually squeezed out of the economy.

Before the 1920s, the AFL under President Samuel Gompers had professed indifference to problems of industrial efficiency. Now, worried about labor's decline, Gompers adopted a new policy. "Labor is not asking for a chance to get," he declared. "Labor is asking for a chance to *give*." A handful of unions, including the Machinists, Ladies' Garment Workers, and Clothing Workers, entered into cooperative agreements with management. Although designed to save union jobs in highly competitive industries, these programs differed substantially in form and content. Nevertheless, they were lumped together with company unions under the general rubric of "union-management cooperation." Since the term originated in the antiunion climate of the 1920s, it always has been associated with "class collaboration" and union weakness. "All in all," write the historians Selig Perlman and Philip Taft, speaking of the 1920s, "union-management cooperation was the program of a spiritually defeated unionism."[6]

The 1930s, in contrast, brought a resurgence of unionism. Although steel organizing had slowed down when SWOC began promoting cooperative programs in 1937–38, the union still was very much on the offensive. The offer to cooperate with management through worker participation became part of an overall strategy to organize workers and extend their collective voice to the highest levels of economic decision making.

The program began, however, in troubled circumstances.

One day in 1937, a committee of SWOC members from Empire Steel in Mansfield, Ohio, accompanied by the company president, appeared at Golden's office in Pittsburgh. Joseph N. Scanlon, the local union president, had previously met Golden and thought that he could counsel the Empire group. The company had just emerged from bankruptcy, was burdened with obsolete equipment, and couldn't pay a wage increase. Did Golden have any advice? Recalling the B&O Plan and his own experience as a machinist, Golden suggested that they tap the ingenuity of the workers. They should interview each employee for suggestions on improving production methods and cutting costs.

When the group returned to Mansfield, Scanlon devised a systematic method of eliciting ideas from the workers. The company put the ideas into practice and substantially reduced costs. Within months, Empire became solvent and granted a wage increase. Later, it inaugurated a productivity bonus designed by Scanlon. This was the groundwork for what Golden later described as "a new and more creative concept of union-management relations."[7]

Impressed with Scanlon's work, Golden brought him to Pittsburgh in 1938 and put him in charge of a new industrial engineering department. Scanlon had worked as a cost accountant before entering the mill during the Depression, and he became an expert on wage rates and incentive pay. Very soon, most of Scanlon's work for SWOC involved helping other locals and financially troubled companies install participation programs. SWOC, however, would not extend this aid, or grant requests for wage cuts, unless the company signed a union shop agreement, requiring that workers join the union after being hired. These agreements helped entrench SWOC in the metal-working industries at a time when the union shop was by no means prevalent.

In early 1938, Golden carried the program a step further. He had discussed SWOC's efforts with an acquaintance, Morris L. Cooke, a well-known industrial engineer who had been preaching the gospel of union-management cooperation since World War I. Cooke also was a disciple of Frederick W. Taylor, who founded "scientific management," a systematic method of extending management control over workers through stopwatch time study, detailed instructions for each task, and incentive wages (see chapter 12 for its application in steel). Cooke, however, was a "liberal" Taylorist, one who urged management to accept and work with unions.

Cooke assigned an assistant to write a pamphlet for SWOC, explaining how workers could help management increase production. Published in July 1938, the booklet *Production Prob-*

lems declares that "many workers in any large-sized establishment...could offhand, as a result of their daily observations, give the management hints as to how it could save money and put out a better or a less costly product." Though written fifty years ago, the pamphlet is astonishingly modern. Like participation agreements negotiated in the 1980s, it sets three conditions for union cooperation. (1) Management must share with the union any benefits obtained in the joint effort to reduce costs and improve quality. (2) No worker should lose his job as a result of improvements in practices or technology. (3) The union must be involved jointly with management at every step in the improvement process. SWOC initially printed four thousand copies of *Production Problems* and distributed them upon request to local unions, companies, trade associations, government agencies, and colleges.[8]

Eventually, SWOC—usually in the person of Joe Scanlon—helped install productivity plans based on participation in forty to fifty companies at the request either of management or the local union. Many, if not most, of the companies had competitive problems. Most were fabricating shops employing from a few hundred to several hundred people, but one large steel producer had four thousand employees. Aside from anonymous examples cited by Golden and Ruttenberg in articles and a book, little data on these cases is recorded. It is not known how long the companies retained the programs.

Steel Labor gave approving mention to *Production Problems*. Indeed, from 1939 to 1942, favorable stories on shop-floor cooperation appeared in at least eight separate issues. This is not to say that the newspaper was given over to writing puff pieces about cooperation. Its general coverage and tone still reflected the labor-management hostility of the day. But the appearance of occasional stories on participation demonstrated that Murray, unlike most other CIO leaders, at least tolerated union-management cooperation.

At this point in the story, the central issue becomes Philip Murray. Was his support an expedient measure or was it based on a deep belief in the values and goals of participation?

Murray's Ambivalence About Cooperation

Murray was part hardheaded Irish miner and part idealist. Born in Scotland in 1886 of Irish Catholic parents, he left school at the age of ten and went to work in the coal mines with his father. The elder Murray was a union official who apparently instructed young Philip in the tenets of unionism. The two emigrated to America in 1902 and took jobs in a mine at Madison in Westmoreland County, Pennsylvania. When Philip got into a fight with a crooked pit boss, he and his father were fired. Murray was driven from the company town. This experience, he later said, committed him to the union cause for life.

He found other mine jobs and settled in Horning, where he was elected president of a UMW local in 1905. Murray took correspondence courses and developed an intellect, though unlettered, of some power. Known as a soft-spoken young man who used conciliatory methods to persuade people, he rose rapidly in the union and in 1915 became president of District 5, western Pennsylvania. He supported Lewis for the UMW presidency in 1920 and, in return, was appointed vice-president. Murray served as Lewis's chief aide during the chaotic 1920s, when coal operators crushed the union.[9]

Lee Pressman, who served for eleven years as Murray's general counsel, told an interviewer that Murray's labor philosophy could be understood only in terms of his searing experiences in the coalfields. "The difference between death and living was a very narrow line in the coalfields, and the ordinary living and working conditions were deuced bad. Over and above that, if you were a union man, your ups and downs, in terms of strikes, whether you lived or whether you starved, was a very frequent issue for most of the people in the coal mines." This background asserted itself, Pressman suggested, in Murray's thinking about employers. Harold Ruttenberg, who became a close adviser to the USW president during the Second World War, added that Murray "could never bring himself personally to cooperate and collaborate with the capitalists and their 'factotems,' as he called them."

Still, Murray had grown sufficiently beyond this past to accept new ideas. His actions indicate that he vacillated between conducting all-out warfare and accepting the possibility of long-term labor-management cooperation. The latter was certainly the more prudent course in dealing with the troubled companies that asked for SWOC's help in the late 1930s. Murray endorsed the cooperative program for two reasons, Ruttenberg told me. "He didn't stop us, because it would facilitate getting a contract and getting more dues-paying members. So he supported us, but not only for that reason. Intellectually and morally, he was honest. He could overcome his [anti-capitalist] orientation in a decision to take a certain action." [10]

In 1939 Murray took his biggest step in support of cooperation. He agreed to be listed as co-author of a book with Morris Cooke, who, though liberal, could in a sense be characterized as a capitalist "factotem." The result of their collaboration, *Organized Labor and Production*, was published in 1940. Golden's biographer, Thomas R. Brooks, says that Golden "instigated" the effort. Golden probably argued that such a book would enable Murray to put a positive light on organized labor as a "responsible" institution. The nature of Murray's involvement in the project, however, demonstrates his ambivalent attitude about labor-management cooperation.

Golden and Cooke recruited more than a dozen writers, including academics, SWOC staffers, management consultants, and government technicians to churn out copy for the book, which Cooke assembled. Murray himself wrote nothing and, according to Golden, read little of what was written. Golden carefully concealed this fact at the time, but years later he candidly disclosed it. "Confidentially, and I *don't* want to be quoted," he wrote to a friend, "Phil Murray just don't [sic] read books to any extent. I doubt whether he ever read more than a few pages of *Organized Labor and Production*. . . . He left the preparation of his contribution to the contents of this book entirely to me. He would not even agree to have it publicized among the members of his own union."

Murray's disinterest put the union in an embarrassing position. Very late in the editing process, after the book had been set in type, Murray seems to have read enough of the manuscript to discover that, philosophically, it espoused principles that were anathema to him personally and organized labor generally. A significant portion of the book was given over to Cooke's proselytizing for scientific management and functional specialization of the kind that American companies did, in fact, embrace, and largely to their own detriment.

Repeating Murray's criticism to Cooke, two months before the book was published, Golden wrote that people in the labor movement "have come to associate his [Taylor's] name with about everything that is detestable in management's attitude toward workers. With this very prevalent feeling of hostility, for Mr. Murray to come out and appear to completely endorse both Taylor and his work in this field places him in a position which he could never explain to his constituents." It is unclear whether this criticism resulted in large changes. In a "dialogue" with Cooke which appears as the last chapter, Murray is presented as criticizing scientific management. But Cooke's argument for functional specialization remains deeply embedded in the book.

Organized Labor and Production did, however, make a good case for union representation. It urged " a wider opportunity for participation of union representatives in matters of plant and business management, as well as in industry economics." The book also called upon management to "invite labor's cooperation in the management processes" because "the goal of all industrial organization must be to make it possible for every worker to give the best that he has in him."

A chapter entitled "Tapping Labor's Brains," written by Cooke, set forth the argument for giving workers a voice in production matters. When early union contracts were being negotiated in the 1880s, Cooke wrote, "workers often expressed a desire to cooperate with management in 'improving the business.' These proffers were discouraged by management on the theory that 'head work' was the prerogative of white collar folk." Cooke obviously disagreed with this attitude, but his discussion contained few practical ideas for involving workers in decision making. The book said nothing about participation that had not been said before. One knowledgeable observer termed it "an excellent restatement of the philosophy of the twenties."

Despite Golden's comment that Murray would not let the book be publicized in the union, *Steel Labor* reviewed it and offered members a special price of $1.60. Golden, however, doubted that as many as two hundred copies were purchased by union officers and members. This remark and many others in correspondence with Cooke during the editing process suggests that the book produced some friction between Golden and Murray. This may have been when the two began to perceive that they looked quite differently at labor-management relations. Ruttenberg confirmed that the book was essentially Cooke's. "Murray bought the idea [union-management cooperation] as a long-term policy and accepted its logic, because Phil Murray had a touch of greatness in him," Ruttenberg said. "But while we got his name, we never got his heart."[11]

Golden and Ruttenberg next collaborated on a book, *The Dynamics of Industrial Democracy*, that turned out to be much the better of the two. Published in 1942, this book drew directly from SWOC's experiments in participation and presented concrete examples of how workers at several firms contributed their knowledge and ideas to help management cut costs, eliminate production bottlenecks, and set work standards. In every case, "efficiency" increased by at least 20 percent, the authors reported. They argued strongly (as the earlier book did not) that participation must be accompanied by the union shop. Golden and Ruttenberg set out thirty-seven "principles" for union-management relations, based on SWOC's experience in the steel and metal fabricating industries. Contending that collective bargaining represented an extension of democracy into industry, the authors emphasized participation and cooperation to an astonishing degree—perhaps unrealistically, given the labor-management hostility that still existed. In many ways, this book presaged the worker participation movement that finally emerged four decades later.

For example, Golden and Ruttenberg laid down a psychological underpinning for worker participation as well as any behavioral science theorist would in the 1960s and 1970s. Workers, they wrote, "crave to be recognized as human beings, to be treated with respect, to be given the opportunity to find satisfaction in their daily work through the free play of their inherent creativeness." This craving could not be fully satisfied by a representative system in which union officials bargain for economic gains, the authors said. Only the "creative participation of workers in making the vital decisions with management" could fulfill the psychological needs that helped create the demand for unions.

The Dynamics of Industrial Democracy remains an important book in the history of American labor. It became a "bible" for students of labor and a new class of industrial relations professionals—arbitrators, mediators, and negotiators—created by the expansion of collective bargaining. It also gave leftist militants a target to shoot at in their war against collaborationists. "It's a cinch that what makes 'sense' to the industrialists must, of necessity, be *non*-sense to the workers," commented a socialist newspaper.

Ten years after its publication, Golden said *Dynamics* "has had a remarkable sale but *NOT* among union members" and implied his disappointment that the USW had not encouraged a wide distribution of the book in the union. It is not recorded what Murray thought of the book, although sections urging the union shop, industrywide bargaining, and "national democratic planning" were fully in accord with his philosophy.[12]

In the area of national planning, Golden and Ruttenberg argued for a controversial Murray proposal to get labor involved in the management of war preparations. The USW leader introduced his idea at the CIO convention in November 1940 upon being elected to succeed Lewis as president. Splitting with most other labor leaders, Lewis had endorsed Wendell L. Willkie in the presidential election of that year and threatened to resign if Roosevelt won. Roosevelt beat Willkie handily with the help of a large labor vote, and Lewis had to make good on his threat.

The new CIO chief demanded an equal voice with management in the federal government's war preparedness program. He urged that production planning in basic industries be assigned to industry councils with labor and management members. A huge leap from shop-floor participation, this concept was more representative of Murray's thinking. Although never adopted, it led to a further expansion of shop-floor cooperation during the war and continued the steel union's strong interest in participation.

Labor's Failed Bid for a Wartime Voice

In times of national crisis, the federal government has unfailingly appealed for organized labor's help—to prosecute a war, to end a strike, to combat inflation, to spur economic growth by working harder—and in doing so it has validated the unions' self-proclaimed assertions of labor's importance in society. At such times, the hopes of union leaders that they will be given a voice in national affairs have swelled like their expanding chests. However, the actual influence gained by labor has been pitifully small in comparison to that granted industry leaders, especially during the two world wars.

From the beginning of war preparations in 1939 until the conflict ended in 1945, union officials played only subsidiary roles in administering the massive war production program carried out by the War Production Board (WPB). Clint Golden, who served on loan from the USW as vice-chairman of the WPB, occupied one of the higher posts given to labor people.

But he complained in 1943 that labor staff members of the board were "thoroughly frustrated" by their "inability to make any very real contribution to the war effort and program." Golden wrote that he could not understand why a labor person "should be any less capable of discharging a public trust than one with a management background.

The internal memorandums and correspondence of the WPB, contained in voluminous files in the National Archives, leave no doubt about the reason for labor's exclusion from policymaking roles. Industry executives, who controlled the nation's production facilities, quite simply refused to share power with labor leaders. They feared, perhaps with good reason, that the unions would attempt to convert wartime authority into postwar power plays against industry. Since government officials needed industry's cooperation, they sided with management.

But a deeper, darker reason underlay the decision to subordinate labor's role. Since the early days of industrialization, a peculiar notion had gained ascendancy in the United States: that wage workers and their representatives lacked the competence to handle complex issues and problems that required abstract knowledge and analytical ability. Only business executives and technical "experts" were qualified to perform society's "head work," as Cooke and Murray termed it.

This attitude—it is part of the business ethos in the United States—has been very destructive of cooperative programs. It represents a deep bias engendered by a too slavish adoration of entrepreneurial success. The nation has paid a heavy price for this bias in terms of lost opportunities to use the full resources of the work force. This became especially evident in the early days of World War II.

In December 1940, Murray presented to the White House Walter Reuther's famous proposal to build "500 planes a day" by converting auto plants to airplane production. Reuther, then a UAW vice-president, had enlisted his skilled-worker friends to survey the availability of machine tools in auto plants. They concluded that existing plans for conversion to defense production would use less than 10 percent of the equipment scattered over many plants. Reuther proposed setting up a board of equal numbers of labor, management, and government officials to supervise the pooling of machine tools and dies among manufacturers. Plane production lagged at the time, and Reuther's idea received much favorable comment. "There is only one thing wrong with the proposal," sarcastically remarked Treasury Secretary Henry Morganthau. "It comes from the wrong source." A government agency eventually rejected the proposal.

Nine months later, with the country reeling from the Japanese attack on Pearl Harbor, the Reuther Plan jumped back into the headlines. The Philippines had fallen, partly because of a lack of planes and arms. Newspaper columnists expressed anger at the initial rejection of the Reuther Plan. Bruce Catton, the information director of the Office of Production Management who later achieved fame as a Civil War historian, wrote a blistering memo to another OPM official. If the Reuther Plan only had resulted in a few days' additional output of planes, they would "have been a Godsend to Mac-Arthur and his boys, to say nothing of the lads on Guam, Wake and points East." But the plan had been "laughed out of court," he said, "because nobody wants to admit that the labor gang might be entitled to a little say on how the job is done."

WPB officials again rejected the Reuther Plan, although one of Reuther's central ideas—the pooling of facilities—was written into a new reconversion plan. *Business Week*, which a year earlier had dismissed the Reuther Plan as a "fancy scheme [that] has fallen flat," now commented: "Pooling of equipment will not be a problem. Corporate lines will be disregarded, and to this extent the basic concept of the Reuther Plan will be effectuated."[13]

That a union man would have the effrontery to suggest a management idea dominated the debate over the Reuther Plan. Its most radical element, the creation of a tripartite board that essentially would run the industry, received less comment. The Murray Plan, also proposed by Murray in December 1940, had the same feature. His industry councils, composed of equal numbers of labor and industry members and chaired by a government representative, would act as a "top scheduling clerk" in determining most production decisions for each basic industry. Seeing the interest accorded Reuther's specific conversion ideas, Murray in early 1941 followed up with a proposal to "speedup" steel production and publicized it in *Steel Labor*.

The industry council idea seems to have arisen from diverse sources and contained a blend of Catholic, syndicalist, and socialist concepts, as Sanford M. Jacoby and Ronald Schatz have shown. Murray must have known that such a proposal would only alarm industry executives. Why did he propose it? Ruttenberg, who worked closely with Murray during the war, said the USW president pressed the idea because "it was good PR. It was a positive thing, and it got attention." Moreover, the Murray Plan represented a message that the CIO leaders wanted to drive home to industrialists and government officials. "We were saying, 'We're big boys now. We're not just a pestiferous fly pecking away at you, that you brush off with your hand.'"[14]

Yet more than this was involved. Some of the CIO leaders, particularly Reuther and Golden, believed strongly in rank-and-file participation in the plants. Murray focused primarily on power relationships at a higher level. "Some labor leaders," observed a WPB scholar at the time, "even hoped that such sharing of responsibility might carry over into peacetime management-labor relations."

Whatever his motivation, Murray did not let up. He sent a "survey" of the steel industry (with his "speedup" suggestions) to the president in January 1941 and vigorously promoted industry councils in SWOC's newspaper. After the war with Japan started on December 7, 1941, labor renewed its bid to win an equal voice with industry in the newly created War Production Board. Its chairman, Donald M. Nelson, a business executive, showed some sympathy for labor's demand. After first meeting Reuther, he commented to an aide: "I can see why manufacturers dislike organized labor. Reuther is much more intelligent than the manufacturers." Nonetheless, Nelson probably could not afford to offend business by *giving* equal status to labor. The Murray Plan and the unions' bid for power in the war agencies were rejected.

On March 2, 1942, however, Nelson announced a production drive to accomplish Roosevelt's goal of producing sixty thousand planes and forty-five thousand tanks in 1942. He called upon companies and unions to form committees at the plant level to discuss ways of increasing output and promoting a patriotic spirit through rallies and so forth. A large outpouring of letters, telegrams, and phone calls welcomed Nelson's plan. Some companies, however, "looked upon it as an insidious device to give labor a share in management."

Murray probably fostered this impression. In a nationwide radio broadcast on March 6, he called upon CIO locals to cooperate with management in setting up such committees.

The committees, he added, "are directly in line with our industry council proposals." This claim significantly distorted the truth: There could be no comparison between voluntary plant-level committees and industrywide councils with statutory authority to direct the production of hundreds of plants. Murray's statement raised the spectre of the dreaded Murray Plan. An official of the National Association of Manufacturers characterized it as "the first step towards sovietizing American industry."

Some large corporations expressed open hostility to the labor-management committee (LMC) approach in public meetings conducted by War Production Drive staffers. When a local union officer in Detroit asked C. E. Wilson, GM president, whether union members on a plant committee would have an equal voice with supervisors, Wilson replied that there would be "none of this equal voice bunk" at GM. As late as 1943, halfway through the war, relatively few LMCs were functioning in the Detroit area because of "the General Motors veiled hostility to the joint committee program."[15]

About five thousand committees registered with War Production Drive headquarters during the war, although no more than about three thousand functioned at one time. Many LMCs either quickly foundered or lay dormant from the outset. Many others never went beyond conducting rallies, blood donor drives, and war bond sales. These activities probably boosted morale among workers. Hundreds of committees reported gains in output and quality. If these reports are believed, it seems probable that the LMCs made an important contribution to the war effort.

The committees, however, did not have a lasting effect on the industrial relations system. Only about five hundred LMCs, 15 percent of the active groups, discussed production-related issues and thus could qualify as dealing with matters that were "outside the traditional scope of employee-management relations," as Jacoby says. Some of the others solicited ideas from workers, usually through a suggestion box system, hardly an innovation in American industry. This probably worked better than normal only because patriotic feelings produced suggestions that otherwise would have been withheld.

Union leaders in some instances refused to cooperate, contending that the plan "was the old speedup system wrapped in the American flag." Overwhelmingly, however, it was management opposition that prevented LMCs from engaging in substantive discussions about improving productivity. This is evident in the reports of WPB staffers who visited plants to assess the LMC program.

The prevailing management theory of the time held that unions must not be allowed to intrude upon managerial "prerogatives" spelled out in the labor agreement, including the sole and exclusive right to manage the business, run the plants, and direct the work force. Companies insisted on this "management rights clause," fearing that the CIO unions intended a socialist assault on private ownership. While the "red takeover" threat was greatly exaggerated, militant rank and filers in many industries challenged management on shop-floor issues such as production standards and job assignment, eroding managerial authority in these areas. In some companies—mostly small firms—union and management struck a new balance of power which gave the union a voice but left management with the final decision-making authority.

Corporate giants like General Motors established the national management ethos regarding labor issues, however, and these companies almost unvaryingly opted for strict enforcement of management rights. Many of the larger firms simply refused to establish LMCs. Ford had only 2 committees in 22 plants; GM 18 committees in 115 plants; and U.S. Steel 74 in 201 plants.

In the steel industry, Murray and his staff continued to point out deficiencies in steel operations in a series of memos to WPB officials. Intensely patriotic, Murray insisted that labor should be heard on these issues. For the most part industry representatives managed to sidetrack his proposals. Nonetheless, and despite criticism from labor's left wing, Murray also scrupulously tried to observe a wartime no-strike pledge and cracked down on locals that led wildcats.

The USW also promoted the LMC concept throughout the war, and about five hundred committees were set up in the steel industry, at least on paper. One report, which admittedly focused on "successful accomplishments," cited numerous instances where LMCs were helpful in reducing absenteeism and lost-time accidents and raising productivity. In July 1943, however, WPB employees visited thirteen steel plants in the Pittsburgh area and found that only one had an effective committee. In October of that year, a WPB visitor to U.S. Steel plants in the same region faulted plant managers for "appointing management representatives [to the committees] with no real authority, refusing to allow meetings on company time, refusing in many instances to set up adequate subcommittee organization or well-planned suggestion systems with adequate cash rewards."

At the end of the war, there was talk of transferring the LMC program from War Production Drive headquarters to the Commerce or Labor Department and making it a permanent government function. But this did not happen. The committees disbanded rapidly and, with the exception of a few that lingered on, faded from sight and memory. The opportunity to experiment in a relatively new dimension of labor relations, to change attitudes, to build a body of technical knowledge about interactions between groups of people with differing interests—this marvelous opportunity was not exploited.[16]

Postwar Confrontations

The USW's commitment to worker participation and plant-level cooperation ended abruptly in 1946. An assortment of factors is responsible, including clashes of philosophy and personalities in the USW's top ranks, the termination of wartime restrictions on labor and management behavior, and a growing determination on the part of management generally to restrain and contain the union movement, legally, economically, and on the plant floor.

As the end of the wartime wage and price controls neared in late 1945, labor and management acted as if they approached a brink over the unknown. The unions had to prove themselves to satisfy members who had grown restive under tight wartime restrictions. There was great fear that a postwar recession would take the nation back to Depression-level unemployment and erode the higher living standards that workers had achieved during the war. Despite the wage freeze, full employment and plenty of overtime pay significantly raised workers' earnings during the war. In 1943 the government established a standard forty-eight hour week, with the last eight hours paid at time and a half, and steelworkers' earnings took a big jump. From 1940 to 1945, according to one calculation, average annual earnings in steel increased by 26 percent, adjusted for inflation and extra hours. However, the earnings bubble burst after V-E Day in 1945, when the companies went back to the forty-hour week. Earnings plunged an average of $52 per month.

Companies had also done well during the war. As one writer has pointed out, corporations gained liberal investment grants, a generous treatment of profits, expenses, and investments under tax laws, and new production facilities that were financed by the government. But they also anticipated a postwar business slump and wished to hold down wages unless the government decontrolled prices.

These attitudes set the stage for massive confrontations. When Phil Murray in September 1945 demanded a $2 a day wage increase at U.S. Steel, the corporation said no. Bargaining was futile, it declared, until the government allowed steelmakers to put through a $7 a ton price increase. In the meantime, GM was saying no to Reuther's "open the books" demand. "Unlike the union in the General Motors case," Murray said, pointedly referring to the UAW, "the steel union . . . has not coupled its demand for wage increases with a flat demand that present basic steel prices be maintained." While implying that price was no concern of the USW, Murray actually was inviting the government to allow steel price rises so the union could win higher wages. This formulation differed significantly from Reuther's attempt to win wage increases without the need for price hikes.[17]

The wage-price standoff also stymied labor negotiators in other industries in late 1945. Amidst this turmoil, President Harry Truman convened a national labor-management conference, hoping to bring forth an agreement on mechanisms to handle labor disputes in the absence of wage-price controls. Labor and management found common ground on some issues but not on the central questions. In an opening speech, Murray declared that "we are confronted with a major collapse in labor-management relations." As if to confirm his remarks, three hundred twenty thousand auto workers went on strike against General Motors on November 21, midway through the three-week conference.

Management representatives at the conference left no doubt that the "right to manage" would be a major issue in postwar bargaining. In a joint committee on that subject, management members noted that bargaining had expanded into "the field of management" and concluded that "the only possible end of such a philosophy would be joint management of enterprise." To prevent this from happening, management insisted that a long list of functions be excluded from bargaining and from the grievance procedure. The list included decisions on products to be manufactured, prices, and financial policies. Management also demanded the exclusive right to decide the organization of work, job content and assignment, plant layout, scheduling of operations, and plant closings—all matters of intimate concern to workers and unions. The only decisions that workers could file grievances on involved disciplinary actions and seniority applications.

The committee's three labor members, one of whom was Clint Golden, rejected this approach. They argued, in what would become a significant irony in American industrial history, that the sharp delineation of such areas *might well restrict the flexibility so necessary to efficient operation* (emphasis added). Forty years later, management would use this same argument in demanding a loosening of union work rules and a reduction in the number of job classifications specified in labor contracts. If the unions went too far in laying out a maze of "legal" boundaries that impeded flexibility in the workplace, it was management that led the way.

The labor-management conference failed to ameliorate the growing antagonism between the two sides. Meanwhile, the White House waffled on decontrolling prices, and U.S. Steel refused to negotiate on wages until the government allowed price relief. In January 1946, Murray applied the maximum possible pressure on the government by shutting down the entire steel industry. It was the only time the USW struck all fabricators along with the basic steel producers; some seven hundred fifty thousand USW members walked out at eleven hundred firms. The strike in basic steel ended on February 17, 1946, when the steel companies were authorized to raise prices $5 a ton. Strikes at other firms ended only when USW staffers went to Washington and lobbied for price increases for these companies (see chapter 4).[18]

In this corrosive climate, with management determined to curb union power, it would have been politically difficult for any union leader to push cooperative programs. The patriotic fervor of wartime had ended. Management's struggle to prevent the union from broadening the scope of bargaining threw up a solid barrier to collaboration at the plant level. Furthermore, the union's ability to shut down most of the industry (a few basic steel plants remained outside the USW's grasp) to win wage increases for hundreds of thousands of workers overshadowed what might be done to increase productivity in individual plants. Bread-and-butter issues had to take precedence, at least temporarily, over shop-floor democracy.

The background is now set for piecing together in more specific terms the story of the USW's dramatic change of direction in 1946. Telling the story now cannot change anything in the past. It can illuminate *one way* in which the steel industry, steel unionism—and, indeed, industrial unionism as a whole—took a fateful wrong turn forty years ago.

The USW Turns Away from Cooperation

In 1986 the only surviving person who had been deeply involved in the steel union's participation program from the beginning was Harold Ruttenberg. I interviewed him on a rainy April evening that year in his apartment near the University of Pittsburgh. At the age of seventy-two, he was tall, balding, and slightly stooped but still had the physical frame of the young man who had been an amateur heavyweight boxer. Although he had had open heart surgery in 1981 and a recent "repair job," Ruttenberg appeared quite healthy. He wore a sporty string tie on a soft gray shirt. We sat in a large, pleasant room whose walls were covered with oil paintings, sketches, and other art pieces. Ruttenberg had left the USW in 1946 to become a steel industry executive. He had done quite well for himself and now was chairman and principal owner of American Locker Company, headquartered in Pittsburgh.

Although solicitous and helpful in our interview, Ruttenberg also displayed some crustiness, reminding me that one writer had described him as "the brash young intellectual of the rank-and-file movement" in the mid-1930s. We talked about the rise and fall of the SWOC cooperative program. Later, I supplemented his story with material from other interviews and the papers of Clint Golden, Morris Cooke, and Phil Murray.

In 1942 Murray sent Ruttenberg to Washington to serve as assistant director of the WPB's steel division. Murray himself spent a large amount of time in his CIO office in Washington and came to depend on Ruttenberg's intelligence and analytical ability. Although Ruttenberg had joined the union as an aide to Golden, he now became Murray's "brain-truster." Toward the end of the war, Ruttenberg said, he had many discussions with Murray about formulating a bargaining policy for the postwar period. During the early SWOC experiments with shop-floor collaboration, "it became very apparent to me, and to Golden, that you have to have union-management cooperation," Ruttenberg said. "You had to tie the economic benefits that flow to workers to increased productivity. You had to develop practical measures of productivity in order to do it. When the war was over, I argued stronger than anyone else, 'Listen, we've got to continue this program, and if we don't we're going to bankrupt our companies.' I said that unless you followed a policy of trying to feed the cow, you're going to milk it to death. You can't just milk it. You've got to help feed it."

What he advocated, Ruttenberg said, was a long-term policy *direction*, not its immediate imposition during 1946 negotiations. "We'd have had to do it a step at a time. If you did it as a long-term policy over the course of ten years, you could accomplish a whole lot. We should have adopted a policy that we would not condone pegged production and stretchouts." By these terms he meant the shop-floor practice by informal groups of workers of setting unofficial production quotas to stretch out the amount of work to be done and thus, theoretically, keep employment levels high. "It was a false idea that you can create jobs by stretching it out," Ruttenberg said.

Murray was not unsympathetic to these arguments, Ruttenberg said. " 'It'll come, Harold, be patient,' " he would respond. In *Self-Developing America*, a little-noticed book published in 1960, Ruttenberg tells somewhat the same story. There, he acknowledges that Murray was correct in arguing that "management itself was dead set against it [collaboration]." According to Ruttenberg, Murray predicted that when management could " 'no longer carry the full load itself, they'll be around asking for the union's help on plant efficiency, but not one day sooner.' " [19]

When it became apparent to him that Murray would not continue the program, Ruttenberg resigned. "I said I'm not going to be part of it. You'll have the house of cards come down." Industrialist Cyrus Eaton offered him a job as vice-president of Portsmouth Steel, and Ruttenberg accepted.

Clint Golden resigned at the same time as Ruttenberg. Murray announced their departures on July 1, 1946. The timing was coincidental and decidedly not the result of joint planning, because their relationship had turned cool and distant.

Golden had developed an intense dislike for the younger man when Ruttenberg served with the WPB. It was a heady position for a person still in his twenties. "Harold's sense of self-importance grew tremendously in that period, and I found him to be a very intolerant, conceited and increasingly aggressive person," Golden later wrote. Ruttenberg acknowledged that when Murray "stole me away from Golden, it created friction. I didn't know how to manage the thing at the time."

This and other personality and political clashes on the USW staff had undercut the participation program. It appears that Dave McDonald, then the USW secretary-treasurer, and Lee Pressman both lobbied against Golden and his program, McDonald because he saw Golden as a potential rival to succeed Murray, and Pressman for ideological reasons. Thomas Brooks, who read through Golden's voluminous papers, quotes a diary entry of May 5, 1946: "It seems that most of the effort Joe Scanlon and I particularly have put forth to both advocate and to build cooperative relationships has been undone by our associates." Brooks concludes that the downgrading of the participation effort "precipitated" Golden's decision to resign.

Golden referred obliquely to his distaste for this political intrigue in a parting address to the union executive board. He "tempermentally" did not like politics, Golden said, adding: "I don't like politics in the government, and I am not interested in them in the union." His main reason for resigning as vice-president, Golden told the board, was that he wanted "to feel free of the compulsions that go along with executive responsibility."

There is no evidence of an outright split between Golden and Murray. The latter, in fact, gave a warm and obviously heartfelt appreciation of Golden's contributions at that final board meeting. For ten years, he said, "I don't think Clint Golden and I have had a cross word." Unlike Ruttenberg, Golden apparently never claimed that he left the union *primarily* because of Murray's bargaining philosophy. Nevertheless, his papers contain many hints that a tension existed between him and Murray on the subject.

In December 1946, for example, Golden attended a meeting of the Wage Policy Committee, which in those days determined USW bargaining policy. "I had a feeling," Golden later wrote, "that his [Murray's] remarks in part at least were directed to me when he said he 'was not among those who believed that any substantial part of management wanted to get along peacefully with the union' and wound up with a heated peroration which included, 'I do not want peace at any price!'" Golden added in an aggrieved tone: "As if anyone had suggested that he did."

The final member of SWOC's participation triumvirate, Joe Scanlon, resigned in the fall of 1946 to teach at the Massachusetts Institute of Technology. Scanlon was invited to join a unique program in MIT's Industrial Relations Section which enabled him to continue his practical experiments in worker participation. That opportunity no longer existed in the USW, according to a close friend, Meyer Bernstein. A hard-working idealist, Bernstein signed on with SWOC in 1936 after graduating from Cornell University and served on the staff for more than thirty-five years. Scanlon, he recalled in 1976, could make no headway on union-management cooperation after Golden left the union. "He had ideas, but he couldn't get support for them," Bernstein said. "Phil Murray didn't care too much for this kind of thing. Clint did. Phil Murray never trusted the employer. Clint had more respect for them as individuals. I think Clint went too far in one direction and Murray did in the other." Ben Fischer, who joined the USW in 1946, added that Golden's "view of cooperation was to serve as an agent of management. In any dispute between a local and a company, he took it for granted that the company was right."

The only question that remains is why Phil Murray, after tolerating the Golden-Ruttenberg-Scanlon program of shop-floor collaboration for years, allowed it to die. The salient point is that to achieve a voice for the union on the shop floor the union had to cooperate, to some degree, with management. Murray's lifelong distrust of employers did not permit him to pay this price. In the late 1940s, he continued to urge the establishment of industry councils. But these would be national planning entities, presumably created by statute, which would give the union equal status with management. Murray would enter willingly a partnership in which power was conferred and protected by law but not one in which the union could gain power only by trusting an employer.[20]

Is it true, as Ruttenberg contended, that Murray explicitly rejected the idea of embarking on a long-term strategy of helping the steel companies raise productivity, "to feed the cow"?

We shall never know for certain. No one who is still alive can corroborate Ruttenberg's account of his discussions with Murray. Everything considered, it is likely that they occurred generally as Ruttenberg remembered. Murray's bargaining leadership from 1946 to 1952 shows no evidence, with one minor exception noted below, that he took steps to inaugurate the kind of policy advocated by Ruttenberg.

It is much less likely that Ruttenberg left the union merely because Murray rejected his advice. In announcing Ruttenberg's resignation to the executive board, Murray quoted the young economist as saying to him, "Mr. Murray, if you say I shouldn't go, I won't go, because my heart is in my work at the Steelworkers." This does not sound like someone resigning in protest. Murray refused to make that decision for him, and Ruttenberg went. Ten years later, he began talking publicly about his disagreement with the USW's bargaining philosophy. In a 1956 interview with *U.S. News & World Report*, Ruttenberg declared that he and Golden had believed that "as the union grew older and matured, it would of necessity have to assume joint responsibility with management for the economic health and success of the company." He continued: "I left the union because they wouldn't buy that kind of program in 1946."

People who knew Ruttenberg are skeptical of that assertion. They remember that in his job at the WPB he had learned much about the inside workings of the steel industry and how money could be made. Fischer, who served under Ruttenberg in the research department, described Ruttenberg's claim as "sheer, unadulterated nonsense." He continued: "Harold left because he wanted to make a million. 'This is the time,' he told me, 'when a son-of-a-bitch who is ruthless can really make a mint. Why don't you join me?'"[21]

History does not lack examples of people who invented reasons for their behavior after the fact.

The Scanlon Plan: A Beginning

Joe Scanlon made important contributions to industrial relations after he joined the Industrial Relations Section at MIT. This group was on the cutting edge of work reforms in the late 1940s and early 1950s, and Scanlon's ideas won the support of faculty members who were, or would become, famous. The two best known were Douglas McGregor, the organizational development expert and management theorist who actually hired Scanlon, and George P. Shultz, a young labor economist who later served in the cabinets of Presidents Nixon and Reagan.

With their encouragement and aid, Scanlon continued to help companies and unions implement participation programs. For years he had experimented with various types of group incentive plans to find a way of correlating bonuses with improved company performance. Finally, he struck upon the idea of relating payroll cost to the sales value of production (sales revenue plus or minus real changes in inventories of finished goods). If a work group with a given payroll produced more than a preestablished norm, a bonus would be paid. Such a formula, unlike a general profit-sharing plan, was not affected by price changes or other external factors. Nor was it a straightforward piecework plan.

The Scanlon Plan, as it became known, encouraged employees to use their ingenuity to reduce costs and wastage, not just to work faster and harder. It enabled those in a specific work group to see a direct link between their efforts and the size of the bonus. Scanlon also insisted that the formula must be accompanied by *real* rank-and-file participation, not only in improving the work process but also in developing and administering the bonus plan.

A modest number of companies, usually small to medium in size, adopted the Scanlon Plan in the late 1940s and early 1950s. After his death in 1956, two of the people he worked with at MIT continued to promote the plan. Frederick G. Lesieur, a former USW local president at a firm that established the plan, formed a consultancy in San Francisco and installed some two hundred Scanlon Plans over a thirty-year period. He estimated in 1986 that "well over one thousand" Scanlon Plans existed in the mid-1980s, although many might have other names. Carl F. Frost became a psychology professor at Michigan State University, where he continued theoretical and practical experiments with Scanlon Plans and later also became a consultant.

Both Lesieur and Frost thought it curious that the organization that spawned the Scanlon Plan, the USW, lost all interest in it. After Golden and Scanlon left the union, Frost recalled, the union "wouldn't even acknowledge that it had been involved in labor-management cooperation." Lesieur was disappointed with the USW. "I thought my union got old awfully young," he told me. "They were way in the forefront, but they let it go."[22]

As a labor reporter, I began covering the USW in 1965. Not once in twenty years did I hear an official or staff member refer either to Joe Scanlon, who remained loyal to the union and never spoke ill of it after resigning, or the Scanlon Plan. I am not suggesting that the union leaders caused history to be rewritten. There existed, however, a sort of unofficial, official silence regarding the entire "union-management cooperation" episode of the 1930s and 1940s.

It never amounted to much in terms of employees affected or the percentage of steel produced in the cooperative mode— that is, where the workers were allowed to take responsibility for the success of the shop. None of the large, integrated steelmakers permitted such bizarre behavior. Steel management gave off fierce sparks of opposition to any notion of labor participation in production decisions. Otis Brubaker, who succeeded Ruttenberg as the USW's director of research, remembered that period very well. "I am very sure," he said, "that Murray was convinced there was not a chance in Hades of getting the basic steel industry working in cooperation with the union in any future period of time." Brubaker himself, however, was an arch foe of cooperation.

And yet, one wonders. While rummaging through the Clint Golden Papers at the Penn State labor archives, I found a copy of a letter written in 1950 by Ben Morreel, president of Jones & Laughlin Steel. He told an intermediary that he would welcome a meeting with Murray to discuss labor and social issues, including participation. The current labor situation was "so bad that we must experiment," Morreel wrote. "I would, therefore, be perfectly willing to try out the Scanlon Plan or something like it in J&L, with the approval of our executive committee." Unfortunately, the trail ends there. The papers I examined show no evidence of whether the meeting occurred. If it did, nothing came of it.[23]

All tragic stories contain might-have-beens like the Morreel letter. The atmosphere of distrust and management's determination to keep its turf inviolate choked off all promising initiatives. The USW lost interest in gaining a voice for workers in decision making—and thus in improving the work process—and started down the road of power bargaining described in chapter 4. As Sanford Jacoby puts it, "The CIO's visions of industrial democracy were discarded in favor of a philosophy that portrayed labor's exclusion from decision making not as a necessity but as a virtue of American union- ism."

Judged on its own terms, labor's drive for money and security cannot be considered a failure, for the unions pene- trated areas, such as pensions, health insurance, and other welfare benefits, that the companies preferred to keep under unilateral control. The rise in the standard of living for blue- collar workers between 1945 and 1980, because of union efforts, was phenomenal.

In a larger sense, however, labor took the easy way out. By accepting management's limiting definition of their role, unions could evade responsibility for the success of the enter- prise without feeling guilt. "We decided," said Ben Fischer, "that management ran the show and that our only course was to demand more of the pie." Murray and Reuther had preached industrial democracy and made it the cornerstone of their early approaches to labor-management issues. In the 1940s and thereafter, "we accommodated to the notion that unions have a limited function in industry," Fischer said. "Labor accepted the residual rights of management theory in the 1940s and established its own downfall."[24]

Attempts to Reform the Bargaining Relationship

Between the early 1950s and the 1970s, the USW and the steel industry made fitful attempts to improve their relationship. After a fifty-nine day strike in 1952, U.S. Steel Chairman Ben- jamin Fairless and USW presidents—first Murray and then his successor, McDonald—made a series of highly touted joint visits to steel plants. The sight of these leaders standing side by side in the glare of open hearth furnaces was supposed to encourage workers and foremen to cooperate in raising pro- ductivity. Little came of this endeavor.

During the long 1959 strike, the industry rejected one USW settlement proposal that included the establishment of a tri- partite committee to recommend "a long range formula for the equitable sharing of the fruits of the industry's economic progress." "I can vividly remember R. Conrad Cooper [the chief industry bargainer] throwing cold water on the idea, say- ing it was absurd, no way to conduct labor relations," said Marvin Miller, a USW negotiator in 1959. Kaiser Steel, how- ever, broke ranks with the industry and accepted the USW's wage proposal, along with the tripartite idea. Such a committee was created, with Kaiser and USW representatives and an outside chairman. It later developed a gain-sharing plan patterned partly on the Scanlon Plan but with a different bonus formula. Put into place in 1963 at Kaiser's plant in Fontana, California, the Long Range Sharing Plan was meant to gain worker commitment to technological change. Kaiser's labor relations improved markedly in the early years. But aside from high bonuses paid in 1963, the plan encountered many difficulties and remained a source of rank-and-file dissatisfaction. Disabled by low-cost Japanese steel landed on the West Coast, Kaiser sold the Fontana plant in 1984.[25]

Miller, who designed the Kaiser plan, also served as union chairman of the Human Relations Committee (HRC), one promising initiative that came out of the 1959 strike. So destructive was the strike that the CCSC and the USW formed this group as a means of avoiding "crisis bargaining." In the early 1960s, the HRC and its subcommittees met continuously to study and recommend solutions to complex problems in preparation for the next bargaining round. To ensure free discussion, the committee adopted a problem-solving approach rather than the position-taking of formal bargaining. From 1960 to 1965 the committee and its subcommittees met many hundreds of times and arrived at some recommendations. For example, a subcommittee thrashed out the details of the thirteen-week extended vacation plan before it was actually negotiated in 1963.

In HRC meetings, the USW tried to generate interest among the major steelmakers in the Kaiser plan but got nowhere. The companies regarded Kaiser as a maverick and were put off by the high bonuses in the early stages of the plan. The concept also drew criticism from many union officials, including I. W. Abel and Otis Brubaker, Miller said. Moreover, the USW's industrywide policy of equal pay for equal work presented a major problem since gain-sharing would produce different bonus yields from plant to plant. Miller and his economic technicians gave some thought to developing a formula that would "result in uniform wage and benefit changes" across the industry. But company opposition made this an academic exercise.

One HRC subcommittee considered "guides for the determination of equitable wage and benefit adjustments." What would be the necessary elements of a long-term wage-productivity relationship? The union wanted strengthened job and income security programs, as well as annual wage increases reflecting long-term productivity growth *and* cost-of-living raises to protect the buying power of wages. But the industry had just rid itself of COLA in settling the 1959 strike and opposed bringing it back. The companies also wanted to modify or eliminate a contract provision protecting old work rules. When a survey of plants turned up little proof that it impeded efficiency, the industry had to drop the idea.

Despite two years of unpublicized meetings, the subcommittee couldn't find common ground on these issues. "They wouldn't give, and we wouldn't give," Brubaker, the union subcommittee chairman, said. R. Heath Larry, U.S. Steel's second-ranking labor relations executive in the early 1960s, confirmed Brubaker's account. Union and management negotiators preliminarily agreed that steel productivity was rising at 2 percent per year, he said, but they could not surmount the COLA obstacle. "We spent hours and many days arguing how to get to a rationale under which both would be comfortable," he recalled in 1986. "But we were never able to come to an agreement."[26]

This failure did not seem important at the time. The companies were still profitable. Although steel imports were beginning to pose a real problem, the union and the companies didn't perceive how serious it would become. No one foresaw that the industry would shrink to a shadow of its former self. Because the HRC process was veiled in secrecy, neither the public nor the great body of steel industry employees ever realized that this critical opportunity had been missed.

The cooperation between top officials of the USW and the industry during the HRC days helped produce low-cost settlements. Steelworkers lost their COLA and received no pay increases between 1961 and 1965. Dave McDonald also acquiesced to the wage restraint blandishments of his friend, President John F. Kennedy. U.S. Steel Chairman Roger Blough, however, did not listen with equal respect to Kennedy's hold-the-line request on steel prices, precipitating the famous Kennedy-Blough price confrontation of 1962. Pressure by the administration forced Blough to rescind a price hike announced after a modest labor settlement.

The Human Relations Committee died abruptly when Abel ran against McDonald in 1965. Abel attacked both the modest settlements of 1962 and 1963 and the structure of the HRC. Although he favored the use of study committees, Abel charged that McDonald used the HRC—whose members mainly were staff people—to keep other elected union officials out of the negotiating process. An innovative approach to solving problems, the HRC *was* flawed politically by the lack of participation of most board members and local union officials. It could not have survived, in its existing form, growing demands for greater rank-and-file involvement in the union.

The HRC enabled the two sides to deal with knotty issues that required extensive study. Although the committee itself was disbanded, the USW and the industry continued the joint study approach in later years by forming task forces on specific issues. Yet the two sides could not bring themselves to grapple seriously with the wage-productivity question or face up to the changing structure of the international steel industry. Looking back on the 1960s, Miller believes both sides lacked leadership in this crucial decade. The industry had no theoreticians studying world steel economics and growing competition. Conrad Cooper was an industrial engineer, not an economist. "The industry had no ideas, none, and there was a real vacuum in the union," Miller said. The USW's influential chief lawyers, Arthur Goldberg and David Feller, left the union, Goldberg in 1961 and Feller in 1965. Miller resigned in 1966. Of the USW's top-notch staff, only Fischer and John Tomayko remained, and Abel didn't allow his staff as much independence as McDonald had.[27]

A final initiative before the ENA years occurred in 1971. The Nixon administration, determined to avoid a steel strike on the eve of Nixon's wage-price controls, pushed the industry to grant a handsome wage settlement to the union. In return, the administration urged the two parties to adopt a productivity program that the government could point to in justifying its support of a hefty wage increase. The result was a provision in the master contract requiring that each local union and plant management set up a productivity committee. The committee should "advise with plant management concerning ways and means of improving productivity ... and also promote orderly and peaceful relations with the employees, to achieve uninterrupted operations in the plants, to promote the use of domestic steel, and to achieve the desired prosperity and progress of the company and its employees."

Although hundreds of committees were set up, at least on paper, all but a few collapsed without accomplishing anything. The provision itself was an afterthought and contained no carefully developed guidelines for ensuring that real worker participation would take place. It called only for local union and management officials to sit on the committees, not rank and filers at the departmental level. It did not specify that the committees could discuss methods of improving working conditions as well as productivity. And applying the name "productivity" to the committees doomed them from the start with steelworkers.

For a few years in the early 1970s, the USW and the major companies made a pretense of operating an industrywide productivity program. They gave out statistics on the number of committees and the number of suggestions made by the local unions. But neither corporate management nor top USW leaders made a personal commitment to participation. Most plant managers, believing they had a mandate to negotiate with the committees on crew reductions, refused to listen to other suggestions for improving production or quality. The average steelworker, the USW itself concluded years later, viewed the productivity committee "as another speedup, crew-cutting process which inevitably meant more layoffs. And, in fact, that is what it was."[28]

Another opportunity had been missed to bring real participation to the steel industry.

Management Proposes LMPT's in the 1980s

Steel management killed the possibility of worker participation in the 1940s, and the USW—as if embarrased for having broached it—drove the idea into deep obscurity. So deeply that when the participation concept finally reemerged thirty-five years later in the steel industry, it was dredged up not by the original proponents of industrial democracy—but by management. Negotiators for Bethlehem Steel and Jones & Laughlin developed and proposed the 1980 contract provision setting up Labor-Management Participation Teams (LMPTs). An additional twist in the story is that the two companies acted in concert to thwart a possible veto by U.S. Steel.

Bethlehem and J&L had concluded independently that steel's lagging productivity growth should be addressed in 1980 bargaining, although the two firms approached the issue in different ways. John Kirkwood's suggestion that COLA pay boosts be linked to productivity had been turned down by Bruce Johnston (see chapter 9). At Bethlehem, meanwhile, Chairman Lewis W. Foy, Trautlein's predecessor, told George Moore that the company should have some sort of productivity program. Moore passed the order on to Tony St. John, who had become interested in problem-solving teams. St. John looked into the UAW–General Motors Quality of Worklife (QWL) programs, which had been under way at some plants since the mid-1970s. The UAW and GM carefully avoided emphasizing the productivity-enhancing aspects of QWL, fearing that doing so would turn away workers. It did not require much examination, however, to discover that in some GM plants QWL discussions had produced significant cost reductions without eliminating jobs. St. John developed a plan for similar groups in the steel industry and obtained Moore's approval.

St. John was aware that J&L's productivity proposal had been turned down by Johnston. He also knew that Kirkwood and A. Cole Tremain, J&L's general manager of labor relations, felt as strongly as he that the steel industry suffered from the top-down approach in labor relations. Too little effort went into involving rank-and-file workers in work improvement programs. Before 1980 bargaining began, the three men discussed St. John's LMPT proposal. Kirkwood liked the idea in lieu of his own, and the two companies agreed to push it, possibly against the resistance of U.S. Steel. As it turned out, Johnston accepted the idea when Moore mentioned it to him and helped draft a proposal. St. John and Tremain introduced it in the problem-solving subcommittee in 1980. Two of the three union negotiators, Ed Ayoub and Sam Camens, a special assistant to McBride, were enthusiastic about the idea. USW attorney Carl Frankel suspected that the companies proposed it for manipulative purposes but later changed his mind. With a few changes proposed by the union, the two sides wrote the LMPT provision into the contract as an "experimental agreement." How much it would be used depended on the commitment of management and union leaders at each company.

McBride appointed Sam Camens to coordinate LMPT activities for the USW, bringing to the fore a veteran staff man who became an eloquent advocate of worker participation. "Steel has so damn many problems," Camens told me at the time of the 1980 settlement, "we realized that unless we replace the adversarial relations with a cooperative attitude, we're both in trouble."

A short, heavy man with white hair bristling from a balding head, Camens was nearing sixty in 1980. He had started working in the old Ohio Works of U.S. Steel (now torn down) in Youngstown and was elected local president in 1945. He attended college for a year, learned steam engineering in night school, and worked as a stationary engineer in the blast furnace department. In 1957, after twelve years as local president, Camens joined the USW staff in Youngstown. Moving to Pittsburgh in 1968, he became Ben Fischer's assistant and a specialist on noneconomic issues such as work practices, contracting out, and seniority.

The LMPT assignment brought out a missionary zeal in Camens. He had always felt that industrial democracy was a necessary building block of political democracy in the United States. In the 1940s, Camens remembered, "management was absolutely hysterical about union encroachment on managerial rights. They wanted no part of the union having any rights in the production process. They wouldn't let us near it. The plant superintendent would always tell me, 'Sam, you run the union, I'll run the plant, and we'll get along fine.' We had good enough relations, but they took the position that everything we got, we had taken away from their rights. When we had a grievance, they would argue against it on a strict construction basis." As a result, Camens said, the USW locals focused their energies on "setting up strict rules in order to protect our membership against autocratic, vicious dictates of the company." The relationship became a legalistic one instead of a cooperative venture of working together to produce a good product at competitive costs.

In the 1980s, Camens traveled from plant to plant, helping locals install LMPTs, visited Japan to study its labor system, and became a sought-after speaker on participation and workplace democracy. His often emotional outbursts against dictatorial managers and Taylor's scientific management became familiar events at labor and behavioral science conferences.[29]

Despite Camens's fervor, LMPTs spread far less widely in the steel industry than he had hoped. It was a *voluntary* arrangement that required the concurrence of the local union and the plant management. A number of locals refused to be involved, either because of ideological opposition to cooperation or unwillingness to take a political risk. Managers at some plants turned down the idea, either because they distrusted the union or because they didn't think the union could tell them anything about making steel that they didn't already know.

The decline of American steel's competitive ability indicates that the all-knowing managers and the uncooperative union officials made a poor choice. As the loss of markets accelerated in the 1970s and 1980s, the quality of working life in the plants also deteriorated. "We're living with forty-year-old union practices and forty-year-old management practices," Mike Bilcsik said in 1983. "The alienation runs clear from the top all the way down."[30]

Chapter 12

By the 1970s, demoralization pervaded the Monongahela Valley mills. This conclusion became clear in dozens of interviews I conducted with workers and supervisors. A corrupt atmosphere had devitalized human relations and the quality of work over several decades. The companies' failure to modernize spurred employee fears about job security and eroded management credibility. Productivity shriveled. A profit-grabbing focus on quantity withered employees' pride in producing good steel. Relations between several groups—employees and foremen, union and company officials, plant-level supervisors and corporate management—turned increasingly sullen and hostile. Management kept the union out of its business, and the union held the companies prisoner in an outmoded structure that management itself had created. Toward the end, as foreign and domestic competition accelerated, it was as if everyone in the mills wore straitjackets on a forced march to industrial perdition.

When did all this begin? I found a surprising unanimity of opinion, given the span of time involved, that many aspects of worklife began to deteriorate in the early 1960s. Perhaps not coincidentally, this is when competition began to eat away at American steel. By then, the giant bureaucracy that had so profitably exploited the assembly line of plants and towns in the Mon Valley could no longer respond quickly enough to head off looming disaster.

Manny Stoupis, who worked in the McKeesport plant from 1947 to 1985, put it this way. "In the 1950s, it was a pleasant place to work. We had a good working relationship with management. Grievances were minimal. No chickenshit discipline. No forced overtime. Good relationship with your immediate supervisor. The people in the personnel office

296

spent their lives there and knew workers personally." The atmosphere changed in the 1960s, Stoupis said. "The company brought in college grads as supervisors and sometimes paid them more than general foremen who had come up the old way. This helped destroy working relationships. In the early seventies, they began hiring attorneys to run personnel services. Most of them used it as a stepping-stone and became tough bastards. 'If you want something, arbitrate!' they told us."[1]

"We were living through the end of an era, although we didn't know it then," said William Schoy, who spent twenty-nine years in valley plants as a rigger. "When I started at National Tube in the 1950s, the whole attitude was different. I enjoyed my work. Then the scene started to change. It wasn't a partnership any more. We were considered stupid. In earlier years, the foremen listened to us and were more on our level. Then there was a widening."

Everything had deteriorated badly by the 1970s, said Allan J. Sarver, a former U.S. Steel superintendent in the Mon Valley. "The way they [U.S. Steel] made steel was crummy, the way they operated was crummy, and the way they handled labor relations was crummy. The only thing that worked well at U.S. Steel was the safety program."

On Wall Street, financial analysts explain the tragedy of the American steel industry in macroeconomic terms. It was swept away, they say, by great structural shifts: a decline in steel consumption in the United States, a rise in excess capacity around the world, the inability of American companies to raise capital for modernization, a wide disparity in labor costs. Equally important, I believe, a creeping rot on the inside eroded competitive ability and stymied innovation in the most vital of areas, human relations.

In the following analysis of how this happened, I deal primarily with the U.S. Steel plants in the Monongahela Valley. Much of what I say about the Mon Valley could be applied in Gary–South Chicago, Birmingham, Baltimore, Youngstown, Johnstown, and Bethlehem. I focus on the Pittsburgh region, however, because I know it and its people best.

"Working" in the Mill, ca. 1950

For the first forty-five to fifty years of steelmaking, most mill workers faced working conditions of unmitigated wretchedness. Before U.S. Steel adopted the eight-hour shift in 1923, steel employees worked seventy-two to eighty-four hours a week. The mills were charnel houses. In one year, 1906, 405 workers were killed in U.S. Steel plants. Under the headline "Making Steel and Killing Men," a magazine reported that 46 men had been killed and up to 2,000 others "merely burned, crushed, maimed or disabled" at the South Chicago plant.[2]

Conditions improved somewhat in the 1920s as steel companies joined the trend to welfare capitalism, establishing safety programs in the mills and social aid programs in the communities. The management style, however, remained as dictatorial and arbitrary as in the older days. Foremen abused their virtually unlimited authority to hire and fire by engaging in favoritism. In the view of Joseph W. Lenart, who started at National Tube in the 1920s and became a roller on a blooming mill, favoritism "was the biggest sin they committed down there." It was the reason he joined the union.

At the age of eighty-one, Lenart was a spry and wispy little man with a mental toughness and assurance that many older people lose upon retirement. He retired from U.S. Steel in 1969 with forty-seven years of service and worked another fifteen years for the American Association of Retired Persons in western Pennsylvania. When I interviewed him in early 1986 at his home in McKeesport, Lenart lived alone (his wife had died) in a narrow frame house just up the street from the Christy Park Works. He had thinning white hair on the back of his head and wore an impish smile. As we talked across a low table in his living room, his memories came tumbling out without much pushing on my part.

"Before the union came, the foremen were bringing their sons and nephews and brothers-in-law in," Lenart said. "After I worked in the National Tube office for a couple of years, they brought a fellow in and said, 'Starting tomorrow he's going to do your job.' I said, 'What am I going to do?' He told me, 'You're a good man, we like you, you can do any job. But I have to take care of this man. He's a foreman's son. You know, we take care of our own.'"

On at least two other occasions Lenart lost a job because of nepotism. "The foremen would bring in their kids and push you out. This happened so many times! You worked your butt to get up there, and when you got there, they'd chop you off. The best thing that happened was that the union came in and everybody was established on their job. Now, when there was an advancement, you bid for the job, and if you had the seniority you got it."

The coming of the union halted favoritism and established seniority rights and a grievance procedure. Foremen no longer possessed unquestioned power over mill workers. The USW negotiated a job classification system formalizing an older management system that slotted workers into narrow jobs, the boundaries of which should not be breached. Management drew the wagons around its managerial prerogatives, and the union, in return, acted on its own territorial imperatives. To protect existing work-force levels and prevent foremen from arbitrarily assigning workers in disfavor to the worst jobs, it aggressively patrolled the job boundaries to prevent crossovers. It was as if the two adversaries were dividing up the workplace forever more.

None of this was apparent to me when I worked at National Tube. The general growth and economic well-being of the late 1940s and early 1950s concealed the deeper struggle. Fat and lazy, the entire steel industry wallowed unconcernedly in a sort of prosperous torpor. My strongest memories of working in the mill involve not mistreatment by foremen but a massive waste of manpower and material.

I first walked into the National Works in 1949 at the age of eighteen and worked for the next year in the office of an engineering firm that was constructing a boilerhouse. On lunch breaks and errands to the construction site, I'd wander around and watch the steel operations. Making iron and steel in the blast furnace and open hearth departments were batching processes that called for short periods of sustained work, the men handling hazardous materials at high temperatures. I could understand why they took frequent breaks, standing around smoking cigarettes.

What I couldn't understand was the pace and nature of work in departments such as the blacksmith shop. This was located in a hoary brick building with high, arched windows. I often stopped to watch the work, singling out a man who operated a hydraulic forge hammer. He sat at the machine and used levers to raise and lower the hammer, banging on a forging that other workers had placed on the anvil. The hammer man himself never went looking for things to do. If there was nothing to hammer, he would sit there and read or doze off. It was the same with men who operated other kinds of machines.

I was amazed. Nothing in my previous experience had prepared me for the reality of work in a large industrial plant, where people were confined to one task or a set of related tasks. From the age of thirteen or fourteen I had worked on weekends and during summers as a beer-truck helper and driver and as an odd-jobs man in a small hotel. People held different jobs, of course, but we all helped one another, and it never occurred to me to draw a line and refuse to cross it. I knew nothing about job classifications, stretchout, or speedup, but I thought unnatural a system of work in which a person could want, or be permitted, to contribute so little as the hammer operator.

No one in the plant seemed to care that an enormous amount of worktime was spent in not working. I'm not referring here to furnace operators or to production crews that operated the blooming mills and pipe mills. Working on incentive pay, these crews worked steadily, if not at break-neck speed, to meet an output standard that would give them a sufficient return for a day's labor. Some mills had larger crews than were actually needed; the least-skilled crew members could double up on duties to spell one another for extended breaks. First-line foremen seldom protested excessive crew sizes, since their power rested in part on the number of people under their command.

Maintenance workers, laborers, and people assigned to other miscellaneous jobs, however, had neither economic nor psychological incentive to do any more than the foremen required that they do. In the 1940s and 1950s, labor gangs were expanded in the summer, made up largely of college kids on vacation and other part-time employees. They had no loyalty to the company or commitment to the work, and no one in charge seemed to care what they did in the slack hours.

In the dark recesses of a mill building, it was easy to find a place to hole out for a few hours to read or sleep. Heaps of rusting steel plates and slabs, jumbled piles of old machinery, crates filled with new parts, and abandoned pits and mill structures littered the insides of mill buildings and storage yards. Some of this industrial bric-a-brac had lain so long in one place that no one could remember why it was there. Whatever the reason, sleeping nooks were plentiful. On the night turn, 11 P.M. to 7 A.M., more people may have been sleeping in National Tube than in all the hotels of Mckeesport. It was a mark of esteem, the "macho" thing of the day, to brag about sleeping on company time.

Bantam Publishing introduced a new form of leisure activity in the mills when it began publishing the popular 25¢ paperbacks in the early 1950s. Many employees carried a dog-eared Bantam novel in their hip pockets, to be read a chapter at a time, behind a stack of pipe, when the foreman wasn't around. Even machine operators would ingest a page or two of mystery or tepid (in those days) sex while sitting at their machines. Barney Joy, a foreman at National Tube from 1954 until he retired in 1976, remembered that time especially well. "It was difficult to break them of that, those Bantam books," he said. "We had to set up rules that anybody who was caught reading would get time off [be suspended]. If they'd sit and read while you were broke down, you wouldn't mind. But trying to read and operate, that created hazards."

I learned first-hand just how lax steel management was in those days. In 1951, home from college, I worked the summer as a vacation replacement at National Tube. I had expected, even rather hoped, that I'd be assigned to a labor gang. I wanted to test my mettle in hot, rough work of the kind I'd heard about ever since the day's my father took me into Musulin's Cafe. A typical laboring job involved shoveling "scale," heavy, oxidized steel chips that had to be cleaned out from under the rolling mills—work that was heroically dirty and exhausting. Instead, I was assigned as a "production checker" in a shop called the Coupling Tap. All summer long I never lifted anything heavier than a pad of paper.

My job was to record the number of couplings produced by each machine operator and fill out a timesheet used to calculate incentive pay based on the output. A timekeeper's clerk gave me a ten-minute lesson in using the form. After entering hours worked and daily output in the first two columns, I should apply various arithmetic functions to these numbers in the remaining columns across the form. I didn't need to know why, the clerk said, just do it. He took me to the Coupling Tap, located in a high-roofed brick building that seemed to reverberate with a continuous metallic clanging. Dust floated in shafts of light that slanted down from high windows. I was installed in the production checker's "shanty," a small frame shack with windows on three sides, a desk, a swivel chair, a telephone, and a hotplate. I found a small pillow and a couple of ratty Bantams in the desk drawers.

In this shop, short sections of pipe were made into couplings, first by machines that tapered and recessed the insides of the "blanks" and then by other machines that tapped (threaded) them. The checker was supposed to make periodic rounds and record the number of finished pieces at each machine. On my first round, I became fascinated by one of the tapering-machine operators, a marvelously dexterous man who chewed tobacco and wore a work shirt with sleeves cut off above the elbow and a soft cap (a hard hat wasn't required in those days). Tending two machines and working on four blanks at a time, he moved back and forth between the machines in a near blur of motion. After placing a fresh blank on a spindle to taper one end, he would cross to the second machine, remove a once-tapered blank from one spindle, flip it around, and insert the untapered end on another spindle. Back and forth he would go, crisscrossing between four spindles, now and then spitting a stream of tobacco juice into the sawdust at his feet. When both ends of a ten-pound coupling had been tapered and recessed, he yanked it off the spindle with one hand and tossed it into a bin with a twist of the wrist that made it line up side by side with the previously finished piece.

I was to count the number of couplings in a bin before an overhead crane picked it up and carried it to the automatic tapping machines. It was then that I discovered how hard it is to count couplings stacked pyramidally. The number alternated from row to row, and my eyes glazed over from trying to count the number of holes—coupling ends—in each row. The operator, grinning faintly, watched me for some time as I peered into the bin, pretending to know what I was doing.

"Hey, kid!" he finally said. "Read this." Bending over with a piece of chalk, he scrawled on a coupling the number of pieces in each row times the number of rows, ending up with some sixty pieces. "Right?" he asked.

"Right, right," I replied, as if he had merely confirmed my own conclusion. I wrote the number on my pad.

From then on, I understood it would be *his* count, not mine, and that this was the accepted way of doing it. All of the operators did the same, chalking up their output, which I copied. The operators weren't about to have a vacationing college kid undercount their production. The shop foreman raised no objection to this practice. Did the operators cheat on the count? Probably some, but not much. Over the course of the summer, I learned how to estimate the number of couplings in a bin (although I never did count them piece by piece) and concluded that the operators' count came *close* to reality.

By the end of my first week on the job, I could complete a round of the machines in about ten minutes, once every two or three hours. Toward the end of each shift I made the final calculations on the time sheet and turned it in at the time-keeper's office as I left the plant. This left plenty of time to spare. I doubt that I put in more than two hours of work per shift. I simply ran out of things to do. It seemed strange that here I sat, a healthy young man, doing nothing except gazing out the shanty window at the tapering machine operator, a grizzled man in his sixties who worked at a demoniacal pace.

In my second week or so, my boss—a timekeeper supervisor—came to my shanty and asked how things were going. Fine, I said, except shouldn't I be doing something else? Was I neglecting other duties? "Nope," he said. "You're doing okay. Keep it up." I never saw him again.

I passed the time by sleeping (although I generally tried to confine my sleeping to the night turn to maintain a semblance of propriety) and reading. At first, I only brought in Bantam books, concealed in a pocket. Getting bored with the deception—and with the books published by Bantam—I graduated to hardcovers, including some quite large ones. I read *Les Misérables* that summer to the din of whirring machines and clanging couplings. I also visited neighboring mill buildings, observed other pipe-finishing operations, and occasionally ran into friends.

Larry Delo, who worked in a labor gang that summer, was sent to the coupling shop several times to run a machine that chopped sixteen-foot lengths of pipe into coupling blanks. The job didn't require any skill, but hefting the heavy pipe and couplings for several hours took its toll. I'd stand there in my clean shirt, chatting with Larry, as he got dirtier and damper, lifting, chopping, and stacking the blanks in a bin. Once or twice, when he fell behind, I offered to help. "You'd better not, John," he said. We both knew that union rules—and company rules, for that matter—forbade the crossing of job boundaries. I could do nothing but watch him get redder in the face. I felt foolish.

On my last day on the job, one of the regular production checkers arrived early at shift change while I was completing the time sheet. He immediately spotted an error. In one of the final columns, I had multiplied instead of divided, or added instead of subtracted (I forget which). The result was a higher incentive yield for the tapering machine operators than should have been the case. But, I protested, I had been computing it this way all summer and no one had said anything. He laughed. "That's management for ya," he said. "But I can tell you who *did* discover it." He glanced out the shanty window to the shop floor. "Those guys must have had a great summer."

The Management Bureaucracy

In the late 1950s, U.S. Steel and other producers, feeling competitive pressures, began paring the work force, and increasing numbers of college-trained men replaced the old foremen. A different managerial problem arose. Al Sarver, who headed iron- and steelmaking operations at the Duquesne plant when he resigned in 1983, was known throughout the valley as a tough but fair and highly competent manager. After serving twenty-one years in supervisory positions, he believed the management system itself had failed. "For the most part," he said in a 1986 interview, "the hourly guys were very poorly directed. They were directed by people who didn't understand what they were doing."

Managing a steel plant is a difficult business. It is not a case of making steel correctly the first time and merely duplicating the procedure endlessly down the years. At the beginning of the process, the manager must marshal enormous amounts of raw materials—iron ore, coke, and lime—which are dumped into a blast furnace and come to a boil at 3,000° F. Men in asbestos suits tap the furnace, guiding the molten iron into a railroad car especially constructed so that it will not disintegrate under its freight's fiery breath. A locomotive transports the "torpedo ladle" to the steel shop, where the molten metal is poured into another gargantuan vessel, a basic oxygen furnace in the most modern steel mills. Furnace operators add scrap steel and fluxes, roil the mixture around for about forty-five minutes at 2,800° F. and pour the contents into a ladle dangling from a crane hook. The crane moves the ladle, sirens screaming, across a wide expanse of shop floor and trickles it into a continuous caster. This is a channel several hundred feet long and shaped like a child's sliding board that curves downward and levels out. As it creeps slowly down the channel, the steel hardens and emerges at the bottom in one of several different shapes, a slab or a bar or a bloom.

If the plant doesn't have a caster, men on the pouring platform pull the ladle plug and allow the molten steel to flow into ingot molds, where it sits cooling for several hours. Workers later strip away the molds and deposit the ingots in soaking pits—sunken furnaces equipped with gas or oil burners—where the steel is reheated to about 2,000° to 2,450° F. to soften it for rolling. When the ingot reaches the right temperature, a crane snatches it out of the pit and lays it on a conveyor line. Snapping, crackling, throwing off scalding chips, the ingot moves into the jaws of a primary mill. The roller, manipulating levers and foot pedals, runs the ingot back and forth between sets of rolls. In a process not unlike squeezing the water out of a garment in an old-fashioned wringer, the steel is compressed and elongated into a slab (six to twenty-one inches thick by thirty-six to seventy-two inches wide) or a bloom (a smaller square shape), sheared off at the end, and passed on to other rolling mills for further refinements before it goes to the finishing mills. It may yet undergo a half dozen more reheating, rolling, piercing, extruding, welding, galvanizing, and chemical-treating processes, depending on the finished product.

In making raw steel, these processes must be repeated from eight to ten times a shift, and in the rolling and finishing operations hundreds of times—repeated without hurting anyone and avoiding undue stress on expensive equipment. Nothing ever happens exactly the same way. No two heats of steel turn out chemically identical; no two ingots have precisely the same properties. To manage such a process successfully, bringing the raw materials and supplies to the right place at the right time, keeping the equipment well maintained and the workers well motivated, dealing with dozens of variables—to make all these things happen at the lowest possible cost, in the minimum amount of time, and yet produce the highest possible quality—to do all of this successfully over and over and over is no small managerial feat.

In reality, it is impossible to manage a steel plant without the aid many times over of employees who operate the mammoth equipment. Nevertheless, like most steel companies—indeed, most manufacturers of any product—U.S. Steel used the "control" method of directing the work force. That is, management assumed that hourly workers had to be closely supervised and controlled, a notion that had persisted since the early twentieth century, when the mills were manned largely by unskilled immigrants. Under the control model, as Richard E. Walton, a Harvard work expert, describes it, workers are regarded as replaceable parts in the industrial system. They respond only to orders, show no initiative, and take no responsibility. (The opposite is the "commitment" model, which starts with the idea that workers tend to be self-directed if management assumes that they want to be and gives them the opportunity.)[3]

The control method went hand in hand with, and was in part necessitated by, U.S. Steel's vertical integration, which had turned the Mon Valley into an assembly line and a series of mill towns that served as locker rooms. Not only must a large corporation dedicated to management by control have battalions of supervisors to command the armies of hourly workers. Having accepted the assumption that hourly workers must be controlled, the management must further assume that low-level supervisors have caught the disease of irresponsibility from the hourly ranks and, therefore, must be supervised by higher managers who, themselves, need to be managed...and on up the line to the "big brain" at the top, who alone requires no guidance other than the yes votes tendered by a board of friendly directors.

The steel industry was the epitome of this many-layered approach to work-force management. By the mid-1970s, U.S. Steel had ten levels of "operating management" in steel (excluding staff functions such as marketing), ranging from the top of the corporation to the production line in a plant. At the corporate level were the chairman and chief executive officer, the president, and a group vice-president of steel. Under the group VP were four steel divisions, each headed by a vice-president and general manager. Within each plant, the hierarchy consisted of the general superintendent, the assistant general superintendent, division superintendents over groups of departments, department superintendents, assistant department superintendents, a general foreman of each department, and—at the bottom—first-line foremen (usually called turn foremen because they worked rotating shifts like the hourly employees).

With all of these managerial posts, one would think that promotion would be frequent. Actually, the reverse was true, as the following example shows. Until the early 1980s, the Duquesne plant had a department known as the primary mill. This included soaking pits, three primary rolling mills, and shipping yards. Each of the three areas had a general foreman and four turn foremen, the number necessary to cover each shift in a seven-day continuous operation. To be promoted, the twelve turn foremen had to wait for a general foreman opening. The three general foremen, in turn, waited for the assistant department superintendent to move up to department superintendent when—and if—the latter beat out three other department superintendents for advancement to divisional superintendent.

Working one's way upward through this maze required a good bit of luck or pull in high places. Indeed, many turn foremen—including college graduates who started as management trainees—remained at that level for fifteen to twenty years or more. Lured to U.S. Steel by relatively good pay and a superb benefit program (a replica of the program won in negotiations by the USW), they would be mired in the front-line trenches for near-life terms. Unquestionably, this had a bad effect on morale. Larry Delo, a Purdue graduate, was not promoted to general foreman until he was fifty, after twenty-four years as a turn foreman. "I regret having to spend all those years working turns when my family was growing up," he said. "Unless you were a real great performer, or made a lot of noise, or the department superintendent took a liking to you, you were really stuck."

Delo was among the first of the college graduates who joined U.S. Steel as management trainees and were slotted into turn foremen jobs. Before the mid-1950s, most foremen came from the hourly ranks and didn't expect to rise beyond general foreman. The purpose of the management trainee program was to give young graduates practical steelmaking knowledge that would be invaluable as they rose to jobs as engineers or higher management positions. Unfortunately, the program contained major flaws from the outset and in the end contributed enormously to demoralization in the plants of the Mon Valley.

The lack of opportunities for promotion and the rotating shift schedule caused increasing dissatisfaction among the college graduates. The management trainees of the 1970s, possibly reflecting the higher expectations of their generation, didn't accept the system as passively as their earlier counterparts. Unwilling to languish unappreciated at the foremen level, many of the later trainees aggressively maneuvered and lobbied for promotion out of the trenches. This irritated foremen who had risen from the ranks.

Barney Joy, for example, made foreman at the National Works in 1954 after twenty-one years as an hourly worker and eventually rose to general foreman of the blooming mill and soaking pits. A rotund man with a completely bald head and a direct way of speaking his mind, Joy was the very image of the gruff, old-style foreman. Many college trainees were assigned to his department over the years. "A lot of them were nice guys," Joy told me, "and I was never jealous of their education. I was proud of the experience that I had that they could never get. But they didn't want turn-foremen jobs. They wanted steady daylight jobs. And they were always looking ahead to the next job up the line. I always told them, 'The question is, are you going to keep our department competitive, or are you simply trying to use this job for your own advantage?'" In too many cases, Joy added, it was the latter.

U.S. Steel's hierarchal system created another problem. The best of the trainees undoubtedly wanted to improve production and employee relations through their own initiative. But the system foiled them. Orders came down the line from above, and if the troops at the bottom didn't march as required, the turn foreman bore the blame. Foremen were expected to exert control over their crews rather than motivate the workers. They had little authority to settle grievances on grounds that they might set precedents that the union would cite elsewhere. These pressures crunched front-line supervisors on the border between the workers and union committeemen below them and the managers above.

The first trainees benefited from the experience of old-time foremen like Barney Joy. In 1961, however, there occurred one of the dark days in Mon Valley steelmaking history. U.S. Steel had seen profits tumble in two, nearly back-to-back recessions, and it decided to shrink overhead costs. On a day in April that is still remembered as Black Friday (although it was no more than a gray day compared with similar episodes in the 1980s), the corporation eliminated thousands of salaried employees through firings and forced retirements. These included middle-level executives and staff technicians in the Pittsburgh office, as well as hundreds of plant managers and supervisors. The Mon Valley lost a high proportion of the oldest, most experienced (and highest paid) supervisors. To replace them, the company stepped up the hiring of young college graduates at much lower salaries.

The deterioration of managerial capability paralleled a similar trend in the production force, according to Al Sarver. His explanation is arguable but probably contains a substantial amount of truth. It starts with the USW's victory on the seniority issue in the 1930s. The use of favoritism and nepotism in promoting, laying off, and recalling employees had to be halted. Seniority was the only objective standard that could be applied. Although a worker bidding for a higher-paid job must demonstrate "ability to perform the work" in addition to his seniority status, this method of promoting people obviously relies less on merit than managers would like.

"In the 1930s, the hourly people in the most important jobs were very competent," Sarver said. "They had some natural ability that got them to the responsible positions. Foremen were picked from the very top hourly people. They made good steel, good quality, for that time." Sarver continued: "The impact of seniority began to be felt in the 1950s. As the top hourly guys from the 1930s retired, they [the company] had to promote the next guy in line. They just had to be able to reasonably perform the job. This didn't mean they were the best available. So the experience in the hourly ranks deteriorated."

At about the same time, the management trainee program began to run into trouble. By the 1970s, Sarver said, "the last thing most of those new college guys wanted to do was get their hands dirty. They had had four years of school, and most of them figured that learning how to do the manual jobs, getting down to the nitty-gritty, was beyond their station. So most of them didn't understand the jobs they were managing. We had a continual deterioration of capability in the hourly ranks and no one in the foremen ranks to assure the continuance of [production skills]. We had an ever descending curve in terms of competence."

William Behare, who worked thirty-seven years in the McKeesport plant, also believed that the pool of foreman knowledge deteriorated over the years. "The philosophy of many foremen was, 'We don't have to know how each piece of equipment operates,'" Behare said. "If a foreman wasn't familiar with how a department operated, the employees would take shortcuts and get around the foreman, and he wouldn't even know it." The trouble with some foremen who graduated from the ranks was that they changed stripes upon promotion and became dictatorial, Behare said. "They'd crack the whip to show the college-trained bosses they could make a go of it."

The most arrogant of the college men often learned the least about steelmaking. Their lack of knowledge showed up when they got into decision-making positions as general foremen and superintendents. Hoping to make a name for themselves, they stormed about the shop, ordering changes in equipment and work practices without consulting people who had worked in these operations for decades. Unfortunately, many of these ideas not only failed to boost production but also turned into costly errors. Practically all of the workers and managers that I talked to in the Mon Valley had a catalogue of tales about such wrongheaded decisions.

For example, Stephen A. Krivda worked for twenty years as a crane operator at Homestead before being laid off in 1985. He related the following story about the soaking pits at Homestead. In 1979 or 1980, a superintendent decided to speed up the heating process by installing new air ducts and redesigning the pits. This only increased the amount of time required for the pits to reach 2,200°. "The heaters [senior hourly operators] told the foremen we weren't getting enough air or gas into the pits through the ducts," Krivda said. "We kept telling the foremen to go back to the old design. But they wasted a lot of money by not listening. In the end, they had to go back to the old way. I told one foreman, 'You have four heaters in this department with forty years' service apiece, and you come in from college and won't listen to them.' He said, 'I don't care how long you've been working here. I'm the boss.'"

The more arrogant managers even ignored the advice of experienced foremen like Barney Joy. "I remember somebody had the idea of putting electric motors in damp conditions under the rolling mills, to move scale along a belt," Joy said. "When I first saw the plan, I said, 'That's crazy. We got people down in there using picks and bars, and they're gonna cut a ground off a motor and somebody will be electrocuted. Put in a big trough with high pressure water,' I told them.

"'Oh, no, look at the cost of water!' they said, and they went ahead and installed the motors. Of course, the motors didn't work, and so they took them out and put in air cylinders. But they had to tear them out, too. They spent a lot of money on all that. They wound up putting in the sluiceways like I suggested. When I retired, they were still under the mills."

No manager could learn all the intimate details of the more complicated production jobs, such as operating a rolling mill. But the best managers learned enough to know when something was going wrong. "When I was a first-line foreman," Sarver said, "I took time to sit down with the hourly guys. I said, 'How do you do this, how do you do that, and why do you do that?' My approach was that every job I got, I was going to retire on it. I wasn't afraid to get my hands dirty."

I could believe that from Sarver's appearance when I interviewed him in 1986. He had his own consultancy by then and also was affiliated with a company that manufactured steel mill equipment. We met in a small plant north of Pittsburgh, where he worked at the time. A thick-chested man as solid as a professional fullback, he had light blue eyes, black hair, and a remarkably square jaw. To see him was to see a *manufacturing* man. He wore a khaki sweater over a work shirt, and his pants were streaked with dirt, evidence that he had been wrestling with a shop-floor problem of some sort.

Sarver's outspokenness got him into trouble with his supervisors at the Mon Valley Works (by 1983, all U.S. Steel plants in the valley came under this designation). He also had an interest in a small firm that sold equipment to U.S. Steel. Accused of conflict of interest, Sarver offered to sever connections with the firm. When it became apparent that this wouldn't satisfy the superiors, Sarver resigned. Many hourly workers and supervisors suspected that he was forced out because he openly fought what was shaping up as a corporate decision to close the Duquesne plant.

After Duquesne's shutdown in 1984, a coalition of political groups and the USW attempted to acquire Duquesne's giant Dorothy Six blast furnace and operate it as a worker-owned enterprise. The coalition asked Sarver to run the plant if the plan worked out, believing that only he could manage it effectively. "He tried to instill an attitude that nothing was impossible," recalled Larry Delo, who served as a general foreman under Sarver. "Any time we had to do something differnt, like handle a product we had never made before, we figured out ways to do it. There was nothing we weren't able to do. He was very tough and very outspoken, and he didn't care what he said to whom. I think the top U.S. Steel people in the Mon Valley were afraid of him."

"A lot of people disliked Sarver because he eliminated some

jobs at Duquesne," said Mike Bilcsik, the last union president at Duquesne. "But he was the only one I could envision to run the plant. I gained respect for him because he dealt with me as a human being."

Labor could be obdurate and wrongheaded about many things, Sarver told me. "But I'll tell you what," he added at the end of a two-hour, taped interview. "American workers have no peers. They are that good. Most workers—I'll bet the percentage is in the high nineties—want to do a good job. All you have to do it give them the opportunity. Make them a part of it. But nobody lets them be a part of it. Everybody thinks they're dumb. Shit! They're good!"

"Scientific Management"

The assumption that workers are dumb, however, has been governing steel management ever since the industry adopted its dominant mode of work organization early in the twentieth century. Michael Maccoby, a researcher, psychologist, and consultant in work reforms since the early 1970s, classifies this form of organization as an "industrial bureaucracy," based in large part on Frederick Taylor's "scientific management" methods.

Taylor taught that there was "one best way" to perform each work task. He saw that the increasing mechanization of manual work in the late nineteenth and early twentieth centuries would enable companies to achieve great economies of scale. But management had to master production and distribution process by instituting systems of labor and inventory control, cost accounting, and production planning. Taylor developed many of his specific work-control methods for machine shops rather than for more generalized production jobs. But his "philosophy" of people and work could be, and was, applied to most industrial occupations. Taylor, says a modern interpreter, believed that "men worked largely or wholly for pecuniary rewards and would work harder for more money." Incentive pay systems, such as those in the steel industry, were conceived as a means of driving workers.

The Taylor approach also dwelt on eliminating human variability from the production process through engineering and method experts. Workers became mere attachments to the machine, feeding it, monitoring it, repairing it. A perceptive Japanese labor expert, Haruo Shimada, explained the meaning of this in a 1986 study comparing Japanese and American auto factories. By minimizing human variability in production, American engineers protected the system from human error and misjudgment. "However, ironically," he adds, "it cannot for this reason take advantage of the potential productivity and creativity of fully utilized human resources." Minimizing the possibility of human error also minimizes the use of human ingenuity to correct a production failure, as well as reform an outmoded labor-management system. American management, like Taylor, put its faith in systems rather than people. "In the past," Taylor wrote, "the man has been first; in the future the system must be first."[4]

Workers came to be regarded almost like spare parts that could be replaced with no damage and little cost to the production process. This policy conformed, not so accidentally, to the social and economic reality of the times. The massive immigration of the early twentieth century provided an enormous pool of replacement labor. The companies paid low wages, invested little if anything in training, and got what they wanted, a pair of hands and a pair of feet that moved in response to orders. Few managers saw the need for motivated and committed workers.

Although a Tayloristic division of labor most readily evokes the visual image of an assembly line (and a hilarious Charlie Chaplin frantically trying to keep pace with a speeded-up line in *Modern Times*), in fact its principles were applied in many industries such as steel which had no assembly lines. Jobs were defined narrowly, wage incentives installed, and detailed shop rules and operating procedures promulgated. New workers were hired at entry-level laboring jobs and given neither training nor basic orientation about the processes and hazards of steelmaking. They learned new production skills by watching, and occasionally filling in for, experienced steelworkers.

Sam Camens of the USW vividly described how scientific management influenced U.S. Steel's employee policies. "You were told when to do something, how to do something, how far you could go. Every job was broken down to its smallest integral part so that, as we said in the plants, 'You could hire any dumb hunky off the boat and put him into the plant, and he could do the job.' ... And it developed a perception in America that you've got to guard a worker, that he can't be trusted, that you've got to have a foreman on his back, that he's irresponsible."[5]

Specialization, the separation of doing and thinking, labor-force control methods, and inviolable procedures characterized the system built on Taylor's principles. Contrary to popular notion, it was management—not organized labor—that introduced as early as the 1920s what today seem to be artificial divisions between jobs—for example, the machine operator who is not allowed under union rules to screw in a light bulb but must wait for an electrician. Rea McKay, a McKeesport steelworker, recalled that even as early as the 1920s, when steel management had complete control over work practices, a sharp division existed between production work and maintenance work.

By the late 1930s, when the union came on the scene, many steel companies had rudimentary job classification systems in which thousands of mill occupations were narrowly—but loosely—defined. Supervisors could move people from job to job, subject to the restrictions of skill, of course. However, pay rates for the same job varied widely within plants, from plant to plant within a company, and from region to region. One of the USW's early goals was to eliminate these "inequities" and achieve equal pay for equal work on an industrywide basis. After two years of joint study and negotiation, the USW and most steel firms signed agreements in 1947 that eliminated inequities and classified all production and maintenance jobs.

Despite the "traditional trade-unionist's distrust of so-called scientific, industrial engineering methods," USW leaders accepted a job evaluation plan developed by the companies. The parties evaluated each job and assigned points for skill, effort, working conditions such as heat and hazards, and responsibility for materials and equipment, and so on. The jobs were grouped according to point total into thirty-two classifications, each of which specified a basic wage rate. A classification manual included a written description of the duties of each job. At U.S. Steel alone, classifications and descriptions were developed for about twenty-five thousand production and maintenance jobs in about fifty plants with one hundred fifty thousand employees.

Management "legally" could rearrange job duties but not by transferring the primary function of one job to another. If the assignment of new duties raised the point total of the job by one full point or more, without a corresponding wage increase, the union would have a good basis for a grievance.

Adoption of the job manual codified and gave quasi-legal status to the preexisting organization of work in the mills, which in turn had been based on faulty assumptions about worker motivations. This increased the rigidity of the work system and the industrial relations derived from it. Obviously, distinctions must be drawn between different kinds of work calling for different skills. The classification systems in most manufacturing industries, however, divide semiskilled production work, with overlapping and related duties, into a hard mosaic of minutely defined jobs. It encouraged workers to think of themselves as "owning" a specific job whose pay was determined by the content of the job, not by the employee's ability.

Job classification, combined with seniority rules and the legalistic grievance procedure, also guaranteed that labor-management relations would be characterized by a continuing war for control of the shop floor. Michael J. Piore and Charles F. Sabel give a penetrating analysis of this kind of work organization in *The Second Industrial Divide.* The American system of shop-floor control in mass production, they write, "produced a parceling of de-facto rights to property in jobs that would bring a smile of recognition to the lips of any historian who had studied the struggles between lords and peasants in medieval Europe."[6]

Labor's interest in categorizing jobs as narrowly as possible has two motivations. One is to prevent bosses from exercising their biases in assigning duties to people, like putting blacks on the meanest job. This is another very real problem, especially when the management system is highly authoritarian, as it was in steel. The second labor objective is to maintain the level of the work force, an understandable goal which, however, can't always be consistent with remaining competitive over a long term. That management and labor can work together to establish a flexible, as well as fair, system is demonstrated in some of the more participative workplaces of the 1980s (see chapter 23).

By the 1980s, steel managers probably no longer supported the more extreme of Taylor's views. But the industrial bureaucracy that had been constructed on Taylor's ideas continued to operate *as if* those principles were still tacked to the bulletin board. As Maccoby says, supervisors were trained to use "administrative methods that are imbued with the theory of economic man."

Application of the theory that employees don't want to be involved in their work left no room for innovation by steelworkers. "We had a lot of very good people, especially in the top hourly jobs," Larry Delo said, "who resented the supervisor telling them what to do. But in most cases they wouldn't try to think of alternative ways of doing something to get us out of trouble if things went wrong. They'd do it by the book. If not, they'd be reprimanded."

In a steel mill perhaps more than any other kind of factory, the employee on the production line should have leeway to make changes when an untoward event occurs. Foremen can't be everyplace at one time. Maccoby and other students of bureaucratic systems note that power to make decisions should reside where problems must be responded to, on the shop floor. The classic studies of bureaucratic organizations tell us what happens when employees are prevented from influencing decisions. They adopt "retreatism" as a personal solution. That is, they "choose to reduce their involvement and to commit themselves as little as possible to the organization." For the alienated worker, a typical work day in a steel mill consisted of "punching in, catching hell, punching out," as the saying goes.

U.S. Steel management also made extensive use of rules and procedures to convince itself that the plants were producing good steel. Written procedures can be helpful in establishing a framework for carrying out the objective of the firm. But when the administration of rules consumes much of the time of middle-level managers, it blurs the real objective—in this case, making steel—and, eventually, *becomes* the objective. "To help maintain order and consistency," writes a critic of corporate bureaucratic systems, "lists of standard operating procedures, employee handbooks, and various policy manuals proliferate in every corner of the organization. When someone does something novel, there is a knee-jerk reaction in which bureaucrats lunge for their manuals, hoping to find guidance therein."[7]

In the Mon Valley plants, first-line foremen had to spend a large part of each shift filling out forms to show that various procedures—production procedures, safety procedures, and procedural procedures—had been followed. Bureaucratic administration also required practically continuous meetings to formulate new procedures. Actually, Al Sarver said, the plants of other steel producers that he visited may even have been more bureaucratic than U.S. Steel. But he particularly remembered the endless meetings at the Duquesne plant.

"I used to start work every day at five o'clock in the morning because I knew that after eight o'clock, because of all the horseshit, all the meetings, I would have very little time to do my job," Sarver said. "So I'd start work at five o'clock in the morning to make sure that everything that was going to be done the next twenty-four hours in terms of making steel was addressed. Because I knew I'd spend the rest of the day in meetings." Employee involvement means even more meetings—but meetings to solve problems, not to create more procedures.

The Safety Program: A Numbers Game

U.S. Steel's safety program, paradoxically, provides the best illustration of the evils of the procedural life. The company had a first-rate safety program and spent considerable amounts of money and energy to ensure safe working conditions, emphasizing preventive safety measures. Frequently it posted the lowest accident rates in the steel industry. While there is a certain unreliability in safety statistics, I am convinced from visiting U.S. Steel plants and talking to employees over a twenty-year period that a real commitment to "safety first" existed throughout the corporation. "If we had made steel the way we handled the safety program," Sarver said, "we'd have made the best steel in the world."

Even the safety program, as Sarver acknowledged, fell victim to proceduralitis and, tragically, became a management weapon to keep workers in line. This came about partly because U.S. Steel's Job Safety Analysis (JSA) program was based on Taylor's "one best method" concept. Written safety procedures described in step-by-step detail the "one safe way" to perform every job in the plant, including operating furnaces and mills, doing skilled maintenance work, and sweeping the floor.

If workers followed the instructions to the letter, it would be "impossible" to get hurt. Foremen issued a safety violation notice to any worker caught deviating from the correct procedure. A serious violation, such as recklessly driving a forklift, called for immediate suspension with loss of pay. A less serious violation, such as working without a hard hat or eye goggles, would earn a warning notice. The accumulation of a certain number of warning slips in an employee's personnel file (five per year in most plants) automatically triggered a suspension.

In the 1960s, I began hearing complaints from local union

officers that foremen had to issue a certain number of safety violations per week or month to satisfy quota systems. To make their quotas and avoid a reprimand, many foremen passed out disciplinary slips for minor, involuntary infractions—a worker might take off his hard hat for a few seconds to wipe the sweat off his forehead—or even for no infraction. The corporate policy didn't establish a quota system, but the practice became common in many plants at the order of general superintendents or department heads. So pervasive was it that the USW raised the issue in companywide negotiations in the late 1960s.

I forgot about the problem until I did a series of interviews for this book and discovered that the quota system had not only lingered through the years but become even more widespread. Everybody I talked to brought it up voluntarily. To workers and supervisors alike the bogey system represented an exercise in absurdity, established to appease the bureaucracy's hunger for meaningless statistics. One of the angriest persons I interviewed was a retired manager named Jack Bergman.

I visited Bergman at his home in West Mifflin on a cold night in February 1986. Driving up the valley on River Road, I had a sense of going through a tunnel in which somebody had switched off the overhead lights. Then I realized, for perhaps the two dozenth time in recent years, that the lights would never go on again. The glow of the Bessemers, the flames from boiler blow-off stacks, the lick of fire sometimes seen atop the blast furnaces, the rows of silvery lights in the finishing mills—these were gone for good. But I caught the last bit of sunset far down on the horizon, west of the river. In the old days, I asked myself, had we ever noticed the setting sun in the midst of the Mon Valley's pyrotechnics? I couldn't remember.

The Bergman home sat on the brow of a ledge overlooking the Steelworkers union hall and the Irvin Works. If you counted the mileage consumed by the highway that switchbacked up the hill, the Bergmans lived two miles above river level. Their bungalow was an updated version of the old steelworker's home—mortgage-free (paid off in 1962), modestly adorned, and crowded with furniture and people (the Bergmans had brought up four children here). It had the lived-in quality of a well-worn bathrobe. We sat in a paneled den, Bergman in a felt-covered rocking chair and I on a couch with a pink slip cover.

Bergman had received an engineering degree at Lehigh and started with U.S. Steel in 1952 as a management trainee. He spent many years in the National Works open hearth shop as a "melter" (a first-line foreman in charge of the furnaces), served as assistant superintendent of the blast furnaces, and moved up to superintendent of the shipping and transportation department. Transferred to Duquesne in 1979 during a period of cutbacks, he had to revert to turn foreman. When the Duquesne plant was closed in 1984, Bergman retired with thirty-two years' service. Although he had a comfortable pension, Bergman couldn't sit still. He took a $5.46 per hour job as a clerk for the state Department of Environmental Resources.

We had known each other fairly well in high school, but I hadn't seen him for years. A fine athlete, Bergman played fullback on two of the best teams ever fielded by McKeesport. Having tried once to tackle him head-on in a scrimmage, I remembered that Jack had an oddly intimidating way of running, his knees pumping high and on the outside, like paddle wheels mounted on either side of a boat. I had a sensation of falling under churning wheels. On his way back from the goal line, Jack helped me get up. After that, I pretty much stuck to writing about football.

At six-foot-two or three, Jack still looked impressive on his feet. For years he has headed a four-man crew refereeing high school football games in western Pennsylvania. He had direct, gray eyes, an angular face inherited from his Alsace-Lorraine forbears, and he talked with a restless intensity, a sort of controlled anger. Although he had enjoyed his job in the early years, Bergman said, he was glad to get out of the mill because of the way relations had decayed.

"What was more aggravating than anything was the safety program," Bergman said. "It was sickening the way it was handled. The brass would start every meeting at divisional and higher levels with all kinds of safety statistics. How you compare with another department, how you compare with another foreman, how you compare with another plant. They had one plant competing with another to see which could show the most safety violations. Down at Homestead they were showing eight thousand violations a month, and we had only five thousand at McKeesport. No matter what the people content was, they'd compare us. Who would want to work in a plant with eight thousand violations?"

At various times, Bergman said, he had to hand out violations under a quota system. "It even got to the point, you'd have to come in on your day off and observe until you got your violations. I always got mine in, except not enough time-off slips. If somebody really earned time off I'd give it, otherwise no. You just put something minor in, just to give them the paper that they wanted. I got so fed up with it, I said one day to a VP from Pittsburgh, 'You know, all these numbers you want generated, all of this stuff you're jamming down our throat about Homestead...I wouldn't want to work in a plant with all those safety violations. I think something ought to be done. You know, correct the problems and have the total going down, not each month trying to break the record of the month before.' No, they'd run up the numbers."

William J. Daley worked for U.S. Steel for thirty-eight years as an hourly employee and had risen to the top job of roller on the slabbing and blooming mills at Duquesne. "When I started in 1948," he said, "safety was hardly anything. But it got better. The company spent a lot of money on safety. But then they got shitty about it, around 'sixty-four or 'sixty-five. Foremen had to spy on people instead of doing their job. The general foreman had to come in on night turn, on his vacation, to write up enough slips to meet the quota. This continued up to the end. I saw guys get a slip for not holding a handrail going up the stairs. I got a slip for taking one glove off to hold a caliper. I was on vacation once and heard that a foreman I liked needed a couple of violations. I saw him on the street and told him, 'What the hell, give me a couple.'"

Bill Schoy recalled a rainy day when he stood on a girder about 65 feet above ground, directing a motor crane operator as he swung a 120-foot high boom. "We probably shouldn't have been working in the rain. You know, the rule in construction is, the first drop that hits you is God's fault. The second drop is your fault. Anyway, I'm looking up, watching the top of the boom to make sure he cleared a high tension power line. The rules were not to take your goggles off. That's a good rule. But the rain was streaking my glasses. I took them off, even though I knew my foreman was on the ground below me. He whistled. 'Put your glasses back on.' I told him I couldn't see the top of the boom. 'Put 'em on,' he said. What was I supposed to do, let the guy electrocute himself? I said, 'Either I take the glasses off, or you get somebody else up here.' He said, 'You get your ticket and go home.' I think I won the grievance, but the whole thing was a mockery."

To many employees, the division of labor mandated by

safety regulations often defied common sense. Manny Stoupis recalled that in the 1950s mill crewmen would help mill-wrights on a repair job, "not to put a guy out of work, but just to give a helping hand." He continued: "We'd have a break-down—say, a sideguard would come out. You needed a mill-wright to put it back in. He couldn't do it all alone. All it amounted to, you'd jump up on the rolling table, lift the side-guard up with a bar, and he'd put a bolt through. If he'd had to call for a rigger, we'd lose a half hour to an hour of production. But later, when the JSAs came in, you weren't allowed to help the millwright. The JSA for replacing sideguards didn't pro-vide for it. If I'd jump up on the table and they'd catch me, it'd be a one-day suspension. They jammed it up your ass. I couldn't even put a pin in a turntable. All it needs is a whack with a sledge. Nope! Not allowed."

The final irony of the safety game was that the reverse side of the disciplinary slip had a form to commend a worker for doing the job right. It was rarely used. Division and plant superintendents "wanted to see the violations and the time off, not the pat on the back," Bergman said. "If there were too many turned in that men were doing the job safely, that was nix."

Corporate managers tended to dismiss the anger over quotas as a minor problem down in the plants. After all, quotas weren't part of corporate policy. The fault lay with the unfortunate bureaucracies that existed at the plant level, the executives said. Moreover, workers complained too much; they would just have to learn to live with the bogey system for the greater good of preventing injury and death.

Sitting in their sixty-first-floor offices in the U.S. Steel Building, the executives could delude themselves with such a patronizing argument. It ignored a basic psychological fact: People don't like to be trifled with. The bogey system con-vinced rank-and-file steelworkers, as well as many super-visors, that upper management persisted in treating them like children. The absurdity of it undermined morale and demon-strated to employees that management couldn't be trusted to make rational decisions about their plants, jobs, and live-lihood.

Quantity Over Quality

U.S. Steel's Mon Valley mills turned out many excellent prod-ucts over the years. The Edgar Thomson Works pioneered steel rail production. Homestead, with its giant plate mills, produced the structural steel used in the Empire State Building, the Mackinac Straits Bridge, and the Sears Tower. Duquesne was known for high-quality bar products used in critical car components such as crank and cam shafts. National Tube in McKeesport led the world in pipe-product innovations during its first one hundred years and produced, among other things, the pipe that was used to construct the Big Inch and Little Inch oil pipelines during World War II. When they were first built, the strip mills at the Irvin Works were considered marvels in the high-speed production of quality sheet steel. Clairton's mills turned out small structural shapes that were used round the world.

Yet this reputation for quality became sullied in the 1960s and 1970s. Why? Even without replacing the old furnaces and mills that lined the Monongahela, U.S. Steel could produce steel at a profit until the early 1980s. "We were content to do what we were doing," said Al Hillegass, who became U.S. Steel's vice-president of steel operations in 1980. "We could sell everything we could make and didn't look for better ways to do it."

Quantity came to have a higher priority than quality. This was especially obvious to highly skilled rollers like Bill Daley of Duquesne. To become a roller, a worker had to advance from an entry-level job up the "line of progression," learning each job on the mill until his seniority graduated him into a vacant roller's job. Generally this took twenty to thirty years. Sitting in a "pulpit" that straddled a conveyor line and facing big sets of steel rolls some twenty feet away, a roller could inspect each new, hot ingot as it passed under the pulpit and headed toward the rolls. One glance would tell him whether the ingot was "green" (too cold) or "soft" (too hot), whether it had the correct carbon content, and so forth. A roller occupied such a pivotal position in the plant—at the beginning of the process in which raw steel was converted into finished products—that, in a sense, he could peer right into the management decision-making process. If orders came down to push the steel through, quantity over quality, he'd know it.

"From the 1960s on, until nineteen seventy-four or 'seventy-five, it was quantity they wanted," Bill Daley told me. "You had the customers waiting for you. When you have the volume, you can do anything. We sent steel out that was so bad it had seams in it. The people in the Pittsburgh office wanted the tonnage. A lot of customers got junk from us. I rolled steel that wasn't fit to be rolled, cold steel, or too hot. The foreman would say, 'Don't worry, the customer has to finish it anyway.' "

Larry Delo was a conscientious man. During one conversation he told me, lowering his voice with embarrassment, that he had to go along with the production-at-any-cost philosophy as a soaking-pit foreman. "The bosses would call down and ask why we weren't getting the ingots out of the pits and into the rolling mill," he said. "The pits weren't heating right, I'd tell them. The steel wasn't hot enough. 'Get it moving!' they'd say. Sometimes I'd find a way to delay, but not always. Sometimes I took green steel out and gave it to rolling. I knew it. Everybody knew it."

As the shipping superintendent at National in the 1970s, Jack Bergman knew the quality of the pipe that he loaded into barges. "The main thing at U.S. Steel always was 'pipe over the pump,'" Bergman said. "Pump" referred to jets of water that washed the pipe as it left the finishing line. "That was what managers were rated on, the pieces of pipe or tons of pipe that passed over the pump. If you didn't get the pieces over the pump on a turn or on a twenty-four hour basis, a lot of hell was raised by upper management with lower management. The wage people were looking for quantity, too, because that pushed up their incentive pay. Some of the stuff wasn't the greatest."

Oil field customers began to complain about quality problems. In 1981 orders came down from Pittsburgh: Quality should be the priority. "All management in the valley had to attend a four-day program in Homestead," Bergman remembered. "It was all on quality, the same program that the Japanese used. A vice-president from Pittsburgh talked to us. 'This time we're behind you 100 percent on quality. We mean it!' As if to say all the other times it was bullshit! We were told this is life and death. If you get steel from a sister plant that doesn't meet specifications, send it back. One of the general foremen at National was getting coils from Gary to make pipe. After the Homestead meeting, he put two bad coils back on a truck and sent them back to Gary. The next day he was called to the office. 'You do that again, you're out of a job.' National had to pay the cost of sending the coils back. Life and death!"

Top management tolerated poor quality far too long. It also

made little effort to break up the bureaucracies at the plant level and no effort to encourage participation by rank-and-file workers. The management system, in fact, militated against obtaining worker commitment to quality and productivity. Corporate executives, by example, established the style of management that would be adopted down through the hierarchy. It was a macho style perhaps best summed up by Joe Odorcich. "U.S. Steel talks through their balls," he told me.

Section 2B

The deteriorating relations in U.S. Steel's Mon Valley plants by no means can be laid solely to poor management of people and processes. The USW's retreat from participation in the late 1940s left a void in union policy toward work itself. While union leaders didn't actively encourage featherbedding, they did little to implant the idea that the union should bear some responsibility for making the plants productive and competitive. This was a major failure.

During my years as a reporter in Pittsburgh, I came to know many of the people at McKeesport Local 1408. Among the long-time officers were Edward Galka, who served several terms as president and vice-president, and Bill Behare, who held various union offices for twenty-six years before retiring in 1985. Both knew the mill and its people as well as anyone, both had been deeply involved in the tangled thicket of USW politics in the Mon Valley, and both had always been candid with me.

In 1986 I asked Behare and Galka to look back on their years at National Tube and assess the union-management relationship and how it changed. My question proved to be an easy one because, in their view, nothing changed. From his first day in office in 1958 until he took early retirement in 1976, Galka said, "management never wanted even to meet with us. The only reason we met," he added, "was because the contract called for periodic meetings. We sat mostly glaring at each other, the union guys putting proposals on the table and listening to management say no."

Behare, a tall, round-faced man with a soft voice, felt that many possibilities went unfulfilled. "The plant managers didn't feel the union was there to help," he said regretfully. "They felt the union was there to stymie progress. So we never settled anything important locally. The rules were set in concrete, and we couldn't bend them. No flexibility. If anything precedent-setting came up in a grievance case, both sides would take it out of our hands and settle it in Pittsburgh."

"The union," Behare continued, "never instilled in the employees' minds that production *was* important. The attitude of a lot of guys was, 'What can I get away with, do the least amount of work that I can?' My dad was a coal miner for forty-five years. When he met with his friends after work, they'd talk about the day in the mine. They were proud of how much coal they put out. At National, if you went to a bar after work, you'd hear talk about what they got away with."

The majority of hourly employees didn't go to work with this attitude, but enough did, especially in the 1970s, to cause real problems in the plants. "I enjoyed working at Duquesne," said roller Bill Daley. "We all felt that way until the younger steelworkers came in, in the seventies. They didn't give a damn. They thought they could get a job anywhere. A lot of them were dopeheads."

Jack Bergman said too many younger workers seemed to feel that "jobs should be created for people where there were no jobs. It seemed that was the American way. You get paid top dollar for doing as little as you can. I had four men in a shipping crew, loading pipe in a bay area, I'd get a report that two would be working, two sleeping. I'd go there. They'd be smoking pot, actually. The stench in the air...people in the cranes would almost be overcome by the stench coming up."

Practically everybody who worked in the plants after the 1950s, that I talked to, expressed unusually strong opinions that the 1970s generation of workers did not measure up to earlier groups. Such claims, of course, must be considered first in the context that it is common for the older generation to regard the younger as less able and less willing. Feelings run deep on this issue, and trying to sort half-formed notions from more objective observations is impossible, barring a full-scale study.

Another labor problem that became one of the most controversial issues in the steel industry involved Section 2B of the basic labor agreement and the question of crew size. Did the union engage in widespread featherbedding by forcing the use of bloated crews? Without doubt, this happened in some departments of some mills.

Section 2B protected certain kinds of past practices, such as the length of break periods, starting and quitting times, and crew sizes, that had been embedded in local custom. The concept that past employment practices should be preserved, whether they redound to the benefit of workers or bosses, is practically as old as employment itself. No one wants to change old ways, especially if it means losing something in the bargain—a few minutes of relief time, an extra hand to ease the workload, or whatever. Moreover, many practices originate as part of a bargaining tradeoff: A manager approves the practice in return for union agreement to drop a troubling grievance.

Steel management originally faced no restrictions on its ability to change local working conditions. In 1945, however, the USW won a vague but limited protection of past practices. This blossomed into a five-point provision in the 1947 U.S. Steel agreement that the union later extended to other companies. Basically, Section 2B says that management can't change or eliminate any existing practice that confers greater benefits than the master contract, except by negotiation. The critical phrase, as applied to work crews, prevents management from changing the crew size unless "the basis for the existence" of the practice—the "underlying circumstances," as arbitrators have construed the words—is also changed.

The USW originally developed the provision to protect practices other than crew size; for example, unusual weekly schedules that existed in some plants. In the early 1950s, however, the union began citing 2B in the grievance procedure to prevent management from unilaterally cutting crews. Ben Fischer developed a "strict construction" strategy, arguing in hearings that the arbitrator must interpret the language of the labor agreement as it exists, not as the company said was intended. This argument prevailed. Starting in 1953, the U.S. Steel Board of Arbitration issued a series of rulings that interpreted 2B as covering crew size. These rulings enunciated the concept that a company could change the number of workers on a particular operation without violating the contract only if it installed new equipment or technology or otherwise changed the "underlying circumstances" of the job.

The effect of 2B often has been misstated. One writer recently asserted that it enabled the USW to "block the adoption of new technology." This is dead wrong. Only the lack of capital, or management indifference, blocked the adoption of new technology in the steel industry. If anything, 2B provides an incentive to management to introduce new technology as a means of increasing productivity. Unquestionably, though, the provision deterred the producers from correcting manning mistakes of the past and led to excessively large production and maintenance crews in some places.

Early in the history of 2B, U.S. Steel realized it had made a dreadful mistake, from a management point of view, in accepting the clause in the first place. The company argued unsuccessfully in one arbitration case that management negotiators simply neglected, in last-minute bargaining, to add a qualifying phrase which would have kept crew size out of purview of the provision. U.S. Steel tried to void the clause in 1952 and declared all-out war over 2B in 1959 negotiations, demanding its elimination. The demands solidified a previously split union and brought on the 116-day walkout. When it ended, 2B remained intact.

While the industry declared that 2B had caused immense harm, it detailed few specific instances. A Past Practice Committee set up by the two parties as part of the 1959 strike settlement launched an investigation but came up empty-handed. Local managers rebuffed attempts by the committee to compile a list of inefficient practices that couldn't be changed because of 2B. Marvin Miller and Philip Scheiding, a lawyer, coordinated the investigation for the union. "We couldn't get anybody to confess that he or she originated an inefficient practice," Miller said. The companies, embarrassed by this failure, never disclosed that the committee had been disbanded without reaching a conclusion.

Employment statistics indicate that the industry reduced manpower in a very substantial way, despite 2B. In 1950 the industry employed an average of 503,309 hourly workers and shipped 72.2 million tons of steel. Twenty years later, in 1970, the AISI producers shipped 90.8 million tons with 100,000 fewer wage employees (403,115). After each recession-induced layoff, the mills resumed production with a smaller work force.[8]

Shifting the spotlight to the plant level, 1,080 people worked in Manny Stoupis's blooming mill department in 1964 but only 600 in 1982. The crew on an individual blooming mill at National declined from 12 to 6 between the 1940s and the 1960s, according to Rea McKay. U.S. Steel eliminated one job by installing a television set, enabling a worker to monitor the scrap chute from a remote point.

The use of TV monitors was one simple way that the companies could change the underlying circumstances. "I thought 2B was the greatest management tool we had," said a retired U.S. Steel labor relations official in a 1986 interview. "If it was really a practice that cost you money and not something that simply offended you, you just put a couple of guys there and had them study it very carefully. Find out the basis for establishing the practice, and then change the underlying circumstance. Used that way, it forced management to carefully analyze its operations that cost it money. I say that 90 percent of the time you could legally, legitimately change the practice and make some money."

Nevertheless, the hassle of dealing with Section 2B undoubtedly delayed productivity-enhancing changes in the steel mills. Over several years, a series of incremental changes in equipment or usage might make the old manning schedule obsolete without qualifying as a change in the underlying circumstances. At the same time, the general educational level of the work force rose steadily from the 1940s. By the 1980s, these and other trends probably did make possible major changes in job content and thus the size of crews. Many skilled crafts, for example, could be combined. But the rigidity of steel's industrial relations system, the adversary nature of relations on the plant floor, and the resistance of management bureaucracies became major barriers to change.

Many locals fought management every step of the way in its attempt to shave employment, filing grievances over obviously legitimate changes as well as questionable ones. Responding to formal complaints can take up a considerable amount of a supervisor's time, and many backed away from making needed changes when threatened with a grievance. Too often, Al Sarver said, management "caved in" on the issue of cutting bloated crews. On the other hand, he didn't blame the USW for trying to save jobs. "If the union doesn't go to bat on that, there's no point in having a union," he said. "What *is* wrong is when they fight for a guy who's been chronically absent for the last thirty years and they want back wages for him being suspended."

In many mill departments in the Mon Valley, this kind of adversary bickering became the rule rather than the exception. Caught in a legal gridlock, the two sides postured and fulminated. The union complained that management stalled

on settling grievances and obeying arbitration rulings. Supervisors complained that, increasingly, local unions elected
unscrupulous grievance representatives.

A decline in the quality of union leadership in USW District
15 of the Mon Valley did occur in the 1960s. This final chapter
in the story of what happened to the mills in the Mon Valley
is one of the sadder ones.

Union Corruption

Holding union office carries many temptations, including the
gratification of desires for money, power, alcohol, and sex. In
the prosperous 1960s, money was the lure. Dues income piled
up at the big basic steel locals, tempting some unionists to
misuse funds set aside to pay officers in lieu of company
wages when they worked on union business. Moreover, under
Director Paul M. Hilbert the District 15 headquarters itself
became a fount for unionists thirsting for money. A one-time
president of the Firth-Stirling local in McKeesport and a Dues
Protest leader, Hilbert was elected as an anti-administration
director in 1957. He held the post for twelve years until un-
seated by a reform movement headed by Joe Odorcich. After
leaving office, Hilbert pleaded guilty to embezzling money
from a union strike fund and received a suspended sentence
from a federal district judge.[9]

During Hilbert's tenure, the District 15 office liberally
doled out union funds to supporters at various locals. There
were more than enough takers. Hilbert's cronies drank at restaurants and taverns where he set up bar tabs, paid for by the
union. The money flowed especially freely during union election campaigns. Odorcich, who had been on the District 15
staff since 1949, said a number of local officers received
payoffs of $250 to $350 a month by filing expense reports for
largely nonexistent union expenses. "It became a standing
joke in the district office to see these fellows come in once a
month to sign the sheets," Odorcich said. "We had yellow
expense sheets then, and we'd say, 'Here come the yellow
birds.' If they needed extra money, they'd say to Paul, 'So and
so is mad at you down at the local, and it's important you
keep the votes in the local.' They'd wind up with another
yellow bird."

Votes were bought and sold in many milltown bars, and

various kinds of ballot fraud occurred in a number of local and district elections, resulting in investigations by the U.S. Labor Department. "The district oozed sleaze," said one government investigator.

McKeesport Local 1408 got caught up in this corruption in the late 1960s. Before that, however, the local under President Tony Tomko had demonstrated that persistent rank and filers could force the USW leaders to listen to their ideas. One of the more colorful local presidents in the USW, Tomko put together a cohesive team of local activists—including Manny Stoupis, Bill Behare, Ed Galka, Dick Grace, and others—and led a movement among basic steel locals to gain influence in the USW's bargaining decisions.

A shrewd politician, Tomko solidified the major ethnic groups at National by including Irish, Croats, Slovaks, and Poles in his leadership group. (Blacks constituted a small minority of workers at the plant until the 1970s.) Tomko wasn't an idealogue, but he had a strong rebellious streak. On one occasion, U.S. Steel suspended him for leaving work unauthorized; in fact, he crawled over a fence to get out of the plant. When he returned to work, the punishment continued. Tomko was put in a labor gang and assigned to the hardest, dirtiest jobs. He began wearing a suit and tie to work. "There he'd be, down in a ditch, digging a sewer line and wearing a suit," Behare recalled. "Everybody would see him and laugh. This made the bosses look bad. It got so hilarious that they took him out of the labor gang and gave him a better job."

Tomko also kept his finger on the pulse of life in the mills. He knew what bothered people and what they wanted in the next contract. In the early 1960s, he and his fellow officers felt that the International had not pushed hard enough for improvements in benefits such as pensions, insurance, widow's benefits, and vacations. This became 1408's agenda. Tomko organized his official family—some fifteen local officers and committeemen—into a disciplined lobbying group and took them everyplace: to negotiations, to Congressional hearings, to conventions, to union summer school at Penn State. Local 1408 practically invented the politics of confrontation in the USW.

"We spent money like there was no end to it," Behare recalled. "Some of the guys maybe didn't like Tony, may even have hated his guts, but going on these trips was like being on vacation. We'd go down to Pittsburgh for negotiations and stay at the Penn Sheraton [as the William Penn Hotel was called in the 1960s], sixteen of us, rent a couple of rooms. We'd take the local's checkbook with us in case we needed emergency funds." The team members demonstrated for their bargaining program, lobbied people from other locals, and badgered the district and International officers when they could get close to them.

Local 1408's biggest victory occurred in 1965. From talking to workers in his plant, Tomko came to believe that early retirement should be ranked high by the union leaders in negotiations that year. But it had not been among the union's original demands. Tomko and his 1408 crew circulated leaflets demanding "30-and-out," the option to retire after thirty years of service regardless of age. Written by Tomko, the message on the leaflets contained some interesting, if unpunctuated, ideas. If workers could retire before their "physical makeup has ... totally degenerated," one leaflet said, they could put their knowledge and craftsmanship to work in the Peace Corps and the National Park Service, thus giving the nation "a clean almost god-like look."

The "30-and-out" campaign spread rapidly to other locals and workers in the mills. A groundswell developed, putting immense pressure on I. W. Abel and his top bargainers. Emissaries from Abel came to Tomko and asked him to stop the campaign (I happened to be present at one such meeting). Early retirement would cost too much, they said, in a year when the USW leaders had committed themselves to winning a large wage increase. Tomko remained unmoved.

Local 1408 won its battle. The settlement package in 1965 contained the first 30-and-out provision negotiated in private industry (twenty-year and thirty-year pensions were common in some government occupations), and it set a precedent that most other unions soon clamored to copy.

The 1408 group split up in 1968. Odorcich decided to run for director to rid the district of corruption. Behare, Galka, and other officers of 1408 sided with Odorcich, becoming his strongest campaign workers. Tomko, once a Hilbert opponent, switched sides amidst rumors of a large payoff. On election day in early 1969, election misconduct occurred in a number of locals, including 1408, and Odorcich didn't gain his board seat until the union, under pressure from the U.S. Labor

Department, conducted a rerun in 1970. Odorcich cleaned up the district. This kind of corruption is rare today in the USW. But the political sleaziness of those years took its toll on member loyalty and public regard for the Steelworkers in the Mon Valley. It also added an unsavory element to the devitalized climate of decline.

Chapter 13

Atlantic City, September 1982

As the recession grew worse in the summer of 1982, un-employed workers began to call for government action. On July 31, two hundred people jammed a "hot and steamy" meeting room at Local 2227 in West Mifflin to demand an extension of unemployment compensation (UC) benefits. The unusual size of the audience on a Saturday testified partly to aggressive organizational work by the Mon Valley Un-employed Committee. The self-help group had expanded its activities since the spring; in July it distributed free food to nine hundred families in the Mon Valley. A food bank started by Ron Weisen's Local 1397 in Homestead also served five hundred families. But for these efforts some people would have gone hungry—in one of the wealthiest industrial regions in the world.

The length of the business slump had exhausted supplemen-tal unemployment benefit (SUB) funds at most companies. Many steelworkers who had been laid off since 1981 soon would be left without any income when their thirty-nine weeks of UC eligibility ran out. Democratic members of the Senate Finance Committee had called a hearing at Local 2227 to collect testimony on a bill to extend benefits to fifty-two weeks. They were unlikely to find opposition in the Mon Valley. But inasmuch as the Reagan administration opposed the legislation, the hearing gave steelworkers a forum for attacking Reagan, the steel industry, and the business estab-lishment. Barney Oursler, a laid-off steelworker and leader of the Unemployed Committee, said his group would take "direct action" if utilities cut off electricity and gas. Andrew "Lefty" Palm, the newly elected director of District 15 in the Mon Valley, declared that the steel companies "want concessions

333

without jobs . . . this is union busting! The White House is their ally in this attack on us." The audience "roared approval of his words," reported the *Pittsburgh Press.*

Little of such militancy was displayed a month and a half later when thirty-four hundred steelworkers gathered in Atlantic City for the USW's biennial convention. For several months the Reagan administration had been predicting a business rebound in the second half of 1982. By September, however, the divine will of the marketplace hadn't responded to the president's wishes. Despite the stimulative effects of the Reagan tax cut, high interest rates still stifled economic expansion. The national unemployment rate crept up to 10.1 percent in September. Steel production continued to fall at a rapid rate. Only 42.5 million tons of raw steel were produced in the first eight months of 1982, a 37 percent drop from the corresponding period in 1981.

A total of 129,634 hourly and 11,442 salaried employees were now laid off in the steel industry, representing 36 percent of the 1981 work force. About 25,340 steelworkers had been furloughed just since the union rejected the industry's wage proposal in July. The magnitude of this decline and the bleak outlook for the next several months had stunned even the stoutest rank-and-file militants, enabling McBride to set the union back on a concession course with relative ease.[1]

Atlantic City was having one of its milder Septembers. With the exception of a couple of drizzly days, temperatures were in the seventies, and the big new casinos enjoyed a brisk trade. For many decades, organized labor had brought conventions to Atlantic City in September to avoid the summer crowds and to take advantage of a surplus of hotel rooms and the enormous space afforded by Convention Hall. Most of the shops would be closed by then, and only an occasional elderly couple would be strolling the rain-swept boardwalk. In this placid emptiness, the oldtime AFL leaders had led a rich man's life by proxy, staying in splendid suites at group prices in the Traymore and Shelbourne hotels and dining on lobster and soft-shell crabs at Captain Stearns's.

The arrival of legalized gambling had converted the staid old resort city into a tawdry, glittering pleasure dome, swirling with crowds. I watched with astonishment as bus after bus pulled up alongside the big gambling palaces—Bally's, Playboy's, Caesar's, Sands—and disgorged loads of people. Practically running from the moment they stepped out of the bus, the people crushed through the casinos' side doors. Every weekday hordes of people, and twice as many on the weekends, poured into the city by bus and departed the same day after a five to eight-hour fling at the tables and slots. The buses were filled with retired people, representing a vast migration of the gray-haired from New York, North Jersey, and Philadelphia, clutching the roll of coins that is provided as an inducement by the bus companies and that will soon wind up in the slot machines. Several hours later, the same oldsters could be seen sitting passively in queues outside the casinos, waiting for buses to take them back home. It was not a happy sight.

Amidst this crush, the USW delegates were just another group of people traveling from one spot to another to be acted on by economic events. In contrast to the bustle on the boardwalk and in the casinos, the 1982 Steelworkers convention was the most subdued I have attended since my first in 1966. It was as if the depression atmosphere of Gary, South Chicago, Bethlehem, Birmingham, Youngstown, and the Mon Valley had been frozen for travel, transported to Atlantic City, and thawed out in Convention Hall.

In the past, most USW conventions had been marked by contentious debates over internal issues involving union democracy. In 1982, however, no such debate took place. The convention passed strong resolutions condemning Reaganomics and calling for broad new retraining programs and restrictions on plant closings and imports. But such issues do not elicit debate; everybody was in favor. Only a resolution on collective bargaining, offered the first day of the convention, stirred a bit of excitement. It authorized union leaders to "balance" positive wage goals with the need to preserve jobs in a "distressed bargaining situation." In other words, to grant concessions where, in the leaders' judgment, this was necessary. Militant delegates made only a token fight to defeat the resolution.

Delegate Joseph Gyurko of Inland Steel Local 1010 stated a concern of many rank and filers in the basic steel locals. He worried, he said, about the union cutting wages at small me-too companies, or "bucket shops," as he called them. No sooner did one firm receive concessions than "along comes the other one, and one right after another one. They're all singing the same song. They're all singing out of the same hymn book, 'Me, too. Me, too. Me, too.' If we don't watch it, they'll get to the big ones, and pretty soon we'll find ourselves back where we started from." David T. Slaney, from the New England region, observed that concessions in the auto industry "have not put auto workers back to work.... It's time for the American labor movement to say, 'No more concessions!'" These delegates were in the minority, however. Bruce Thrasher, who chaired the convention during this debate, called for a voice vote in response to shouts from the floor, and the ayes had it overwhelmingly.

The next morning, September 21, McBride and his executive board met at breakfast with some three hundred local presidents who belonged to the Basic Steel Industry Conference. The group discussed the July negotiations and the deterioration of the industry since then. Joe Odorcich later said that complaints by jobless steelworkers apparently had had an effect since July, for he noticed "a real change of attitude" among the presidents. After the meeting, McBride reported he had detected "a very clear consensus" to "keep the door open" for a resumption of bargaining. Only Ron Weisen spoke against the idea.

It was by no means evident, however, that a majority of the BSIC members would now vote for wage cuts. Perhaps all that had been demonstrated was that those opposed to concessions did not wish to be openly identified as being unreasonably against further talks. Even the Bethlehem presidents, who had been so staunchly against negotiations in July, failed to speak up. The increasing steel layoffs worried them. The companies' attempt to turn the rank and file against union leaders, in letters sent to employees' homes following the July rejection, also had had some effect. McBride's office did receive a flood of mail from members. The letters ran in favor of granting relief to the companies, McBride had told the industry.

Management's hoped-for revolt of the unemployed against their local officers didn't occur, however. Dick Grace, president of McKeesport Local 1408, an anticoncession man in July, acknowledged that he had received complaints from many members. "When U.S. Steel put out that letter... what garbage!" he said. "But the phone calls increased. They [the company] were playing on the sympathy of the wives. One lady said, 'What the hell are you doing, Dick?' A lot of people looked at it with smoked glasses. After I explained it to them, they understood. But that letter brought the phone calls, and then the unemployment situation got worse and made the mood of the local presidents change."

McBride made no attempt at the convention to mask his belief that the union must reduce labor costs in the steel industry. "Our successes," he said at a news conference on September 21, referring to USW contracts, "have created some problems. [The steel companies are] not playing any games. They're in serious trouble." He repeatedly said that he wanted to "keep the door open" to negotiation, an obvious signal to the industry. "Unemployed people is my biggest concern," he told reporters on September 20. "At fourteen," he said, "I was the breadwinner in a family with a father who wanted to work and couldn't get a job. We didn't have a home to lose. I can imagine the frustration of people who face losing their homes."

The president studiously avoided anticompany polemics. McBride's speech writer, Michael Drapkin, told me that when McBride reviewed a draft of his keynote convention speech earlier in September, he singled out one sentence. Instead of saying "the industry wants to take us on," McBride told the writer to soften it to, *"it would appear* that the industry wants to take us on." He looked up at Drapkin and said: "They don't want to take us on, and we don't want to take them on."

McBride must have recognized, however, that rising unemployment created a bargaining problem. Because of the downward spiral of steel production, the bargain that was possible in July—a union give-up of 50¢ per hour in future wage increases to provide jobless pay to most steelworkers for a year—was *passé* by September. With fewer steelworkers working fewer hours, a contribution by the companies of 50¢ per hour worked would no longer build a fund large enough to provide that benefit. "The call on SUB funds has increased beyond anybody's imagination only three months ago," J&L's John Kirkwood told me at the time. "We're scared as hell." It was imperative, Kirkwood said, that the union and the industry reach an agreement before November 1, when another COLA payment was due.[2]

Some of the more militant delegates kept sniping at McBride. Dennis Shattuck, a Local 1010 grievanceman, compared the USW's situation *vis-à-vis* the steel industry with that of European countries which had tried to appease Hitler in the 1930s. "Every time he wanted something, they went and they talked to him and they gave him some more, and then he wanted some more, and they talked to him again. And the result was a tragic disaster for this country—Pearl Harbor." Weisen also attacked the International on various matters. McBride struck back at these "unreasonable, illogical criticisms that are offered...obviously for the purpose of political aggrandizement." His war with the militants continued.

On the Friday before the convention started, I had interviewed McBride in his sixth floor suite in Bally's Park Place. Everything about the interview typified McBride's simple directness. His wife Delores opened the door for me and quickly disappeared into a bedroom. McBride and I sat down in the living room of a very mundane, no-frills suite, with $20 prints hanging on the wall. This was not the Traymore of olden days. There were no aides hurrying in and out, no hangers-on or cronies waiting on him, and no proffers of drink or food. McBride wore a light blue polo shirt and blue trousers. He seemed stoically at ease with himself and the world, an outward appearance that may have hidden intense pressures boiling inside.

During a long conversation about Round One negotiations, McBride said that he had rejected the industry's proposal because he felt that the companies "went too far." He added: "That feeling may have been shored up by the conviction that if I couldn't buy it, how the hell could I sell it to our membership? It wasn't even borderline the way I saw it."

If this statement accurately depicted McBride's thinking in September, he would undergo a remarkable change of mind by November.

Wage Cuts and Profit-Sharing

A gorgeous Indian summer came to Pittsburgh for a couple of days in late October. Temperatures rose to the low seventies, and the sky looked as if it had been dipped in Mother's laundry bluing. I drove up and down the Mon Valley, looking desperately for signs of revival. Instead, on October 28, one of those spectacular sunny days, U.S. Steel put its only working blast furnace in the Mon Valley, the huge Dorothy Six at Duquesne, on forced draft and stopped casting iron for several days. Aside from strike periods, it was one of the rare times in more than a century that U.S. Steel, or its Carnegie Steel predecessor, stopped smelting iron ore in the Mon Valley. In

1950 U.S. Steel operated no less then twenty-five blast furnaces in the valley. Many of them still existed, towering
above the mills at McKeesport, Duquesne, Braddock, and
Rankin like stern despots. But they were hollow reminders of
what used to be. In 1982 only four furnaces were capable of
making iron, two at Duquesne and two at Braddock.

But that is nostalgia. Technology changes, markets change,
and people must change what they have been doing and do
something else. This was easy for me to say. I met Larry Delo
for dinner on October 29 at the last non–fast food restaurant
extant in downtown McKeesport, a motel on Lysle Boulevard.
Two years later, it too would be closed. He had just survived a cut in management forces at the combined National-
Duquesne Works. Twelve percent, or 82, of the 680 supervisors
and managers at the two plants, were severed permanently
from the payroll in September. More cuts were expected.
"Everybody is waiting for somebody to tell you, 'You're
finished,'" Delo said. "I've decided if it happens, it happens. I'll
find another job." He had been with U.S. Steel for twenty-six
years.

Relations in the mills were getting worse. I told him I'd
heard union complaints that supervisors transferred workers
from one department to another to avoid recalling laid-off
employees. Even Delo, one of the mildest of men, bristled.
Recently, he said, Duquesne had received an order from a
sister plant for a small shipment of semifinished steel. As
shipping foreman, he needed a crane operator, a hook-up person, and a posting clerk for a few hours to fill the order. If he
called in people on layoff, they would get four hours' pay for
the week but lose their UC benefit for the entire week. "If you
wanted to be a bastard, you'd call them in," Delo said. Instead,
he recruited three people from another department, loaded the
steel, and shipped it. The local union filed a grievance. If the
union won, the three affected workers who were not called in
would be awarded back pay for four hours.

"My boss vowed the next time this happens we'll call the
guy in just to screw him," Delo said. He shook his head.
"Some people get the impression that management despises
wage workers. Jesus! My father was a mill man. My father-in-
law and my brother were mill men."

During the same week, McBride announced that the union
would defer payment of a 9¢ COLA due on November 1
pending the outcome of negotiations. Raising wages when the

steel companies' losses were soaring would have been like a gratuitous insult. McBride had resumed discussions with the industry negotiators on October 15, and they had given him some numbers. In 1982, according to projections, the eight companies of the bargaining group would lose at least $3 billion. U.S. Steel's losses were mounting at the rate of $3 million a day.

The companies in July had miscalculated badly the amount of relief they needed and would have put themselves in grave financial jeopardy if the union had signed a three-year agreement under the July terms. When Bruce Johnston and George Moore met with McBride the first week in October to set up a negotiating schedule, the industry men told him, not so joshingly, "You may have saved our ass by not buying the July deal." Now, the industry was determined to get "true concessions"—that is, actual wage cuts instead of the wage freeze discussed in July.

On the morning of October 21, Johnston and Moore received their marching orders from the chief executives of the Coordinating Committee Steel Companies (CCSC). In the afternoon, they told McBride and his fellow negotiators, Joe Odorcich and Bernie Kleiman, that the companies needed a $5 per hour reduction in labor costs. Instead of specifying how to achieve the $5 cut, the industry men took an unorthodox approach. They handed McBride a list showing the components of hourly employment costs (wages, COLA, shift pay, vacations, pensions, etc.), averaged for the eight companies. The union should choose the items to be reduced to reach the $5 total. This way, the union would decide what was "doable," the industry negotiators said. With the $5 cut, McBride was told, the industry would advance money for the unemployed. It would also offer a profit-sharing plan.[3]

The profit-sharing idea had come up in previous meetings when the industry men complained about large wage reductions granted to non-CCSC firms. A pact negotiated with Northwestern Steel & Wire reduced labor costs by $7 to $7.50 per hour. McLouth Steel, despite concessions ordered by a bankruptcy judge in late 1981, had faced liquidation in September 1982. The USW saved the firm by negotiating further reductions with a new owner, Tang Industries. This time, total employment costs were taken down to $18 to $19 per hour, well below the $23 paid by McLouth's CCSC competitors. Interlake Steel and Colorado Fuel & Iron also received hefty concessions. McBride tried to justify these deals by saying some of the firms had given the union profit-sharing. The CCSC negotiators said they also would offer profit-sharing if the union gave up COLA.

It is almost certain that McBride had expected a demand for a wage cut, although not of the amount asked for. "I'm not even going to tell you how much money Bruce is asking us to give back. It's an outrageous amount," McBride said at a meeting of his think-tank advisers. Nevertheless, he told them to begin preparing a concession proposal. Now McBride could contend that management had insisted that the union take the initiative. He also indicated that he was willing to replace the COLA provision with a profit-sharing plan.

Ed Ayoub took sharp issue with McBride. If the union gave up COLA, and if double-digit inflation returned (which seemed quite plausible in 1982), steelworker earnings would be devastated, Ayoub said. Other staff people, including Bernie Kleiman and Jim Smith, also spoke against abandoning COLA, though not as vehemently as Ayoub. McBride thought about it over a weekend and relented. He would not give up COLA, he said on the following Monday, but he would link it to profit-sharing. He ordered Ayoub and Smith to find a way to do that.

The profit-sharing concept, although long anathema to corporate America, had begun to spread in 1982. In earlier negotiations, companies such as Ford, General Motors, Pan Am, Uniroyal, and International Harvester had for the first time negotiated profit-sharing plans with their unions. It did not even require a skeptical mind to conclude that the sudden popularity of sharing profits with workers found its source in a lack of profits to share in 1982. But management also had begun to view profit-sharing as a method of coping with the changing economic environment. They wanted to retain the stability of long-term union contracts without being locked into pre-set annual wage increases and high COLA payments for three- and four-year periods.

One way to do this was to substitute variable pay plans based on company performance for fixed rises in compensation. Indeed, the whole nation might benefit if profit-sharing or other gain-sharing plans (such as wages based on productivity or value-added in manufacturing) were widely used. The New York Stock Exchange noted in a 1984 study that "with gain-sharing general throughout the economy, prices and wages would be downwardly flexible."

Organized labor traditionally had opposed profit-sharing for a number of reasons. The concept was viewed as a management device to win worker loyalty. Moreover, management could easily manipulate reported profits according to their treatment of depreciation and inventory, among other items, and still adhere to accepted accounting principles. Even in successful companies, rank and filers disliked giving up fixed pay boosts in return for an uncertain share of the profits.

In October, for example, Chrysler workers had voted down a settlement containing profit-sharing but no immediate wage increase. Rebounding from its near-bankrupt situation of three years before, Chrysler almost certainly would have to pay handsome bonuses in 1983 through a profit-sharing plan first negotiated with the UAW in 1981. But its employees hadn't had a raise since early 1981 and wanted "up front" money. They accepted a 75¢ per hour wage boost and gave up profit-sharing. As it turned out, they would have netted more money under profit-sharing, but the "bird in the hand" concept is a powerful motivator.

Many large corporations, on the other hand, had ideological objections to the concept. In 1958, for example, when Walter Reuther proposed profit-sharing at GM (the thought of sharing the auto maker's enormous profits easily offset the traditional union antipathy to profit-sharing), the corporation termed it a "radical scheme" and refused to discuss it. In 1982, however, GM readily granted a profit-sharing plan to the UAW.[4]

The steel companies hoped to dangle profit-sharing as a lure (although it could not have been very alluring in an industry that was on its way to losing over $3 billion in 1982) and as a substitute for COLA. In the eyes of the industry negotiators, proposing it was no trick. They thought of a profit-sharing offer as a legitimate method of weaning the union away from its attachment to the cost-of-living principle.

On November 1 McBride made a dramatic offer that overshadowed the profit-sharing issue for the moment. Usually it is management that takes on the unhappy job of actually proposing a wage cut. Convinced that the industry had a "serious emergency," McBride shouldered that task and offered to reduce steel wages by an average of $1.50 per hour. With compounding, the companies would actually save $2.25 an hour. The union also offered concessions in the benefit area and estimated total hourly savings at $3.43. Johnston welcomed the offer but indicated it was not enough.

Many staffers, however, thought McBride had gone too far. When Ayoub heard the $1.50 proposal mentioned in a pre-negotiation think-tank meeting, he was shocked. Although in July he had urged a concessionary settlement based on the cost-moderation concept, the economist objected to reducing wages on grounds that it would hurt the union as an institution. Ayoub asked for a private meeting with McBride to plead his case. Because it was one of the most fateful meetings of his career, he had no difficulty recalling the conversation.

"Lloyd," he said, "a wage cut is not going to solve the problems of this industry. If you cut wages, you in effect are telling the membership, 'If you give something back, you will get something in return'—that they will save their plants and their jobs. But the economics are against you. There's no way you can do that. The economics say you're going to lose 30 million tons of capacity."

Ayoub referred to the studies he had presented over the past two years. He pointed out that the industry could not sustain the current steelmaking capacity of 155 million tons, that steel demand had slowed down, and that capacity inevitably would shrink at least to 130 to 135 million tons (by mid-1987, it had declined to 115 million tons). "Anybody that knew anything about the industry knew that the kind of technology that was coming on stream, such as continuous casting, needed fewer workers," Ayoub later told me. "And the numbers on steel intensity were declining because of smaller cars and a growth in services and high tech, which don't use steel. Anybody who understood the steel industry knew that 155 million tons was simply out of the question."

"What I'm saying, Lloyd," he told McBride, "is that the industry is going to shut down plants. If the members take a wage cut and discover that they don't save their jobs and don't save their plants, then you've got the beginnings of disaffection."

The session lasted for about forty-five minutes, and both Ayoub and McBride raised their voices. Finally, McBride stood up and said crisply, "You've been helpful."

Ayoub left McBride's office, realizing that his usefulness to the president was coming to an end. The economist had differed strongly with McBride before other advisers on the wisdom of granting concessions to the less efficient me-too companies (see chapter 3). He had urged a settlement in July at a time when McBride was uncertain. The president also had rebuked him for talking to reporters about steel's economic situation. "I had pretty much compromised my involvement in negotiations," Ayoub recalled. "We had failed in July, and I blamed myself partly for that, and I didn't want to be part of a process that I felt so strongly would not work. Maybe it was cowardice, and maybe it was a sense of despair about the whole thing. But I wanted to withdraw, and I did."

After that meeting, McBride no longer relied on Ayoub as a principal adviser. Other staff members noticed the change and later verified that Ayoub had argued vociferously against the $1.50 wage cut and profit-sharing. Many shared his concerns but didn't voice them as loudly. Ayoub continued to act as a technician, working on the profit-sharing plan and other matters, but his heart wasn't in it. He remained on the staff, however, until 1985, when he took early retirement. Still loyal to the union, he didn't seek to publicize his differences with McBride and told me about them only after he retired.[5]

Ayoub conceded that he could not offer a viable alternative to wage cuts in November 1982. If the USW had followed his advice and signed a three-year agreement in July, he maintained, it would have been in a position of control when the industry asked for further concessions, as it probably would have been forced to do before the end of 1982. With nearly three years to go before the pact expired, the union wouldn't have to worry about hedge-buying or being forced into a strike situation. McBride *did* worry in November 1982 that the union might be crushed if it were forced to strike in 1983. "One of the things that bothered Lloyd was that if you pushed this thing to the breaking point, you would have steelworkers crossing picket lines," Ayoub said. "Not many people believed that, but Lloyd did." (A number of other staffers confirmed McBride's fears about strikebreaking.)

Under Ayoub's scenario, bargaining would have taken place in a noncrisis situation. Both sides could have studied their needs in depth. The union could have exacted a real price for additional concessions, such as modernization of competitive facilities, the elimination of poor management practices, and other reforms. In short, it could have used its leverage to acquire a voice and decision-making power in the firms. The USW also would have had to change work practices that impeded productivity growth.

An idealist, Ayoub believed in rational behavior. Unfortunately, the union and the steel industry hadn't progressed beyond a primitive relationship at the plant level. Since 1959, the two sides had put too much reliance on preventing strikes and not enough on reforming the workplace to keep up with technological, social, and attitudinal changes. Strikes are only one manifestation of flawed labor relations, and by no means the worst.

The Round Two Bargain

By early November, the profit-sharing issue dominated the steel negotiations. The industry said it would grant profit-sharing only if the USW cut costs by $5 an hour; or it would settle for a $3 reduction but without profit-sharing or COLA. The companies had retained Howard Pifer of Putnam, Hayes to devise a conventional profit-sharing plan. Johnston asked Pifer to find a way to combine profit-sharing and COLA. Aided by financial experts from the eight companies, Pifer went to work on the new concept.

Over a two-week period, the company technicians negotiated frequently with a union team headed by Ayoub and Smith and including lawyers Jim English and Carl Frankel and a research assistant, Edward Ghearing. The result was a complicated arrangement called the COLA bonus plan. It provided that each company would place 25 percent of its profits in a pool every six months. Any increase in COLA over the period would be paid as a bonus from the pool. If the amount pooled was not sufficient to pay that accrued cost-of-living, the unpaid amount would be carried over to the next six-month period. The plan defined profits as revenues from steel sales minus employment costs and other operating expenses.

Ayoub and Smith insisted that the plan include a mechanism to prevent the companies from inflating costs and understating profits. Along with Pifer, they devised a "proxy" method of estimating increases in each cost category between the base year of 1977 and the year in which earnings were being calculated. This made the plan even more complex (see notes for explanation). As Moore noted, "Jim Smith and Ed Ayoub wanted finite equity instead of rough equity, and the more finite they got, the more complicated the plan got."

The union economists worked hard on the plan, because McBride wanted it, but both disliked the concept from the beginning. "I thought it was a catastrophe," Smith said. "It was not susceptible to comparison against any of the published financial data of the companies. No worker could pick up a newspaper and read whether or not U.S. Steel had made a profit and reach a judgment on whether he was going to get anything." Added Ayoub: "The major trouble would have resulted from the fact that it became so damn complicated. It was cumbersome to begin with, and it became more cumbersome as we did patch work."

The final plan was one of the more bizarre arrangements to emerge from collective bargaining in steel. McBride had misgivings, but he thought the plan workable. He had been persuaded, he said at a bargaining session on November 13, that "the union has to tie its fortunes to the general performance of the company involved and not to the economy." The USW's wage demands had always been based on factors outside the steel industry, including national productivity growth, national inflation rates, and wage movements in other industries, especially autos. By agreeing to the COLA bonus plan, McBride signified that the USW would depart from its uniform wage policy. For the first time since the 1940s, hourly wage rates for the same job would vary from company to company.[6]

Despite progress on the COLA bonus plan, many issues remained unresolved by November 16, three days before a BSIC meeting scheduled by McBride. In addition to the wage cut, the union had offered to give up two holidays and reduce Sunday premium pay from time-and-a-half to time-and-a-quarter. Johnston and Moore wanted further cuts. "You are going to get hell, Mac, no matter what you do," Johnston had said at a critical point. "So you might as well get hell for the right reasons." Added Moore: "This is our one opportunity to do something meaningful. We don't want our employees to think they have sucked up their guts for a short time and then everything will be all right."

The industry especially wanted to eliminate the thirteen-week sabbatical or extended vacation (EV). Doing so would save the companies 50¢ an hour, as well as spare the chief executives embarrassment when they went to Washington and asked for import relief. Their critics always cited the EV as an example of the industry's largesse. What the critics ignored was that the steel companies first granted it in 1962 as a low-cost (at the outset) benefit in lieu of a wage increase that year. So far, McBride had offered to modify the EV, but it would still cost the companies 36¢ an hour.

McBride told the industry negotiators he could only go so

far. Fifty dissidents, he said, had met the previous weekend in Detroit to plan a strategy for "taking over" the industry conference meeting and defeating a concession agreement. Indeed, a meeting vaguely fitting this description had occurred, but McBride exaggerated the threat either as a bargaining ploy or because he believed it *was* serious.[7]

At the November 16 meeting, however, McBride offered to give up the EV if the industry gave back something of equal value. This initiated some old-fashioned trading. Johnston offered to trade the November COLA payment of 9¢, which the USW had agreed to defer, for the 36¢ EV. The teams caucused, after which McBride replied that he didn't like that trade but would accept the COLA payment plus the two holidays he had previously agreed to wipe out. The parties split up, conferred quickly, and this time Johnston and McBride met in the corridor. The union could have the COLA payment plus one holiday, in return for the EV, Johnston said. Done, said McBride. At 3.10 P.M., the five negotiators reassembled in the conference room and shook hands. They had a deal.

The $1.50 wage reduction would be restored in 50¢ increments starting August 1, 1983. But an additional 75¢ an hour also would be deducted from the paychecks of active workers to finance a so-called SAFE program providing weekly benefits for unemployed steelworkers. This would enable the companies to realize an immediate increase in cash flow. They had won permanent reductions and converted COLA from a fixed wage cost to a variable bonus outside the wage system. But they failed to achieve the major goal of cutting labor costs by $3 per hour. Hourly employment costs would actually be slightly higher at the end of the forty-four-month agreement than at the beginning.[8]

Many details remained to be cleaned up, however. In traditional "crisis" bargaining, the existence of a time deadline—like the imminence of the hangman's noose for Samuel Johnson—focuses the mind wonderfully on making a deal to avoid disaster. And it is exhilarating for those involved. But the steel contract had become so complex that verbal deals on the brink of a deadline had become a dangerous undertaking.

The drafting of contract language to implement the agreement typically wasn't finished by the time the USW and the industry held their ratification meetings. Both sides had long followed the practice of submitting a summary of the settlement agreement, rather than the document itself, to their constituents. Each depended on the "good faith" of the other to complete the bargaining after the agreement had been approved.

In this case, the negotiators continued arguing over details of the COLA bonus plan until November 18. The union agreed to it before knowing how it would work in practice. Smith requested data that would enable the union to calculate payouts for each company if the plan had been in effect the past five years. But the industry side didn't respond with computer printouts until the evening of November 18, only several hours before the BSIC would meet. Smith asked Ayoub to look over the data.

"I started playing around with the numbers," Ayoub said, "and, lo and behold, I discovered that some companies would have paid out an enormous amount of COLA money under the assumptions of the plan. U.S. Steel would have gotten off virtually scot-free."

The extent of the variation startled Ayoub. From 1978 through 1981, the plan would have generated profit-sharing bonuses totaling $9.17 per hour at Armco Steel, $4.06 at Republic, $3.44 at National, $3.06 at Inland, $1.52 at J&L, and only $0.42 and $0.33, respectively, at Bethlehem and U.S. Steel.

By the time Ayoub finished his calculations, it was close to midnight. Early the next morning, he, Smith, and Carl Frankel gathered in Ayoub's office and went over the numbers once again. "Holy Christ!" Smith said. "We can't go into the meeting with these results."

Smith and Ayoub hurried to McBride's office and showed him the data. He told the economists to meet with Pifer and see what could be done. Only about two hours remained before the 10 A.M. BSIC meeting. Smith and Ayoub took a cab across town to the U.S. Steel Building and began intensive negotiations with Pifer. It soon became apparent that it would be impossible to identify the source of the problem and revise the formula within a few hours. Nevertheless, McBride decided to go ahead with the presidents' meeting while the technicians continued to talk.

At about 2:30 P.M., Smith later said, "we were sitting there arguing about the damn thing when the vote was taken [in the BSIC]." The agreement was voted down, for reasons other than the COLA bonus. Only one delegate even mentioned the plan. If the delegates had discovered the degree of variation in the payments, Odorcich speculated later, "our membership would have killed us."

At the industry ratification meeting on the morning of November 19, the chief executives were given projected payouts in the future based on volume assumptions. Armco and Inland protested heatedly. Roderick of U.S. Steel and Trautlein of Bethlehem even stepped in to defend the integrity of Johnston and Moore. Despite this outburst, the CEOs "felt this was the best deal they could get, and they accepted the agreement," said John Kirkwood, who sat in the meeting. How CEOs of companies at the top of the payment list, such as Armco, Republic, and National, would have acted upon seeing numbers based on actual past performance is open to conjecture.

"It wouldn't have been attractive for us," admitted Stan Ellspermann, National Steel's chief negotiator, in a 1986 interview. "But we weren't all that disturbed, because I didn't believe in my head that it would fly anyway. I don't think you can have a profit-sharing plan, or a stock plan, or an employee security plan, that involves eight different companies with eight different sets of problems. To assume that everybody's needs and desires are the same is ridiculous."[9]

It was not lost on the USW negotiators that the two companies that would have paid least under the plan, U.S. Steel and Bethlehem, had negotiated it. Was this purposeful? Moore and Johnston themselves didn't see the computer run on past performance until the union did. They knew, however, that U.S. Steel and Bethlehem came out well on projected payments. But they could argue legitimately that the companies had approved profit-sharing as a bargaining goal knowing that the payouts would vary. U.S. Steel and Bethlehem, on the other hand, agreed to a benefits plan for the jobless that would have required USS and Bethlehem to increase their SUB funding significantly more than some other firms.

Considering the results of Round Two only from a management perspective, Johnston and Moore did a masterful job of drawing the eight companies together, despite the many unorthodox issues. If the test of a management labor negotiator in a depressed industry is how much he or she extracts from a union leader, Johnston must be given high marks. The $1.50 wage reduction *proposed* by McBride topped by far the pay cuts accepted by other major unions up to that time. Johnston's persuasive ability and his close relationship with McBride had paid off.

The BSIC Votes NO

The storm signals went up on Thursday night, November 18: Round Two's tentative agreement would encounter serious, perhaps fatal, opposition at the BSIC meeting the next day.

The warnings came at a meeting of several dozen local presidents. Called into session by McBride, these seventy to ninety presidents (estimates of their number varied) represented employees of CCSC companies who did not produce basic steel products. They worked in limestone quarries, steel warehouses, and fabricating and specialty steel shops such as U.S. Steel's Christy Park plant in McKeesport. McBride gave them the bad news that their members no longer would be covered by the industrywide agreement. Instead, they would be cut loose to bargain individually with the companies and, very likely, be subjected to further concession demands. The news infuriated the local officers.

Typically, McBride had attacked a long-festering problem head on. For twenty odd years, the companies had urged the severance of these units, called "List 3" operations because they were listed separately from basic steel plants (List 1) and iron ore mines (List 2) in the industrywide agreement. In most cases, the List 3 plants competed with small firms outside the CCSC, including nonunion minimills, which paid their workers as much as $7 per hour less than the CCSC rate for wages and benefits. Involved were seventy-five bargaining units, including thirty-three at U.S. Steel, fifteen at Bethlehem, twelve at Republic, eleven at Inland, three at Armco, and one at J&L.

Previous union administrations, unwilling to allow any erosion of the uniform wage policy, time and again had rebuffed the companies on this issue. As the years wore on, competitive pressures forced the closing of many List 3 plants. USW membership in these units had dropped from forty thousand to nine thousand, McBride later told the BSIC, and only about four thousand remained at work. Because the List 3 employees numbered a small fraction of the members covered by the industrywide agreement, the USW could sacrifice these plants without large political repercussion. On no other issue had the union been so guilty of placing the interests of the institution over those of minority groups of members.

One of the nation's best known design-engineering companies, Bethlehem's Fabricated Steel Construction subsidiary, had gone out of business partly because of the union's inflexible wage policy. This was because it had to pay CCSC wages in shops where steel was fabricated for construction. The Bethlehem unit, which designed and built the Golden Gate and George Washington bridges, among others, paid $3 to $8 more per hour than competitors by the early 1970s. In 1976, after the USW rejected Bethlehem's request for a 10 percent pay cut and a two-year wage freeze, the company closed or sold six fabricating plants employing six thousand people.[10]

In 1980 McBride's intention to remove the List 3 units from the master agreement was thwarted by his illness at a critical point in negotiations. Determined to face the issue in 1982, he included a severance provision in the tentative settlement. He told the presidents on November 18 that the employers had threatened to close the plants. But the local officials objected strongly to McBride's decision, either refusing to believe the shutdown threat—which had been made before—or fearing even more the prospect of engaging in an unequal bargaining battle with a large corporation. It was a tumultuous session.

McBride, however, refused to back down, despite indications that the issue could jeopardize ratification of the entire contract. When I reviewed this event in 1986 with Jim English, a USW lawyer who attended the November 18 session, he explained why McBride would take the gamble of rejection. "Lloyd was the kind of man who could be absolutely convinced that something was correct but who didn't have his ego involved in it," English said. "He could live with rejections."

On Friday, November 19, the Round Two negotiations went down in smoke and fire at the stormiest steel ratification meeting in the USW's recent history.

The local presidents had only to glance at a summary of the agreement to see the extent of the concessions. Disgusted and angry, they began lining up at the microphones in the William Penn ballroom before Lynn Williams finished reading the summary aloud. When McBride opened the meeting to debate, the first speaker launched into an emotional tirade that set the tone of the meeting. "I can't understand how the Steelworkers who are supposed to be so proud and who have worked so hard for forty to fifty years to be united [can] now throw us away as soon as the heat gets hot," said delegate Anderson of Local 1566, a List 3 unit. "I hope the hell it doesn't pass."

McBride replied that the union was not abandoning the List 3 locals but acting in their best interest. To keep them under the basic steel umbrella would be like making "funeral arrangements" for those plants. "I wish somebody else would have handled this problem before I got here," he added, making an unusual, critical comment about previous officers. "But nobody did, and now it's got to be handled and it will be handled one way or the other."

Responding to another critic, McBride said that "if these people [List 3 presidents] want to ... say that we've got to have basic steel wages, then they're writing their own epitaph." But Roland Pigeon of Local 3734 at Bethlehem's Buffalo Tank Division, a List 3 plant, charged that the List 3s were being "thrown to the wolves." If his plant had to bargain separately, it might end up with wages of only $5 or $6 an hour. Even so, he said, Bethlehem would close the plant if it lost money. "So," he concluded, "we're going to end up in Potter's Field, with no epitaph at all."

The opponents criticized other features of the tentative agreement. Charlie Grese of the Clairton local, who was not noted as a dissident, complained that unemployed people with less than five years of service would not receive jobless pay under the SAFE plan. The presidents, he said, needed two or three days to consider the proposal. "We have to make U.S. Steel remember," he said in a fiery peroration, "that we are the United Steelworkers of America, and we're not going to fall down and just lie down and play dead to them."

Mike Bonn had been standing at a mike for an hour and a half when McBride recognized him. Looking up at the balcony packed with people, mostly local unionists who could not vote, he likened the delegates on the floor to Christians in the "old Roman pits" (forgetting that fellow unionists in the balcony then would become Romans). He continued: "You know I'm a Christian, Brother Lloyd, and I'm damn proud of that. But if we vote in favor of this proposal, the score is going to be six hundred for the lions, zero for the Christians. Let's circle the wagons and fight!" Loud applause rang out.

McBride tried to generate some compassion for unemployed steelworkers. Under the SAFE program, he pointed out, jobless members with at least fifteen years of service would receive $200 a week for one year, or more than $10,000, to help them feed their families and save their homes. "We have an obligation to do something about that," the president said.

Although no delegate directly attacked the plan to take 75¢ an hour out of active workers' paychecks to aid the unemployed, no one praised the idea. The total pay cut of $2.25 per hour would have cost a full-time worker over $400 a month, not counting incentive-pay loss. Tom Duzak, the USW's chief benefits negotiator, explained that even though the workers' contributions would total $140 million to $170 million, this amount was insufficient to cover the low-seniority jobless.[11]

"I don't object to helping the laid-off workers, and I don't think anyone in here does," said one delegate. "But if we're just prolonging the inevitable, then I don't know." By "inevitable" he apparently meant the permanent severance of furloughed workers from the steel industry. By November 1982 the belief was widespread that many of the unemployed would never be called back to the mills and that concessions would not save jobs.

One of the more militant local presidents, David Sullivan, president of Local 6787 at Bethlehem's big Burns Harbor plant, made this point more directly. Steel's financial troubles, he said, came about because of Reaganomics, changes in the world economy, and the tight monetary policy of "greedy bankers and the Federal Reserve." Concessions had not saved jobs in other industries. Instead, they had undermined the general principles of unionism. "What we are doing is just competing with ourselves," Sullivan concluded.

McBride replied that Sullivan's attitude was like that of U.S. Steel regarding the me-too companies that had received USW concessions. Big Steel's position was "let McLouth go down the drain. Don't let them be around to survive when this depression is over...bury them. That's exactly what you are saying, and I disagree with you."

At no point did McBride criticize the steel companies, apparently feeling that this would have been irrelevant to the proceedings. As Carl Frankel once said of McBride, he "did not massage people. He was not political. It might have been better if he portrayed himself more as a defender of the workers. But that just wasn't his style. He was a firm and very direct, blunt man."

The meeting lasted four-and-a-half hours, and for most of

that time McBride stood at the podium, catching brickbats, and throwing them back. Not once did he retreat in the face of unrelenting attacks by the militant sections of the union, as company officials had accused him of doing in July. Few other union presidents would have subjected themselves to such a storm of criticism. Not until the tenth delegate rose to speak did anybody make an approving comment about the agreement. Only three of twenty speakers urged its adoption.

McBride failed to employ an old USW tactic for dealing with such rebelliousness at a meeting. When an administration proposal received rough handling, one of the union's better orators—someone like Joe Molony, a vice-president under Abel, or Jim Griffin, a gifted district director—would take the podium and make an emotional appeal for solidarity. The eloquence would have an effect, and the speech also would halt the momentum of the dissident cause. McBride, however, made no effort to enlist the other four officers who sat at the head table. Joe Odorcich later said he offered to speak but that the president didn't call on him. Although the executive board had approved the agreement unanimously, no board member rose to defend it, perhaps believing—to be generous—that McBride wanted to take full personal responsibility for the pact.

Finally, Joseph Samargia of iron ore Local 1938, a well-known dissident, demanded a roll-call vote. McBride flatly ruled out the idea, saying it would subject the delegates to "too much peer pressure." Ron Weisen protested the decision, but McBride dismissed him. (Actually, the agreement might have had a better chance *with* a roll-call vote.)

Several delegates shouted, "Question! Question!"

Acceding to their call, McBride asked for a standing vote on closing debate. Company officials, who tend to believe that union leaders can always impose their will on subordinates, later suggested that McBride mistakenly identified the question callers as pro-agreement delegates and should have kept debate going. No one knows whether this is true. When the sergeants-at-arms reported that 245 presidents stood in favor of the motion and 163 opposed, McBride closed debate and ordered balloting on the contract to begin.

It took nearly an hour to conduct the secret ballot vote. In a startling repudiation of McBride's policy, the presidents voted down the agreement, 231 to 141.[12]

"The decision has been made," McBride said laconically. "I hope it's for the best." He adjourned the meeting at 2:40 P.M.

As in July, the dissidents were ecstatic. I remember meeting Mike Bonn in the corridor outside the ballroom. He was so full of himself that he couldn't get the words out fast enough. In my notes, I find him saying, "McBride has lost his credibility with us. He's a dictator. The local presidents are going to start meeting to organize alternatives to concessions. We're going to take it [control of the negotiating process] away from him." This promise never came to fruition.

If the most important thing was to defeat McBride, the dissidents and their allies had done that. But how long would this triumph last? "When the shock sets in, it will be tremendous in the locals," board member Bruce Thrasher told a group of reporters. "I don't see how a local union president can tell an unemployed member, 'I have just futzed you out of a home and a car and education for your kid.'" But he also conceded that back in the plants, "there is complete mistrust of the companies, which is justified. Our people don't believe them or trust them."

At a news conference a few minutes after the meeting, McBride seemed disappointed but stoical. It would be "futile," he said, to try a third time to reach an early agreement. He did not expect to bargain again until May 1983, the normal starting time for negotiations leading up to the contract expiration of August 1. A strike at that time was a distinct likelihood.

The Reasons Why

Why did 231 local presidents reject a contract that would have helped thousands of jobless workers and possibly (though nobody could be assured of this) save *some* jobs? The dissident point of view was summed up by one writer who termed the pact "probably the worst contract a union leadership ever offered to its membership since the formation of the CIO in 1935."

A comprehensive analysis of why people voted for and against the agreement is impossible because of the secret ballot. However, a study of the meeting transcript, combined with many interviews, provides important clues.

The structure of the Basic Steel Industry Conference was an important factor. Some weeks after the meeting, the USW made a rough count of delegates certified as voting members of the BSIC. The union's 29 executive board members had voting rights, as well as about 150 presidents from basic steel locals and 75 or so from List 3 locals belonging to the CCSC companies—a total of 254. One must assume that practically all of these delegates were among the 372 who cast ballots on November 19. The remaining 118 must have represented workers in another category, the me-too companies such as Wheeling-Pittsburgh Steel, McLouth, and many others. They had always been allowed to vote on the CCSC agreement on the theory that its terms would be extended to their firms.

Since the USW had put wages back into competition by granting concessions at the me-too's, the makeup of the BSIC created a not-so-theoretical conflict between locals. The me-too presidents could help their own firms competitively by voting against the industrywide contract. According to Odorcich, more than one Wheeling-Pittsburgh president admitted casting "no" votes on November 19.

Another important structural point is that the largest groupings of local presidents, by company affiliation, were from U.S. Steel and Bethlehem. Since these companies had relatively worse relations with local officers than other firms, the result must have been influenced accordingly. The presidents of at least five U.S. Steel locals in the Mon Valley voted against the agreement. On the other hand, the leaders of two large locals at the Gary Works supported the agreement. On balance, however, the poor relations probably had a negative impact on the vote. An even larger proportion of Bethlehem delegates voted against it. Four presidents representing locals at the firm's large Lackawanna plant cast "no" ballots, although (or because) they were virtually certain the plant would be closed by the end of the year. It was.

Aside from the List 3 problem, relatively few provisions of the tentative agreement were actually cited by opponents at the meeting. Undoubtedly, the size of the wage cut caused consternation. According to one estimate, a steelworker who worked full time at the average wage would lose $12,800 over the forty-four-month contract. Elimination of the EV was attacked, as was the lack of strengthened protection against contracting out. The SAFE plan, however, drew the most critical comment after the meeting.[13]

The plan had two major defects, according to the opponents. It provided only meager benefits to idled steelworkers who were still receiving UC and no benefits at all to workers with less than five years' service. In Pennsylvania, for example, the typical UC payment of $198 per week would be subtracted from a $200 SAFE benefit, leaving a payment of $2. Only in Alabama, where stipends averaged $90, would the SAFE benefit have much value for workers still collecting UC. The less-than-five-year category included about twenty-six thousand people, or one-quarter of the ninety-eight thousand unemployed in the eight companies.

The USW ran into a cost barrier, Tom Duzak later told me. Given the magnitude of unemployment in the industry, the negotiators had to draw the line somewhere. The companies had rejected the union's original idea of providing at least short-term benefits to all of the unemployed. Instead, the employers wanted to earn the good will of workers who had the greatest chance of being called back to work, those with more than five years' seniority. To have included the under-five-year group in the SAFE program would have added another 50¢ per hour to the cost, Duzak said. It was doubtful that active workers would have been willing to divert $1.25 per hour of their earnings to help the unemployed.

The payroll deduction aspect of SAFE did not make the plan any more appealing to the active employee. Each paycheck would show gross earnings, net earnings after taxes and regular deductions, and finally a deduction of 75¢ times the number of hours worked in the pay period. "When we think about it in retrospect, that was really stupid on our part," Duzak said. "The worker would be reminded every payday exactly how much he was giving up for the so-called unemployed."

Besides these negative aspects, employed workers lacked McBride's altruism. In the Mon Valley, for example, U.S. Steel had agreed at the request of six locals to devise a payroll deduction plan for active workers who volunteered to contribute money to food banks. Only two locals—at Homestead and Duquesne—eventually got into the program. Less than 3 percent of the workers volunteered, and the average deduction was only 17¢ per week.

It would be unfair to conclude that steelworkers in general were abnormally selfish. A partial explanation for the seeming niggardliness is that even those who still had jobs came to feel, as the steel depression deepened, that these jobs were by no means safe. Many probably also felt that they were already contributing to the unemployed by way of union dues and tax payments.

The seniority system, for all its positive contributions, also tended to breed a kind of institutional narrow-mindedness that enabled some people—probably a small minority—to rationalize their unwillingness to help the unemployed. I have heard older workers say that *they* had lost income in many layoffs in the past. Why should it be different for the younger generation? The youngest workers never would be called back to the mills, said some older members. It would be wrong to encourage them to wait for jobs that would never return.[14]

The lack of job guarantees also helped sink the agreement. If workers agreed to a wage cut, many felt, they should be guaranteed a job. At the very least, the companies should agree to "reinvest" the cost savings in steel operations rather than in other businesses. Critics implied that McBride failed to ask for such a guarantee because of faint-heartedness. One critic wrote that "forcing the companies to invest in the steel industry . . . could have been a point of pride for the union."

Actually, USW negotiators had discussed such a provision with management. "The companies showed us that they weren't anywhere near having enough cash flow to use for anything other than to pay their bills and spend on the plants [to keep them operating]," Bernie Kleiman told me shortly after the November failure. The companies probably would have pledged to use the savings only for these purposes, but this would only have confirmed the obvious. "Lloyd became convinced that whatever you might say would be window-dressing, and Lloyd's not much for window-dressing," Kleiman added.[15]

The Aftermath

In the first few weeks after the vote, union and industry leaders expressed profound discouragement. Since the 1959 strike, the steel industry and the USW had become adept at solving, or circumventing, difficult issues in top-level bargaining. With the second negotiating failure in four months, that vaunted expertise seemed to have disappeared.

Several elements of the relationship seemed to have broken down. Either the companies had failed to communicate the gravity of their plight to their employees, or the BSIC presidents didn't truly reflect the will of the rank and file. Another factor involved the very nature of multiemployer bargaining. This approach works best when all employers have roughly equal interests. In prosperous times, differences in interests could be overlooked. Adversity, however, magnified the differences. The List 3 issue is a good example. National Steel and Allegheny Ludlum had no List 3 operations, and J&L had only one. Yet it could be argued that they were denied financial relief largely because of a problem they didn't have.

Similarly, the friction between U.S. Steel and Bethlehem managements and their locals to some extent penalized other companies. Watching TV news in Pittsburgh on the night of November 19, the chief negotiator of another steel firm saw interviews with some of the Mon Valley presidents, including Mike Bonn and Ron Weisen. The unionists expressed extreme hostility toward U.S. Steel management in remarks that bordered on the scurrilous. The vehemence of the attacks astonished the company man. "You've got to examine the labor relations that causes presidents to express rage and hate against the company," the vice-president later told me. "We probably have some presidents [in his company] who voted against the agreement, but they didn't express a sense of rage against us."

In the immediate aftermath of the vote, some company officials and USW leaders began to question whether the industry conference would approve *any* agreement with substantial wage concessions. In an interview with me on November 22, Joe Odorcich planted the idea that some companies should think of bargaining individually with the union, especially if they had good relations with their locals. "We know some of the major companies are hurting," Odorcich said. "If I were such a company, I'd go to my locals and the district director and ask them to talk. It would be the end of industrywide bargaining, but our guys put us in that position." Printed in *Business Week*, this comment drew wide attention.

This approach, however, posed two major problems. To retain wage uniformity, the USW would have to set a pattern with one company and extend it to the other firms. Only U.S. Steel and, conceivably, Bethlehem, were large enough to make such a pattern stick. But these companies had too many enemies among their local presidents. The second problem was that McBride disagreed with Odorcich, who had made the comment while McBride took a brief vacation away from Pittsburgh. When the president returned, he quickly demonstrated that he had no intention of abandoning industrywide talks.

On December 16, the USW executive board, acting on a McBride proposal, changed the BSIC's voting procedures. No longer could me-too locals vote on a Big Eight agreement. The change remedied the unfairness of the old procedure. Stripping the me-too's of the vote also acknowledged the reality that a many-tiered wage structure was taking root throughout the steel industry.

Even as the board acted, USW negotiators met with Wheeling-Pittsburgh Steel, which had appealed for further wage cuts. To grant more relief to W-P and still allow its local presidents to vote on a pact covering its competitors was hardly tenable. "I might become selfish and vote against it," one W-P president told the *Wall Street Journal*.

Meanwhile, three of the Big Eight companies, J&L, National, and Republic, sent out feelers to the union on the subject of individual bargaining. McBride was not inclined to explore the idea. Individual bargaining, he said, would lead to "competition for the lowest wage rates [which] can do serious harm to industrial relations in the industry and create some pretty serious situations in terms of strikes." With that alternative dead, no one knew at mid-December how the USW and the industry would avoid a confrontation on the picket lines in the summer of 1983.[16]

Chapter 14

For the second time, GM chairman Roger Smith stepped on stage as the *deus ex machina* of the steel labor drama. In late December 1982, Smith telephoned McBride to give a friendly warning. GM must know by March 1, 1983, whether steelworkers would be producing steel as of August 1, or whether they would be on strike. The giant automaker followed a practice of letting supply contracts in March for steel and other materials to be used in producing cars for the new model year, which began in August. If a domestic steel strike on August 1 were still a possibility in March, Smith said, GM would be forced to sign contracts with Japanese and Canadian suppliers.

"He was very pointed to let me know that this was not something that he wanted to do," McBride said. "His alternatives were somewhat limited. Their system would not let them stockpile enough steel to carry through a possible strike situation. He could not afford to let his plants close down because of a steel strike. He was very, very specific that he did not want to get into a posture of using any kind of foreign steel." He would do so, however, to protect his company's interests.

The GM chairman undoubtedly gave the same message to steel management. It was highly unusual for an executive in one industry to interfere in the affairs of another. But Smith's warning prodded McBride and Johnston. They agreed to make a third attempt to reach a peaceful agreement. By January 3, 1983, when I talked with McBride, he had already set March 1 as the deadline for the new talks.

In addition to the GM threat, the economic climate gave more than enough reason to resume bargaining. Nationally, the key "economic indicators" showed that a recovery had started. But 10 million Americans, or 10.7 percent of the civilian labor force, were out of work in January 1983. In the four-county Pittsburgh metropolitan area, joblessness peaked at 15.9 percent in January, with 168,500 people seeking work. McKeesport registered a 20.4 percent rate, a decline for some reason from the 23 percent of October.[1]

In mid-January the steel companies announced financial results for 1982, taking huge writeoffs for closed plants. The full list constituted a catalogue of ruin unparalleled in American industrial history. Companies affiliated with the AISI lost a total of $3.2 billion in 1982.

U.S. Steel permanently shut the legendary open hearth shops at Homestead and the blast furnaces at Rankin, where one thousand people had worked a few years before, at a cost of $123 million. The corporation's steel operations had lost $852 million in 1982, but profits from Marathon and other operations reduced the year's net loss to $361 million. Steelmaking contributed only 28 percent of sales revenues in 1982.

Bethlehem Steel said it would eliminate seventy-three hundred jobs by closing most of its eighty-two-year-old Lackawanna plant and reducing operations in Johnstown. These writeoffs totaled $800 million, producing a 1982 loss of $1.47 billion, one of the largest ever compiled. Armco discontinued operations at three plants in Missouri, Ohio, and Texas, permanently cutting twenty-two hundred jobs. National Steel wrote off $60 million worth of idled facilities. Republic lost $239.2 million in 1982 and desperately tried to stay out of bankruptcy. Mesta Machine Company, the once prosperous manufacturer of mill equipment in West Homestead, filed for bankruptcy.

The industry was in the middle stages of a profound shakeout, one that started in the mid-1970s and would continue through the 1980s. One steel analyst, Peter Marcus, predicted that the USW's rejection of a contract in November would "speed up the closing of 20 million tons or so of capacity." The number turned out to be correct, but the linkage with steel's labor problems seemed tenuous. A more accurate perception had been voiced by Ed Ayoub and many local presidents: The industry would continue to shrink even *with* labor concessions.

Steelmakers shipped only 61.6 million tons of steel in 1982, the lowest since 1958. But optimism about the industry's future still reigned in the U.S. Commerce Department, which forecast a 25 percent rise in shipments in 1983, to 80 million tons. An analyst at Data Resources put the figure at 74 million tons. Steel executives, finally wary of planning for sharp upturns that never came, agreed on an appropriately more cautious forecast of 70 million tons. They weren't cautious enough. By the end of 1983, the industry had shipped only 67.6 million tons.[2]

On December 30, Wheeling-Pittsburgh employees ratified a settlement reducing labor costs by $2.85 an hour, including a $1.53 wage cut. W-P's hourly employment costs went down to $20.65, or $3 to $5 below the industry average. The company agreed to put 90¢ an hour into a fund to help the unemployed. It also guaranteed that "any savings resulting from a moderation of its labor costs will stay in the steel industry." Meanwhile, U.S. Steel cut the pay of twenty-eight thousand management and salaried employees by 5 percent, the second such reduction in seven months. Only the USW "has refused to contribute to the sacrifices made by everyone else in the steel family," David Roderick said in a letter to employees. Bethlehem slashed the salaries of fourteen thousand salaried and nonunion workers by 2.5 to 8 percent.[3]

Day after day, bad news piled on top of bad news.

A Groundswell for Concessions

As steel region economies deteriorated, UC and SUB benefits expired, and queues of people at food distribution places lengthened. Steelworkers lining up for free food! They may have held the "premier wage position" in America, as Bruce Johnston described it in a letter to employees, but increasing numbers had to apply for food stamps and welfare payments. A total of 148,691 hourly and salaried steel employees were still laid off, and more than 20,000 worked short weeks.

In many locals, the unemployed far outnumbered the active workers. Some local presidents who had opposed concessions began having second thoughts. Whose interests were they representing in continuing to reject aid for the unemployed? Public attention had been focused on the local officers when they voted down the November agreement. The companies sent letters chastising the presidents to all employees. Newspaper and magazine editorials pointed out that the jobs of most presidents who voted against the November pact were secure. Under a superseniority provision in the contract, local officers could bump more senior employees in a cutback. It was a matter of note that the International did not rush to the defense of its local officers.[4]

This criticism caught the presidents' attention. Many began to promote various lines of action. In District 31 (Chicago-Gary), a group of USW activists called the Committee to Save Our Union circulated a petition opposing "one-sided concessions." But in the same district, Phil Cyprian, the president of twelve-thousand-member Local 1014 at the Gary Works, estimated that his membership supported his proconcession stand by about seven to one. A USW committeeman in Lorain, Ohio, however, wrote in his local paper that "labor has learned that concessions will kill, not help, our nation and our economy."

In Duquesne, Mike Bilcsik, who voted against the November settlement, said that U.S. Steel had "convinced me they're going to get out of the steel business unless we take a cut." He regarded himself as a militant, but about two thousand of his three thousand members were laid off. To make concessions palatable, Bilcsik said, the companies must guarantee that cost savings be used for "modernization or improvement of existing steel facilities." He passed out questionnaires to about seven hundred and fifty steelworkers at plant gates and food distribution places, asking whether they would have approved the November proposal if it contained an investment guarantee. Eighty-three percent checked the "yes" box.

Several other local presidents began to focus on the investment issue. Even Mike Bonn said such a *quid pro quo* might win his vote. McBride had told me early in January that he recognized that the issue should be addressed. He was skeptical that such a guarantee would be meaningful. The steel companies could document that "they've been putting more money into steel than they were taking out of steel," he said. To win ratification, however, he would ask the companies for the "window-dressing" that he had earlier declined. On January 10, the USW's Wage Policy Committee adopted a statement calling for a company investment pledge, among other things.

At the time of the November rejection, some USW leaders had talked of removing the contract ratification responsibility from the BSIC and adopting a referendum procedure. The rank and file, including unemployed steelworkers, would be more likely to approve a contract than the BSIC, some board members felt. Nothing happened then. But the issue suddenly sprang to life in January. At Local 1066 in Gary, activists collected twelve hundred signatures in four days on a petition for the right to ratify. The purpose was not to assure passage of a concessionary agreement. Rather, according to one activist, Valerie Denney, "On a question this serious ... people feel they have the intelligence and the interest to vote on their own contract."

The idea also arose in the Mon Valley. Bilcsik proposed that the District 15 local presidents adopt a resolution calling for a membership vote of the next contract. Twenty-six presidents met on January 15 at the district office and rejected Bilcsik's resolution by twenty-one to five. This result commented powerfully on the sincerity of anticoncession militants like Ron Weisen who had demanded rank-and-file ratification in the past but who voted against it now. "He wanted the membership to have the vote when he thought the membership would support him, but not when he thought they wouldn't support him," Bilcsik said. Mike Bonn split with Weisen on this issue; Bonn favored voting on contracts in local meetings, a halfway measure that is also susceptible to leadership control.

By mid-January, a groundswell of opinion favoring concessions rippled through many steel communities, although it wasn't highly visible or audible. Many mill workers who responded yes to concessions in a questionnaire were unwilling to shout yes in a meeting. They didn't want to be condemned by the militants for "running scared." Recalled Bilcsik: "Not one single person, including my closest supporters, would admit openly that they voted yes on the questionnaire. They put one thing on paper and said the opposite in the meetings. Some of the most outspoken against concessions were the unemployed. They knew they would never get back to work, and they wanted to hurt the company. They hated U.S. Steel. Ninety-five percent of the membership shared that feeling. I shared that feeling. How can you despise a corporation? You had to work for them to understand. People voted against their own jobs, that's how deep the feeling was."

All this turmoil and turnabout did not speak well for the political system that had developed in the USW. The left-wing militants and all-season dissidents established the terms of debate in the mill towns: No one could be a strong unionist unless he or she agreed with the idealogues. Other members were afraid to stand up for what they believed. The USW establishment had long ago created this polarity by branding all who disagreed with it, on any matter, as a dissident to be shunned. At a time when the union most needed a responsible democratic system, it didn't exist. "The membership *was* irresponsible," Bilcsik later reflected. "But the reason the membership was irresponsible was because they didn't have responsibility for all those years."[5]

The right-to-ratify movement was short-lived. It lacked McBride's support. The BSIC, meeting in Pittsburgh on February 2, voted to retain the authority to ratify. It also adopted a statement listing major goals for a third round of negotiations, scheduled to begin February 15.

Company and union strategists began to prepare for negotiations. McBride acknowledged to management bargainers that he must find a different way to approach the List 3 problem and that he had misread the sentiment for helping the unemployed. Since he had made such a public issue of the latter, however, the USW strategists concluded that the union must save face by finding some way to raise pay for the jobless. Having decided that the November negotiations failed in part because of lack of local participation, McBride also proposed that bargaining over local issues be resumed at the company level.

Industry negotiators agreed to schedule the company meetings, although with reluctance. "We had pled our case over the last twelve to fourteen months to the top of the union, to the middle, to the rank and file, by letters, newspaper ads, slide presentations, and so on," said George Moore. "What more do we have to do?" The industry strategy was fairly simple. At a meeting with their negotiators, the chief executive officers declared "vehemently that they were not willing to take anything less" than the USW officers had agreed to in November, said one company official. The phrase of the day became "November-plus."[6]

USW strategists, meanwhile, tried to find a combination of give-backs and *quid pro quos* that would produce a yes vote in the Steel Industry Conference. How could they determine what the local presidents—and by extension, the rank and file—would give up most willingly and what they would accept in return? Jim Brown, president of Local 1066 in Gary, supplied the answer to that question. He polled the rank and file.

What the Rank and File Wanted

U.S. Steel's vast complex of furnaces and mills at Gary is split into two parts. The iron- and steelmaking or "hot" end of the plant comes under the jurisdiction of Local 1014. The finishing end, which in known as Gary Sheet & Tin and which includes cold rolling mills, pickling lines, and an eighty-four-inch hot strip mill, is covered by Local 1066. Unlike almost every other U.S. Steel local, 1066 had a large majority of its members working in January, thirty-six hundred, compared with only four hundred on layoff. The atypically high proportion of actives made 1066 an excellent place to find out just how militant the actives were.

Since the previous July, Brown had been a staunch supporter of McBride's attempt to negotiate relief for the steelmakers. "I'm just as mad at U.S. Steel as the other presidents," he told me on January 19, "but I see no reason to cut off your nose to spite your face." He felt that Round Two foundered because the union bargainers didn't know with precision which wage and benefit items the members considered most important and which they regarded as expendable. Brown drew up a questionnaire and mailed it to all members in Local 1066. Forty-five percent, or 1,750 members, returned the forms. Sidney P. Feldman, professor of marketing at Indiana University Northwest, tabulated and analyzed the results for Brown in late January.

Were the workers willing to strike to avoid concessions? Sixty-three percent of those who answered this question said no, a significant finding in that only 7 percent of the replies came from jobless employees. As for benefit preferences, high percentages of the members wanted either "improvement" or "no change" in provisions involving job security (which could be construed as a demand for job guarantees or reinvestment pledges), pensions, medical and hospital insurance, and life insurance. Wages ranked next on the "improvement or no change" list, followed by SUB, overtime premium, COLA, and aid to the unemployed. The nonpriority items included dental and optical plans, paid holidays, shift differentials, extended vacations, funeral leave, and vacation bonus.

When Brown showed his data to union negotiators in Pittsburgh, they realized its importance and decided to follow up by polling the local presidents. By this time, McBride had been hospitalized with a heart problem and Joe Odorcich had assumed command of negotiations. At a breakfast meeting on February 18, the presidents filled out a lengthy questionnaire. The results profoundly disturbed Odorcich and his top staff people. Of the 140 presidents who responded, 79 (56 percent), said their locals were not willing to strike on August 1 to avoid concessions.

This finding seemed to support McBride's fear that the union could not mount an effective industrywide strike in August. "This bothered both of us," Odorcich said. "We had both come to the conclusion that if there was a long strike, they'd walk over us." Some unionists strongly dispute this conclusion, viewing it highly unlikely that steelworkers in the mill towns of the Mon Valley, or anywhere else, would ignore a strike call or attempt to act as strikebreakers. Odorcich understood the psychology of mill town workers as well as anyone. However, as a boy in the 1920s he had seen hungry miners cross UMW picket lines in the coalfields of western Pennsylvania. He had seen what desperation will drive people to do.

Because of the sensitivity of the strike finding, Odorcich withheld it from all but five or six top union people. When Bruce Johnston heard about the BSIC poll and asked for a copy, the USW delayed giving it to him until after the talks, but even then without the strike data.

The presidents' survey provided other extremely useful information. Like the 1066 poll, it showed that pensions and medical and hospital insurance were high on the list of "no concessions." The presidents, however, placed more importance on retaining COLA (87 percent) than the 1066 members. In the most revealing finding, one that enabled the USW finally to obtain ratification of an agreement, 83 percent of the presidents put top priority on an early retirement incentive as a tradeoff for concessions. Only contracting-out protection ranked higher at 84 percent, and investment of savings in existing steel plants polled third at 73 percent.[7]

In Round One the industry had brushed aside a USW proposal to spur early retirement, and the USW didn't raise the issue in Round Two. It surfaced again in January, when local presidents of the Lackawanna and Johnstown plants asked for a special early retirement incentive to help deal with their plant closings. This gave pension negotiator Tom Duzak the idea of putting the early retirement question in the survey.

The concept proved so popular that the USW decided to push for it in the third round.

Before negotiations began, Allegheny-Ludlum pulled out of the CCSC (Coordinating Committee Steel Companies), reducing the industry bargaining group to seven companies. The specialty steel producer withdrew largely because of a special competitive problem involving Jones & Laughlin, and the loss didn't weaken the CCSC.[8]

A more important loss, to both sides, was that of Lloyd McBride's leadership. Few people in any occupation could have been under the kind of pressures that had weighed on the USW president over the past year—the long stressful weeks of negotiations, the worries about laid-off members, a growing concern for the future of the union. He had "wrestled with [his] conscience," as he put it at a staff meeting in November 1982, in deciding how to deal with the industry's dilemma. In January 1983, another major decision faced him.

The USW's dues-paying membership had dropped from slightly over 1 million in June 1981 to about 750,000. The loss of income forced a reduction in the 1,460 people employed by the USW. Although 157 older employees accepted an offer of early retirement, Treasurer Frank McKee urged further payroll cuts. McBride and his closest aides studied the problem for weeks and came to the painful conclusion that 200 staffers must be laid off. "This decision preyed heavily on his mind," said Harry Guenther, a personal assistant.

In late January, when McBride announced his decision to the executive board, he encountered overwhelming opposition. The reduction would strip some district directors of onethird to one-half of their staffs of field representatives, who serviced local unions and performed organizing duties. One staff member who attended the meeting, Ed Ayoub, had seen board members disagree with, and vote against, USW presidents in the past. But this meeting, he said, "was the only time in twenty-five years that I really saw the board take out after a president." Laying off loyal district staff people could damage a director's political future. All but one of twenty-three directors argued against the cut and demanded that McBride reconsider his decision.

A few days later, the USW president, who had a history of heart trouble, suffered chest pains. After two days of hospital treatment, he returned home. But within hours his wife summoned an ambulance. Rushed to another hospital, McBride went into cardiac arrest and had to be revived with electric paddles. Fortunately, there was no brain damage, and doctors installed a pacemaker on February 9. McBride began to recuperate, but he could not return to work for several weeks.

There was no question which of the four remaining international officers should take over the negotiating responsibility. Joe Odorcich had stepped in as chief bargainer during McBride's 1980 illness and had participated in the two 1982 rounds. Some management people worried that political maneuvering within the USW would leave it leaderless. McBride's term would end in 1985, and rumor had it that Frank McKee and Lynn Williams would vie for the post. McKee and Williams, however, did not allow political ambition to intrude at this point. Odorcich, McKee, Williams, and Leon Lynch met to discuss the division of responsibilities in McBride's absence, and Odorcich emerged as the chief negotiator. A vote was not necessary.

Odorcich presided over the BSIC meeting on February 2. When a few dissidents tested his leadership, he displayed a far different style than McBride. Ron Weisen and Mike Bonn moved to unseat him as lead bargainer. But Odorcich refused to accept the motion. He referred to the two presidents as "hard-headed hunkies" and added: "If you want to unseat me, the next guy to take my place is Lynch, and he agrees with my position. The next guy is Williams, and he agrees. The next guy is McKee, and he agrees with me. Now, how many people do you want to unseat?"

Weisen shouted, "Don't forget, you work for us."

"I work for you," Odorcich replied, "but I don't have to be a horse's ass to work for you."

The delegates laughed, and Weisen and Bonn sat down. Odorcich went on with the meeting.

"I'm not going to be a gentleman," Odorcich told me that night in a phone interview. "I don't mind getting into the gutter with them. Quit depending on integrity and character and faith in the human man and tell it like it is. Our guys in the Mon Valley would just as soon spit in your eye as say hello to you. But if they know you aren't bullshitting, you'll do all right."

Although Odorcich enjoyed projecting this image of toughness, a deep-lying sensitivity belied the image in some respects. But his hard-bitten realism was by no means all show. Odorcich had experienced more of labor's elemental struggles with miserable living conditions, grueling work, dictatorial bosses, and narrow-minded union superiors than any other USW leader.[9]

Odorcich's Odyssey

In 1982, at the age of sixty-six, Odorcich looked hearty and fit, with a tanned, leathery face, a graying moustache, and a nearly bald head. A golfer and jogger, he sometimes still ran as much as ten miles at a time, explaining his penchant for movement in this curious way: "I'm like the average ethnic. I think better when I'm either working or running." A sly smile frequently played under Odorcich's moustache, accompanied by dry witticisms clothed in the earthy language of a miner.

Odorcich was born in Johnstown, one of fifteen children of Croatian immigrants. His father moved back and forth between factories and coal mines in western Pennsylvania and eventually settled the family in a coal patch called Mather, near Waynesburg. In 1927 the mine where his father worked blew up, killing scores of miners. Odorcich can remember watching rescue crews bringing up bodies "stacked like cordwood" on mine cars. His father survived that blast only to be killed a few days later when a large chunk of slate fell on him in the mine. The company evicted Mrs. Odorcich and her children from the company-owned house. "We were sitting at the edge of town," Odorcich recalled. "The kids were around her like peepies with a chicken, and she was crying her heart out." A Jewish store owner took pity on the family and let them stay in his barn for a few months.[10]

Later, the family moved to Fredericktown. Odorcich left school before he turned sixteen, lied about his age, and got a job as a miner. He joined the UMW and became a mine committeeman at eighteen. Like many union activists in those days, he served time in jail. On one occasion, he and other miners paraded through Fredericktown on a miners' holiday. "The state constabulary was there, and we didn't think they looked good on horses," Odorcich said. "Another fellow and I took a couple of them off the horses. I woke up about six hours later with a cracked skull. I was in jail for about ten days. Phil Murray came and bailed us out. The other time I was in jail was in an organizing drive. Somebody questioned my antecedents, and I didn't like it."

Odorcich's last day as a miner came in 1937. A cave-in

occurred at the J&L mine where he worked on the afternoon shift. He saved himself by crawling under a haulage car, but it took two days for rescue workers to dig him out. "The flashlight battery pooped out after eight hours, and I sat there in the dark. Every once in a while a little more roof would fall," Odorcich recalled. "I died a thousand deaths. Finally, they shined a light in and said, 'Wait, we'll make the hole bigger.' I said, 'Get out of my way, it's big enough,' but I lost a yard of skin getting out of there. The pit boss, a fellow named Hines, said, 'Joe, take tomorrow off, you'll feel better.' I said, 'Mr. Hines, the next time I come into a mine, the goddamned thing is going to have windows on it.' And that was the last day I worked in a mine."

Odorcich went to McKeesport, where a relative helped him get a job at a foundry, Fort Pitt Steel Casting. Odorcich helped organize the shop during the SWOC (Steel Workers Organizing Committee) campaign and became a committeeman. Over the next few years, he participated in other organizing drives. Many steelworkers lacked enthusiasm for the union, he said, and SWOC used intimidation in some cases to collect dues. Setting up picket lines outside the plant gates, union activists kept employees out of the plant unless they handed over 50¢ and had their names checked off.

Odorcich served on a number of dues picket lines and remembered in particular an episode at McKeesport Tinplate Company, located on the Youghiogheny River bank across from McKeesport. He and two other men lined up in the middle of the Fifteenth Street Bridge and demanded dues from people as they crossed the bridge to work. One black worker refused to pay, pulled out a pistol, and discharged it between Odorcich's legs. The three pickets "beat the living hell out of him and threw him over the bridge."

"But," I said, interrupting the story, "the Yough is pretty shallow. He could have got killed."

"There was enough water where we threw him," Odorcich replied.

Fired in 1942 for beating up a foreman who cheated on an incentive plan, Odorcich was out of work for several months until Phil Murray once again came to his aid and helped him get a job at Pittsburgh Steel Foundry in Glassport. Odorcich served as the local president for several years and joined the District 15 staff in 1949. Although he had never worked in a basic steel plant, Odorcich thought that he received a better education as a union officer in the small locals. The International didn't allow officers of large steel locals much autonomy in handling their affairs. "I find that better unionists, better staff guys, came out of the smaller fabricating plants," he said. "You negotiate your own contracts, you take your own cases to arbitration, and you don't have the people upstairs do your thinking for you."

The first of three disillusioning experiences in union politics—episodes that turned Odorcich into a bitter, angry man by the time he retired in 1986—occurred in 1969. By openly supporting I. W. Abel when he ran against Dave McDonald in 1965, Odorcich defied his district director, Paul Hilbert, a McDonald man. Odorcich detested Hilbert and his corrupt administration and ran against him in 1969. Abel also disliked Hilbert and, Odorcich says, encouraged him to mount the challenge. Other officers, however, persuaded Abel not to violate the one-for-all and all-for-one "official family" concept by endorsing Odorcich over an incumbent board member. Abel stayed out of the District 15 race.

Although feeling betrayed, Odorcich refused to back down. I first met him during the 1969 campaign and learned that there was no empty bravado about him. In the rough and tumble business of USW politics, one did not lightly challenge one's own director. Odorcich received threats on several occasions and endured some bumping around by Hilbert stalwarts. A cadre of rank-and-file followers, particularly the McKeesport unionists, accompanied him as he made the rounds of local union halls. Odorcich promised to bring financial and other reforms to the district.

The union's election officials declared Odorcich the winner over Hilbert and two other candidates, with a plurality of 522 votes. Two weeks later, the executive board reversed that decision, threw out the Local 1408 vote, and pronounced Hilbert the winner by 72 votes. Local 1408 President Tony Tomko, who supported Hilbert, admitted to USW officials that he violated election rules by campaigning in the balloting area. This "tainting" of the local vote obviously was meant to void the results in the pro-Odorcich local. Instead of ordering a rerun of the election, the board chose to keep Odorcich out of office. Wanting to be rid of Hilbert, Abel forced him to resign and praised his "courageous action" in doing so. The USW established an administratorship over the district.

It was one of the more shameful episodes in the USW's

political history. Odorcich reluctantly appealed to the U.S. Labor Department. "Before that," he said, "I thought any union man who takes his union to court is a fink. But when I got to thinking that they [the board members] thought I was such a nonentity that they could do that to me, I got like my old man—I took them on." After an investigation, the Labor Department threatened to seek a court-ordered rerun. The union agreed to hold a new election, with the same names on the ballot, and Odorcich won by 6,000 votes.[11]

Odorcich finally took his board seat in 1970 and, for a year or so, acted as an anti-Abel rebel. Finally, the two declared an uneasy truce, but Odorcich never forgave Abel, or the union establishment. His second disillusionment came when he tried to form his own slate and run for president in 1977. Odorcich received the support of a number of board members, only to see it suddenly melt away. He believed that Abel was also at the bottom of this move. Later, he grudgingly accepted an offer to run for vice-president on McBride's ticket. His third experience occurred late in 1983 following McBride's death (see chapter 15).

For all of his close contact with the rank and file, Odorcich rarely indulged in back-slapping or standing drinks at the bar. A curiously aloof man, he discouraged unionists who sought friendlier ties and went his own way. One high-level staff man in the USW felt that Odorcich "threw up a lot of defenses, and it was difficult to know him."

Although Odorcich had started his career as a militant, even a radical, he eventually adopted a middle-of-the-road economic philosophy. "I started out wanting to kill all companies," he once told me, "but it didn't take me long to understand one thing. To have a union you got to have a company, and the company got to make bucks before you can get pennies."

Odorcich Versus Johnston: A Deal Is Struck

According to one school of thought, economics and politics dictate the outcome of a union-management struggle, not personalities. There is some truth in this. If the USW had been headed by the most rabid anticoncession person, a Weisen or a Bonn, say, it could have struck the steel industry in 1982–83 to avoid wage cuts. But in the end it could not have prevailed over the economic forces that were dismantling the domestic steel industry, mill by mill, company by company. Eventually, with or without the union, many steelworkers would have accepted work at a lower wage.

In the long sweep of time, the traits and characteristics of individual leaders seem to be ground up in the maw of much larger forces. In February 1983, however, a striking contrast in the personalities of the two lead negotiators at the steel bargaining table *did* make a difference.

Bruce Johnston, a sophisticated bargainer with a strong grasp of economic issues, had a head for numbers and a large and colorful vocabulary. He could think quickly and impose his agenda on a negotiation through skillful filibustering.

Odorcich, a blunt-talking coal miner, left "g's" off the end of "talkin'" and "negotiatin'" and sometimes set his mind against understanding the detailed arithmetic of an economic issue. This limitation, in fact, would cause a problem for the union in the aftermath of Round Three bargaining. Odorcich, however, realized his weakness and respected Johnston's abilities. "He's much smarter then I am," Odorcich volunteered. "He's cunning. He almost psychoanalyzes you. One for one, on the figures, he'd wipe me out." Odorcich's strength was his stubbornness and aloofness, the impenetrability of the Croatian serf, which made him less vulnerable to Johnston's persuasiveness than McBride had been.

Odorcich told an anecdote about the difference between him and McBride. "Bruce liked to start out meetings telling jokes and talking about extraneous things. In one of those episodes, he said, 'It's too bad you gentlemen are from a different social strata because, God, how nice it would be if we could socialize together.' I look at Bernie and Bernie looks at me, and we look at Lloyd, and Lloyd has a great big grin on his face. He fell for that bullshit. What he didn't know was that it was Bruce's job, like it was our job, to use the people on the other side to see your point of view. One of the best ways you do that is to get them to like you. The only time I liked Bruce was when I walked away from negotiations. In negotiations he was like a board to me."

The cost-of-living issue dominated Round Three, as it had the two previous negotiations. The industry still wanted to eliminate it, or combine it with profit-sharing. Odorcich disliked the COLA bonus plan and was determined to keep COLA, even if modified. "I'll fold it, I'll bend it, I'll twist it, but in the end we will emerge with COLA," he told the local presidents at a breakfast meeting.[12]

Odorcich held forth at three of these breakfasts over the

negotiating period. It was his way of giving feedback to the presidents and keeping them on his side. He felt that McBride did not report back to the local presidents often enough and when he did "it seemed like he spent more time worrying about saving the industry" than about the union's problems. Odorcich intended to "carry the message back [from the bargaining table] but jab the company once or twice to let the rank and file know you're still with them."

On another occasion, Odorcich told the presidents about a conversation he had had with Johnston. The U.S. Steel executive urged the union bargainers to give up COLA. "Things will change," he told Odorcich. "Let the next [USW] president get COLA back, and he'll be a hero."

"He may be a hero," Odorcich shot back, "but I'll be dead by then." The BSIC delegates enjoyed this kind of repartee. It gave them the feeling that the union had a goal other than merely following management down the concession trail.

The positions of the two sides, however, created a standoff that persisted through two weeks of bargaining. Johnston and Moore kept saying they would settle for nothing less than "November-plus." When the union indicated it would give up slightly over $2.00 per hour, they demanded an additional $1.00 to reach the $3.11 McBride had agreed to in November. But the USW adopted stalling tactics in subcommittee meetings, hoping to whittle down the November give-backs.

The List 3 problem cropped up again. Odorcich believed he couldn't obtain ratification of a pact that severed the List 3 units from the industrywide contract. Johnston likened the union's position on List 3 to Lucy in the "Peanuts" comic strip when she holds a football for Charlie Brown to kick but pulls it aside at the last moment. "You've been doing that to us with List 3 since at least 1968," Johnston said. Three days before the March 1 deadline, however, the company men relented on the issue, hoping to get the talks moving. They reluctantly accepted the union solution: Although the subsidiary locals would be covered by the master agreement, they would be free to negotiate separate pacts with the companies. Odorcich later told the industry conference that the List 3 plants would "go on the chopping block" if they did not bargain new pacts within 120 days.

Although retreating on this issue, the industry negotiators continued to insist on the COLA bonus plan. Odorcich argued against it, calling it a too-complicated "mishmash." Apparently, McBride also had decided to abandon the idea. In January, Jim Smith told McBride that he "wasn't going to be a party to any more of that [COLA bonus]," and McBride also withdrew his support, Smith said.

So intense was Smith's dislike of COLA bonus that he took an extraordinary step on his own to kill it. The industry had revised the plan so that the disparity in payments would not be nearly as great as in November and asked the USW bargainers to look at it. Odorcich agreed at least to do that much. Smith knew that Odorcich disapproved of the plan and believed that the chief negotiator had given him a signal to build support against it. Through union intermediaries, he arranged for the chief negotiators of Armco, Republic, and J&L to see the November calculations (even though the plan had been revised), which showed that these companies would have paid substantially more than U.S. Steel and Bethlehem.

Johnston heard about Smith's plotting and complained strongly to Odorcich, who in turn reprimanded Smith. "If I was president," Odorcich told the industry men, "I'd fire him."[13]

It is questionable whether the incident undermined the industry's cohesiveness. Johnston kept insisting on COLA bonus. Odorcich kept saying no. He would accept deep cuts in future cost-of-living payments but not the COLA link with profit-sharing. He dug in his heels and refused to budge. "Joe's stubbornness was a very valuable tool," Kleiman recalled in 1986. "I think Bruce had an awful lot of trouble figuring Joe. I don't think he ever did figure Joe."

On February 27, after several hours of meetings and with less than forty-eight hours left before the March 1 deadline, Johnston broke the logjam. Saying he didn't want to be so rigid as to ignore other ideas, he asked the USW negotiators for their proposal. Kleiman said the union would forego COLA in the first year of a contract and accept a 4 percent "corridor" in the second year. Under this concept, suggested by Ed Ayoub, steelworkers would receive cost-of-living payments only if inflation rose more than 4 percent annually. The USW, however, wanted to return to the full COLA in the third year of a pact. Kleiman also proposed reductions in other benefits.

Bargaining over this proposal continued into the late evening, with Johnston and Moore complaining that the union still hadn't boosted the proffered concessions by the $1 per hour needed by the industry. Close to midnight, Johnston made a last appeal to Odorcich, saying that he could not obtain the chief executives' approval on anything less than a corridor of at least 5 percent in the last two years.

At this point, Kleiman wrote in his notes, rather incredulously: "Odorcich just sat." When he read his notes to me, Kleiman laughed and repeated, "He just sat there. Johnston and Moore didn't know what to do. They asked whether there were other areas where the union would compromise. Joe just sat for a while." Moore suggested lowering the corridor in the third year to 4 percent. Steelworkers would have a better chance of being recompensed if inflation shot up, but the contract would end without a restoration to full COLA.

Still, Odorcich said nothing. "They waited and waited, but he sat there," Kleiman recalled. "Then they said, 'What *can* you do?' He sat a long time again. Finally, Joe said, 'If you can't sell it and we can't sell it, we just can't do it.'"

The meeting broke up, and it appeared there would be no settlement. Odorcich put on his hat and coat and left with Johnston, while Moore and Kleiman stayed in the conference room, commiserating about this failure. "Can't you do something to get Joe to move?" Moore said.

Meanwhile, Johnston and Odorcich walked some two hundred feet to a glass door that led to the elevators. "It's too bad after all the work, we can't put this thing together," Odorcich later quoted Johnston as saying.

"Well, Bruce, I have a job to do, and you do, too, and I think we done it," Odorcich replied.

Johnston returned to his office. "I was waiting for Bernie," Odorcich said. "Finally, he came and said that maybe with a little nudge we could put this thing together. He suggested what we might do. I said, 'Okay, but let's make this the final proposition.'"

Moore picked up the narrative. "We [he and Johnston] were in Bruce's office for five minutes or so, standing around, looking at one another and bemoaning the fact that we were so close and didn't make it. Suddenly, 'Bang, Bang, Bang,' somebody was knocking on the door. It was Odorcich. Bruce opened the door and said something like, 'I knew it had to be either you or Roderick, because you are the only two who bang on my door like that.'"

"Goddamnit, this is my last offer," Odorcich said. Instead

of insisting on a full COLA in the third year, he would accept a 3 percent corridor.

The company men exchanged glances. Not having to pay full cost of living in the final year made the proposal more digestible. But Johnston thought he might squeeze out an additional concession. "What about the Sunday premium?" he asked.

The union had agreed to reduce the premium from time-and-a-half to time-and-a-quarter but said the higher figure must be restored after the first year. Moore and Johnston wanted to make the cut permanent. It was a costly benefit because continuous steel operations required many hours of Sunday work.

Odorcich still wore his hat and coat. "Hell, no!" He said. "Sunday premium must be restored." Kleiman suggested that it could be restored toward the end of the contract. "Okay," Odorcich said, "I'll back it up to May of 'eighty-six. But that's it!"

The union negotiators left the office while Johnston and Moore caucused. Once again someone knocked on the door. This time it was Kleiman. Odorcich had made a mistake on the COLA proposal, he said.

"I held my breath," Moore said, "and thought, 'Christ, what now?'"

The union wanted a corridor of 1.5 percent for the first two quarters of the final year and full COLA for the last six months, Kleiman said. Johnston and Moore said they'd consider it, but they realized this was the best they could get. Kleiman brought up a number of subsidiary issues. Finally, at about 3 A.M. on Monday, February 28, Johnston said, "I've got to go back to my principals, but I think we have a tentative agreement."

The Last Industrywide Ratification

As the small band of people at the top acted out this old bargaining ritual in the U.S. Steel headquarters, hundreds of local unionists and company technicians swarmed in the William Penn Hotel. The union and the industry had started plant negotiations in mid-February and later imported their people to Pittsburgh to work under the stern eyes of the top negotiators. They met in suites and conference rooms throughout the hotel, talking about local and companywide problems. About one thousand plant-level issues had been raised by the locals and, one way or another, they had to be disposed of by the March 1 deadline.[14]

The multitiered bargaining involved in the industrywide negotiations was an amazing process, one that created an impression, if nothing else, that democracy flourished in the steel industry. Logistically, it was like channeling one hundred lanes of cars into a tunnel from which would emerge a single, orderly line of bumper-to-bumper traffic. When I first observed the process in the 1960s, presidents of the United States still trembled at the thought of coping with a nationwide steel strike. A huge corps of reporters would descend on the William Penn in the closing days of the talks, its members peeping through keyholes and staking out listening posts in the many nooks and crannies on the ballroom floor.

By 1983 the possibility of a steel strike had become much less important to the nation. The press corps had dwindled to half its former size. Nevertheless, the scene was still impressive, a colorful slice of industrial Americana: Big Labor versus Big Management, circa 1946–75. By 1983 industrywide bargaining had survived beyond its economic life. Although we didn't know it then, we observed the phenomenon for the last time.

The local and companywide bargaining process always had a theatrical aspect, giving the *show* of rank-and-file participation without a lot of substance. The issues were limited to those not negotiated at the "main table." The companies' policy of maintaining standard practices from plant to plant also discouraged innovation in the local talks. Moreover, the top bargainers sometimes used the lower-level discussions as tactical weapons. In 1971, for example, just as the main table negotiators approached the deal-making point, the USW's top negotiator at U.S. Steel, Jim Griffin, injected a large number of militant demands on working conditions into the talks to set up a trade for a large economic offer. The tactic worked.

In the last few days of February 1983, it appeared that U.S. Steel turned the 1971 tactic on the union. Many local presidents from U.S. Steel plants complained that the company not only slowed the process of settling union demands but also escalated management demands. Local management negotiators called for changes in work practices which would enable them to combine production duties, merge craft jobs into supercrafts, eliminate precedents set by arbitration, wipe out all past practices that had been protected by the 2B provision, and contract out work performed by the bargaining unit. In many plants, the company wanted to stop providing safety equipment such as work gloves, eye glasses, and ear plugs.

Most locals resisted the sweeping demands at this time. But U.S. Steel's tough approach left no doubt in the minds of local officers that the company would continue an aggressive campaign to change what it considered to be inefficient practices, with or without union approval. The locals also believed that U.S. Steel intended to play one plant against another in the local concessions game. If two plants produced the same product, the one that conceded more on local practices would be most likely to continue in business. The companies knew that most locals were too weak to strike. Odorcich, in fact, had ruled out local strikes. The right to strike locally carried with it the management right to lock out, and USW leaders worried that some firms would exercise that right as an excuse to close marginal plants.

A profound pessimism settled over the delegates in the William Penn on February 27 and 28. As we waited for news about the top-level negotiations, I chatted with Bill Behare of McKeesport Local 1408. Management at his plant had told local officers that "the bottom fifteen hundred jobs have fallen out at National." Without substantial changes in work rules, the plant would lose pipe orders to U.S. Steel pipe plants in Lorain (Ohio), Baytown (Texas), and Fairfield (Alabama). Only two hundred people now worked at National and four thousand were laid off. Older and younger workers were sharply divided on taking concessions, the younger being more amenable.

Local 1408 started a food bank by holding a fundraising dinner. "We made about $1,300," Behare said. "But I don't think ten people came from the plant to support their own people. They're worried about themselves. Nobody sees no hope anywhere. They're looking out for themselves. The company forced us to take a week's vacation in January. When the laid-off people got their vacation pay, they saw that union dues were taken out. People came down to the union hall, screaming at us." The local asked the International to stop collecting from employees who worked less than five days a month. The International refused, saying it would lose too much income.

In the early afternoon of Monday, February 28, Odorcich briefed the executive board on the tentative settlement. The chief executives had met that morning at Pittsburgh Metropolitan Airport, where U.S. Steel owned a hangar with a conference room, and approved the agreement. The executive board voted unanimous approval. Although the agreement contained significant concessions, it was a more attractive package than the pact rejected in November. Not only had the USW beaten back the industry's effort to win deeper cuts; it also managed to reduce the size of the give-ups. In addition to modifying the COLA provision as outlined above, the tentative forty-one-month agreement called for a $1.25 per hour wage cut. It also eliminated one holiday and the thirteen-week vacation, cancelled one week of vacation for a year and a vacation bonus for the last two years, and reduced the Sunday premium until mid-1986. Although the companies lowered the wage cut of November by 25¢ per hour, they gained a cash-flow benefit for six months longer, since the first 40¢ restoration would not occur until February 1, 1984.

On the positive side for the union, it emerged with the COLA principle intact, though modified. The most important "positive" accomplishment—and the union's one brilliant stroke in three rounds of bargaining—was the early retirement incentive. Under the negotiated plan, all employees aged sixty or over, who had thirty years of service—about ten thousand workers at the seven companies—could retire on a regular pension with a $400 monthly supplement that would be paid until the normal retirement age of sixty-two. Workers had to exercise this option between March 1 and May 1, 1983. The plan would appeal to four classes of workers—young, old, active, and unemployed. The more people who opted to retire, the more jobs (theoretically) that would open up for employees on layoff. The November proposal had created some resentment among active, older workers by taking cash from them and giving it to younger, unemployed members. The new plan, Tom Duzak pointed out to me, enabled the presidents to "go back to their people and say, 'The older people aren't doing all the giving and the younger people all the receiving.' Here is something that would help the unemployed by giving something to the older people."

Money to finance the early retirements would come from the same source as money to pay for new benefits for unemployed steelworkers, from the SUB fund. This meant that members wouldn't see the deduction on each paycheck. The companies agreed to increase their SUB contributions by 50¢ per hour worked, enabling them to provide a portion of a regular SUB benefit to most unemployed steelworkers with at least two years of service for three months. Employees with longer seniority would be guaranteed benefits until early 1986.

The industry also agreed to a "side letter" in which it pledged to apply all cost savings "exclusively to the needs of the existing facilities covered by this agreement." The savings could not be used in businesses other than steel, and each company would provide data annually to the union to prove it had complied with the provision. Finally, the USW won a "dignity and justice" provision which said that workers must be proved guilty of wrongdoing (such as excessive absenteeism, tardiness, drunkenness) before they could be suspended or discharged. The new procedure would be used on a trial basis in one-third of each company's plants.

The two sides also agreed that local unions and managements could continue to negotiate over contracting out. They set up a potential tradeoff: By agreeing to combine skilled craft jobs, a local could regain some work currently contracted out by the company.

The local presidents filed into a 5 P.M. BSIC meeting in a resigned mood. Opening the session, Odorcich said that "there's no question we took a step backwards." Union and industry negotiators had had "a hell of a fight" over COLA, he said, but the union had maintained the principle. Although the industry had demanded a $5 an hour cut in November, the union bargained "them down a hell of a long way." Odorcich added: "They did get some pieces of your hide. But you'll find out that by the end of the contract, if you accept it, you not only growed a little hide back on, but maybe a little thicker hide."[15]

The BSIC delegates gave Odorcich a much easier time than they had McBride in Round Two. Indeed, it might be said that by taking the beating at the November meeting, McBride softened up the presidents for the Round Three ratification conference.

A number of delegates commented favorably on the positive aspects of the agreement, especially the retirement incentive and the jobless aid provision that provided some benefits for *most of* the unemployed. Many presidents criticized one or more elements of the agreement, but in two days of debate only five delegates said they would vote against it. Four of the definite noes came from presidents who had been the most vocal in opposition from the beginning: Ron Weisen, Mike Bonn, David Sullivan of the Burns Harbor plant, and Joe Samargia of the iron ore range.

Bonn angrily listed issues that he thought the agreement should have addressed but did not: job guarantees, a moratorium on plant closings and department shutdowns, an equality of sacrifice provision, elimination of forced overtime and vacations, a revised ENA, and a procedure for further local-issue negotiations. Odorcich responded indirectly. He said that Johnston had given him a pointed warning about the plants in the Mon Valley. Following the November 19 rejection, U.S. Steel executives had watched District 15 local presidents criticizing corporate management on TV, using rhetoric that "wasn't very constructive." It was then, Johnston suggested, that the company decided against building a continuous caster in the valley. (The presence of a caster would ensure the continuance of some steelmaking in the valley.) With excess capacity at its many plants, U.S. Steel would have to decide which plants to modernize and which to close. Odorcich quoted Johnston as saying, " 'We're going to [keep] those plants where we get people who cooperate.' "

When Weisen got the floor, he said these comments had been directed at him. "We used to negotiate out of power," he said. "Now we're negotiating out of fear. I'm ashamed it's happening here." He said his local opposed concessions "because U.S. Steel is not putting none of the money back into the Mon Valley plants." Nevertheless, he said, the Homestead plant had more active workers than any other plant in the valley. "So, Joe," he ended, "you better get Bruce to educate you a little better."

Weisen's remarks gave Odorcich the kind of opportunity he liked. "Ron," he said, "I don't even remember mentioning your name. The wicked fleeth where no man pursueth." The delegates and observers applauded.

Sullivan urged the BSIC to vote down the proposal, saying his members were willing to give up a $40 annual allowance to buy safety shoes but not paycheck earnings. Odorcich pointed out that Sullivan's plant, the most modern in the industry, was not likely to be shut down. "You would be against whatever we did," he said.

Among the ten delegates who specifically said they would vote for the agreement were Mike Bilcsik of Duquesne, Charlie Grese of Clairton, and Don Thomas, president of Braddock Local 1219. The dissidents were routed. The tenor of the meeting so clearly indicated a yes vote that Odorcich adjourned the session at about 7:30 P.M. to allow the presidents to talk about the contract with the members back home. After two hours of debate the next morning, the conference ratified the pact by a vote of 169 to 63.

Union and company officials signed the agreement in the afternoon, and it became effective that day, March 1, 1983, almost a year and a half after Bruce Johnston and George Moore decided to ask the USW to reopen the steel contract.

Results of the 1983 Settlement

The industry failed to achieve its goal of cutting labor costs by $5, or even $3, an hour. Labor-cost reductions of about $2.65 an hour in the first year would be offset by 50¢ in increased contributions to the SUB fund, bringing net hourly savings down to about $2.15. Actual figures complied by the industry in 1984 put the average savings per hour worked at $2.19 for the seven companies (see below). Because of the wage restorations and escalation in health care prices, the terminal cost of the contract in 1986 would be close to $2 higher than in 1983.

Why did the industry accept a settlement so short of its goal? A failure in this round of bargaining presumably would have driven General Motors into the arms of foreign producers. It also would have increased the chance of a strike on August 1, when the industrywide agreement expired. Even without a strike, a series of negative events would occur leading up to the deadline. Hedge-buying would begin anew, attracting yet more steel from abroad. The steel producers would have to begin the costly business of shutting down furnaces and mills days ahead of the deadline. And bargaining against a contract expiration deadline, in contrast to a mid-term deadline, creates uncontrollable situations in which strikes are apt to occur even with the best of wills. None of the companies wanted a strike. The most troubled firms, Bethlehem, Republic, and Jones & Laughlin, could not have survived a long walkout.

U.S. Steel could have absorbed a strike, and Chairman David Roderick was no laggard about accepting challenges.

Very likely, Roderick already knew that eventually he would have to take on the Steelworkers, but not in 1983. Too many things were left undone. The oil business hadn't yet improved to the point where Marathon Oil produced enough cash to offset strike losses, and Roderick also wanted to diversify further into energy. An industrywide strike, while it would be much less damaging than twenty years before, still could cause enough domestic economic problems to turn the White House and Congress adamantly against any measure to protect the industry from foreign steel.

Finally, another factor undoubtedly influenced Roderick and Johnston in accepting what they viewed as an inadequate wage cut. After the economic settlement, they intended to initiate a broad assault at the plant level to cut the work force. Johnston believed that he had McBride's approval to negotiate locally without interference from the International. One provision in the new agreement called for plant talks in which the local union should consider combining skilled craft jobs in return for a "good faith" effort by management to avoid contracting out construction work. A skillful management could turn these discussions into a broader-based effort to cut jobs, leveraging on widespread fears by steelworkers that more plants would close. The USW also had given a go-ahead for local bargaining at the List 3 plants.[16]

In the months and years after the 1983 settlement, many unionists and some news articles reported that the companies, U.S. Steel in particular, didn't reinvest the labor-cost savings in steel. The evidence indicates that they did. As the top USW negotiators had expected, the seven companies spent far more on new equipment than the total amount of savings.

Under the reinvestment guarantee, the CCSC agreed to provide annual data to the USW so it could verify that the firms put the labor savings back into steel operations. The first report supplied figures showing that the seven companies realized savings of $521 million from March 1, 1983, to February 28, 1984, and invested $1.5 billion in steel. Because USW members worked a total of 238 million hours at the seven producers, investment per hour worked was $6.32, compared with the $2.19 in savings. Indeed, these companies reported total operating losses in 1983 of $1.246 million. The USW research department concluded that without the labor concessions, the firms would have lost $1.767 million. The cost savings provided working capital to companies that they would have had to borrow.

The projected savings over the forty-one-month period of the contract totaled $1.3 billion. It was impossible to estimate the extent to which new equipment projects would have been carried out without the concessions. The USW concluded that most of them probably had been planned before 1983 and that they would have been cancelled or the companies would have borrowed money to carry them out.

The report also enabled USW technicians to compute the amount of savings and investment on a plant-by-plant basis. This analysis disclosed that, as many Mon Valley employees suspected, U.S. Steel did not reinvest the total savings from those plants in the valley. The USW estimated that while USS workers in the valley gave up a total of $38.9 million in wages and benefits in 1983, the company spent only $7.5 million in major capital expenditures in those plants. The reinvestment provision didn't specify that the companies put the same amount of money generated at each plant back into that plant. Union bargainers believed such a requirement wouldn't be realistic, given the fact that a major piece of equipment installed at one plant would absorb the savings from several plants. In 1983 USS spent $675.5 million, or 87 percent of total capital expenditures, on a new pipe mill and other equipment at the Fairfield Works.[17]

The USW Gave Up More Than It Intended

While the companies didn't get as much as they bargained for, the USW in the end lost more than it *thought* it had bargained for. Some weeks after the pact was signed, the union discovered that it contained one sentence which hadn't been discussed during negotiations. In an appendix listing the new, lower wage rates for each job class appeared this harmless-looking sentence: "The above adjustments shall be treated for all purposes, except as provided in the pension agreement, as general wage changes or increases."

Industry negotiators contended this sentence meant that all benefits affected in the past by wage increases—vacation and holiday pay, overtime, and SUB benefits—must be affected in the same way by wage decreases. If wage boosts had a compounding effect on the way up, they must have the same effect on the way down. Each 1¢ of cut in the hourly rate would actually cost a worker 1.6¢. The industry argued that six items in the contract were affected this way.

Finally, it had happened: In the crisis atmosphere that enveloped the last two days of negotiations, groggy negotiators had made a mistake that would escalate into a major dispute. While Odorcich presided over the BSIC meeting, lawyers on both sides worked on the "settlement agreement," a document that expresses in contractual language what the negotiators agreed to. When union attorneys read a draft of the agreement submitted by the industry, they overlooked the crucial sentence. Never having drafted language in the industrywide agreement to reflect a wage *decrease*, the lawyers simply missed the significance of the sentence. They accepted blame for the mistake but also felt the industry didn't deal fairly with the union.

The sentence hadn't been mentioned at the bargaining table and, therefore, said Bernie Kleiman, industry negotiators should have pointed out to the union that it was there. "Somebody slipped the damn sentence in," Kleiman said, "and we didn't see the significance of it. But the person who slipped it in did. Fair dealing requires that they tell us they are changing the substance of the agreement we bargained." Kleiman contended that on many occasions in the past the union had informed the industry of a change in substance in the drafting process.

The industry, however, argued that USW negotiators should have understood that the benefits would be impacted by wage decreases. Indeed, the union's pension negotiators made certain that the pension portion of the agreement specifically excluded pensions from being affected by the wage cut. Why didn't they do the same for the other benefits? Moreover, said George Moore, "it behooved them to give as much as they could. To come in and say these six items should be excluded was the opposite of what we were trying to do, save money."

Unable to settle the disagreement in negotiations, the two sides agreed to submit it to arbitration. In the summer of 1983, the negotiators went to Los Angeles and testified in hearings before Benjamin Aaron of the University of California Law School, Los Angeles. Only top negotiators and lawyers participated, and both sides agreed not to release a record of the hearing. If it got into the hands of USW dissidents and political enemies of the McBride administration, it could prove highly embarrassing. At one point the industry offered an "out of court" settlement in which the union would win on one point. But Odorcich turned it down, fearing that such an agreement could appear to the rank and file like a conspiracy between the top negotiators. On the other hand, the union could claim that an arbitration ruling was wrong, though binding.

Aaron ruled in favor of the industry on all points but one. It was a humiliating loss for the union and added to the bad feelings that had begun to accumulate between the USW and U.S. Steel.[18]

Blast furnaces at U.S. Steel's National Works in McKeesport being demolished in 1985. *John Beale/Pittsburgh Post-Gazette*

(Left) The U.S. Steel Building, Pittsburgh, shown in 1984. USS sold the building in 1982 to help pay debts after acquiring Marathon Oil Company, but continued to maintain its headquarters there. *USX Corporation*

(Above) The Ann blast furnace at J&L's plant on the north bank of the Monongahela River, shown in 1983 before it was torn down. In the background are houses in Pittsburgh's Southside, across the river. *Mark Murphy/Pittsburgh Post-Gazette*

Aerial view in the 1950s, showing a new open hearth furnace at J&L's Southside plant. Pittsburgh's Golden Triangle is in the background. *Pittsburgh Press*

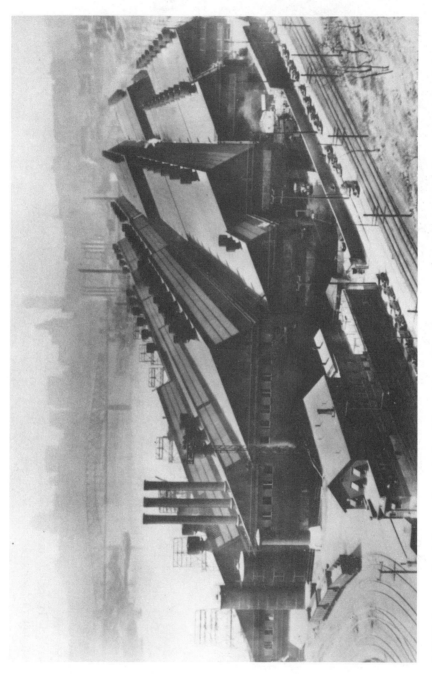

A 1937 photograph of the newly completed hot strip mill at J&L's Pittsburgh Works on the north bank of the Monongahela. *Pittsburgh Press*

By 1986 the hot strip mill had been demolished, leaving this bare riverbank, a 48-acre site on which will be built the Pittsburgh Technology Center. In the background are skyscrapers in downtown Pittsburgh. *Pittsburgh Press*

U.S. Steel's Duquesne plant in 1984, viewed from a residential area on a hill behind the plant. *Darrell Sapp/Pittsburgh Post-Gazette*

Shift change at the Locust Street gate of U.S. Steel's National Works in 1968 when the pipe plant was going strong and employed some 4,000 steelworkers. *Pittsburgh Press*

The railing along Locust Street in McKeesport, where steelworkers gathered before and after a turn in the mill, shown in 1970. The B&O tracks that ran through the center of town are across the street, along with a signal tower (left) and the passenger depot. *Pittsburgh Post-Gazette*

A view northward on Fifth Avenue in the heart of McKeesport's business district in 1951. The narrow brick street with trolley tracks and two- and three-story buildings was typical of main streets in Mon Valley mill towns. *Pittsburgh Post-Gazette*

In 1987, three years after U.S. Steel shut down the Duquesne Works, deteriorating walls and fences at this and other closed plants provided easy access to
vandals and thieves. But they faced hazardous conditions inside the plant
grounds. A trespasser was accidentally electrocuted at Duquesne the day before
this picture was taken. *John Beale/Pittsburgh Post-Gazette*

On Aug. 5, 1986, shortly after the beginning of the work stoppage at USX, pickets at the Clairton Coke Works hear that their final paychecks will be delayed. USX was retaliating for the pickets' refusal to allow managerial employees to enter the plant. *AP/Wide World Photos*

Both the Steelworkers and the steel companies blamed imports for the decline of the domestic industry, as this 1982 sign on a fence outside U.S. Steel's Homestead Works indicates. *Morris Berman / Pittsburgh Post-Gazette*

Aug. 28, 1987: the day that U.S. Steel closed the last section of the National Works, 115 years after the National Tube Company opened the plant. Charles Tice, a 40-year veteran and the last security guard at the plant, stands in the passageway of the Locust Street gate. *Mark Murphy/Pittsburgh Post-Gazette*

The Minimum Wage Recovery

The steel labor settlement of March 1, 1983, removed a cloud of discord that had overshadowed the Monongahela Valley and other steel regions for nearly a year. The agreement, of course, could not, and did not, cause the steel industry to rebound. Wage cuts of themselves do not stimulate business. The industry kept dwindling over the next four years, and so did other basic industries, permanently contracting the once prosperous world of the American blue-collar worker. The Steelworkers, meanwhile, went through an agonizing political battle and elected a new leader, Lynn Williams, who would respond to the steel industry crisis and the changing economic environment in a very different manner than Lloyd McBride.

The climate in which these changes occurred remained dismal. Steel always lagged behind other industries in a business upturn. In 1983 the accessibility of foreign steel, as well as the weak growth of demand in the United States, further delayed the steel recovery. Unemployment in the Pittsburgh metropolitan area did not show a marked improvement until July 1983, when it dropped to 13.9 percent and thereafter declined slowly to 12.7 percent at the end of the year. A similarly slow recovery occurred in steel-producing areas in eastern Ohio, northern Indiana, the southeast side of Chicago, Johnstown, and elsewhere.

Even when economic activity finally picked up in the steel regions, the steel business had little to do with it. Many mill gates remained closed through 1983, some permanently. Steelworkers returned to work by the scores or hundreds rather than by the thousands. Starting in 1984 and extending for about two years, the Mon Valley experienced what Paul Lodico, co-

390

director of the Mon Valley Unemployed Committee, sardonically referred to as "the minimum wage recovery." Most job openings occurred in shopping malls, wholesale and retail stores, fast food restaurants, supermarkets, warehouses, and business offices. The manufacture of hard goods, the valley's staple for a century, dropped to a minimum. Nor did employers build anything. To the degree that any structural work went on, it involved demolition—the tearing down of abandoned mills—rather than construction.

The low-paid, service jobs occurred in endless variation. In the Aliquippa area, jobless steelworkers worked four hours a day driving shuttle buses to the Pittsburgh airport. Temporary and part-time work abounded. Many employers, themselves hurt badly by the recession, offered twenty hours a week, or less, thereby escaping a legal requirement to provide benefits. Pay ranged from the minimum wage of $3.25 per hour to $4, sometimes going up to $5. As might be expected, the minimum wage recovery was a nonunion movement. Temporaries and part-timers, including many ex-steelworkers, didn't raise a clamor for a union, and little organizing occurred. Indeed, supermarkets in McKeesport and Monessen managed to oust their unions by engaging in a paper transfer of ownership. The Steelworkers and the United Food & Commercial Workers waged a bitter battle to represent clerks and butchers at other markets.

Even obtaining a part-time, minimum wage job proved difficult for many former steel employees. Jack Bergman, a one-time department superintendent at U.S. Steel, retired on a pension when the Duquesne plant shut down in 1984. At the age of fifty-four, he felt much too young to sit at home. He answered the want ads for minimum wage jobs but discovered that employers didn't want former mill workers or supervisors. "I was 'overqualified,' they told me," Bergman related. "I went for a job at Murphy's warehouse, a laboring job. I put down that I graduated from college, worked for thirty years in the mill, foreman, and so forth. Turned down. Overqualified. That was a helluva thing! Rather than hire somebody who can do the job, they come right out and say they wanted a man who knows nothing. A couple months later I went back, put in a new application. This time I'm a high school graduate, labored in the mill. The same woman interviewed me, but didn't remember. I said, 'I worked hard all my life. That's the American way. I listen to what the boss says. I do everything he tells me.' The next day I got a call to report to work. No guarantees, no benefits. I signed a paper showing I had no rights. We got maybe three days a week, usually four hours a day."

Something revolutionary was happening, and not just in the Mon Valley.[1]

Decline of the Blue-Collar Worker

In Houston during the early 1980s, bumper stickers read, "Drive 90 and Freeze a Yankee." In part this exercise represented harmless interregional razzing, but it also expressed the contempt felt by cowboy entrepreneurs of the free-wheeling Southwest for the growing armies of the jobless in the Rust Belt states. President Reagan had said that jobs were going begging, and yet the cry-baby Yankee steelworkers sat at home, unemployed, resigned to a fate of economic uselessness.

Then came the sudden decline in demand for oil. Petroleum prices fell, drilling slowed down, and unemployment shot up in Texas and Oklahoma. By 1986 Houston alone had lost one hundred fifty thousand jobs. Up to that time, the *Wall Street Journal* reported in March of that year, "Houston had maintained the self-assured swagger developed during a generation of undeterred growth. But the latest collapse in oil prices has finally proved that Houston lies largely at the mercy of forces beyond its control." The paper quoted a woman executive whose responsibilities included corporate entertainment. "I once asked my boss how long the boom would last, and he said, 'The rest of our natural lives,'" said Robin Stanier. "There was never any question of budget. The only limit was my imagination."

At roughly the same time, I sat in the home of former steelworker Richard A. Pomponio in McKeesport. He had felt the effects of the oil decline even before Houston did, having been laid off from National Tube in July 1982 *because* of the drop in demand for pipe. "In late 'eighty-one," he said, "I asked my boss how are my chances of working? He said, 'Richie, we have enough work for ten years,' we were so backlogged with orders. So I went ahead and put in new windows in my home."

Pomponio was laid off at the age of forty-nine after thirty years of service. A high school graduate, he had learned bookkeeping in night school and worked as an office clerk at

National, earning $22,000 in his last year. But although Pomponio submitted more than two hundred resumes and had countless job interviews in the first three years after his layoff, nobody would hire him. "How many times I heard, 'You're overqualified. I can't pay you what you were getting.' I said I don't need that. But I was too old. What were they going to do with a fifty-three-year-old? If you're over forty-five, you don't have a chance." In 1985 he finally got a job keeping the books for a school bus company, earning $10,000 a year, less than half his former income. The Pomponios managed to keep their home only because the Commonwealth of Pennsylvania temporarily assumed the monthly mortgage payment under a new law suggested by the Mon Valley Unemployed Committee (see chapter 6).[2]

As the Houston-McKeesport connection demonstrates, the regionally arrogant are likely to have their comeuppance. More important, it shows the interrelatedness of economic activity in a world dominated by giant corporations whose decisions, zipping around the world on the wings of information technology, are converted almost instantly into actions that help some nations, regions, and groups of people, but hurt others. For economists, of course, economic interrelatedness is an old concept, one that is clearly evident in textbook equations. In the 1980s, this phenomenon became painfully apparent to ordinary people, causing a change in perception about life and work that has profound implications for the way companies deal with employees.

Consider the following activities and business relationships, which once occurred behind the facades of financial institutions but which in the 1980s regularly made headline treatment on television news. With improved communications and the concentration of wealth and power, the natural volatility of some parts of the economic system, such as the stock market, began in the 1980s to make every other part almost as volatile. Speculation on the gold market in London changed the value of the dollar *vis-à-vis* other currencies, immediately affecting the cost of imported and exported goods, which had an almost immediate impact on decisions of manufacturers to make or buy components abroad. The deregulation of airlines in the United States prompted fare wars and an explosion of carriers, followed within a few years by an implosion and a rapid coalescing into a relative few survivors.

The Federal Reserve, meanwhile, manipulated the basic

money supply to control inflation, causing the interest-rate rises that put the price of homes beyond the reach of an entire generation of young people. Pension funds and other institutional investors, engaging in computer-programmed buying and selling, forced wild, sometimes daily swings in stock market prices. A few Wall Street arbitragers indulged in illegal insider-trading on a scale so massive that most of the corporate mergers of recent years would have to be taken apart to restore the status quo ante.

In the 1980s the strong dollar and the rise of Japan and Developing World competitors moved enormous amounts of manufacturing out of the United States. In light-manufacturing industries, management could almost carry the assembly line on its back to a low-wage country. But finally, after a great deal of damage had been done to basic industries, the dollar began slipping, and management began pulling those assembly lines back into the United States. It was like spinning out a yo-yo and making it "walk" the floor before reeling it back in.

Takeover artists and company raiders, battling in an arena far removed from the ordinary worker's small world, destroyed thousands of jobs with the wave of a hand at corporate auctions. Between 1981 and 1986, the number of takeovers worth $1 billion or more quintupled, but many of the mergers merely shuffled assets on paper and created nothing of value. The $179.6 billion involved in mergers and acquisitions in 1985 exceeded combined expenditures for research and development and net new investment. While American managements toyed with their financial shell games, they had little time to build new plants, bring out new products, create jobs, and learn how to relate to their employees. Billions of dollars that might have been spent to make America more competitive went to legal fees and investment banking commissions. Even when high-level executives found themselves on the wrong side of a hostile takeover, they could float to earth under golden parachutes and land standing up.

At the bottom of the financial pyramid, meanwhile, workers who produced the products that created the wealth so cavalierly traded at high altitudes found themselves vulnerable to all kinds of economic storms. In the new environment of a swiftly changing world economic order, they dangled forlornly on slender threads controlled at the top, or fell with a thud. Shoe workers in Lynn, auto workers in Flint, rubber workers in

Toledo, agricultural implement workers in Moline, refinery workers in Houston, steetworkers in dozens of communities, machine tool workers in Cincinnati, iron miners in Minnesota and copper miners in Colorado and New Mexico, dozens of varieties of fabrication employees throughout the Midwest—the list goes on and on. In the great grain-growing states—Iowa, Kansas, Minnesota, Illinois—thousands of farmers who had been urged to borrow money to expand production suddenly found themselves without a market, partly because of a foreign policy turnabout in Washington (no more wheat to be sold to the Soviet Union). When they failed to pay off the loans, farmers then found themselves without farms or homes. In the year ended June 1986, financial problems caused the loss of one farm every eight minutes. Nearly two hundred thousand rural Americans had to seek new lives.

Victims are generally the first to understand the revolutionary nature of a catastrophe, if not the reasons for it. One of the failed Iowa farmers, Dean Hagedorn, had to sell insurance for a living. "I loved seeing things grow, being my own boss and working together as a family," he told the *New York Times*. "Nowadays it seems everybody goes their own way. We're getting a different life in this country and it's rough to adjust."

In the Pittsburgh area, Alexander Stright, forty-one, worked for three years at the Duquesne Works before he was laid off in 1982. "I really like the job," he told the *Pittsburgh Post-Gazette*. "I was a welder and moved all over the mill. It was interesting." But after two years of unemployment, he moved to Spring, Texas, just north of Houston, to become a carpet installer. "Down here," he said after moving, "you don't even know what the seasons are. They don't have any trees—just pines, all you see are pines. If you keep thinking about it you get depressed. I try to block it out. Life's a bitch but it goes on. It won't be the same. It will never be the same. But we'll give it our best shot."[3]

Despite widespread criticism of American unions for granting "too little, too late" in reduced wages, they demonstrated amazing resiliency. In 1981, major contracts negotiated by unions provided annual wage increases averaging 7.9 percent over the life of the contract. The average fell to 3.6 percent in 1982, stayed under 3 percent for the next three years, and dropped to an all-time low of 1.8 percent in 1986. Adjusted for inflation, average production-worker wages actually declined 6.3 percent from 1978 to late 1987.

Even in 1986, four years after the end of the recession, 32 percent of workers covered by newly negotiated pacts had their pay cut or frozen in the first nine months. Despite the impression created by bargaining difficulties and contract rejections in some major industries, the five-year period of wage-cutting did not produce unmitigated strife. The years 1982 through 1986 witnessed the lowest level of strike activity in the forty years that the Bureau of Labor Statistics had kept figures on work stoppages. Generally, the American blue-collar worker responded to economic change the way that the gods of free markets would have wanted him or her to respond.[4]

As the wage-cutting phenomenon played itself out, or continued in collapsing industries such as steel, a new issue emerged from the turmoil of the early 1980s. American industry clearly had lost its competitive edge. To regain it, labor and management had to abandon uncompetitive work practices and improve performance and the concern for quality. Labor especially had to understand that change had become the norm.

American workers could meet this challenge, and in many cases would be forced to do so. But force does not secure long-term commitment. The question was, would management bring labor into the game as a partner, sharing information and power? In the new, fast-changing world, with information technology playing an ever greater role in the workplace, workers must commit their minds and hearts to the job, as well as their bodies.

No American industry had to change more in the mid-to-late-1980s than the steel industry. The steel companies had to starve themselves to health by abandoning plants and product lines and cutting the labor force. They had to develop new markets, find innovative ways of holding on to old ones, live within scarce resources, and reform—perhaps revolutionize— their management and industrial relations systems. The participation and involvement of workers would not only be helpful in this undertaking; it would be crucial.

Some steelmakers understood the need for greater participation and invited the USW's cooperation in taking this direction, while others moved more hesitantly or, as in the case of U.S. Steel, tended to emphasize change by force (see chapter 16). On the union side, Lynn Williams, an articulate Canadian who succeeded Lloyd McBride as USW president in 1984, advocated participation more boldly than most American labor leaders, challenging companies in a way they rarely had been challenged before.

A Change in Leadership

By the summer of 1983, Lloyd McBride was tired and ailing. He had returned to work soon after his operation in February and worked too hard, putting in ten-hour days. "He's very intense," said his assistant Harry Guenther. "He pays attention to the minutest detail and wants to do everything himself." When I last interviewed him on August 19, 1983, he looked wan and vulnerable but had energy enough to talk for nearly two hours. Among other things, he said that pattern bargaining from industry to industry could not be a viable strategy for organized labor at a time of increasing economic differentiation between industries. "To bargain beyond an industry's ability to pay is simply not reasonable or logical and is doomed to failure in the long run," he said.

In late September, his strength declining, bothered by a kidney stone and other problems, McBride returned to the hospital. He stayed for several weeks for tests and underwent multiple bypass surgery on October 18. This time he accepted the possibility that he might not be able to recover sufficiently to resume his duties. If this was the case, he told his friend Buddy Davis, director of District 34, he would retire early instead of clinging to office as a part-time president. He was released from the hospital but died in his sleep a few days later, on Sunday, November 6, at his apartment in a South Hills suburb of Pittsburgh. He was sixty-seven years old and would have retired when his term expired at the end of 1985. His prediction to Davis that he would not have a normal retirement had come true.

Perhaps McBride's greatest contribution was in forcing the Steelworkers to face economic reality. Finding this a politically difficult undertaking, he had made mistakes of judgment. He had failed to develop a broad policy for dealing with concessions, but few leaders on the cutting edge of significant change can invent strategies of brilliant foresight. McBride had moved the union in the only general direction, given the circumstances, it could move, which was backward. If his leadership lacked imagination, it reflected more than the average person's portion of courage.

The union had been leaderless for the entire period of McBride's final illness and was deteriorating rapidly when he died. The combination of a continuing attrition in membership (from 1.2 million in 1979 to slightly over 700,000), McBride's absence, and the reduction of staff by 230 members since early 1983 had seriously affected the morale of the remaining employees and local officers. Some board members had talked of appointing an interim leader while McBride recuperated. But the USW's constitution contained no provision for filling the post if the president was incapacitated. None of the officers or directors wanted to be perceived as pushing an ill man aside, and so nothing had come of this talk.

On the night of McBride's death, the four remaining officers—Leon Lynch, Frank McKee, Joe Odorcich, and Lynn Williams—discussed the situation. Naming a successor as soon as possible was imperative. A political fight seemed certain. It had been rumored for years that Williams and McKee had presidential ambitions and likely would face each other in the 1985 election. McBride's death also gave Odorcich an opportunity to fulfill his long-held ambition to be president. Although he would be too old under USW rules to seek election in 1985, he wanted to fill the vacancy as a "caretaker" president until then. Out of respect for McBride, the officers postponed action and pledged not to engage in politicking until after the funeral.[5]

McBride was buried in a rainswept cemetery outside St. Louis on November 10. The officers seemed—and undoubtedly were—unified in their sorrow for McBride's passing. There was a sense that he had suffered a lot for a less than gratifying presidency. A large contingent of industry executives attended the ceremonies, including Johnston, Moore, Kirkwood, Tremain, St. John, and others.

Shortly after the funeral, the USW officers agreed to settle the question of successorship at a board meeting on November 17. The union constitution specified that a referendum must be held to fill an unexpired presidential term if the vacancy occurred more than two years prior to the next scheduled election. McBride had died two years and twenty days before an election slated for November 26, 1985. Because it would take at least three months to go through a nominating period and make other preparations for a special election, the board would have to name an interim president. Whoever it was would be able to consolidate power and enhance his chances of winning the election.

Williams began lining up support. McKee, however, decided to postpone his own candidacy until later and endorsed Odorcich for the interim job. Winning the post had become a consuming passion for Odorcich. He felt that the union owed it to him for reasons of seniority and experience. He had served on the board longer than the other officers. Twice he had stepped into the role of chief steel negotiator when McBride became ill. Everybody knew he had sought the board's support to run for the top job in 1977, only to be foiled by what he thought was a conspiracy led by I. W. Abel. His turn had come, Odorcich contended, and a number of board members agreed.[6]

Between November 11 and November 17, Williams and Odorcich and their staff and board supporters spent hours each day on the phone and in meetings, cajoling and negotiating with board members for their votes. Almost immediately the board split nearly down the middle, and for an interesting mix of reasons.

Initially, Odorcich was supported by McKee and directors in older steelmaking areas like the Mon and Ohio valleys, Buffalo, Cleveland, Bethlehem, Birmingham, Detroit, and Baltimore, as well as directors in Atlanta, Los Angeles, Worcester, and Indianapolis. Many members of this group backed Odorcich because of personal friendship and a shared concern for the steel portion of the union. Odorcich talked of concentrating the union's resources on solving the problems of steel industry members and spending money out of the USW's treasury, or diverting its $200 million strike fund, to help jobless steelworkers. Ethnicity also played a part, though perhaps the least important part, in Odorcich's campaign. He saw himself as representing the USW's "ethnic" members—those of Eastern European nationality—and five of his original supporters had such a family background.

McKee's support of Odorcich also attracted followers who favored McKee as the next full-term president. Although the Odorcich group represented a somewhat more conservative union philosophy than the Williams partisans, the ideological split was not clearcut. McKee had made it clear that he opposed wage concessions, particularly actual cuts in pay, as a hard principle. Odorcich and others in the group didn't share this conviction. For example, the Atlanta director, Bruce

Thrasher, who served as the USW's chief aluminum negotiator, favored wage moderation and had accepted a pay freeze
and other restraints in 1983 aluminum bargaining. But he had
hopes of joining a McKee ticket in the next regular election.

The most divisive issue separating the two factions was a
dark strain of nationalism. Some Odorcich supporters felt
uncomfortable with Williams's Canadian background and the
ideas he brought from it. While Canadians had constituted a
significant portion of USW membership since the 1930s and
accounted for 20 percent in 1983, Williams was the first Canadian to hold a top office in the union. Many American steelworkers regarded the USW as "our" union.

In the Canadian union tradition, Williams had a special
interest in international labor matters and headed that area of
USW activity. He advocated closer cooperation with organizations such as the International Confederation of Free Trade
Unions, the International Metalworkers' Federation (IMF),
and the International Labor Organization (ILO). Odorcich
termed Williams an "internationalist," adding: "I'm concerned with here, not Zimbombuie (sic), the ILO and the IMF.
We got so much problems with our own people, that all this
intellectual, high-flying stuff just leaves me cold."

Williams, in a sense, had been preparing for the presidency
for many years. By far the most intellectual of the USW
officers, he had been underutilized by McBride and had stayed
in the background. Now, it became clear, he was a strong-
willed, politically shrewd man who projected a sense of knowing what he wanted, how to get it, and where to take the
union. "Joe would concentrate on the Steelworkers union, the
steel industry, and even the [Mon] Valley," Williams told me.
"But our problems are not problems of administering our
traditional base. We have to find a place in the new world, and
we won't find that by looking inward. We have to look outward." He didn't regard himself as an "internationalist," but
rather as "a North American trade unionist."

Since 1977 Williams had occupied what was in some ways
the choicest office from which to launch a presidential bid. As
union secretary, he had a strong grip on important internal
functions, such as the personnel, education, and local union
departments, and also supervised public relations, organizing,
and international affairs. "You never make anybody mad in
that position," one director noted. The secretary had few
negotiating responsibilities that could raise the potential for

angering district directors and local officials. But Williams hadn't set out to insulate himself in the secretary's office. Originally slated to be a vice-president, he had switched spots on the McBride ticket with Odorcich because the latter didn't like paper work. While Odorcich could claim more negotiating experience in the steel industry, Williams was regarded the better administrator.

Supporters of Williams included Lynch and basic steel directors in Canton, Chicago, and Philadelphia. On his side also were directors in Milwaukee, Duluth, St. Louis, Houston, the national director in Canada, and two of the three Canadian district directors. The third Canadian director, Dave Patterson, known as an anti-McBride, anti-Williams militant, endorsed Odorcich.

Typically in Steelworker politics, the longing for a clean line of succession from president to president, almost as strong as royal succession, played an important role. It was a natural inclination in a group of politicians who hoped at all costs to avoid divisions within ranks. McBride never had announced publicly his choice of a successor, but a number of people—including Buddy Davis and McBride's personal assistant, Harry Guenther—said he had told them privately that he preferred Williams. Odorcich disputed these assertions, but he couldn't overcome a widespread feeling among board members that this was true. He and Williams had got along well in the six years since they took office together, and Williams gave Odorcich credit for doing a good job in steel negotiations. "But that's not necessarily a reason to be a president," Williams said. "That's just giving service to the union."[7]

With the two sides fairly evenly divided, the fight went down to the last day. Odorcich and Williams both rented suites in the Pittsburgh Hilton the night before the November 17 board meeting and relentlessly talked to board members who might be wavering. "Up until 2 A.M. on the seventeenth, I thought I had sixteen and Williams had twelve," Odorcich later said. "Then one of my guys spotted two defectors in the Williams suite. I knew we were in trouble." Five minutes before the board meeting started, a third defector, Paul D. Rusen, director of District 23 headquartered in Wheeling, told Odorcich he had concluded that Williams would win and therefore would switch to his side.

Rusen's decision to change allegiance probably was typical

of the others. A one-time golfing partner of Odorcich, Rusen originally had backed him out of friendship. In the Odorcich group, however, Rusen found himself thrown together with people he disagreed with on fundamental issues. Frank McKee, for example, had criticized Rusen for negotiating wage concessions with Wheeling-Pittsburgh Steel. Late at night on the sixteenth, Rusen had a long conversation with two Williams supporters, Jim Smith and Edgar L. Ball, the Houston director who Williams later named to the vacant post of secretary. They convinced Rusen that Williams would be better for the USW in the long term. Rusen also had hopes of running for a top office on the Williams slate in 1985.

Odorcich presided over the board meeting. He had hoped to avoid a special election so that he could spend two years as a "caretaker" president. But the board overruled this idea and set a date of March 29, 1984, for the balloting. After a three-hour debate on the merits of the candidates, the board voted, sixteen to twelve, to name Williams as temporary acting president until the special election. In a final vote to make Williams's election unanimous, only Odorcich, Lefty Palm of District 15, Frank Valenta of District 26 in Cleveland, and the Canadian rebel, Patterson, refused to go along.

In addition to the differences in philosophy and ability of the two candidates, the fifty-nine-year-old Williams seemed to offer greater long-term stability. If he was elected for the twenty-month interim term until the election of 1985, he would still be eligible to serve two more four-year terms, carrying the Steelworkers into the 1990s. This factor swayed many directors. "If Joe had got in, we could conceivably have ended up with three presidents in ten years," said Buddy Davis. "We can't keep operating on the basis of changing policies and plans. We need long-range planning, and Lynn is a far-reaching kind of person."

"They've [the union establishment] kicked me in the ass two or three times," a bitter, somber Odorcich told me the next day. "Nobody can say I haven't handled responsibilities. I'm most bitter about the directors who shook my hand and said, 'We're with you,' and then turned against me. In my opinion, they're less than men." Along with Rusen, Thermon Phillips of Birmingham, Harry E. Lester of Detroit, and David Wilson of Baltimore had changed sides.

Odorcich said he would run against Williams in the March referendum but changed his mind about a week later. The board vote had convinced him that the lame duck issue damaged his candidacy. He told McKee he would pull out of the race and support the treasurer. McKee gladly accepted the offer.

Despite the internal split, Williams quickly gained the board's unanimous approval of a wage-concession policy. Meeting on December 13, the board adopted a policy prohibiting any director or local officer from granting concessions in wages, benefits, or work practices that would be inconsistent with the industrywide steel agreement. It was an important move to prevent locals and districts from undermining one another—if it had been observed. In the end, the International allowed many exceptions.[8]

Williams Versus McKee

Williams and McKee announced their candidacies in early December 1983. McKee, who turned sixty-three on December 31, 1983, would be eligible to run for a four-year term in November 1985. Tall and heavyset, a former open hearth worker at a Bethlehem Steel plant in Seattle, McKee had served as a local president, district staff man and, from 1971 to 1977, director of District 38 in California. As one director described him, McKee was "a good old boy." He got along well with people, had many friends in the locals, and prided himself on being a hard-line, bread-and-butter unionist. He simply did not believe in the dawning of a new economic order that would require nontraditional union responses.

In a letter announcing his candidacy to all local unions in early December, McKee staked out a tough, anticoncession position. "I am openly declaring war on concessions, vacillation, timidity, discrimination and apathy in our union," he wrote. "I am sick and tired—as I know you are—of policies that have brought our union and its local unions to the brink of disaster." The union, he added, had allowed management "to walk all over our hard-won rights."

In this statement, McKee implied that he had opposed concessions to the steel companies. The record didn't support fully this assertion. The executive board had voted unanimously in the last two rounds of steel bargaining to endorse recommendations for wage cuts. According to participants, McKee didn't raise a word against those agreements. "He voiced his opposition in the bar," Rusen remembered. McKee justified his lack of opposition in the board meetings on grounds that he was not involved in steel negotiations. "I don't second guess anybody else's bargaining," he said.

His personal opposition to wage cuts, however, led to a spectacular defeat for the union in the copper industry. Since 1977, McKee had been chairman of a coalition of unions that negotiated contracts at companies that mined, smelted, and refined copper and other nonferrous metals. In 1983 bargaining under McKee's leadership, the coalition accepted modest new pacts with a wage freeze at Kennecott, Asarco, Magma, and other companies. Phelps Dodge Corporation, however, refused to acknowledge the existence of a "pattern" and insisted on suspending COLA and reducing some benefits. The company argued that it could not compete with foreign producers during a period of severely depressed prices and high imports without lower costs. During a long, violent strike that started on July 1, 1983, Phelps Dodge continued producing at its large mine in Morenci, Arizona, with strikebreakers—including many former union members—under the protection of the Arizona state police. This might have been anticipated, since only a little over 50 percent of the Morenci workers belonged to the union in the first place.

It became clear within a few months that the strikers had lost their leverage and could not prevent the company from operating. McKee, however, refused to compromise on his principle that Phelps Dodge must accept the industry "pattern." He might have been able to negotiate a pact limiting COLA and making other changes, but this would have spurred demands for renegotiation by the other companies. As the strike wore into the fall months, even McKee's friends at USW headquarters in Pittsburgh believed that he had made "a big mistake," as one board member told me, in not perceiving the reality of the situation.

Although the strike became a *cause célèbre* for the labor movement and political progressives, the cause had been lost by early 1984. In February, the company worked its mines and smelters flat out with 947 new employees and some 1,300 union members who crossed picket lines. In the fall of 1984, these workers voted to decertify the USW and other unions in the coalition. The unions' intransigence brought about their own downfall. Prostrike sympathizers deplored tactics employed by the company, such as evicting strikers from company housing, and the sometimes violent methods of the state troopers. But deploring could not alter either the economic circumstances that forced the company to take its tough stand or the labor laws that enabled it to hire permanent replacements for the strikers.

The Phelps Dodge disaster also exposed an old deficiency in the USW's decision-making process. McBride always had allowed McKee a free hand in copper and made no attempt to interfere in the first three months of the walkout. After that, McBride's illness left the union leaderless. Other board members, despite their apprehension about the strike, also muzzled themselves. If they kept out of McKee's affairs, he would keep out of theirs. Finally, in 1985, some directors openly attacked McKee for his administration of the strike. Williams, then president, removed McKee as chief copper negotiator and replaced him with Ed Ball. By then, the strike was long since lost.[9]

When the Williams-McKee campaign started in December 1983, this outcome wasn't apparent. McKee's reputation as the leader of the Phelps Dodge strike and his campaign pronouncements against wage cuts gained him many adherents among militant union members. Some militants supported an early bid by Ron Weisen to get on the ballot. But he failed to obtain the required 111 nominations by local unions. Williams won far more nominations than McKee, 1,593 to 984, and had the edge from the start. Of the approximately 1 million members eligible to vote, 195,000 belonged to the USW's Canadian districts, where Williams could be expected to do very well. Williams also retained the support of sixteen district directors, plus Leon Lynch and E. Gerard Docquier, the national director of Canada. Only eight directors, along with Odorcich, endorsed McKee.

In terms of what policies the USW should pursue at a time of basic restructuring of American industries, Williams and McKee offered steelworkers a choice—although their differences were not made entirely clear in the campaign. Wedded to traditional USW policies, McKee was more likely to fight to retain pattern bargaining in the big metals industries and resist workplace changes. Williams did not advocate reversing the old policies. His campaign literature noted that he "has publicly drawn the line on wage and benefit concessions in the steel industry." To many union insiders, however, Williams had demonstrated his willingness to shuck old concepts and rigid union behavior if the times called for it. As chairman of the union's "futures" committee, he had developed some very specific ideas for trying new organizing techniques and adapting internal union procedures to changed circumstances.

One of the earliest USW supporters of worker participation, Williams wanted to expand the use of Labor-Management Participation Teams and similar mechanisms. Shortly after becoming temporary president, he quietly lifted a moratorium on LMPT activity at U.S. Steel imposed by McBride to retaliate against the company for planning to import semifinished steel slabs from Britain. Williams also opposed the British proposal, but he believed a company and a union could disagree in one area while cooperating in others.

But he avoided the LMPT issue as much as possible in his campaign. He felt, probably correctly, that the election should not hinge on his attitude toward worker participation. Very likely, a demand for more LMPTs in the steel industry would invite charges that he was "in bed with management." Instead, he issued a general call for "involving workers in the economic processes of this society."

Williams campaigned hard on the theme that the labor movement must defeat Ronald Reagan in 1984. His second priority was rebuilding the USW. McKee stressed his anticoncession approach and, as the campaign wore on, focused on Williams's Canadian background. "It would just not be right to have our union dominated by a Canadian," McKee said, implying that Williams would favor Canadian steelworkers to the disadvantage of U.S. members on matters such as trade legislation.

Williams labeled this talk "jingoistic" and partially disarmed McKee on the issue by supporting a bill pending in Congress to establish quotas that would limit steel imports—from all nations, including Canada—to 15 percent of the U.S. market. This action rankled some Canadian members but not enough to change the pro-Williams sentiment in that country.

When steelworkers voted on March 29, the Canadian support put Williams over the top by a vote of 192,767 to 136,264. McKee beat him in the United States by nearly 15,000 votes and may have narrowed the gap even more but for a low voter turnout. Unemployed steelworkers who had little chance of regaining steel jobs were apathetic. "Nobody could care less," Arthur J. Sambuchi, a local president at Bethlehem's largely closed Lackawanna plant, told the *New York Times*.

"We're dead. We're just waiting for the ground to be shoveled over us."

McKee complained of vote-cheating in Canada. It was suspicious, he said, that Williams garnered over 90 percent of the Canadian vote. But USW election officials rejected this complaint for lack of evidence, noting that McKee had won some districts in the United States by more than 80 percent.[10]

After such a divisive election campaign, the healing process took a relatively short period of time. Most of the ill will was cleared up at the USW's convention in Cleveland in September 1984, when Williams met with seven of the eight directors who had opposed him. One of the seven, Bruce Thrasher, proposed the meeting and set it up. "We asked for no commitments," he said. "We recognized that he won and we lost, and we were there to offer our cooperation. Whatever differences we had with the administration, we couldn't solve them on the outside."

Neither McKee nor Odorcich attended the session. McKee later made a short speech to the convention, wishing "Lynn Williams and his reign in this union the very best." McKee and Williams shook hands as the delegates stood and applauded. But the rapprochement was more symbolic than real. For the remaining eighteen months of his term of office, McKee performed his treasurer's duties largely from an office in California.

Odorcich, however, did not pass so easily from view. Asking to address the convention on a point of personal privilege, he made an emotional speech filled with the typical Odorich combativeness and stubbornness. It was like the Odorcich of fifty years before, trapped in the coal mine, unable to free himself but determined to hold out. This time, however, there was no escape, and he must have known it. The delegates had demonstrated overwhelming approval of Williams, but Odorcich wanted to leave his mark. He complained that Williams had stripped him of his duties. If *he* had nothing to do, he said, the convention could reduce the number of top officers in the union and save money at a time when dues income was declining.

But the emotional core of Odorcich's speech concerned four staff members who had supported McKee in the election and had been demoted by Williams from their positions as chiefs or assistant chiefs of departments. "We are supposed to be a union, for God's sake," Odorcich said. "We fought these kinds of tactics against the companies." He would not shake the hand of the new president, he declared, until Williams "makes whole those four little people that we kicked around."

It was a unique moment in USW political history. An opponent had been allowed to attack the union administration from the podium. Williams made a speech in response, contending that the four staffers had been demoted, but not fired, because he lacked confidence in their ability to carry out their responsibilities. The president opposed the idea of reducing the number of top officers and said that Odorcich surely could find things to do in a union as large as the USW. Williams left no doubt that he had won the election and had no intention of deferring to Odorcich.[11]

The two never did shake hands during the next eighteen months. Occasionally, they crossed paths on the way to and from their twelfth-floor offices in the USW headquarters, and Odorcich carried out the occasional assignments given him by the president. Essentially, the union did without the services of two of its principal officers, Odorcich and McKee, until their successors took office in 1986. Odorcich attended the 1986 inauguration of new officers and wished the new team good luck. McKee didn't bother attending.

The Making of a USW President

Lynn Williams, the fifth president of the Steelworkers, was in several ways the most interesting of them all. Born in 1924 in Springfield, Ontario, he grew up in a family that emphasized social issues and concern for people. His father, a United Church of Christ lay preacher, ministered to the poor in working-class neighborhoods during the Depression in Sarnia and Hamilton. As a teenager, Williams worked in the YMCA youth program and came in contact with older men who were attracted to the Steel Workers Organizing Committee as it began signing up members in Ontario. The labor movement in Canada, says writer Robert Kuttner, combined idealism, service, and struggle, an irresistible attraction for a young man of Williams's background. He received a liberal arts degree from McMaster University in Hamilton in 1944 and served in the Navy for a year.[12]

After the service, Williams volunteered to help the USW in 1946 conduct a bitter strike at Stelco, the big Canadian steelmaker. He never forgot how Stelco's resistance to unionization divided families and the community. "One brother struck and the other scabbed," he recalled. "A father struck and a son scabbed. It was a struggle for survival, which the union won. We fought for every inch of ground." This struggle took place at an elemental level, one of determining whether the union would exist or not. But it convinced Williams that even at this level "collective bargaining means moving areas that had been exclusively the company's into some kind of joint decision-making process." He believed this applied with equal validity to most decisions that a company makes.

Having decided to make the labor movement his career, Williams worked for a brief period on an assembly line in a Toronto plant. In 1947, at the behest of a USW official he had met during his YMCA years, Williams signed on as a full-time staffer in an attempt to organize the large Eaton's department store in Toronto. Although the drive in the end failed, Williams learned all the facets of organizing from the ground up—talking to employees one-on-one, writing leaflets, raising money, developing strategy. He also fought Communist influence in these campaigns.

After four more years of white-collar organizing, Williams in 1956 joined the staff of USW District 6 in Ontario at the invitation of Director Larry Sefton, an early Williams mentor at the YMCA. This became a showcase position for Williams in the USW, because Sefton was one of the most respected and influential of district leaders in the Steelworkers, and he regarded Williams as his heir apparent. For the next eighteen years, Williams served on Sefton's staff, negotiating contracts, servicing locals, engaging in political action, and organizing. He personally planned and conducted many organizing campaigns, spending weeks at a time with "bushwhackers" in northwestern Ontario's gold and copper mining camps. "He didn't like to go someplace and just attend a meeting," recalled William F. Scandlin, a Canadian USW staffer who accompanied Williams on some trips. "Nobody could talk as effectively to people as Lynn. He was fairly unique as a college graduate in the labor movement. But people never felt he was talking down to them. He made you feel you were on the same level."

When Sefton died in 1973, Williams succeeded him as director. He quickly made a mark on I. W. Abel's executive board as a man of obvious intellect and ideas but one who didn't rock the boat as a rebel. Williams had no need to be a rebel; he was "in" from the beginning.

He inherited and carried forward the Canadian labor movement's long attachment to international unionism. This arose naturally in a country that depends to a high degree on exports for jobs. Because many large employers in Canada are branches of U.S. corporations, Canadian unions had to form alliances with U.S. unions. Despite occasional lapses, the Pittsburgh office of the USW has generally allowed the Canadian wing to operate autonomously. Nevertheless, nationalistic factions in the Canadian USW sometimes attracted intense support on specific issues. Williams was always on the side of staying in the International, instead of pulling out, as the Canadian branch of the UAW did in 1985.

Skeptics argue that attending meetings of the IMF and ILO in Europe is an exercise in unreality that seldom accomplishes much in terms of solving specific problems. Even the countries of the industrialized West are so unlike in political, legal, and economic structure that international labor solidarity at the grass-roots level has seldom developed, except between the United States and Canada. However, the process of learning about other labor movements and economies and developing foreign contacts can be a broadening experience. By the time Williams moved to Pittsburgh from Toronto in 1977, he had developed a worldwide perspective on political and economic problems.

Williams believed in free trade up to a point. He didn't think that labor movements in individual countries should stand aside and allow a massive drain of jobs to other nations. "The old idea that you leave trade to the free market isn't viable now, if it ever was," he told me in 1983. "We have a world where technology is mobile. Corporations take advantage of cheap labor in the Third World. We need to insist on the right of American workers to have jobs." Williams did not consider it "protectionist" to advocate bilateral negotiations between nations to determine the extent of trade in competitive products such as steel.

One of the founders of Canada's New Democratic party (NDP), Williams became skilled at close-in political maneuvering and fought there against radical elements as he had in union organizing. He would be classified in turbulent Canadian politics as slightly right of center. When he moved to the United States, Williams retained his Canadian citizenship and couldn't vote in elections. But he took a strong interest in the Democratic party and the AFL-CIO's attempt in 1984 to elect Walter Mondale as president. Indeed, he attracted considerable attention in high union circles in October 1983 when he represented McBride at the meeting of the federation's general board that made the historic decision to endorse Mondale long before the Democrats had nominated him. In the midst of a worried, low-key discussion about surveys and polls of union members, Williams electrified everybody with a drum-beating speech about the need to defeat Reagan.

Perhaps the most important leadership trait Williams brought to the USW involved a matter of emphasis. The four previous presidents had followed in the American tradition of emphasizing, first, the provision of economic benefits to members through collective bargaining and, second, the protection of these gains, and the protection of unions as institutions, through political action. Williams didn't totally reverse these roles. He by no means downgraded the importance of bargaining. But he seemed to feel most comfortable flying about the two countries to influence political processes, serve as a spokesman for the USW and labor at large, and develop broad stratcgics and coalitions to create an economic and political environment in which bargaining would yield more for the members.

Williams believed it more important for a union president to create the context for social and political equity than to negotiate the division of equities at the bargaining table. The latter role could be delegated to subordinates, though Williams himself participated when he felt it necessary. All these statements of emphasis are relative, of course, because the best labor leaders always did some of each. More than most American union presidents, however, Williams tended to follow the model of European labor leaders in expending his energies at a more cosmic level than the bargaining table. This tendency would draw sharp criticism from U.S. Steel negotiator Bruce Johnston in 1986 (see chapter 19).

A lean, bespectacled man of medium height, Williams kept in shape by jogging frequently. He did not have the commanding presence of people who think they are always on stage and often could be seen standing around with hands in trousers pockets. His mild outward demeanor would enable him to pass for a friendly high school English teacher. Williams enjoyed talking to rank-and-file workers and would drink a beer or two. When the occasion demanded it, he changed his persona into a strong leader who delivered forceful, extemporaneous speeches.

Williams liked to sit and chat with staffers, commenting with a dry wit, asking questions, figuring things out. He called this process "fussing things through." More than McBride, he consulted his fellow officers and staff members before making strategic decisions. "He frequently told us that if he had problems, he liked you to share them," Paul Rusen remembered with some amusement. In his quiet, patient way, the preacher's son could make people believe that all sorts of things might be accomplished if they only had faith—faith in the union as an institution, faith in its processes, and faith in his ability to make the processes work. "He spends an awful lot of time immersing himself in the facts, wrestling around in there, and saying, 'How do we solve this problem?'" said Bernie Kleiman. "He is always looking for a large, hopeful, solution. He's a very constructive thinker in terms of being able to solve some of these problems that a lot of people have given up on a long time ago."

This is key to Williams's thinking. He tried to avoid reacting to short-term economic trends and management initiatives. One example involves the issue of plant shutdowns. At one of his earliest news conferences after becoming interim president, Williams was asked what the union would do about U.S. Steel's decision in late 1983 to close down several plants and portions of plants, permanently idling more than fifteen thousand employees. McBride, the realist, probably would have said that nothing could be done, that a steel company could not be expected to keep all plants open in a deteriorating market flooded with steel. Williams had a much different response. "I think they [U.S. Steel] should keep them all open, and we should figure out ways to put people to work," he told reporters on January 13, 1984. The government, he added, "must put in place policies that help the steel industry move ahead, not continue its gradual erosion and destruction that erodes the industrial base of America."

Williams preferred to put issues such as world overcapacity in a long-term perspective. At some point, the shrinkage of the steel industries of Western Europe, Japan, and the United States would bring supply into balance with rising demand. There was always a chance that the U.S. government would act to increase the demand for steel (through programs to

rebuild bridges and other parts of the nation's infrastructure)
and restrict both direct steel imports and the amount of foreign steel used in other imported goods. Instead of accepting
plant closures as inevitable, Williams tried to keep them
open, using employee buyouts as a last resort.

Emphasizing the long-term enabled Williams to enunciate a
politics of hope, not of despair, a policy that would enable him
to restore institutional pride in the USW and individual pride
in members. To some degree, this approach conflicted with
what the steel companies saw as their immediate needs.

In American labor circles, the term "intellectual" can be
the kiss of death when applied to a union official. Although I
never asked him about it, I doubt that Williams would like to
be branded an intellectual. In truth, he didn't act elitist in any
sense and usually avoided speaking from a rhetorical or
polemical podium. But he followed a decidedly intellectual
approach in formulating a rationale for union policymaking.
After becoming USW president, for example, Williams decided that the union must position itself, in plain view of the
members and the corporate world, on the complex issue of
union-management cooperation. The problem was that while
some companies were begging for cooperative endeavors,
others were fighting the union tooth and claw. This reality
made it difficult for union leaders to adopt a public posture
other than one of belligerence. Williams wanted to devise a
policy that straddled the extremes of corporate behavior but
dealt individually with them.

Williams addressed the problem in an unorthodox way. He
constructed a philosophical framework, based on his assessment of the business environment in the United States and
Canada, to put the issue in perspective. Then he formulated a
two-pronged policy and discussed it with his executive board.
Finding a positive reception there, Williams outlined the
policy at the 1984 USW convention, leaving out the philosophical girding. He talked about the latter several weeks
later at a small conference of labor and management people. It
was one of the most creative speeches by a union leader that I
had heard.

The American entrepreneurial spirit, Williams said, has
contributed much good to the material success of the society.
It also has produced harmful authoritarianism on the part of
management. "The challenge is how we can unleash that
entrepreneurial energy and tradition from the authoritarian

mold." It would make great sense, he said, to combine the entrepreneurial spirit with the creativity of American workers "to find a new dynamism that is appropriate to the real world in which we live." Therefore, he strongly supported worker participation and involvement in companies that accepted the union. This would be the union's preferred approach.

At the same time, Williams continued, "we are letting corporate America know that if they want a row with us, they are in for the damndest fight they have ever seen.... We will be just as adversarial as we know how to be—if that is the way they want to play the game."[13]

In fact, Williams *did* play this double-edged game in his first four years as USW president. Under his leadership, the union started or strengthened participation and other cooperative efforts at many companies. He also led the USW in bitter struggles with other firms, including Wheeling-Pittsburgh, U.S. Steel, Danly Machine, and Phelps Dodge. Some of these clashes were unavoidable, but Williams's decision to mount a corporate campaign against PD in 1984 seemed of questionable judgment. The USW urged depositors at PD's creditor banks to withdraw large amounts of money in an attempt to turn the banks against PD. By then, the strike was all but lost, and the campaign could have had little purpose other than to punish the company after the fact, out of an acute sense of institutional embarrassment. It accomplished little.

Williams, however, did not often indulge in such wasted efforts. A cautious man, he considered every step very carefully and rarely boxed himself in with unrealistic promises. But once committed to a course, he showed himself capable of taking very large risks. Not untypically of leaders who act boldly, one of Williams's major strengths—the tendency to plan patiently and optimistically for the long term, to look for "large, hopeful solutions," as Kleiman put it, and to take risks in carrying them out—also could be viewed as a weakness by his opponents. In the USW's response to events that continued to worsen American steel's prospects from 1984 through 1986, some industry leaders argued that Williams put the long-term needs of his institution above those of the industry and its employees.

Chapter 16

Roderick's Tough Leadership

David Roderick was not a man who lived on political hope. A hard-boiled former marine sergeant, the U.S. Steel chairman drove his corporation like a tank across the pitfalls of high finance and through the barriers of union resistance. In the mid-1980s, Roderick gave a daunting performance on steel's desolate landscape, slashing prices to gain market share, fighting off wily corporate raiders, and trading the Steelworkers' Lynn Williams bold move for bold move. Doughty Andrew Carnegie would have applauded.

Much of what happened in this period stemmed from a decision by Roderick. When the White House in 1982 and 1983 turned a deaf ear on the steel industry's importunings for more relief from the import deluge, Roderick took it as a message. "If we have no desire as a government to have a viable, strong steel industry...then we will not have one," the U.S. Steel chairman told an interviewer in mid-1983. Under his tough-minded leadership, the corporation embarked on a number of strategies aimed at shrinking American steelmaking capacity—in other companies as well as his own—and cutting costs with Carnegie-like relentlessness in U.S. Steel's remaining steel operations. Roderick intended that his company would take charge of the industry and its people and emerge the dominant force. It was a legitimate business goal but one that bespoke a politics of confrontation.[1]

The Worldwide Restructuring of Steel

On June 24, 1986, PaineWebber greeted participants in the Wall Street firm's first Steel Survival Strategies Forum with this description of the steel industry: "The industry as we

415

knew it five years ago is no more; the industry as we knew it last year is gone. The industry is now in a perpetual state of change; it will be at least a decade before it stabilizes into a new configuration. The battle for market share among the major mills, reconstituted mills and minimills is awesome. The next decade is sure to be a wild one."[2]

At the time this summary was written, the U.S. steel industry had been in a depressed state for four years. Average steel prices dropped nearly 10 percent between 1982 and 1985, preventing most domestic producers from returning to profitability. Two trends contributed to the price decline, according to a study by a union consulting firm, Locker/Abrecht Associates, sponsored by the Steelworkers. The policies of the Reagan administration produced a rapid rise in the value of the dollar against foreign currencies in 1980–84, lowering the price of imports by 40 percent and boosting the price of exports by 70 percent. This effect was exacerbated by deep price-cutting on the part of U.S. steelmakers, especially U.S. Steel, from 1983 to 1985. Unable to make money and weighted down by unused steelmaking capacity, the companies slid inexorably toward a financial black hole. In early 1983 about 151 million tons of capacity remained on the books, some 20 million more than analysts, at that time, said were needed. By 1987 U.S. capacity had declined to about 112 million tons, and analysts declared that another 15 million to 20 million tons would have to be eliminated.

But the American steel industry was just part of a worldwide contraction of older steel industries and expansion of newer ones in the 1970s and 1980s. The other traditional steelmakers for the Western World—the Europeans, Canadians, and Japanese—also were forced to close plants and cut work forces in the face of growing competition from developing nations. From 1975 to 1986, Japan reduced employment in its steel industry by 23 percent, West Germany 32 percent, and France 41 percent. But steel employment (hourly and salaried) shrank most in the United States during this period, falling 48 percent, from 548,200 to 283,200.[3]

Meanwhile, producers in Brazil, South Korea, Argentina, Mexico, and elsewhere were building state-subsidized, technologically advanced steel mills and aggressively scouting for markets. Low wages in these countries, combined with the strong dollar, enabled them to ship steel into the United States under the cost of domestic production. European imports were subject to some limitations starting in 1982. But the virtual open-borders policy of the United States with respect to steelmakers in developing nations invited them to feast on the most lucrative steel market in the world until some restrictions were imposed in late 1984.

Much of the new steel-producing capacity in the Developing World was financed by loans from banks in the United States, Europe, and Japan. Eager to make money at interest rates of 20 percent or more, American banks set up a vicious circle—and put themselves in jeopardy—by loaning huge sums of money to Brazil in particular. Although the U.S. steel industry had trouble getting loans from domestic banks, Brazil's inability to meet interest payments forced the same banks to extend loans to that country and oppose protection for the American steel industry so that Brazil could continue to ship into the U.S. market. The deeper that Brazil got in hock to American banks, the more those banks, in effect, had to help the Brazilian steel industry at the expense of the American. From 1981 to 1986, Brazil's steel exports to the United States more than tripled, from 547,900 tons to 1.7 million.

In 1983 the domestic industry and the USW began a massive campaign for import restraints. Toward the end of the year, U.S. Steel and Bethlehem split over tactics. Big Steel, the most vocal critic of government trade policies on steel, filed suits alleging unfair trade practices against Argentina, Brazil, and Mexico. Bethlehem's Donald Trautlein, strongly supported by the USW, adopted a broader approach by petitioning the International Trade Commission (ITC) to investigate all carbon steel imports entering the United States under section 201 of the Trade Act of 1974. Bethlehem had criticized U.S. Steel earlier in the year for attempting to import semifinished steel from Britain (see below). Even so, the "201 filing," as it became known, was a notable break in the unity usually displayed by steelmakers on trade matters and indicated a pulling-apart that eventually would lead to the demise of the industry's solidarity in labor bargaining.

Roderick worried that the Bethlehem-USW action, supported later by Inland and Armco, would damage the restraint systems already in place for European steel producers. But the Bethlehem gamble worked. After conducting its investigation, the ITC in summer 1984 recommended strict tariffs and quotas on steel imports. At the same time, the USW and the steel firms were pushing for legislated quotas, putting more pressure on the White House to take a position in this presidential election year. On the eve of the election in September, the Reagan administration announced that it would negotiate voluntary restraint agreements (VRAs) with the Common Market countries in Europe. The negotiations later produced VRAs limiting steel imports from European nations to 20.2 percent of U.S. consumption of finished and semifinished steel for five years. U.S. Steel kept pressing for similar arrangements for countries like Brazil, South Korea, and Spain and eventually succeeded.

Despite the VRAs, imports soared to 26.4 percent of the market in 1984 and dropped only slightly to 25.2 percent in 1985. Even though steel demand rose during this period, foreign producers skimmed off most of the increased market. By the time the restraints began working more effectively in late 1986, two major steel makers—LTV Steel and Wheeling-Pittsburgh—were in bankruptcy proceedings.[4]

The United States, of course, could not hope to retain its early postwar position as steelmaker for the world. Nor could it expect that poor countries with burgeoning populations should forever remain poverty-stricken communities of farm peasants. The industrial world had to help these nations build the capacity to make and export products so that they could raise living standards. The question was, should American steel communities pay the total price for this development, and for the overextended loan policies of American banks?

Faced by the onslaught from abroad and excess capacity at home, American steel companies desperately tried new strategies. U.S. Steel, National, and Armco diversified to escape dependence on the deteriorating steel business. With more cash at its disposal and more assets to sell than the other firms, USS was the most successful in taking this course. In early 1986, Roderick snapped up Texas Oil & Gas, and joined it with Marathon Oil to form an oil and gas segment that provided 60 percent of USS sales in 1986, compared with only 25 percent for steel.

Inland and Bethlehem concentrated on whipping their steel operations into shape. Jones & Laughlin took the merger route which held some promise of consolidating the industry into more efficient units. A merged company could more effectively eliminate duplicate lines of production. In early 1984, J&L's corporate parent, LTV Corporation, acquired Republic

Steel and merged it with J&L to form LTV Steel. The new firm displaced Bethlehem as the second largest producer. However, the Justice Department refused to allow U.S. Steel to purchase National Steel from its parent, National Intergroup, on antitrust grounds.

National converted itself from an old-line steel producer to a diversified holder of steel and financial institutions. In 1980 National also began disposing of its steel facilities by closing old furnaces and mills and selling its large Weirton plant to employees who formed an Employee Stock Ownership Plan (ESOP). To reflect this change, the company in 1983 renamed itself National Intergroup Incorporated, and later sold half of the remaining steel business to Nippon Kokan, a large Japanese steelmaker.

None of these moves, however, solved the companies' competitive problems. After the merging and restructuring, they were still left with many inefficient plants and, in large part, industrial relations systems that stymied change on the plant floor, where the competition had to be met. However, in the 1970s J&L had adopted an innovative labor approach, followed later by National Steel and to a lesser extent Bethlehem. A brief description of what happened at J&L will help put U.S. Steel's more confrontational course in perspective (see chapter 17 for the National Steel story).[5]

Labor Reforms at J&L

The irony is that at J&L, Thomas Graham achieved a notable improvement in labor relations and productivity with very different methods than he later would institute at U.S. Steel. In both cases, Graham's goal was to reduce the man-hours required to produce a ton of steel without major investments in new technology. There could hardly be a more difficult task in a declining industry, for it meant cutting an already dwindling and worried work force, by one means or another, so that fewer people did the same amount of work, or more.

The traditional, adversary way of approaching workers under these conditions is to promise swift retribution, in the form of plant shutdown, if they refuse to bend. John Kirkwood, J&L's top industrial relations executive in the late 1970s, took the position that "threatening the workers is not constructive labor relations." He and his chief labor negotiator, Cole Tremain, developed another method. They reasoned that local union officials could not be expected to bargain away jobs without winning something in return.

Kirkwood and Tremain offered to negotiate with the locals on this basis: If a local union agreed to combine duties so that positions would be eliminated, the company would grant early retirement to older crew members under the so-called 70/80 "mutual" pension provision in the USW contract. Workers in their fifties, whose years of service and age totaled seventy in some cases and eighty in others, would receive a regular pension, plus a $400 per month supplement, until they became eligible for Social Security at age sixty-two. "After 1977 there was never a time we weren't negotiating something with a local somewhere," Tremain remembered. "It became a way of life in our plants."

The company conducted this kind of "problem-solving" negotiation in several plants, eliminating about twelve hundred jobs over several years. The effort achieved its greatest success at the Aliquippa Works. Between 1980 and 1983 J&L and Local 1211 negotiated force cuts of about five hundred people, reducing man-hours per ton from 6 to 3.5, and the company granted early retirement to five hundred employees. The existence of a good LMPT program at Aliquippa helped this bargaining. While the LMPTs could not be used for negotiating wages or work rules, the cooperative spirit established by the process spilled over into all areas of union-management relations.

James Anderson, the Aliquippa plant manager in the early 1980s, strongly supported the participation teams. He and his superintendents, along with the leaders of Local 1211, sat on a steering committee that directed the LMPT program and became a forum for exchanging information and solving plant-wide problems. Local President Peter J. Eritano headed a strong, united leadership team. With officials on both sides committed to LMPTs and the idea of making the plant competitive, negotiating the work rule changes became one more form of cooperative behavior.

"The company took the approach of, 'how can we do this together' rather than, 'you *will* do this,'" Eritano said in 1983. "They asked for the workers' cooperation, explained why sacrifices had to be made, and gave us the data necessary to make the decisions. They didn't say they would shut the plant down without concessions. We made sure the new jobs could be done safely without turning the plant into a sweat shop."

Working jointly, company and union officials interviewed

groups of workers from each production unit to decide how to combine duties and eliminate unnecessary jobs. They then developed a "greenfield manning" plan which specified the number of workers that would be needed on each unit if the plant were starting anew. The two sides also agreed to compress many maintenance skills into two multicraft classifications for mechanical repair and one for electrical repair. In most cases, the addition of duties led to a higher job classification and thus increased hourly and incentive pay. No change was implemented unless the affected workers voted approval.

J&L figured that the exchange of pensions for jobs yielded a net cost reduction. The average annual, amortized cost of funding a 70/80 pension in 1983 was $8,000, compared with the $40,000 cost of employing an active worker (paying benefits as well as wages). Other steel companies, however, looked skeptically on J&L's "70/80 bargaining," as it came to be known. "They were concerned," Kirkwood said, "that if they gave pensions in a particular area of a particular plant, the union would demand that they do it everywhere. Our attitude was, 'Fine. Anywhere we can get people out of the work force under an agreed-upon 70/80 pension, as opposed to an adversarial arbitration situation, we're going to benefit economically.'"

Other companies also pointed out that allowing people to retire early raised the cost of funding pensions at a time when most steel firms were short of cash. Bruce Johnston argued heatedly against the practice for that reason and said that U.S. Steel would not jeopardize its pension fund, one of the healthiest in the industry, by negotiating mutual pensions. He later changed his mind. U.S. Steel eventually adopted the practice, starting in 1983, and so did Bethlehem and National. There was no other way to induce people to eliminate their own jobs.

Did J&L put its pension fund in jeopardy? All of the 70/80s granted by J&L increased pension costs by only one-tenth of 1 percent, Kirkwood contended. One friendly critic, however, believes that J&L may have "gone overboard" with the number of pensions granted. J&L's parent, LTV, didn't like the early retirement program and complained about the cost. Graham, however, refused to back down, according to Kirkwood, because the 70/80 bargaining produced major cost savings in steel operations. Unfortunately, LTV for several years contributed only enough money to its pension funds to meet the minimum standards of federal law. Eventually, the under-funded plans helped push LTV into bankruptcy.[6]

By 1983 J&L and National had moved significantly toward a new way of dealing with their employees. Bethlehem had made a strong effort to implement the LMPT process in many plants, although a decentralization of management in 1984 badly damaged this progress. Wheeling-Pittsburgh, though in deep financial trouble, was trying to stay afloat by using the knowledge of workers. Inland and Armco made some progress in this direction. Allegheny-Ludlum, a well-managed spe-cialty steel producer, had good relations with its employees. However, after dropping out of the industry bargaining group in 1983, AL had little impact on the rest of the industry.

Of the major steelmakers, only U.S. Steel operated largely in the old way. The corporation would argue that its strategies for surviving in a world steel industry in the throes of great struc-tural change dictated the course of its industrial relations pol-icies. That may be so. But the way U.S. Steel carried out those strategies, which were essentially the handiwork of David Roderick, made enemies of large numbers of employees.

Confrontations at U.S. Steel

A native of Pittsburgh and son of a postal worker, Roderick grew up on the North Side and entered the Marine Corps upon graduating from high school in 1942. He served as a noncom-missioned officer in the marines' security forces (guarding prisoners, among other things) and saw action in the South Pacific during World War II. After the war, Roderick graduated from Pitt with a degree in economics and finance and worked for Gulf Oil and two railroads before joining U.S. Steel in 1959. From 1962 until 1973, he held top-level accounting posts, became vice-president of international operations, and lived several years in Paris. Returning to Pittsburgh, he began his ascent to the top of the corporation by serving as chairman of the finance committee. He became president in 1975 and chairman in 1979.

Roderick's financial background, and the fact that he never worked a day in a steel plant, led some employees to speak of him as the "top bean counter." That was unfair. In a profile of Roderick published in the 1986 book, *The Big Boys*, Ralph Nader and William Taylor say that the chairman "bristled" at the implication that he was merely a bookkeeper. Instead,

Roderick told the authors, in his accounting posts he specialized in the financial analysis of operations.

His lack of production experience, however, made Roderick immune to the kind of emotional attachment to steel mills that drives many former steel managers. One of the latter was Roderick's predecessor, Edgar Speer. Another was William Roesch, the USS president under Roderick until 1983. Roesch's efforts to make U.S. Steel's plants more competitive by increasing productivity sometimes conflicted with Roderick's inclination to shut them down. One highly placed staff man at USS says that by 1983 their conflict "was tearing the corporation apart."

Although not a "steel man" in the sense of one who delights in crawling into scale pits and designing blast furnace boshes, Roderick looked and acted every inch the typical U.S. Steel executive. A balding, stocky man of slightly less than medium height, he projected power, not the monied power of Texas oilmen or the articulate puissance of certain politicians, but the crude, institutional power inherent in *the* basic industry in a nation of goods producers. Roderick talked in forceful terms, giving the impression that he rejoiced in his bottom-line toughness and ability to punish employees who did not behave according to his concept of economic realism.

This trait came out clearly in February 1987, when Roderick announced the shutdown of several plants in the aftermath of a labor dispute that idled USS's steel plants for six months. He pointed out that he had warned many months ago that the stoppage might cause plant closings. "And now the economic hammer has dropped," he said, obviously relishing his choice of words. "And it's unfortunate, but that's the way it is." Dave Roderick was an intelligent, formidable man who seldom hesitated on the brink of confrontation.[7]

His first action after the 1983 bargaining period immediately soured relations with the USW. The plan involved the Fairless Works, located thirty miles north of Philadelphia on the Delaware river at Fairless Hills, Pennsylvania. Opened in 1952, Fairless still produced steel in open hearth furnaces. To make the plant more competitive with imported steel coming into East Coast ports, Roderick wanted to close the hot end of the plant, eliminating about one thousand jobs, and import semifinished steel slabs from Britain for $20 to $50 a ton less than the Fairless cost. Some five thousand people would still be employed in the Fairless finishing mills, rolling the slabs into finished product. Otherwise, Roderick said, Fairless would be closed by 1990, "a victim of terminal obsolescence," with the loss of over six thousand jobs.

The chairman had been negotiating for several months with British Steel, a government-owned corporation. When a British union official disclosed the plan in March 1983, the USW protested vigorously. The union feared that, with the excess of steelmaking capacity around the world, foreign producers would scramble madly to offer slabs and other semifinished steel at reduced prices to American companies.

Lloyd McBride charged that U.S. Steel demonstrated "bad faith" inasmuch as the union recently had made large wage concessions. Roderick declared that the deal should be no surprise to McBride, since Bruce Johnston had told him about the British Steel negotiations in November 1982. McBride said he had understood that USS would go through with the deal only if the union struck the industry in 1983.

Whatever the facts of the Johnston-McBride conversation, the corporation and the USW became embroiled in a major public squabble. Roderick and McBride presented their arguments to the Congressional Steel Caucus, a body of senators and representatives from steel-producing regions, and debated each other on a national television program.

Other critics, meanwhile, assailed U.S. Steel for trying to make a deal involving precisely what the corporation itself had long complained about—importing subsidized foreign steel. Yet U.S. Steel was by no means the only domestic producer that imported semifinished steel. By the end of 1983, other importers included Lukens, Republic, Sharon, McLouth, Kaiser, Wheeling-Pittsburgh, and Jones & Laughlin. The USW filed grievances against the practice in some of these cases. But U.S. Steel executives could rightly complain that the union protested the British Steel proposal in a much louder voice than it did the actual importation of slabs by other companies.

McBride formed a task force of staffers and local presidents to seek support for the union position. Russell Gibbons and Gary Hubbard, USW publicists, organized a media campaign, which included placing full-page ads in newspapers and national magazines: " 'The British Are Coming, The British Are Coming...And America's Steel Independence Is Going.' " The USW filed complaints with the Commerce Department and the National Labor Relations Board. McBride also ordered a halt to the LMPT program at U.S. Steel.

The USW campaign didn't stop U.S. Steel, but the negotiations petered out. British Steel offered to invest only $300 million in the project, while Roderick had asked for $600 million. He announced termination of the talks in December 1983. The confrontational feature of the British Steel imbroglio typified relations between U.S. Steel and the USW from 1983 to 1987.[8]

Another flare-up involved the corporation's South Works in South Chicago. In 1982 the corporation promised to build a rail mill at this plant, one of its largest, if the USW met two conditions. Local 65 satisfied the first requirement in late 1982 by ratifying a manning agreement which called for fewer crew members than the conventional rail mill. The USW also lobbied the Illinois legislature and won passage of an exemption to the state sales tax applying specifically to rail mill products.

In early 1983 USS added another condition, asking the union to help the company avoid spending $33 million to control water pollution at the Gary Works. The union complied. In August the Illinois attorney general announced a plan to defray the obligation. A few months later, USS returned to local 65 with a third set of conditions. This time, the company demanded additional concessions, including wage and benefit cuts valued by USS at $1.40 per hour.

Claiming "enough is enough," local officers said they would consider the new demands only if USS actually built the mill. "No decent union could accept what they wanted," said Local 65 President Donald Stazak, who previously supported concessions.

The company canceled the project and attacked the union in newspaper advertisements written by Bruce Johnston in his pungent style. The ads pointed out that two major rail makers, Colorado Fuel & Iron (CF&I) and Wheeling-Pittsburgh (W-P), both had won larger concessions from the USW than U.S. Steel and the coordinated bargaining group. Therefore, the union should be blamed for cancellation of the South Works mill.

But another factor was involved. In the two years since USS first approached Local 65, the market for rails had shriveled and could not have supported a third large mill at the South Works, even with concessions, unless USS intended to take business away from CF&I and W-P. U.S. Steel chose not to make the effort. It used Local 65's refusal to meet the new demands as an excuse to cancel a decision that it believed no longer made economic sense.

Business conditions, of course, do change. But the company created a climate of skepticism and mistrust by postponing the project and demanding ever more concessions. This approach made union cooperation practically impossible. Instead of accepting this fact and pulling back from what had become a possibly uneconomic project, U.S. Steel tried to score publicity points against the USW.

The USS decision and ad campaign provoked outcries from state and city officials, as well as the staunchly conservative *Chicago Tribune*, which editorialized against the company. And business writer R. C. Longworth, in a blistering commentary, wrote that USS "violated every standard of decency and broke every obligation to the workers and the community that made it rich." The company's politics of confrontation negated whatever it had hoped to gain.

The Steelworkers filed suit against U.S. Steel for breach of contract but lost in a jury trial. The jury found that the company had made an oral contract to build the mill but concluded that the state had not fulfilled U.S. Steel's request to defray the cost of the pollution control in exactly the way the company had prescribed.[9]

Cutting People at U.S. Steel

In early 1983, William Roesch, U.S. Steel's president and chief operating officer, took leave from his duties because of a terminal illness and died later that year. Roderick, the financial expert, needed a strong production man as second-in-command. He chose Tom Graham, the Jones & Laughlin president who, former associates say, had been unhappy with LTV's treatment of its steel subsidiary. Graham joined USS in May 1983 as vice-chairman and chief of steel operations and related resources.

The corporation's steel operations were wallowing helplessly when Graham entered the picture. USS lost $154 for each ton of steel shipped in 1982, $60 more than the industry average. This was partly attributable to high overhead costs stemming from the company's bureaucratic organization. It had twenty-nine vice-presidents in 1983 and was awash in assistant managers and technical specialists. The latter included fifty-four staffers assigned to the nearly nonexistent steel export business. The proliferation of bosses in the plants had produced supervisory *rigor mortis*. Orders passed down the hierarchy but few ideas from below penetrated the layers of bureaucracy surrounding the decision makers.

Given Washington's denial of greater import relief by 1983 and the overcapacity situation in the United States, the course facing Graham was clear: Massive numbers of people had to be severed from employment and antiquated plants shut down. The per-ton cost of making steel had to be reduced in the plants that remained.

Roderick had picked the right man. Graham moved quickly to put an end to the "civil service atmosphere" in the salaried ranks. He initiated a program to halve administrative expenses and reduce other costs. In September 1983, the company announced the permanent layoff or retirement of 4,000 salaried employees, and further cuts occurred over the next few years. Between 1980 and mid-1986, salaried, nonunion employment plummeted from 20,837 to 7,736.

Force reductions were not new at USS, but Graham shortly developed a special reputation among employees in the Mon Valley as a "people cutter." In late 1983 he decreed a reduction of 50 percent of the supervisors and staff people in the valley plants. Positions such as assistant superintendents and general foremen were eliminated and the duties of turn foremen doubled or tripled. Graham believed that hourly workers should assume more responsibility for operating their production units—an idea that came to the Mon Valley plants twenty years too late. Some operations lost all shift supervisors except the daylight foreman who, however, was subject to call on a twenty-four-hour basis. Jack Bergman frequently reported to the Duquesne plant in the middle of the night to handle an emergency problem, without extra pay.

For those terminated, there was no two-week, nor even a two-day, notice. Instead, superintendents handed out pink slips on the spot, sometimes at the beginning of a shift but often at the end so the company could get a final day's work out of the foreman. Scores of demotions accompanied these changes, with department superintendents becoming general foremen and the latter moving down to turn foremen.

Larry Delo, for example, was transferred from Duquesne to Homestead as a general foreman in 1983 and after a few months demoted to first-line supervisor. He suddenly found himself working rotating shifts, five years after he thought he had escaped that practice for good. Transferred for the second time within a year, he went from Homestead to Braddock in April 1984 and took charge of preparing a slabbing mill, soaking pits, and shipping yard for reopening after two years on standby status. Because of the work-force cuts, he covered an area that once had four first-line foremen.

"I went to work on August 30 [1984] on the noon to eight shift," Delo later told me. "My boss called me into his office and told me that was it. I was finished right then. I was paid through September. I guess it wasn't unexpected." Delo, fifty-four, had spent nearly thirty years with U.S. Steel and received a pension in addition to severance pay. Many younger supervisors, however, fell short of pension eligibility. "The company has changed considerably," Delo said, referring to his abrupt termination. "They don't do things with compassion. They really don't care about your feelings. I know one guy who needed two months for a pension and another who needed four months. Didn't matter. Both were out anyway."

These changes, made in an atmosphere of chaos rather than reform, destroyed what little morale and incentive were left in the plants. Everything caved in at the end—the work ethic, personal relations, even morality. Jack Bergman accepted his termination with relief. "They actually did me a favor getting me out of there," he said. "I couldn't work under the new regime. I couldn't fathom it. Inefficiency... no care... don't give a damn. Thievery was going on. Management was ravishing the plant. I had trucks under my supervision. Superintendents and some of the general foremen, they'd order a truck, load it with stuff, tools, machinery, and out the gate. The guards wouldn't stop them. A helluva lot of brand new railroad ties went out of that mill. A lot of slag went out of that mill... good fill. The stuff went out by the truckload."

The reduction could have been accomplished without the bitter feelings, though it would have taken longer. Graham was impatient for results. Wall Street always applauds work-force cuts as the solution to financial problems, regardless of the effect on production ability. And U.S. Steel's financial performance did improve. From March of 1983 to the end of the year, USS's operating loss dropped from $93 per ton shipped to $56, despite an increase in quarterly shipments of only 300,000 tons. In the first nine months of 1984, the company made money, $18 per ton on 9.2 million tons.

By 1987, Graham had chopped off a third of USS's steel-making capacity and sharply improved output. With a total employment of 48,600 people in 1983, the company shipped 11.3 million tons of steel. After plant closings in 1987, USS said it could turn out just as much steel with only about 18,000 to 19,000 employees. "Tom Graham is a brilliant manager," a former plant superintendent told the *Wall Street Journal* in 1987.

Not all managerial employees agreed with that assessment. In 1987 I interviewed a high-level manager who had worked for the company since the 1950s. He had survived all the cuts since 1983 but felt bitter about the company's "lack of human concern" and Graham's methods of managing people. "U.S. Steel has always been highly politicized," the manager said. "The higher you go in management, the less they listen. The management style didn't change when Graham came. Still the blatant arrogance and not confiding to people with truth and openness. Graham is an autocrat, and he listens to no one. If you speak up to him, give him your best judgment, you're gone."

Of the seven superintendents who ran U.S. Steel's primary steel plants when Graham took over in 1983, none still worked for the corporation in early 1987.[10]

The "Graham Revolution"

Cutting union workers posed a more difficult task than terminating unrepresented employees. Graham, however, maintained his autocratic stance, paying only lip service to contractual commitments to consult with the union. Restoring U.S. Steel to profitability—indeed, ensuring that it survived in the rapid erosion of the steel industry—took precedence in his view over complying with rules that had been drawn up for use in normal times.

The times were abnormal. Even with the wage concessions which took effect on March 1, 1983, the company still employed steelworkers at about $23 per hour, compared with $10 to $11 in Japan and less than $2 in South Korea. No one expected American wages to fall to the Korean level, which was below the federally mandated minimum wage. Nor did

USS management really expect that the USW would reduce employment costs to the $17 to $18 per hour considered necessary to compete on wages with Japan and the minimills. But there was an alternative.

Since employment costs comprised about a third of the total cost of making steel, the company could become competitive if it slashed the number of $23 hours needed to produce a ton of steel. In 1983, each ton of steel shipped by U.S. Steel required 7.5 man-hours of work. Graham ordered his plant superintendents to reduce the number to 4.0, nearly a 50 percent cut. Later, he lowered the goal still further to 3.2. The superintendents were to reach the goal by combining jobs, cutting crew size, and—especially—by contracting out as much work as possible. Under Graham's policy, they should seek union approval to do these things but not hesitate to take unilateral action if the USW balked. The quickest way to reach the 4.0 target was to hire outside contractors, because USS didn't include their hours worked in its own total hours. Indeed, U.S. Steel reached that goal in 1985 by contracting out the equivalent of two man-hours per ton, according to a company official. Jim Smith estimated that USS eliminated five thousand jobs this way between 1983 and 1986.

This improvement came at a high price. Contracting out aggravated strains already existing between the USW and the corporation and was a primary cause of the long work stoppage in 1986-87. So pervasive did the practice become, with such large consequences for the company's labor relations, that some U.S. Steel executives referred to it as "the Graham revolution."

The USW based its "legal" argument against contracting out on the so-called recognition clause in its steel contracts. In this section, management accepted the union's claim to represent all persons (except supervisory people) "employed in and about the company's steel-manufacturing and by-product coke plants." If a firm hired others to do this work, the union argued, it infringed on the right of USW members to hold these jobs. The union, on the other hand, accepted the concept that the company "retains the exclusive rights to manage the business and plants." The clash between these sets of rights frequently produced arguments over subcontracting work.

The USW generally did not dispute the contracting out of major construction, for which steelworkers lacked some necessary skills, and some types of service work for which the companies had no equipment, such as hauling slag out of the plants. Before the 1980s, however, management rarely, if ever, contracted out production work. Most arguments over what could and could not be farmed out involved building and equipment maintenance, whether it was performed inside or outside the plant.

The companies contracted out work for a wide variety of reasons, including cost and availability of equipment and manpower. Outside shops with special tools and lower-paid workers could repair some kinds of equipment cheaper than steelmakers themselves. For small construction jobs and in-plant maintenance, many steelmakers preferred to import contractors who employed skilled, AFL building tradesmen, such as bricklayers and electricians, rather than assign the work to their own journeymen. Even if the tradesmen earned a higher hourly wage than steelworkers, they in many cases could offer a cheaper overall cost. As construction work became scarce in the 1970s, the outside craft workers often agreed to relax jurisdictional rules. In some mills, however, USW craftsmen refused to cross job boundaries; for example, an electrician might refuse to do minor welding and instead insist that a welder be called.

The steel companies also could hold costs down by keeping a slim maintenance force in the plant and hiring contractors for extra work. Using contractors enabled a firm to turn the cost on and off as needed, whereas it was expensive to recall steelworkers from layoff for a job of short duration. Many locals hurt their own cause by "giving up" repair jobs during periods of full employment when no one made an issue of contracting out. But the issue understandably became an inflammatory one for the union at times of heavy steel layoffs.

A significant number of jobs always had been subcontracted. A joint study by a union-management committee in 1979 estimated that from 1971 to 1978 an average of fifteen thousand contractor employees worked full-time each year in the steel industry, representing 15 percent of the companies' own craft forces.

In the past, the USW and the industry had addressed the issue in many negotiations, constantly refining a provision first adopted in 1962 to limit contracting out. Short of giving the companies a free hand to contract out any and all work, or imposing a strict prohibition on the practice, the contractual language had to remain somewhat vague, leaving many gray areas that each side could interpret differently. And there was logic in both sides' position.

Management argued that to limit its flexibility in finding the cheapest, most efficient way to perform work would undermine the economic viability of the plant. The USW contended that it must have restrictions to prevent management from whittling down the size of the bargaining unit. The two sides concluded in the 1979 study that they should refrain from taking these extreme positions. The contracting out problem, said the joint commission, "is insoluble so long as it is seen as a monolithic subject to be dispensed with in head-on fashion."[11]

The question of which work should be contracted out could only be settled locally, perhaps with companywide guidelines to prevent management from setting local against local in a continuous war of attrition. In 1983 negotiations, the top bargainers decided to allow local unions and managers greater latitude in dealing with the subject. The USW authorized its locals to negotiate craft combinations. For example, many of the duties of the boilermaker, welder, and rigger could be combined for structural steel jobs. Such a tradeoff had worked well at Jones & Laughlin's Cleveland plants. Between 1983 and 1986, some locals refused to negotiate, but many others agreed to create "supercrafts," including several at U.S. Steel. In return, companies were supposed to make a "good faith" effort to assign new construction work to steelworkers and give them back some work that had been subcontracted. USS locals, however, complained that the company reneged on this agreement.

Clairton Local 1557 agreed in February 1986 to eliminate 112 maintenance jobs by merging crafts into an electro-mechanical repairman position. The company brought back 75 jobs that had been subcontracted but soon began hiring outside maintenance contractors. "The company told us if we became competitive, we'd have the work," said Charlie Grese, two months after the pact took effect. "They now claim everything is an emergency job. It's a never-ending problem. They will contract out everything they can get away with."

U.S. Steel continued to extend contracting out to more and more areas of the mills. As the steel business slowly picked up from 1983 to 1986, the company in many instances hired contractors to handle the increased workload, particularly in custodial services and maintenance, instead of recalling laid-off employees. Production work such as shearing steel coils into shorter coils and performing other finishing operations also disappeared from some plants. At Gary Sheet & Tin Works, the company brought in three foremen employed by a contractor to supervise USW members on janitorial jobs.

The company also used leasing arrangements to take work out of the USW bargaining unit. The Irvin Works, for example, leased land to a company that built a gas-processing unit and took over a twenty-four-hour-a-day operation that once employed eighteen steelworkers. USW locals cited other leasing examples and many instances of substandard repair work and overcharging by contractors.

USS's willingness to subcontract work induced many foremen to take early retirement, set up shop as contractors, and turn their contacts in the plant and knowledge of the mill into lucrative businesses. At the Geneva Works in Utah, five supervisors retired one day and returned the next day as contractors. Steve Kecman, a local official at Irvin, counted eight "consultants" working in the plant at one time in 1986. All were former supervisors.

Sam Camens, the USW's expert on contracting out, visited many plants to study the practice. He contended that frequently the hiring of contractors did not save U.S. Steel money; it was a way to avoid the time-consuming managerial duties involving employee scheduling and administrative paper work. "It became a style of managing, not of saving money," Camens said. "They [management] didn't even bother to find out if we could do the work. They just sent it out. The easiest way to manage a plant is to evade the responsibility of managing. A maintenance supervisor might not have enough people at work to handle a repair job that comes up. His manager tells him, 'Do what you have to do.' The supervisor gets on the phone and calls XYZ Company—and don't minimize this—he calls his cousin, his uncle, his aunt... The conflict of interest going on in U.S. Steel is amazing."

Company officials conceded privately that some nepotism occurred but argued that, on the whole, contracting out was the most cost-efficient way to perform some kinds of work. Union members must recognize that this strengthened the long-term viability of their plant, they said. However true, this assertion was undercut by the workers' perception of widespread nepotism, conflict of interest on the part of supervisors, wasteful spending, and shoddy repair work.

The battle over contracting out at U.S. Steel reflected in microcosm a larger trend in American industries toward adopting the Japanese technique of creating two tiers of workers. The higher tier, or "core" group, consisted of the company's most skilled workers, who held the primary production jobs. A secondary group consisted of temporary and part-time workers and contractor employees who performed ancillary or seasonal work at much lower wages. The Japanese steel industry employed one hundred fifty thousand such workers. This practice became increasingly popular in U.S. manufacturing industries, but it caused particular trouble in steel because the work force was diminishing rapidly for other reasons.

The grievance backlog, usually a measure of the state of union-management relations, expanded rapidly at U.S. Steel. By November 1983, Local 1066 at Gary Sheet & Tin had 200 cases—mainly involving contracting out and job combinations—pending arbitration and another 150 specifically charging contracting out violations at or above the third step level in the grievance procedure (arbitration is the fifth step). As with other types of grievances, the mere filing of a union complaint could not stop the company from acting as it saw fit. Management, of course, had to balance the advantages to be gained against the risk of losing a case in arbitration and being forced to pay back wages to steelworkers who should have been assigned to the job.

From 1983 on, under Tom Graham's impatient lashing, plant operators increasingly took that risk, even in work situations that clearly belonged to the USW because of past practice. As grievances soared, the company began losing cases in arbitration. At the Fairless Works alone, USS paid $360,000 in back wages because of contracting out during 1985, said the local president, Al Lupini. Between 1983 and 1987, awards totaled $8 million to $10 million for the entire company, according to one union estimate.

By no means did U.S. Steel lose all arbitration cases involving contracting out. But a management that pays out $8 million in avoidable fines and provokes widespread anger in the work force is either amazingly inept or shrewdly competent. The race to cut man-hours per ton probably produced many wrong decisions locally. On the other hand, Graham must have valued the results of contracting out more than the cost of labor turmoil and back pay awards. Management gained considerable flexibility by replacing bargaining unit employees—and union control of the shop floor—with temporary workers who brought no such encumbrances to the job. There was an important symbolism in this. Management demonstrated to employees that the union no longer had monopoly control of either labor supply or work practices.

To some extent, U.S. Steel may have balanced the cost of prounion arbitration rulings by avoiding the "trailing benefits" involved in recalling workers from layoff. The big item was medical coverage. Laid-off steelworkers retained medical benefits for periods of time determined by seniority, ranging from six months for two to ten years of service to two years for over twenty years of service. Recalling an unemployed steelworker, even for a single work shift, would retrigger the full range of medical benefits. The degree to which this internal calculation influenced U.S. Steel's decision to risk arbitration is not clear. Unquestionably, it motivated some management decisions on recall. Bruce Johnston told me the company had decided against bringing employees back from layoff in some situations because this would "recharge the benefit battery for up to two years."

Steelworker anger about contracting out exploded at a three-day meeting convened by the union and U.S. Steel at Clearwater Beach, Florida, in December 1984. The idea was to have frank discussions by local and national officials about problems in USS plants. As Tom Graham, Bruce Johnston, and other executives listened with stony faces, local president after local president voiced strident complaints. Jim Brown of Gary Local 1066 even borrowed the management technique of presenting a slide show to give a documented account of contracting out at his plant. In many cases, the company hadn't notified the union beforehand that it intended to subcontract the work, as required by the contract. In a ten-month period, he said, Gary management notified Local 1066 of 180 instances of contracting out and omitted more than 100 other instances. Phil Cyprian of 1066's sister local, 1014, engaged in an emotional tirade, calling Graham a "hypocrite" and "double-talker."

The company executives didn't apologize for their actions. Graham made a tough speech, urging the union people not to "cling to a nostalgic past that is rapidly becoming irrelevant."

The company must improve productivity, he said, and it could no longer be competitive if it had to pay USW wages for maintenance and service functions the other companies contracted out at lower wages. It made little sense, he said, to pay $23 an hour for a janitor. "We will not want to do in a steel plant anything that can be more economically done outside the steel plant," Graham said. "We will try...to get to that objective first by negotiation. Failing that, [we will contract out] unilaterally within the framework of the contract. That is my philosophy on contracting out."

Graham was being honest, but he also threw out a challenge that most union leaders would find difficult to evade. John Moody, the *Pittsburgh Post-Gazette*'s labor writer, reported from Florida that the challenge was taken up by Lynn Williams, the new USW president. In a talk to the local presidents after company officials had departed, Williams said repeatedly that he sought to cooperate with the company. But he also warned that "in circumstances in which people seek confrontation, we [the USW] have had a lot of experience in that, and we know how to do that."

From this time on, talk of a strike against U.S. Steel in 1986 spread through the corporation. "I told my members to prepare for a strike," said Don Holdaway, president of Local 2701 at USS's Geneva (Utah) Works, after returning from the Florida meetings. "We have no choice. Tom Graham left no room open."

Comments at the Florida meeting also indicated that the contracting out problem may have been exacerbated by a split within U.S. Steel. In a reorganization earlier in 1984, USS had changed the chain of command for industrial relations (IR) managers at the plant level. In the past, they had reported directly to the plant superintendent as representatives of the corporate IR staff headed by Johnston. The superintendents were obligated to consult with their IR people before making decisions that would affect the administration of the labor contract. But there was little coordination at top levels between the IR staff and line managers. Under Graham, the superintendents had to make aggressive operating decisions without corresponding changes in IR policy.

To give Graham control over industrial relations, USS ordered local IR managers to report to Thomas W. Sterling III, a newly appointed vice-president of employee relations who reported to both Graham and Johnston. Asked at the Florida meeting who had the final word on contracting out, Graham replied that line management had final responsibility for running the plant. The same question was asked of Johnston after Graham left the meeting. If the contracting out issue "comes on up the line, we [the IR staff] have the responsibility for labor contract administration," Johnston said.[12]

There had been rumors of conflict between Graham and Johnston ever since the former joined USS. Little concrete evidence existed of personal conflict, but the strains that developed in the company's labor relations argued that operating policy and contract administration policy failed to mesh. Both Graham and Johnston sat on the company's eleven-man corporate policy committee, USS's top decision-making body under the board of directors, and both had the strong support of Roderick. If Johnston carried out his job to the letter, however, he would attempt to intervene when operating decisions went contrary to the corporation's labor policy. The continuing battle over subcontracting indicates that Johnston and his staff didn't intervene very forcefully.

Graham had been an aggressive manager at J&L but without spurring thousands of grievances and arousing the animosity of a large part of the work force. Former staff members at J&L told me that John Kirkwood and his industrial relations people acted as a restraining force on Graham. They persuaded him that the work force could be cut and productivity raised without making enemies of the workers. The facts indicate that no such restraining influence existed at U.S. Steel. If they did, Graham chose to ignore them—and this is a possibility given the breadth of the task facing him in dealing with the corporation's labyrinthine bureaucracy.

The 1984 Plant Shutdowns

Sometime in 1983, U.S. Steel management made a key decision. The demand for steel had not returned to prerecession levels, preventing the company from gaining enough volume to reach the breakeven point. The cost of hanging on to unused operating units constituted a severe financial drain. Wall Street analysts, who influence the decisions of big investors, were urging American steelmakers to cut capacity, increasingly singling out U.S. Steel as the target of their criticism. And the politics of obtaining import relief seemed to demand that the companies themselves dramatize the seriousness of their plight by closing plants and eliminating more jobs.

Under these pressures, Roderick and his top managers decided to take a big write-off of assets at the end of the year by closing plants in 1984. Which plants would be affected depended on many factors, including degree of obsolescence and cost of modernizing, the location of the plant *vis-à-vis* markets for its products, comparative production costs, and the status of labor relations. If it came to a decision on keeping units that ranked equally on the other factors, Johnston had warned during 1983 negotiations, the company would favor those "where we get people who cooperate." The definition of "cooperate," of course, would be management's.

U.S. Steel negotiated with many locals from 1983 to 1986. Johnston felt that he had McBride's implicit support in doing this. In fact, McBride didn't lay down any guidelines for the plant-level talks, apparently feeling that the locals should be allowed to decide their own fate. When company negotiators talked to local officials, the message came through very clearly that their plants, or portions of them, would be closed unless they acceded to changes. Johnston denied that the company used threats or "blackmail" in this bargaining. "We have not threatened any basic steel mill with shutdown, except South Works," he insisted.

Union people saw it differently. Louis Kelley, a USW staff representative assigned in 1984 to help the USS locals in negotiations, participated in many talks. "The way the company man would put it was, 'We must make crew size changes and put in another craft system if that plant is to operate or survive.'" Kelley said. He advised locals to negotiate. "Some of our guys told them [USS] to go ahead and shut it down. I told them, 'Don't tell U.S. Steel to do something, because they'll do it.' They never threatened to shut down a plant that they didn't shut down."

Many locals took the threat seriously. Local 1557 at Clairton agreed to reduce crews for each coke battery from twenty-one workers to fifteen, representing some sixty jobs and a large savings for U.S. Steel in the cost of making coke. This was the first instance in which USS granted 70/80 pensions in return for job eliminations. The company also cited crew levels being considered by locals at its Gary, Fairfield, and Lorain coke plants, thereby involving the four locals in simultaneous negotiations. "We were whipsawed," Charlie Grese acknowledged. Partly as a result of the Clairton concessions, the Fairfield and Lorain coke batteries were shut down. Whipsawing could work just as well for the company on the downside as it had for the union on the upside.

Johnston later complained that local unions from 1983 to 1986 "stonewalled" the company on its demands to reduce crew sizes and combine crafts. Some did, especially locals at the Gary, Fairless, and South Works. However, locals at Lorain, Geneva, Braddock, McKeesport, Clairton, and Fairfield did agree to work-force concessions. The Fairfield agreement, negotiated by staff representatives rather than by the local, contained especially extensive changes which—for the first time in a USW-represented steel mill—actually wiped out all past practices regarding manning, crew sizes, job assignments, and related items and gave management the unilateral right to establish new practices. It also canceled many incentive pay plans and authorized the company to contract out some services previously performed by USW members.

The USW's district director in Alabama, Thermon Phillips, implemented this agreement without holding a ratification vote, an action he felt was justified by a USS promise to reopen closed steelmaking units and build a slab caster at Fairfield, securing some thirty-four hundred jobs for the plant. This many jobs never materialized, largely because the anemic steel demand did not justify restarting some steelmaking units. Because the Fairfield pact gave management great flexibility, it became U.S. Steel's model for negotiating at other plants. But it infuriated unionists at many locals who complained that the manning provision of the agreement, in effect, abrogated the 2B clause in the union's master contract. If a single local could simply ignore the industrywide agreement, many USW officials worried, the union could disintegrate into dozens of competing locals.[13]

On the company's side, the effort by local managers over decades to keep the union out of managerial areas of decision making had left a vacuum. When management really needed the union's help, there was no established basis for cooperation, merely distrust of management's motives. The story of negotiations at the Duquesne plant illustrates this problem.

In early 1983, U.S. Steel officials told Mike Bilcsik that Local 1256 must agree to reduce crew sizes to make the plant competitive. Bilcsik, who believed that work practices *should* be flexible, said he would negotiate on the issue but only if the company guaranteed investment in his plant. By mid-1983 he, as well as many other USS employees in the Mon Valley, had surmised that only one plant with blast furnaces and basic oxygen furnaces would survive in the valley, either Duquesne or Braddock.

Bilcsik charged that U.S. Steel attempted to exploit this competition by pitting one local against the other. There is no evidence that the company actually told either local that it must yield the greater concession to stay in business, but Bilcsik perceived it that way. Moreover, the attitude of his members regarding the Duquesne plant was, "'Let 'em shut it down.' They don't care. They're tired of being jerked around."

Still annoyed that Duquesne management had turned down his request to form LMPTs, Bilcsik also accused the company of reneging on a promise to buy certain equipment for his plant. "We will not agree to these concessions because they won't gain a thing for us," Bilcsik wrote in the August 1983 issue of his newsletter, *The Voice of 1256*. "We will not be able to put this mill back on its feet until they start treating their employees as equal partners," he concluded.

The idea of sharing power with the union ran counter to U.S. Steel's ideology and concept of good management practice. Some local leaders, like Charlie Grese at Clairton and Don Thomas at Braddock, responded to U.S. Steel's demands by doing what they had to do to ensure their plants' future. The Braddock local accepted work-rule changes in 1984.

Bilcsik thought he stood for a principle, but the company merely viewed him as intransigent. After some negotiations early in 1983, few meetings followed at the local level. Bilcsik said none occurred after May. He didn't press the matter, although it seemed clear that U.S. Steel would announce plant shutdowns at the end of the year. Finally, in a letter dated December 15 and received by Bilcsik on December 24, the company said it soon would announce those plans. "We have been doing our best to persuade you and the local union leadership to make every possible accommodation on seniority arrangements, job manning, work rules, and related activities," the letter said. It urged Bilcsik to "secure your members' understanding that the responses made in this area represent the best hope of the future."

Bilcsik maintained he didn't have a chance to reply. Three days later, on December 27, Roderick held a news conference and announced the company would permanently shut part or all of twenty-eight plants and mines, eliminating five million tons of steel capacity, or about 16 percent. Up to then, it was the largest shutdown notice in steel history. The closings would cut 15,436 employees, 10,846 of whom were already laid off. Portions of all Mon Valley plants were affected, including Duquesne's two blast furnaces and the basic oxygen furnace. Some of the operations scheduled for closing, Roderick said, could have been kept open if local unions had granted concessions. "We are willing to make a place process-competitive," he said. "We think labor has the obligation to be labor-competitive, and, by God, [the plants] are not going to get process-competitive if they are not willing to become labor-competitive, just that simple."

Under the USW contract, the union locals had sixty days to offer further concessions to save the plants. "If they want to stop the freight train," Johnston added later, "they'll have to come to us. There's still time, but there has to be a spirit of wanting to do it. If there isn't, *c'est la vie.*" Roderick, however, left it unclear which plants could be saved by further concessions. On the other hand, he said that the "greater part" of the jobs eliminated could not be retrieved, even with more relief from the union. But later in the news conference he said that the closing decisions were not irreversible, that "those plants could be saved if the employees at the plant wish to do it." This semantic confusion left the impression that all of the shutdowns should be blamed on labor obduracy.

Actually, the corporation had no intention of operating the blast furnaces and steel furnace at Duquesne. It had made the decision some time ago to use the Edgar Thomson Works at Braddock as the raw steelmaker for the valley. My interviews with management people such as Al Hillegass, Al Sarver, Jack Bergman, and Al Voss make it clear that top management had favored Braddock for some time because of practical considerations.

Although ET's two small blast furnaces were not as productive as Duquesne's huge Dorothy Six, Braddock had a more modern steelmaking shop. Dorothy could have been kept to supply iron to ET, but this would have meant transporting molten metal constantly across the Monongahela on a railroad bridge. This had been done many times in the past, but USS viewed it as too expensive and time-consuming for a permanent practice.

So Local 1256's intransigence almost certainly did not result in the closing of the Duquesne furnaces. However, the local might have avoided the company's later decision to halt operations at the plant's high-quality bar mill. In 1984, a company man later recalled, Bilcsik and other local officials met with USS negotiators to talk about concessions needed to keep the bar mill open.

"The meeting started at two o'clock," the company man said. "When the three o'clock whistle blew, the union guys got up and started to leave. That's when they stopped being paid. Management had said it could have been a competitive bar mill, but the union people couldn't wait fifteen minutes to listen to what we had to say. I'll never get over that."

Bilcsik didn't remember that specific meeting, but conceded it could have happened that way. "A lot of our guys didn't like to hang around after three o'clock. If they walked out, I would not have stayed behind. You have to show cohesiveness."[14]

U.S. Steel halted all Duquesne operations in 1984.

Following Roderick's announcement, a few locals offered to negotiate but no agreements eventuated. In addition to Duquesne, U.S. Steel closed six entire plants, all categorized as List 3 locations: a wire-rope plant in Trenton, New Jersey; the Cuyahoga Works (a fabricating shop near Cleveland); the Johnstown plant, which made equipment for USS's rolling mills; and fabricating plants at Ambridge, Elmira, and Shiffler, Pennsylvania. The company thus made good its earlier threat to close List 3 operations. At Trenton and Cuyahoga, the locals rejected company demands for extensive wage and benefit cuts, partly because many workers had enough seniority to claim shutdown pensions.

In some of these cases, union members did not adopt a realistic attitude about the competitive situation of their plant. The president of the Johnstown local, Dean Bracken, told me many of his members refused to believe that USS would shut the plant, although the company said it had lost $18.5 million in two years. The Johnstown plant competed with other companies where USW-represented employees earned about $4 per hour less than Johnstown workers. The company asked for a $5 per hour cut and offered 150 early retirements in return. The logic of this offer left much to be desired, noted USW representative Lou Kelley, inasmuch as 250 employees would receive pensions if the plant shut down.

On March 1, 1984, as Johnstown workers voted on the proposal, word came from U.S. Steel's Pittsburgh office that the company would offer forty additional early pensions. It was too late to stop the balloting, which resulted in a contract rejection by sixty-nine votes. The forty additional pensions might well have swung the vote the other way, but the company closed the plant the next day. Several months later, U.S. Steel sold the plant to a company which reopened it with many former employees and under a USW contract that provided for wages of about $5 per hour less than U.S. Steel had paid.

Conclusion: Both Sides Failed

Both sides could have acted differently in the plant shutdowns of 1984. The International failed to play a strong role in educating its members about the faltering steel economy and urging the locals to loosen up on work rules. The experience also revealed a depth of hostility to management that communicated a message to U.S. Steel's Pittsburgh headquarters. Roderick, Graham, and Johnston chose to read the message as one of union failure to see the economic writing on the wall. This perception contained some truth, but it wasn't the whole message.

U.S. Steel's "control" method of managing people, its fight to keep local unions out of management business, perpetuated feelings of disbelief and distrust. Where locals accepted changes, they did so on the basis of perceived threats, not out of a sense of entering into a partnership with management to contend with a new economic environment. The idea of expanding the range of issues on which local managers should consult with union officials, even of sharing some decision-making power, ran counter to the company's management system and style. Workers were treated as contract employees, hired and paid by the hour, who had no interest in the firm beyond a bi-weekly paycheck. The real core of U.S. Steel consisted of the upper-level managers.

In response to this argument, Roderick, Graham, and Johnston would ask why management should give up power to encourage a union to behave in an economically reasonable way. This was a valid position. But was it the correct position for the fundamental changes that had to occur in the new world economic order?

U.S. Steel tried overnight to change workplace practices that resulted from the application of Taylorism and the "economic man" concept for nearly a century—to change those practices *without changing the management culture that spawned the uncompetitive rules and attitudes.*

U.S. Steel's attitude toward LMPTs exemplified the lack of participatory feeling at the top of the corporation. In making this point, I am not implying that Labor-Management Participation Teams or any other form of worker participation, acting alone, would have reversed the forces of worldwide competition and saved all the Mon Valley plants. A true participative process, however, starting at the top and extending to the bottom, might have saved some plants, certainly some operations. It would have helped U.S. Steel initiate needed changes without provoking the enmity of thousands of employees and residents of steel communities.

Another caveat is necessary: LMPTs should not be established with the ultimate objective of persuading workers to negotiate away their jobs. Nor should the participative process be confined to problem-solving committees. It should be expanded into many joint activities, including the creation of autonomous work teams and consultations by management with workers and union officials on decisions that will have a direct impact on the work force. In such a setting, management and union can address competitive challenges and worker concerns in a spirit of solving problems rather than fighting over outmoded contractual rights.

Unlike J&L, Bethlehem, and National, USS officials never really pushed participation as a priority. "There was never any commitment from the top of the house," said a former labor-relations executive who left USS in the early 1980s. "I think there was a disbelief that the program would work. It was literally a struggle to get the two programs going where we had them. And after the 1983 industrywide settlement, we were directed to negotiate locally with the union, and we didn't make many friends." U.S. Steel installed LMPTs at only two plants, Irvin and Edgar Thomson, in 1982. Both of these proved to be successful initial efforts, carried out by people on both sides who sincerely believed in participation. The teams at ET survived and still operated in 1987.

The program at Irvin collapsed in 1983. U.S. Steel began combining jobs in the plant, though not as a result of LMPT discussions. But Mike Bonn and his committee halted team participation. "They felt they could hold the program hostage to stop the combining of jobs," said an informed observer. At about the same time, McBride withdrew his support for LMPTs in retaliation against Roderick's proposed deal with British Steel.

U.S. Steel management often cited McBride's action in 1983 to explain why the company did not move more aggressively in setting up cooperative programs. But this argument begs the question. The key period was between 1980 and 1983, when the company could have got the process started at many plants. Some discussions occurred but nothing resulted, not even when local union presidents like Bilcsik and Jim Brown at Gary Sheet & Tube proposed LMPTs to local management. On the other hand, some locals—Local 1557 at Clairton for one—rejected management requests to get involved.

As a reporter, I talked to Johnston on several occasions about LMPTs. His feelings about participation ranged from cool to lukewarm, compared to executives I have spoken to in other steel companies and other industries. Rather than summarizing his attitude at the risk of being unfair, I shall quote Johnston at length from our longest discussion about LMPTs.[15]

He would not answer directly why the company hadn't moved farther, faster with LMPTs. Rather, like the debater he was, Johnston listed many reasons why advocates of participation claimed too much for it. He noted that participation teams "have become a favorite of university people," a sardonic observation reflecting Johnston's contempt for academics with liberal views on labor issues. "I am not against participation teams," Johnston said repeatedly, claiming that he himself conceived and wrote the 1980 contract provision that first authorized LMPTs in the steel industry. Actually, Bethlehem Steel proposed the concept, although Johnston helped draft the provision and, as chief negotiator, approved its inclusion in the industrywide contract (see chapter 11).

Some people, Johnston went on to say, urge participation because they "don't want to come up against the real tough problems." He gave the following example. "No congressman will take the risk of moving money out of the federal spending sector to private investment. When we complain about depreciation and transfer [welfare and unemployment] payments, they say, 'Hey, how about quality circles, isn't that how the Japanese do it?' Quality circles and participation teams don't begin to answer these huge transfer payment problems we have created. They don't begin to answer the huge differences in how Europeans and Third World countries have been allocating capital resources to their steel industries and how we do it. Many people use quality circles to veil the big problems we have to solve."

Some studies, Johnston pointed out, had concluded that participation produced mixed results, some good and some indifferent, and that no one had developed an "assured formula" for improving motivation and creating a "successful corporate culture." Johnston noted that LMPTs at the Irvin Works "taught us some useful things. There were some good changes made in customer satisfaction and quality. The training helped change attitudes. While we got out of the formal program, we have an investment in that cultural change." He added that "the big gain [in participation] is in worker satisfaction. But [that can be better accomplished] with employee benefits." (Behavioral experts would dispute this view.)

Johnston continued: "It's difficult to work it [participation] in an industry which is contracting, laying off people." This was a good point. Asking for the cooperation of the work force only to close plants would surely bring charges of betrayal. On the other hand, employee involvement groups have continued to function well in many declining plants. LMPTs were still working in the one remaining department of LTV's Aliquippa plant in 1987, long after the rest of the works had been closed. Moreover, U.S. Steel's big closures occurred after 1983, not in 1980 to 1983, when the company still tried to save plants.

As for those companies that embraced LMPTs, Johnston concluded, "I think a lot of people in the industry will have to go to bed with the union. It's a survival approach."

All of this bespoke a credible point of view, one that very likely the majority of American management still accepted in the late 1980s. For most companies, worker participation had nothing to do with establishing labor-management partnerships. It was just one more company program to help motivate people. U.S. Steel was not exceptional in its position. It simply did not want union involvement in any significant way, except on the company's terms.

Chapter 17

Monessen is a gritty little town pressed between a steep, walled-in hill and the brownish Monongahela, just north of Lock & Dam No. 4 and a horseshoe bend in the river south of Donora. The city lies only a short distance as the bird flies, but several tortuous driving miles on congested, two-lane roads from the nearest expressway, Interstate 70. Founded in 1897 as the site of a tinplate mill, the town received its name from a land developer who foresaw the little settlement exploding into an Essen-on-the-Mon, like the big German industrial city in the Ruhr Valley.

On a Saturday afternoon late in July 1985, I drove up the valley to Monessen. Coming out of the final curve as Route 906 parallels the river's bend, one leaves a tunnel of dense foliage and is immediately confronted by the giant red and green structures of the Wheeling-Pittsburgh Steel plant. It is an unexpected sight in this rural setting. The Monongahela is narrow here, and the tall red blast furnaces look as if they are painted on the side of the soaring green bluff across the river. My eye next fell on four men sitting stonily on a bench near a plant gate, looking like they'd been painted there, too. Big signs adorned their lean-to: "Federal Court Unfair to Steelworkers" and "Carney Must Go," the latter referring to Dennis J. Carney, the chairman and chief executive officer of Wheeling-Pittsburgh.

It wasn't the first time that Monessen had known industrial strife, but it could be the last.

I drove through town on Donner Avenue, which bordered the two-mile long plant. It was a hot, lazy day, and I seemed to be the only thing moving in Monessen. I passed another mill gate, where picket signs leaned against a row of empty

447

wooden chairs. Taking refuge from the sun, the strikers lolled under the marquee of an abandoned movie theater across the street. A little farther on, another couple of pickets in T-shirts and baseball caps barred yet another gate. In each case, the guys had taken careful note of me and my car but hadn't moved, just sat there staring, as I coasted through town on a street otherwise devoid of traffic.

The plant grounds had become a graveyard of old mill buildings, most of them unused for years, rusting away. The plant had grown like Topsy sixty and seventy years ago, when the Pittsburgh Steel Company was expanding along the banks of the river and moving inland. But the company fell on hard times in the 1960s and in 1968 merged with another troubled firm, Wheeling Steel. In recent years, the firm had abandoned large areas of the eighty-five-year-old plant as the machinery wore out and demand dropped. The dozens of triangular roofs were headstones for old Monessen, famous for its production of barrel hoops, nails, rod, wire, tinplate, and the home of more than twenty thousand people as late as the 1940s. In the 1980s fewer than twelve thousand people lived here, and the future of the Monessen Works depended solely on its one modern feature, a "world-class" rail mill built in the early 1980s. But it wasn't operating now.

Within a few minutes, it was all behind me—drowsy Monessen, the staring pickets, the silent mill. But the tableau remained etched in my mind. As the steel industry lay dying in the Mon Valley, these workers, as if trying to recapture the glorious days of union power in the 1950s, were engaging in the first major work stoppage in the steel industry since 1959. More remarkably, the seventy-five hundred Wheeling-Pittsburgh employees in the Mon and Ohio Valleys *had shut down a company in bankruptcy proceedings.*

Nobody in the union could (or did) make the remotest guarantee that the company would survive the strike—not the USW negotiators who recommended the action, not Lynn Williams who approved it, and certainly not the union members who carried it out. Yet it went on for ninety-eight days. Why?

The answer had to do with the struggle for power between union and management in a reorganizing industry. At "Wheel-Pitt," as people called it, Dennis Carney tried to circumvent the union by using a tactic employed often in the first half of the 1980s by impatient chief executives. He gained the approval of a bankruptcy court to nullify an existing labor agreement and impose in its place—without union consent—an 18 percent cut in pay and benefits. Carney had done a superb job of modernizing W-P's plants and keeping the company alive, but he overestimated his ability to make people do things by management decree.

The workers' attitude was stated concisely by Joseph Elliott, a machine operator at the Monessen plant. Explaining why he went on strike despite the threat of company liquidation, he said: "The strike will put me into personal bankruptcy, but I'd rather be unemployed than a slave."

The seemingly perverse nature of the Wheeling-Pittsburgh shutdown captures the spirit of the steel industry's chaotic labor relations during the middle years of the 1980s. These were transition years, in which the industry and its employees were forced to abandon their old monopoly ways and begin to forge new relationships. During this time, 20 percent of the steel industry closed down for good, two major steelmakers went into bankruptcy, the industry's bargaining group broke up after thirty years of negotiating industrywide contracts, and the Steelworkers under Lynn Williams experimented with new policies. All in all, it was a period of turmoil in which desperate people did desperate things but also a time when people put aside old belligerencies and began to build new structures.

The latter was particularly true at National Steel, where management initiated thorough-going reforms in the management and industrial relations systems and in 1986 negotiated one of the most forward-looking labor contracts in U.S. industry. The outstanding success story of the mid-1980s occurred at Weirton Steel Company, the 100 percent employee-owned plant at Weirton, West Virginia. Executives of old-line steelmakers had scoffed at the notion that employee ownership could succeed where traditional management could see only failure ahead.

But by late-1987, Weirton had put together a record of fourteen consecutive quarters of profitability. The company's first president, Robert L. Loughhead, established a style of management that converted financial ownership by workers into a real partnership that retained only technical distinctions between managers and workers. Loughhead described his style of management in deceptively simple terms. "The key," he said in 1986, "is listening. It has the potential to change a person, company, and maybe even a country."

The steel industry's agonizing period of adaptation to a competitive world economy could have used more Loughheads.[1]

New USW Policies

In January 1984, the steel industry still writhed in financial spasms as it underwent a forced contraction. U.S. Steel had just announced its massive plant shutdowns. Only the most optimistic analysts could see an end to the industry's shakeout before the mid-1990s. Reflecting this turmoil, industrywide solidarity in the USW, already weakened by steel's post-1981 plunge, had splintered dangerously in the last half of 1983. With no central guidelines or leadership, district leaders pursued divergent policies in bargaining over wages and work practices. To shocked steelworkers in the Mon Valley, Illinois, Indiana, and Texas, the industry seemed to be spiraling downward out of control while the union wallowed in confusion.

On January 13, the USW convened a meeting of twenty-nine U.S. Steel local presidents at the Shoreham Hotel in Washington to counsel them on shutdowns. The meeting was part of Lynn Williams's effort to dispel the confusion, give the locals advice and a sense of leadership. Union technicians instructed the local officials on applying for shutdown benefits, seeking government retraining aid, and using the grievance procedure on contracting out problems. Williams reminded the locals that the executive board a month previously had adopted a policy against making concessions that would undermine the industrywide agreement.

One outsider addressed the meeting that day. He was Eugene Keilin, a senior partner of the Lazard Frères investment banking firm. He had come to Washington to get acquainted with the Steelworkers, bringing an assistant, Joshua Gotbaum, whose father Victor was head of the powerful District Council 37 of the American Federation of State, County & Municipal Employees, which represented more than one hundred thousand employees of New York City. Keilin, who designed the Employee Stock Ownership Plan (ESOP) at Weirton Steel and served on Weirton's board of directors, briefed the USW officials on employee ownership. This first contact would expand into an unusual link between Wall Street and organized labor. Lazard Frères became the USW's investment banker.

It was Jim Smith's idea. By the end of 1983, it had become

plain that the restructuring of the steel industry would challenge the union's ability to make decisions in financially complex situations. The USW's research department, perfectly adequate for doing routine economic projections and examining a company's books, had never had to engage in the kind of detailed economic and financial analysis required by the current state of the industry. When Smith looked around for help, the name Lazard Frères leaped out at him.

Felix Rohatyn, a Lazard Frères partner and an architect of New York City's financial recovery, had demonstrated an unusual sensitivity to union concerns. Instead of adopting the typical Wall Street disdain of unions, he and Gene Keilin had worked pragmatically with Victor Gotbaum to solve the city's problems. Rohatyn also demonstrated in his writings and appearances before Congress that he believed that government, labor, and management should form tripartite partnerships to solve the problems of cities and industries, a concept dear to the heart of USW leaders.

Rohatyn's reputation led directly to Keilin's involvement in setting up the Weirton Steel ESOP in 1983. The Air Line Pilots' Association also retained Lazard Frères for special projects. In the airline industry, as in steel, massive restructuring occurred in the mid-1980s, involving friendly mergers, hostile takeovers, and plant shutdowns. To protect the interests of members of these transactions, unions had to buy expert knowledge of financial markets and investment sources and the ability to put together financial proposals. For its part, Lazard Frères took on the union clients, despite criticism on Wall Street, primarily for commercial reasons, tinged with the centrist-liberal political philosophy of Rohatyn and Keilin. The union business was good business—the firm charged the same fees to unions as to companies ($75,000 a month in bankruptcy cases, according to a *Fortune* profile of Keilin)—and one that likely would expand. Indeed, several other investment houses entered this business in the late 1980s.

Moreover, the firm's developing knowledge of unions would help it in dealing with corporate clients. Rohatyn and Keilin saw that it would be increasingly difficult to calculate the value of a company without assessing its labor relations and labor costs. This was especially true with regard to the increasing cost of pensions and health insurance benefits. Many companies did not adequately finance these funds, and pensions liabilities often exceeded a company's net worth. "The only way to deal with these liabilities," Keilin said, "is to find a way of dealing with the workers to bring them into some kind of partnership to reduce costs and be competitive."[2]

The USW also acquired other types of outside expertise. In analyzing company books, the USW began using insolvency accounting experts from national firms such as Arthur Young and Company. For economic and feasibility studies, the union retained Locker/Abrecht Associates, a New York firm headed by Mike Locker. He and Randy Barber, director of the Center for Economic Organizing in Washington, represented a relatively new kind of union consultant. To help labor cope with capital mobility and the threatened loss of jobs, they devised political and organizational strategies aimed at giving unions a broader role in areas such as financial management, marketing, pricing, plant location, and board membership. Sidney P. Rubinstein, president of Participative Systems of Princeton, helped the USW install LMPTs.

Under Williams, the Steelworkers also turned increasingly to Employee Stock Ownership Plans to cope with financial weakness in the steel industry. This required a fundamental change in bargaining policy. Like most unions, the Steelworkers traditionally rejected employee ownership. The concept seemed to go against the confrontational philosophy of American trade unionism in that employee owners would tend to identify closely with management. The USW believed that management installed ESOPs, a specific form of employee ownership used with growing frequency since the early 1970s, to manipulate workers. Moreover, many companies with ESOPs have no pension plan but award stock upon retirement. Most unions contend that the vagaries of stock values cannot assure a secure retirement.

As mills failed in the 1980s for lack of outside investment and a decline in borrowing power, the USW concluded that employees themselves had to provide investment capital to save jobs. This could be done through an ESOP, which is defined under federal law as a special type of employee benefit plan that can borrow money. An ESOP is useful in arranging leveraged buyouts and it also receives many tax breaks.

In 1980 the USW began changing its collective mind about stock ownership after Jim Smith commissioned a study of ESOPs by a Washington lawyer, Stephen Hester. With the backing of Jones & Laughlin, the USW pushed for and gained a provision in the 1980 steel contract authorizing individual companies to negotiate a stock ownership plan. But the provision remained unused by the large steel companies until 1985, when Bethlehem formed an ESOP in return for wage concessions at a plant in Johnstown.[3]

In early 1982, McBride and Smith initiated a policy of asking for stock in return for wage concessions at small companies. The USW also negotiated partial equity positions for members in some situations to help a new owner take over a threatened or closed plant. While these efforts began under McBride, Williams made it a more consistent policy, using Lazard Frères and Locker/Abrecht to study the feasibility of forming ESOPs to buy threatened plants. Buyout efforts failed in some cases, including the Gadsden (Alabama) plant of LTV Steel and U.S. Steel's South Works and the Dorothy Six blast furnace at Duquesne. But they succeeded elsewhere, and the USW began to win significant equity positions in return for wage concessions. The number and scale of these efforts was amazing, considering that only a few years earlier the USW had regarded ESOPs with extreme hostility.

By early 1988, USW members owned substantial stock, though less than 15 percent, in Wheeling-Pittsburgh, Bethlehem, LTV Steel, and Kaiser Aluminum. Through buyouts and negotiations, USW members and nonunion employees also owned 40 percent of CF&I Industries (formerly Colorado Fuel & Iron), 70 percent of Copper Range Company in northern Michigan, 42 percent of Northwestern Steel & Wire in Illinois, 87 percent of McLouth, 67 percent of Oregon Metallurgical (Oremet) in Wyoming, and 100 percent of three small firms, E. W. Bliss of Salem, Ohio, Republic Storage Systems Company of Canton, Ohio, and Republic Container Company of Nitro, West Virginia.

Studies show that employee ownership in and of itself neither significantly increases worker motivation nor changes a company's management style unless a comprehensive participation process is put into place. In most companies where the USW negotiated an ESOP, it also set up employee involvement programs at the shop-floor level and higher, where possible. By 1988, the union had representatives on boards of directors at Wheeling-Pittsburgh, Kaiser Aluminum, CF&I, McLouth, Northwestern, Copper Range, Oremet, E. W. Bliss, Republic Storage, and Republic Container.[4]

A few other unions demanded board representation in similar cases, including the Air Line Pilots, Machinists, Teamsters, and Auto Workers (at Chrysler). But the USW spread the concept more pervasively through its major industry than any other union. Gradually, the Steelworkers accepted ownership and decision-making responsibilities in an effort to maintain as many steel jobs as possible. In some cases, gaining stock ownership could not have been considered a great victory, since the companies were teetering on the edge of bankruptcy.

While Williams was not shy about taking on these non-traditional responsibilities, stock ownership, shop-floor participation, and board representation by no means formed the core of the USW's collective bargaining program. Wages, benefits, and job security remained the most important issues. Williams pointed out in several speeches that he did not push employee ownership or board representation as the principal means of saving the steel industry. To save jobs, the USW frequently had no choice but to give concessions and demand various mixtures of ownership and power-sharing. This being noted, the movement toward participative relationships in the steel industry was quite remarkable under Williams.

Still, the 1986 round of bargaining indicated that none of the major steel companies had any intention of accepting the USW as a partner at the highest levels of decision making. And Williams's policy of saving jobs through employee ownership contained the same paradox as McBride's ad hoc decisions to allow varying levels of wage concessions at troubled companies. To the degree that worker ownership enabled weaker companies to survive in a steel marketplace with excess capacity, the policy jeopardized the jobs of USW members in stronger companies.

The union's willingness to finance feasibility studies for worker buyouts also created unrealistic expectations in some situations—at Duquesne and the South Works, for example—where the chances of survival were not good. Williams acknowledged that the union would make mistakes as it tried new approaches. He believed, however, that the USW's overriding goal should be to save as many jobs as possible and that it was impossible to calculate prospectively the competitive effect of keeping marginal companies in business. "A great many communities and people are dependent on the so-called marginal companies," Williams said in an interview. "I think the union as a workers' organization has an obligation to be responsive to the needs of our people and those communities. Our purpose has been all along to try to stop the shrinkage, to reduce the base of it, to prevent it, and to preserve as significant a steel industry as possible in America."[5]

Trouble at Wheeling-Pittsburgh

Williams's willingness to take large risks was demonstrated by events at Wheeling-Pittsburgh Steel. To many onlookers, the USW's 1985 strike at Wheel-Pitt, then the seventh largest steelmaker, seemed suicidal. A good case could be made to uphold that view. But beyond that, the origin, conduct, and outcome of the work stoppage illuminated a number of new issues and tactics generated by the changing balance of labor relations in the mid-1980s.

No one could say that the USW had ignored W-P's deteriorating financial position in the 1980s. Three times between 1980 and 1983, the union granted wage concessions to the company, the first time at just one plant but the last two times at all plants. The steelmaker had gotten into trouble largely because its hustling chairman, Dennis Carney, tried to accomplish what few other steel managers attempted—a thorough modernization of Wheel-Pitt's steel plants in western Pennsylvania, West Virginia, and Ohio. But the company suffered major losses during and after the recession and was unable to make interest and principal payments on $540 million in loans.

In late 1984, on the advice of accountants from Arthur Young who examined company books, the USW agreed to defer scheduled restorations of the earlier wage cuts. Two months later, Carney asked for a fourth round of concessions. During this period, Carney's relations with the union's chief negotiator, Paul Rusen, turned very sour. Carney had the reputation of being an autocratic manager, and now he refused to listen to union ideas about restructuring the debt, according to Rusen.

Rusen and his bargaining team—Lefty Palm of District 15 (W-P's Monessen and Allenport plants were in Palm's district) and Jim Smith—offered to cut total employment costs from $21.40 an hour to $19.50. But the union had negotiated on the condition that Carney gain concessions from W-P's lenders. Instead of deferring and rescheduling payments on the interest and principle, the banks only offered to provide an additional $40 million line of credit. Most importantly, they insisted that Wheel-Pitt grant them a lien on all current assets, valued at $300 million. If the company were forced into liquidation by another recession, the banks would take the $300 million, leaving little of value for the workers despite their ownership of a considerable amount of stock.

Carney granted the lien, despite the USW's disapproval. Two Lazard Frères advisers, David Supino and Josh Gotbaum, concluded from a study of W-P's books that without relief from the banks the company would go into bankruptcy. If it had no assets to use for loan collateral, it would run out of cash, and the banks would be in control. By convincing Rusen and Palm that this likely would mean the death of the company, the Lazard Frères men stiffened the union's backbone. The USW negotiators refused to make a deal with Carney as long as the lien existed.

Joseph Scalise, the company's senior vice-president of industrial relations, felt a sense of personal rejection. As W-P's chief negotiator for ten years, he had bargained innovative provisions to make previous wage cuts palatable. "We always had the ability to reach tough concession agreements," Scalise said in 1987. "We explored a lot of new fields. What was different about this time was that outsiders got involved. I still feel we could have reached an honorable agreement without them."

On April 16, 1985, Wheeling-Pittsburgh filed for chapter 11 protection and petitioned the bankruptcy court for permission to cancel the existing labor agreement and unilaterally lower wages. Rusen wanted to avoid such a ruling. On the day before court hearings began, he had breakfast with Scalise, and made an unofficial offer. Supino had advised the union privately that W-P could reorganize successfully with employment costs of $18 per hour.

Rusen told Scalise he would try to sell his members on reducing wages and benefits to $18.50, a number that was negotiable down to $18, Rusen later disclosed to me. Unpublicized at the time, this offer might have led to an out-of-court settlement. But the company turned it down. "It was too late for $18.50," Scalise said. "And the union's advisers were asking for things that Carney just wouldn't agree to, like the union right to approve purchases." In addition, Carney, backed by the banks, apparently believed that the bankruptcy judge would approve an even lower number.

He was correct. After several days of hearings, Judge Warren W. Bentz ruled on July 17 in management's favor, authorizing the company to put in place an hourly labor rate of $15.20. Realizing that a $6 per hour cut was a bit Draconian, Wheeling-Pittsburgh tried to entice workers to stay on the job by announcing it would institute a $17.50 labor cost for six months. Jim Smith, however, contended the terms actually added only to $16.59.

The union quickly appealed Bentz's decision, thus initiating the first significant test of recently adopted amendments to the federal bankruptcy code. The use of the bankruptcy tactic by employers had spiraled into an issue of national significance in the 1980s. Led by Texas Air and Wilson Foods, a number of firms went into chapter 11 proceedings and renounced their labor contracts. The bankruptcy code, developed at a time when going into bankruptcy seemed almost immoral, hadn't anticipated that the law would be used primarily for the purpose of nullifying labor contracts. The code's language on this issue made it relatively easy for companies to meet the standard for showing that existing labor costs were too high, and the U.S. Supreme Court upheld the lenient standard in early 1984. Dissenting justices pointed out that this reading of the bankruptcy law conflicted with the duty to bargain over wages contained in the National Labor Relations Act.

Shaken by these events, the AFL-CIO mounted a strong lobbying effort which led to passage of an amendment to the bankruptcy code in June 1984. The new section spelled out procedures that a company in bankruptcy, or "debtor-in-possession," must follow in seeking to void a labor contract. Among other things, new labor terms proposed by a debtor-in-possession must be sufficient only to "permit the reorganization of the debtor," and they must assure that the creditors, employees, and debtor "are treated fairly and equitably."

In May 1986, a federal appeals court overturned the bankruptcy court's decision in the Wheeling-Pittsburgh case. This court said that the firm's proposal to modify the contract didn't treat the employees "fairly and equitably." It would have frozen wages and benefits at $15.20 an hour for five years under a worst-case scenario. It allowed no possibility of raising wages if the company recovered more successfully than anticipated, although payments to creditors would have been increased in that event.

Although USW lawyers felt confident their arguments would eventually prevail, Rusen couldn't wait for the lengthy appeals process. He and Lefty Palm had to make the monumental decision of whether to strike a bankrupt company, possibly forcing it into liquidation, or acquiesce to management imposition of wage terms. Meeting on July 18, the two negotiators considered all the options. Because about 80 percent of the Wheel-Pitt members never had been involved in a strike, Rusen and Palm worried that some might cross the picket lines as their money ran out. The union and the company had been engaged in a publicity war for several months, and it was unclear which side's propaganda had made the bigger impression.

On the other hand, the USW leaders believed that meekly accepting a compensation cut of nearly $5 an hour, from $21.40 to $16.59, would "break the back of the union," as Rusen later put it. Average hourly wages alone would go down by 28 percent, to about $8. Although they worried about liquidation, the bargainers thought the company's cash holdings of $100 million would keep it intact until mid-November. Furthermore, they thought they could exploit reported disagreements between Carney and two major shareholders. These were Allen E. Paulson, chairman of Gulfstream Aerospace and the owner of 34 percent of W-P common stock, and Japan's Nisshin Steel, which had a 10 percent stake. Carney might be less deeply entrenched than it appeared on the surface.

"It was a tough situation to be in," Rusen said later. "We had the whole principle of unionism riding on what we would do. The $16.59 cost would never be an acceptable principle to establish in the steel industry. It could have destroyed our union. We decided to call the strike, play Carney as the villain, and 'take him out.'" By this he meant devising a strategy to blame the strike on Carney and force his resignation.

Rusen and Palm met the same day with Lynn Williams and told him their recommendation. Following his policy of appointing other officers and district directors to serve as chief negotiators at individual companies, Williams had stayed out of the Wheel-Pitt negotiations and allowed Rusen and Palm essentially a free hand. Now, he backed their decision to strike. Over the next two days, Rusen and Palm held a series of meetings with members in Monessen, Steubenville, and Wheeling and conducted strike authorization votes. Meanwhile, a committee of nineteen local presidents voted 18 to 1 to walk out.

While these preparations went on, the union sent a letter to

Carney offering to continue working under the terms of the old contract. It may have looked to Wheel-Pitt management like just another bit of union propaganda. In fact, the offer concealed a legal ploy suggested by a USW lawyer, Jim English, that could add enormously to the union's strike leverage. While keeping track of unemployment compensation (UC) awards in Pennsylvania, English had noticed that court interpretations of the state UC law consistently favored unions on one issue. The law prohibited UC payments to workers on strike. But if an employer turned down a union's offer to remain on the job under an existing agreement, the ensuing work stoppage should be defined as a lockout and the workers should be granted UC.

The union delivered the letter to Carney's office two days before the strike began. Wheel-Pitt fell into the trap and rejected the offer out of hand. But the outcome would not be known until several weeks after the shutdown began.

The strike started on Sunday, July 21, 1985. The union immediately stepped up its anti-Carney campaign. Richard Fontana, a union speech writer, had already written several letters to W-P members, stressing the theme that, "We're going to save this company with or without Dennis Carney." During the strike, he wrote weekly letters, hammering at the idea that "Carney wants to destroy the union" and also giving facts about the strike. "The members got a great deal of information but also a great deal of rhetoric," Fontana candidly admitted.

This was a risky strategy. While Carney had the reputation of being an autocratic manager, many employees regarded him as a hero for holding W-P together for as long as he had. In 1984 some five hundred employees and local businessmen had honored him at a testimonial dinner in Monessen, hailing him in speech after speech.

"Dennis Carney," Rusen said, "came to believe that to the employees of Wheeling-Pittsburgh he was some kind of messiah. He did save the company. He did tremendous things. But he came to dictate to the people what was good for them. He thought everybody would follow him. He was on an ego trip. In our last meeting prior to the strike, Carney said, 'We'll find out whether these people will follow your or me.' I said, 'I'm a hillbilly from West Virginia, but I'm going to teach you a lesson.'"

In an interview with the *Pittsburgh Post-Gazette*, Carney

denied that his personal qualities caused the strike. "I don't question that I'm aggressive, but that's not a fault that hurts most companies," he said. "I've never seen an aggressive chairman hurt his company by being aggressive." Scalise thought that Carney was "probably the most brilliant manager in the steel industry. Yes, he was tough, but those were tough times."

The union's propaganda campaign was highly effective. During the course of the strike, an independent polling organization conducted three surveys of Wheel-Pitt employees under commission from the USW and found support for the strike rising to 95 percent, an abnormally high support for a walkout. About 80 percent of the members blamed Carney or "management" for the strike.

Meanwhile, in the third week of the strike an Ohio state agency granted UC to employees at W-P's Steubenville plant on grounds that they had been locked out. Pennsylvania, as English suspected, also ruled in favor of the union, authorizing payments of nearly $200 a week to workers walking the picket lines in Monessen and Allenport. However, West Virginia's Employment Security Board of Review denied UC for some eight hundred Wheel-Pitt workers in that state.

At this point Rusen demonstrated that unions can still wield enormous political power where they have numerous members. He went to Charleston and had an hour-long meeting with Republican Governor Arch Moore. Rusen had taken a neutral position in the previous gubernatorial election, ensuring Moore's chances of obtaining a healthy vote from the thousands of USW members and their families in the state. In the meeting, the USW leader complained about the UC decision, noting that one state supreme court ruling supported the union's position. "I never said what would happen if he didn't change the UC decision," Rusen later said. "I just said we needed a change and needed it badly."

Moore didn't have to be told what would happen. On September 30, he fired two persons on the three-member board, appointed new members, and ordered that the benefits be paid. The board complied. Moore made certain that he received political credit for his action. When the first UC checks were ready for distribution a few weeks later, he appeared at the Garibaldi Club in Follansbee and personally handed out some four hundred checks as television cameras recorded the scene.

On another front, the campaign to unseat Carney moved

ahead. One of Carney's strongest supporters on the W-P board was Robert E. Seymour, retired chairman of Pittsburgh Brewing Company, which produced a popular light beer, IC Light. The union couldn't legally call for a secondary boycott, but Rusen easily overcame this barrier. He was quoted in newspapers and on television as saying that he was an IC Light drinker but "because of Seymour's outspoken pronouncements in favor of Carney, I wasn't going to drink it any more." IC Light sales plummeted in the region. Seymour resigned from the Wheel-Pitt board in September.

Rusen and Smith also met several times with Allen Paulson at his headquarters in Savannah and with an official of Nisshin Steel in New York. Paulson's and Nisshin's large investments in W-P were deteriorating as the price of its stock went down. "We told them that as long as Carney was chief executive officer, the people would see the company liquidated before they went back to work," Rusen said. To back up their statements, the USW men showed the poll results to Paulson and the Nisshin official.

It was all over for Carney. Paulson asked him to resign. Several board members still supported Carney, but he faced a personal decision. He realized that the USW would not settle as long as he remained in charge. Eventually, the company would run out of cash and be liquidated and his share of stock would be worth little because of the bank lien. On September 20, Carney and five other board members resigned after Carney negotiated a $1.5 million severance payment for himself and $400,000 each for Scalise and W-P's general counsel. Paulson took over temporarily as chairman and brought in George A. Ferris, a former Ford Motor executive, as chief executive officer.[6]

A Unique Settlement at Wheeling-Pittsburgh

Negotiations for a new contract resumed and the two sides reached agreement on October 15, 1985. As they had intended for months, Rusen and Palm settled for an hourly employment cost of $18, but W-P workers continued to receive $20.33 worth of wages and benefits. This curious outcome resulted from the insistence by Wheel-Pitt's lenders that the company terminate its pension plans, underfunded by $470 million, to reduce costs. This action dumped the responsibility of meeting all past pension obligations for 13,000 retirees and 8,500 workers on the Pension Benefit Guaranty Corporation (PBGC), a federal agency that insures retirement plans. The PBGC assumed pension costs estimated at $2.33 per hour. Thus, employees only had to give up a little more than $1 per hour in wages and benefits to reach the $18 employment-cost level.

The W-P labor agreement was unique in other ways. The union agreed to eliminate six hundred jobs in an effort to boost productivity by 20 percent. Reaching back to an old concept used extensively by Andrew Carnegie, the company agreed to tie wages to prices by paying a "price escalation bonus" of up to $1 per hour, depending on whether, and how much, steel prices rose during the agreement. To ensure that the USW would have a major role in reorganizing the company, W-P agreed to set a contract expiration date of ten days after the bankruptcy court approved a reorganization plan. The USW later gained membership on an unsecured creditors' committee.

The USW won a significant voice in managing the company. A Cooperative Partnership Agreement gave the union the right to nominate a candidate for the board of directors. It also set up an eight-person Joint Strategic Decision Board, with equal numbers of union and company representatives, to discuss capital investment, new technology, the use of facilities, and other business issues. Similar boards were established at each plant both at the department and plantwide level.

In some respects, the Wheeling-Pittsburgh strike/lockout was a resounding victory for the USW and the employees. While it almost caused liquidation, it may have *saved* the company from liquidation at the hands of the banks. And the banks may have hurt themselves. If they had not insisted on the lien on assets, W-P might have escaped bankruptcy and eventually paid the banks in full. In bankruptcy, however, the lenders' unsecured loans are paid off at the same discounted rate as other creditors'. "The banks got greedy," concluded Josh Gotbaum of Lazard Frères. "They were convinced that the union would cave in." The Wheel-Pitt settlement also indicated to other steel firms that the union would demand "a piece of the action" in return for further concessions. By solidifying the union and demonstrating Williams's willingness to take risks, the stoppage served an important institutional purpose.

The decision to strike a bankrupt company, however, drew sharp criticism from the business community generally. The union's action reduced the value of the company, thereby hurting shareholders, creditors, and employees. It also severely punished one of the few chief executives in the steel industry who had tried to undergird the long-term viability of the company and its well-paying jobs by investing in modern equipment. For a union to insist on "excessive" wage increases or resist work-rule changes is taken for granted in management circles, though it is difficult to deal with. But the campaign to oust Carney was seen by many executives as a violation of an unwritten code of behavior between labor and management—as a personal betrayal and a willingness to sacrifice the entire company and its employees to protect the union as an institution.

Two years after the Wheeling-Pittsburgh strike, I pressed Paul Rusen on this issue. Rather emotionally, he rejected the argument. Far from having no interest in the company, Rusen noted, he had many relatives on the payroll and lived in a suburb of Wheeling that depended on W-P's survival. "I intend spending the rest of my life in this community," he said. "Ten of my neighbors work for Wheeling-Pittsburgh. My name and face are well known everywhere I go. I wouldn't give this company up to serve some other purpose. But we couldn't let Dennis Carney break the back of this union."

There was a personal sequel to the W-P episode for Rusen. He had wanted to run for a top union office on the Williams ticket in November 1985. Disappointed when Williams picked other candidates, Rusen decided to take early retirement at the age of fifty and become a consultant. Typically, Williams found a political solution for this political problem. Deciding that somebody other than an elected USW official or company employee should represent the USW on Wheel-Pitt's board to avoid possible conflicts of interest, he nominated Rusen. The latter was elected to the W-P board in early 1986. Later, he became a sort of professional board representative for the USW, serving also in that capacity at Kaiser Aluminum and Bliss-Salem. In addition, the International Union of Electronic Workers put him on the board of Demco-Gray, a small, employee-owned plastics manufacturer in Ohio.

New Approaches I: National Steel

Out of conflict at Wheeling-Pittsburgh came the beginnings of a new labor-management alliance and fundamental changes in the industrial relations system. Reforms of a similar nature, though they sprang from different causes, began to transform the labor picture at National Steel and Weirton Steel in 1984–87.

The story begins at National, the parent company, which had been formed in 1929 by the merger of three companies, including the original Weirton Steel headed by Ernest T. Weir, one of the most antiunion of the old generation of steel bosses. Although National had diversified to some extent, steel accounted for 85 percent of its assets in 1980. That year, steel markets for most products except pipe deteriorated markedly, and National's profits declined by 33 percent. Howard M. (Pete) Love, who became chairman and chief executive officer in 1981, eliminated steelmaking capacity, accelerated the diversification program, and formed a holding company, National Intergroup Incorporated (NII), to oversee its autonomous businesses. Steel operations were placed in a subsidiary of NII under the name National Steel Corporation. Love became chairman of NII.

Like the rest of the steel industry, National Steel had high labor costs and low productivity growth. Relations with the USW were very poor at two of National's plants, at Granite City, Illinois, and at the Great Lakes Division plant at Ecorse, just outside Detroit. Union and management officials got along better at the Midwest Division plant in Portage, Indiana. Employees at a fourth plant, Weirton, were represented by another union, the Independent Steelworkers Union.

Unlike many chief executives, Love attributed many of the productivity and labor problems to management failures, not to union behavior. His years as a foreman and superintendent in steel plants had convinced him that the system had to be changed, though he lacked a conceptual framework for organizing a new system.

In 1981, the company asked a consultant, Lee M. Ozley, president of Responsive Organizations of Arlington, Virginia, to find out why a quality circle program at the Weirton plant wasn't working. After looking into it, Ozley met with Love and told the chairman that the quality circles would never produce much of value unless the company changed its way of managing people, from dictatorial to participative. Such a change could work only if top executives were committed to it. Love listened carefully. "All of a sudden," Ozley recalled, "he kind of brusquely stopped the conversation and said, 'The problem has nothing to do with the hourly people. The problem starts here with me and the managers under me.' I didn't tell him anything he didn't know. I just gave him a framework to organize his thoughts."[7]

From that time on, Love converted what he called an "old-line hierarchical organization into a more participative company from the executive suite to the shop floor." This was a long, slow process of changing the company "culture," a term that gained popular business usage in the 1980s. When used correctly, it referred to the way managers and workers interacted, as well as the beliefs and assumptions that guided their behavior.

For the first two and a half years, Love concentrated the change process at the executive level among thirteen to fifteen managers, mainly in the firm's Pittsburgh headquarters. As in most corporations, people managing the various functions and divisions at National tended to form cliques, deal in power politics, and fight for Love's favor. Unless these executives learned to be participative and open with one another, the company wouldn't be able to persuade workers on the shop floor to change their behavior. Love himself had to prove that he would communicate with his subordinates and listen to their counsel. Sometimes his actions belied his supposed commitment to participation.

For instance, Love kept secret his talks with U.S. Steel on selling National Intergroup's steel segment to the corporation. When the NII board announced its approval in early 1984, morale plunged throughout the company. Employees from top to bottom felt betrayed by the surprise announcement, Love later conceded, because they didn't want to be a part of USS. By then, U.S. Steel was widely perceived to have "the worst labor relations in the business," charged Buddy Davis, the USW's chief negotiator at National. At National's Midwest plant, Local 6103 President Bob Paster said: "We have good relations here, and we're worried it'll go down the drain under U.S. Steel."

The Justice Department's disapproval aborted the merger plan. A few months later, NII sold 50 percent of the steel segment to Nippon Kokan, the Japanese company. It is a curious comment on the times that the Japanese deal "was more favorably received" by National's employees than the U.S. Steel proposal, Love said. Nippon Kokan was highly regarded for its expertise in steelmaking technology. Love didn't mention it, but some National employees were relieved to have escaped U.S. Steel's heavy-handed management.

Stan Ellspermann had been named in 1981 to succeed National Steel's veteran labor-relations vice-president, George Angevine, who retired. A blunt, sometimes tart-tongued man in his early forties, Ellspermann had spent twelve years in the auto industry before joining National in 1975. He rose to chief of industrial relations at National's Great Lakes plant. As vice-president, he became involved in the top-level change process and helped develop a company "mission statement," a crucial element in the process.

No longer would the company's objective be, as many employees assumed, "to make and sell steel" or "to maintain market share." It became, instead, "to provide a superior and consistent return to our stockholders." If any business segment, including steel, failed to contribute to that goal, it would be reduced or divested. One of several operating principles accompanying the mission statement said that "every employee has the opportunity to participate to the extent possible in the decision-making process."

Ellspermann put the statement in more colorful terms when he translated it to USW officials and workers. "We don't want to be a bunch of pot-bellied, cigar-smoking dictators," he said. "We really want to step into the twentieth century and figure out a better way of doing business. We were convinced that with our old attitudes we were not going to be able to achieve the kinds of productivity gains that would save us. If we quit treating people like cattle and started treating them like human beings, maybe we had a chance. The problem was that, after twenty-five years of us acting like Attila the Hun, the people weren't prepared to accept the fact that we'd suddenly got religion."

One skeptic was Buddy Davis. A lean, hard-looking man with a rasping voice, Davis had been named the USW's chief negotiator at National because the Granite City plant was located in his St. Louis territory. "We didn't see any change at first down in the plants from this new approach," he recalled in a 1987 interview. "People were still angry about the 1983 concessions, and we were used to being at war with the company."

Ellspermann approached Davis in 1984 and suggested that they try to improve relations. Davis tentatively agreed. By no means an "early blooming disciple" of cooperation, Davis later credited Lynn Williams, a friend as well as union associate, with persuading him to try the new way.

Before moving much further on this front, however, Love

decided to take a step that Ellspermann had been urging for a year. In May 1984, National Steel withdrew from the Coordinating Committee Steel Companies, the industry bargaining group. Although the next round of negotiations was still two years away, Ellspermann wanted as much time as possible to prepare for bargaining one-on-one with the USW. By leaving the CCSC, Love and Ellspermann hoped to convince the union that they were serious about negotiating in a different spirit, and with drastically different results. Under CCSC rules, the firms involved were supposed to decide jointly on a bargaining strategy. Had National remained in the group, Ellspermann could not have begun informal discussions with Buddy Davis about innovative provisions that would differentiate National from the other companies. "If we'd stayed in," Ellspermann told me, "we wouldn't have had the same sense of urgency, the same candor. We wouldn't have had the same ability to say National Steel is committed to employment security, to the viability of the union. We wanted to demonstrate to our employees that we were different, and therefore we wanted to act different, and by acting differently, we expected to be treated differently."

The USW at first disapproved of National's withdrawal, because it threatened to undermine the industrywide contract. However, Davis agreed to continue exploring a more cooperative approach with Ellspermann. A turning point came at an annual meeting of managers and local union presidents in December 1984—at roughly the same time that U.S. Steel's local presidents were castigating Tom Graham and Bruce Johnston at their Florida meeting. Lee Ozley and another well-known participation consultant, E. Douglas White of the American Productivity Center in Dallas, helped National's union and management people learn how to work together. When the two sides separately drew up priority lists of concerns that should be addressed, Davis recalled, "we were amazed to discover that we had a lot in common." This is a standard procedure used by consultants to demonstrate that adversaries have mutual interests.

Deciding to move ahead with a cooperative program, the two groups formed joint committees at the company's three steel plants (Weirton had been sold by then) and a companywide committee headed by Davis and Ellspermann. Discussions continued through 1985 at both levels. These were not formal negotiations but problem-solving sessions dealing with issues such as safety, contracting out, job security, and productivity. The company gave Davis information about finances, markets, and capital expenditures in much greater depth than ever before.

Ellspermann began to lay out National's game plan. To remain competitive, the company needed to reduce man-hours of work by a substantial amount, cutting its twelve thousand five hundred salaried and hourly workers by about three thousand over a five-year period. If the USW helped to meet this goal by negotiating smaller crew sizes, Ellspermann said, National would be willing to guarantee that nobody would be laid off during the term of the next contract.

National actually had put an informal no-layoff plan into place at the Midwest plant, where managers established a good relationship with Local 6103. In mid-1984, Ellspermann offered to sign a formal no-layoff agreement at the plant in return for two things. The company wanted the union's help in making the plant more productive by combining jobs and ending restrictive practices. In addition, National proposed diverting its SUB fund contribution of 50¢ per hour to a new fund which would help pay for the cost of keeping surplus employees on the job. With a no-layoff guarantee, jobless benefits wouldn't be needed.

But the experiment at Midwest didn't get beyond the discussion stage—for a strange reason. In meetings with employees, Ellspermann and his aides discovered that many "didn't want to give up their voluntary right to be laid off," Ellspermann told me in a shocked voice. "Isn't that unbelievable?" Apparently, the fact that the company *hadn't* laid people off for several months and had invested in new equipment convinced workers that the company was in good financial shape. There was no need for change. "We'd sent them the wrong signals," Ellspermann said.

Despite this experience, National and the USW continued moving toward a more cooperative relationship. In mid-1985, Ellspermann convinced Davis that they should engage in early negotiations to replace the companywide contract that would expire the next summer. Consultants Ozley and White met often with Ellspermann, Davis and their aides, working on problem-solving skills so the bargainers could avoid the traditional posturing that leads up to conventional contract talks. "I didn't think we could really bargain that way," Davis said. "But I was wrong, because we found ourselves arriving at decisions by consensus even when we negotiated."

But Davis couldn't begin formal talks without Lynn Williams's approval, and the USW president first had to develop a strategy for the entire industry. National was put on hold until then.

By early 1986, National Steel had started down a labor-relations road far different from the one followed by U.S. Steel. But the big test was yet to come. When I interviewed Ellspermann in February 1986, just before he entered formal negotiations with the union, he posed a very explicit difference in the two approaches and raised a question so crucial that I shall quote him at length.

"Our experiment is clearly 180 degrees from the way U.S. Steel is approaching their labor-management relations," Ellspermann said. "U.S. Steel may be right and I may be wrong. We'll find out. There are two very big issues riding on this contract negotiation. One is the future of labor-management relations for the next twenty-five years. If the Steelworkers, despite all their rhetoric, reward the guy they publicly view as the bad guy with a good contract and stick it in the ass of the people who have been willing to be open, candid, and participative, that's going to send a signal that will take a generation to overcome.

"The second issue is an economic one. If the company doesn't get what it needs in terms of short-term economic help and the long-term productivity commitment that is necessary for survival, we'll quit investing. The union can't just talk about all their wonderful values and when push comes to shove, throw it all away. I'm willing to guarantee that during the term of the new contract, we'll never lay anybody off. I don't know any other steel company that is prepared to say that. I'm going to find out how important employment security is to the union."

New Approaches II: Weirton Steel

When National Steel began to scale down its steel operations in 1980, the flagship plant at Weirton became a prime candidate for sale or abandonment. Its principal product, tinplate, competed in a declining market for tin cans. The company didn't have enough cash to upgrade the facilities. It couldn't simply shut down the plant because pension liabilities exceeded the value of the assets.

To keep the plant out of the clutches of the USW, National over many years had granted high wages and benefits to the Independent Steelworkers Union (ISU). By 1982, hourly compensation totaled about $26, or $3 to $4 more than the industry average. If National shut the plant, the cost of severance pay, pensions, and health care continuation would have totaled about $450 million, according to one estimate. That was about $56,000 per employee. The liabilities exceeded the value of the plant by $295 million.[8]

In March 1982, Love offered to sell the plant to the employees. If such a plan couldn't be arranged, he said, the company likely would shut down the hot end of the mill and operate finishing facilities to the end of their normal life. A study by McKinsey & Company indicated that the plant could be a viable operation with a 32 percent lower labor cost and a change in product mix. National agreed to absorb 12 percent of the cut in costs by retaining the obligation for all pension costs as of May 1, 1983. In September 1983, the union employees voted by a nine to one margin to take a 20 percent pay cut, form an ESOP, and take over the plant.

To most institutional lenders, the ESOP was a little-known creature in 1983. ESOPs existed at thousands of small companies by then, but in most cases management owned a majority of the stock and controlled the firm. Some elements of the business community had talked disparagingly about worker ownership, predicting that where employees were the sole owners—as at Weirton—militant workers would take over the executive offices and run the company into the ground. A case in point, the disparagers said, was Rath Packing Company, which had turned to employee ownership in the late 1970s but had failed. The reasons for Rath's failure had less to do with who owned the company than with the inefficiencies inherent in its ancient, multistoried plant and mistakes made by previous managements.

Because of this bias against worker ownership, the Weirton Steel plan may have been required to pass somewhat more stringent tests of fiscal soundness than a conventional company seeking loans. The structural features of Weirton's ESOP were designed to demonstrate to lenders that this was no fly-by-night proposition. The workers not only accepted a 20 percent pay cut but also agreed to freeze wages at that level for six years. Although stock would begin accruing in each employee's ESOP account from the outset, the workers would not be able to vote their shares and elect a board of directors before 1989. Until then, the board would be controlled by eight outside directors who were approved by the lenders.

These features enabled the ESOP to borrow $161 million at the start. On January 11, 1984, Weirton Steel Company, with about eight thousand employees, began operating as the largest American company wholly owned by workers. The steelmaker quickly put to rest many of the old concerns about worker ownership. But one of the first problems facing Chairman and President Robert Loughhead was a poor labor-management atmosphere left over from the decades of operation under the old National Steel style of management. Irving Bluestone, the retired UAW vice-president who was one of three union representatives on the Weirton board, said the labor relations were "god-awful" when he arrived at the plant. "There was a very deep bitterness on the shop floor."

It was the kind of problem suited to Loughhead's experience. He had resigned the presidency of Copperweld Steel to join Weirton primarily because "a large employee-owned company was the perfect environment to prove that employee participation could work." The company retained John Kirkwood, who had left Jones & Laughlin to become a consultant, to design a worker participation program. It encountered the usual start-up problems because workers were still wary of managers. But after a year or so, the process began to spread rapidly.

In June 1987, when I visited Weirton, the company had an extensive participation program in place with full-time coordinators and facilitators. One hundred seventeen Employee Participation Groups (EPGs), similar to LMPTs, met weekly in the plant to solve problems. Some twenty-two hundred employees had voluntarily attended three-day training seminars to qualify for EPGs, and 20 percent of the employees were actively involved in some participation activity. A total of $10.6 million in cost improvements resulted from the EPG process in 1986.

Loughhead also realized that communicating with employees can be one of the best tools for gaining commitment. He himself visited a different section of the plant each week, often accompanied by Walter Bish, the ISU president. He would review developments in Weirton's business and answer questions. The company set up more than one hundred television sets throughout the plant for broadcasting fifteen-minute, weekly programs bringing workers up to date on the latest sales figures, orders, customers' comments. Employees could catch up on daily developments by dialing into a hot line.

Before the end of Weirton Steel's first year, customers began complimenting the company for meeting delivery schedules and providing good quality products. By late 1984, Weirton was earning $41 per ton, higher than any of the six largest steel companies. The company increased sales and shipments in its first three years, although profits dipped in 1986 to $45.1 million, down from $61 million in the previous two years because of price erosion. Workers began receiving a tangible benefit of this progress in March 1986 when the first profit-sharing payments, averaging between $1,500 and $2,000 for hourly workers, were made. The bonus decreased slightly in 1987 but was expected to rise significantly in 1988.

In short, Weirton was not merely a limping survivor but a strong and viable company in 1988. It still had many hurdles to cross. A modernization plan called for generating $90 million by 1991 to invest in new equipment. ISU officers and top managers, meeting as members of a Long-Range Planning Committee, had agreed to do this by eliminating thirteen hundred jobs. The two sides could reach this agreement only because they had learned how to work together.

Perhaps the biggest unanswered question for Weirton is what will happen in 1989 when the work force will be able to elect the board. Will it toss out the outsiders and managers and elect an all-worker board? That wouldn't necessarily be bad, although corporate boards work best with a mixture of talents and skills. Workers already have a significant voice in decisions, both individually on the shop floor and through union representatives at higher levels. In addition to Bluestone, the ISU's board representatives are the local president, Walter Bish, and a Weirton lawyer, David L. Robertson. According to company insiders, the three union representatives have never voted as an interest bloc against the rest of the board.

Weirton's experience should dash the old bugaboo about the workers taking over and ruining the business. In a sense, the company is governed by a coalition of worker representatives, managers, and outside businessmen. The basis of this coalition, according to R. Alan Prosswimmer, the firm's vice-president and chief financial officer, is "a clearcut recognition that management is hired to manage the company. As long as we communicate effectively, our obligation is not to let the workers vote on each decision or issue. We have to be willing to explain why we did like we did. But we can't get into the kind of contest where we must defend every management decision."

One of the things that has made this coalition viable is that the professional managers are highly committed to the participatory process. Loughhead, the leader in this effort, retired in 1987 and was succeeded by Herbert Elish, a board member and former senior vice-president of Dreyfus Corporation. He promised to continue the participation process.

Prosswimmer, the second ranking executive and holder of an MBA from Harvard Business School, resigned from ITT in New York because he wanted to participate in the Weirton experiment. "I feel strongly," he said, "that this is an opportunity to try to create a new structure for American industry and revitalize it. I wanted to be part of it. I could probably get 50 percent more in salary somewhere else. I'm working harder than I worked in the ITT organization, and I love it."

The success of Weirton undoubtedly had an impact on the thinking of the people who ran the big integrated steel companies. "The idea is galling to them that Weirton is profitable and they are not," said a financial consultant with deep knowledge of the steel industry. "There is a view among chief executives that Weirton is a reproach to them."

Weirton also presented a competitive threat to the integrated producers. At the beginning of its life as a worker-owned firm it produced good quality steel at a labor cost of $19.70 an hour, about $3 to $4 under the industry average. In 1987 Weirton's profitability could not be attributed solely to its labor rates, Prosswimmer said. Hourly labor costs had risen, though Weirton still had a $3 advantage over some competitors. But efficiency and quality now contributed to the firm's success.

By May 1985 the large steelmakers also had come to regard Weirton's labor costs as a competitive factor. They noted, too, that Wheeling-Pittsburgh had lower than average labor costs and was bidding to reduce them even more in bankruptcy. These two situations added to the companies' increasing disillusionment with their method of negotiating contracts. It was time for a change.

The End of Coordinated Bargaining

Industrywide bargaining came to an end on May 2, 1985. In a short statement, Bruce Johnston announced that the chief executive officers of the companies remaining in the CCSC—U.S. Steel, LTV, Bethlehem, Armco, and Inland—voted unanimously to dissolve the group. The decision, Johnston said, reflected a number of factors that had brought unprecedented change to the steel industry: "abandonment of pattern bargaining" by the USW, government-subsidized foreign competition, joint ventures between American and foreign steel companies, rising use of semifinished imported steel, and sustained financial losses by member companies.

The commonality of interests among the members of the group had diminished considerably. U.S. Steel, LTV, and Armco were conglomerates. Inland produced only steel and Bethlehem very little else. The earlier departures of National Steel and Allegheny-Ludlum from the group also had a bearing on the decision. Armco competed with Allegheny-Ludlum in stainless steel products and through plant shutdowns had reduced the USW-represented portion of the hourly work force to 50 percent. National, primarily a producer of flat-rolled steel for the auto industry, presented a particular problem for Inland, which specialized in the same product and operated in the same region.

U.S. Steel, the bargaining leader, presented problems for all four of its partners because it had split further from the pack than any firm. With its large revenues from Marathon Oil, USS could withstand a long strike that would sink the other four, especially LTV and Bethlehem. The corporation's growing reputation as the "bad boy" of steel labor relations bothered some of the companies. Should they keep their fortunes linked to a company whose local union presidents had talked openly of striking because of contract violations?

Indeed, Donald Trautlein of Bethlehem went to the CEO meeting with a mandate from his board of directors to pull out of coordinated bargaining, partly because of the strike possibility. The board had accepted the recommendation of Tony St. John, Bethlehem's top labor executive in 1985, to get out of the CCSC. St. John gave two reasons: (1) the diverging interests of the companies, and (2) the highly visible tensions between U.S. Steel and the Steelworkers. Bethlehem didn't want to follow the leader blindly into a strike.

James H. Wallace, Armco's vice-president of industrial relations, and Cole Tremain, LTV's chief negotiator in 1985, also recommended leaving the CCSC. The diversity of interests loomed high in their thinking. Tremain had proposed in late 1984 a method of giving his company, Armco, and Inland a larger say in the conduct and outcome of industrywide bargaining. "It was summarily rejected by U.S. Steel," he said. "U.S. Steel didn't want to collectively bargain but to decide themselves what the determination of collectively bargained issues would be."

One reporter with inside contacts in the industry, Gloria LaRue of *American Metal Market,* also cited a dislike of U.S. Steel's policies for the breakup of the committee. "Some officials have gone so far as to say they would rather be dead than go to the bargaining table with U.S. Steel," she wrote after the breakup decision.

Johnston often downplayed criticisms by other companies as small talk by firms that didn't have the responsibility of negotiating the contract. "We could do it either way, with them or without them," he said in 1986. It wasn't as though U.S. Steel desperately tried to hold the group together.

But Johnston was the only vice-president who attended the CEO meeting. Neither Tremain nor St. John could say who did the breaking up, though they welcomed it. The unanimous decision indicates that all companies realized the time had come to put an end to an oligarchic practice in the absence of oligarchic market conditions.

Unhappy about the prospect of negotiating individually with the companies, the USW tried for a while to make the case that the union had nothing to do with the cessation of industrywide bargaining. Williams noted the breakup of the CCSC in a speech in December 1985 and said the companies apparently believed "that level employment costs, guaranteed in the past by multi-employer bargaining among the major producers, is no longer a desirable thing."

Johnston, however, scoffed at this statement and pointed out that U.S. Steel didn't initiate the lower-than-pattern settlements negotiated by the USW starting in 1982. "We had witnessed the abandonment of pattern bargaining by the Steelworkers," Johnston testified at an unemployment compensation hearing; "it was obvious to us that companies that were not inside coordinated bargaining were getting a more flexible and faster response [by] the union to their labor cost problems then coordinating committee companies were."

In 1956, the major steel producers and all other companies

that more or less adopted the industrywide labor pattern produced more than 95 percent of the steel sold in the United States. By 1985, with the growth of minimills and imports, the figure had declined to 60 percent. The USW had lost its commanding position—its monopoly, so to speak—of the steel market. Not only could the union not make the pattern stick across most of the industry. The USW itself had accepted pay concessions of varying levels at many non-CCSC companies.

The USW even had granted extra concessions to companies within the bargaining group. To prevent the shutdown of Bethlehem's Bar, Rod & Wire Division operations at Johnstown, the USW in April 1985 granted a cut of $4.91, or 20 percent, in wages and benefits for twenty-one hundred workers. In return, the employees received a profit-sharing plan and creation of an ESOP in which the company would place $3.91 in preferred stock for each hour worked by each employee in the first year. This was one of the first times the USW agreed to tie wages to the profitability of a single product line. Workers in Bethlehem's more profitable sheet, plate, and structural mills would continue to work at a higher labor rate.

U.S. Steel had reasons for wanting to continue coordinated bargaining. As the chief negotiator, it could bargain and enforce the labor pattern most advantageous to itself, possibly helping weaker companies drop by the wayside. It could also prevent the Steelworkers from picking off the other companies one by one and forcing acceptance of a pattern disadvantageous to U.S. Steel—on contracting out, for example. It would have been most unlike U.S. Steel, however, to beg the other companies to stay united. Roderick, Graham, and Johnston had reason to be confident in the corporation's strength as a stand-alone bargainer.[9]

Chapter 18

Winding Down at McKeesport and Aliquippa

It takes a curious empathy for smoke, fire, dirt, roaring machines, and the people who tend them to become fond of a steel mill. I confess to feeling that way about two particular steel plants. One is the old USS National Works in my home town, a place where I, tangentially, helped build a boiler house and, literally, spent one summer doing mostly nothing. The other plant is LTV Steel's Aliquippa Works, which I had a unique opportunity to observe from the inside. Important things happened in both of these plants in 1985.

In McKeesport, the old guard in Local 1408 began to pass from the scene as the plant dwindled to a few hundred employees. Manny Stoupis, the grievance committee chairman, retired after thirty-eight years, more than twenty of them spent as a local officer. He was still only fifty-five years old. In 1982 he had finally worked up to the top job as roller on the blooming mill. He liked the roller's job because it required considerable judgment and skill and paid well, as much as $160 a day including incentive pay. A week after he took the job, the company closed the blooming mill. "I almost cry when I think about it," he said. "They had electrified the mill, replacing the old steam engines with a really sophisticated motor room, all solid state materials. I told management they didn't know what they were losing, the mill had a great potential. But they wanted the writeoff."

Stoupis continued working in a labor pool. He held the respect of many people on both sides of the fence because he was a reasonable man who didn't press frivolous grievances and dealt fairly with bosses. But Manny disliked making concessions. When management proposed eliminating many jobs by cutting crews and merging seniority units, he refused to go

477

along—and his approval as grievance chairman was required under the by-laws. Manny felt the changes would violate Lynn Williams's 1983 guidelines calling for adherence to the master contract. He and Dick Grace, the Local 1408 president, split over this issue, and the local rejected the proposed changes.

In 1985, Stoupis took early retirement when his section of the plant was permanently closed. Grace also retired after losing his bid for reelection. Plant management approached the new officers with the crew-cutting plan, declaring the plant would be shut down if it were rejected. The proposal called for radical changes, including a reclassification of all employees into three general groups. Production workers would perform maintenance work, and skilled craftsmen would operate machinery. The plan abrogated sections of the master agreement, but the International did not try to block it. A few hundred jobs might be saved.

Stoupis's successor as chairman had been the loudest anti-concession man before. Now, he and two other grievancemen offered to go along with the proposal if the company gave them mutual pensions. Other workers in the affected units had higher seniority, but the grievance representatives stepped in front of their brothers. "We had to buy them off," said Albert Voss, the general superintendent at the time.

What the grievancemen and company did was not illegal, but it violated union principles. Although probably a large majority of union officials refused to compromise themselves, similar deals were made elsewhere.

Management, meanwhile, made a decision that deprived hundreds of National workers of severance pay. The plant had laid off more than two thousand workers in March 1982. Most of the production lines and shops where they had worked never operated again. But the company waited until April 1, 1984, before declaring the units permanently closed and then refused to make severance payments to eligible workers because they had been laid off more than two years. Before retiring, Stoupis took the issue to arbitration, arguing that employees retained recall rights for five years under the USW contract and therefore remained eligible for severance pay during that period.

After an arbitrator ruled in the union's favor, U.S. Steel took the position that it would pay the severance allowance only to workers who had signed a grievance—a few dozen out of the hundreds of eligible employees. The company contended that everybody should have known about the permanent shut-down because it had made a public announcement. The company was not moved by Stoupis's argument that many had moved out of the region to look for work. Nor did U.S. Steel agree that it was obligated to send shutdown notices to all affected employees. It refused to give last-known addresses and phone numbers to Stoupis. In the end, he estimated, about two hundred and fifty employees lost allowances that would have ranged from four to eight weeks of pay per year of service, averaging about $3,500.[1]

What happened at the National Works in its last few years of life could not serve as an ethical model for phasing out a plant. National was permanently closed in 1987 (see chapter 20).

The Aliquippa story provoked a personal reaction in a different way. In addition to normal reportorial excursions to Aliquippa over the years, I had spent a week in the plant as an observer in 1971. It was an unusual sort of thing. Most steel companies wouldn't let a reporter roam free in a steel mill. in those years, however, Jones & Laughlin displayed an openness to the press previously unknown in the steel industry. I spent eight hours each day in the plant, moving from department to department, talking freely to workers and supervisors, trying to understand what the steelworkers of that time experienced in the mills. I came to know many people there, and later I saw an exciting growth of the participation concept at Aliquippa.

In 1971 the Alquippa plant seemed impregnable. Some ten thousand people worked there. Retirees were still so attached to the company that every day one could see half a dozen or more sitting on a bench outside the plant, old fellows with arthritic hands and wrinkled faces for whom the mill was the center of life. It should not have been, of course, because the union had tried to give workers the wherewithall to do other things—travel, set up a small business, move to Florida as so many steelworkers did. But this older generation, men who had started their working lives in the 1920s or before, couldn't pull themselves away from the mill.

Nearly ten thousand people were still employed at Aliquippa in 1981. But less than twenty five hundred remained as of May 17, 1985. On that day, LTV Steel announced it would shut down most of the plant, including iron and steel furnaces and coke ovens, and put thirteen hundred workers on indefinite layoff. Only a tinplate operation and a light structural mill with nine hundred to one thousand employees continued working. In

Pittsburgh, the historic Pittsburgh Works of old J&L had also shut down, except for a coking operation.

This closing of yet another chapter in the history of steel-making in the Pittsburgh region was by no means the end of LTV Steel's financial troubles. Ever since the Dallas-based LTV Corporation acquired Republic Steel in June 1984 and merged it with Jones & Laughlin, the steel subsidiary had been going downhill. The idea of the merger had been to combine the resources of two sick companies to make one healthy producer. At the beginning, the two firms together produced for 15.7 percent of the American steel market, making LTV Steel almost as large as U.S. Steel. But merging the two staffs and integrating the well-worn procedures of two old companies proved more difficult than anticipated. The company moved four hundred fifty executives from J&L's Pittsburgh office to the new subsidiary's headquarters in Cleveland at a cost of $50,000 each. Although LTV instituted a cost-cutting program by, among other things, eliminating more than two thousand salaried jobs, it had to contend with other problems.

The combined companies had a large excess capacity in hot-rolled bars, for which the market was weak. A very high ratio of retirees to active workers boosted its hourly labor costs to the highest in the industry. In the first quarter of 1985, LTV Steel lost $156.4 million, or $44 for every ton shipped. As steel prices continued to decline, LTV fell deeper into the hole.

Other steelmakers also suffered in 1985 from anemic steel orders, low prices, and a high level of imports. The only source of relief in sight was the union. Existing USW contracts called for further restorations of the 1983 wage cuts, and therefore rising labor costs, until they would expire at the end of July 1986. If the steel market did not improve—and few experts thought it would—some of the domestic producers might not last that long. Freed of their group bargaining obligation, a number of firms approached the union for early relief. LTV lined up at the head of the queue.[2]

Union-Management Efforts to Save LTV

In the summer of 1985, LTV Steel's president, David H. Hoag, began talking seriously to Lynn Williams about renegotiating the LTV labor agreement. Cole Tremain, labor relations vice-president of LTV Steel, arranged meetings at least quarterly between Hoag and Williams. The two got along well. The company opened its books to the union's outside consultants, and in September, Gene Keilin, Josh Gotbaum, and Ronald Bloom of Lazard Frères began to study LTV's financial situation. Insolvency accountants from Arthur Young later joined them.

As these studies went forward, Tremain met periodically in the fall with two district directors who would head the bargaining at LTV, Tony Rainaldi, the chief of District 20 in the Upper Ohio and Beaver Valleys of western Pennsylvania, and Joseph M. Coyle, the District 27 director in eastern Ohio. Tremain kept urging the USW to set a date for negotiations. To slow the company's cash drain, he said, wages and benefits must be reduced. Tremain and Hoag indicated that LTV would be willing to discuss *quid pro quos* such as profit-sharing and stock ownership.

Bethlehem, National, and Armco also had asked for early bargaining in 1985. But the Steelworkers' quadrennial election of top officers and district directors was scheduled for November 26, 1985, and it became apparent that Williams didn't want to start a concession bargaining round in the middle of an election campaign. The concession issue could have affected a race for director in district 28, Cleveland, where LTV had combined J&L and Republic mills on opposite sides of the Cuyahoga River into the giant Cleveland Works. A local president, Albert Forney, ran on a "no concessions" platform against incumbent Frank Valenta, who supported flexibility in negotiations. Valenta defeated Forney by 56 percent to 44 percent.

Williams also wanted to avoid divisive issues while bringing two newcomers into top offices vacated by Joe Odorcich and Frank McKee. They were James N. McGeehan, the head of District 7 in Philadelphia, who ran for treasurer, and George Becker, Williams's assistant, who sought a vice-presidency. As it turned out, no challengers won enough local nominations to get on the ballot for any of the five top offices. Becker was one of the few men in the history of the union who jumped into a principal office without serving as a district director. It was a measure of Williams's political control that he could manage this shift without causing an uproar.

Hoag and Tremain had presented strong arguments in favor of early bargaining, and LTV appeared to be in worse financial shape than Bethlehem, National, or Armco. Another consideration that later influenced the USW's decision to negotiate first at LTV was the union's respect for the company's ability to negotiate problem-solving solutions to tough issues.

Industrywide bargaining had provided no incentive for the smaller companies to develop strong labor relations staffs; they depended on U.S. Steel and Bethlehem to negotiate for them. Following the breakup of the industry group, the steel industry was not surfeited with bargaining expertise. USW staffers, however, regarded Tremain as one of the top professionals in the industry. During industrywide talks, he and Tony St. John had frequently been called upon to negotiate with the union on the toughest noneconomic issues. But St. John had resigned from Bethlehem in June 1985. Apart from Bruce Johnston, that left Tremain as the most experienced company negotiator in the industry. LTV, said Bernie Kleiman, "had the most creative bargaining setup in the industry. It's easier to solve problems with them than anybody else."

Employee involvement also had progressed much further at LTV than at all other companies except National Steel. At the end of 1985, about one out of every five workers at each of twelve LTV plants was actively involved in a Labor-Management Participation Team. Conventional thinking holds that the shutdown of a plant with active LMPTs generates employee anger and kills the process elsewhere. At Aliquippa, where 40 percent of employees were active in LMPTs when the hot end shut down, twenty-six teams still met in the surviving operations as of 1987. Team activity across LTV in 1985 resulted in cost savings of $10.3 million, split about fifty-fifty as one-time or recurring savings. Lower than average levels of lost-time injuries, absenteeism, and grievances in plants with LMPTs enabled supervisors and union representatives to solve other problems in a nonadversary atmosphere.[3]

In one new LTV plant, USW members worked under one of the most innovative labor pacts in the United States. This important breakthrough requires some explanation.

In the early 1980s, steel researchers developed methods of producing corrosion-resistant steel by an electrolytic process in which a zinc coating is bonded to steel sheet. The electrogalvanizing process turned out higher quality rust-proof steel, with a smoother surface for painting, than older galvanizing methods. The Big Three auto makers eagerly snapped up the idea and announced they would sharply increase their use of galvanized steel—but only that made on electrogalvanizing lines—in 1987 models. To meet this demand, a number of steel companies began to install electrogalvanizing lines (EGL).

LTV formed a partnership with Sumitomo Metal Industries, which had developed an EGL technology in Japan, and started the L-S Electro-Galvanizing Company (LSE) to produce corrosion-resistant steel. The company decided at the outset that the new business could not be competitive in cost and quality if it had to operate under the same compensation terms and with the same work practices as LTV's other steel operations. What was needed was a highly flexible workplace, employing well-trained, skilled people who, instead of being confined to one narrow job, could shift from task to task and, in a large sense, manage themselves.

This new work system would depart so radically from the traditional industrial relations setup in a steel mill that it had to be located separately from existing LTV operations. Mixing the new and old cultures would be like placing a group of twentieth-century industrial workers in a Stone Age village. For logistical reasons, the company wanted to install the new operation in four unused mill buildings in the Cleveland Works. The area would be fenced off from the rest of the plant and have its own gate.

Cole Tremain asked Lynn Williams for the union's cooperation in planning the new work system and negotiating an agreement to cover it. In return, the company would recognize the USW as bargaining agent for the new employees even before they were hired. If the union refused, Tremain said, the company would consider locating the plant "on a mountain top in Arkansas" and run it nonunion, if need be. But Williams had no objection to trying something new, as long as the union was involved from the outset.

Tremain and Sam Camens headed the bargaining and planning process, reaching agreement on a pact in 1984, long before anybody was hired. Meanwhile, the company built the new line in one of the refurbished mill buildings. On August 1, 1985 the company began a six-month training period for forty-five carefully selected workers who became members of a new USW local upon being hired. All but four were steelworkers who had been laid off by Republic. LTV flew them to Japan for a two-week orientation period at a Sumitomo plant. Back in the United States, all were given extensive training in electrical and mechanical skills, problem-solving techniques, and group dynamics.

At LTV's invitation, I visited the new LSE plant in November 1985 and observed workers and managers together fashioning a form of work organization that would stand Taylorism on its head. Borrowing from production team concepts developed in Britain, Sweden, Norway, and the United States, the plan called for grouping the employees into work groups of about ten persons each. The teams would function autonomously, each taking full responsibility for operating the electrogalvanizing line during its work shift. The function of some fifteen managers was to provide technical expertise and assistance to the work teams, but not to "boss" the workers with the old control methods.

The electrogalvanizing line was an 800-yard long mass of machinery, dressed up in gleaming green paint, consisting of many computer-controlled processes that would transform an ordinary coil of steel into a zinc-coated coil at a maximum rate of 650 feet per minute. In traditional steel mills, production workers would have been assigned to functional jobs along the line, with a different hourly wage for each job. The production employees would know little if anything about the technology and next to nothing about sales, profits, and business prospects. They would serve as machine tenders, subservient to the technology.

The extensive training given the EGL employees, as well as the ability to work autonomously, wouldn't qualify them as chemists or engineers. But the training demythologized the machine, and the new work system put the human factor first. Each employee learned how to perform most of the production and maintenance tasks, although at the beginning only a few could handle the more complex electrical or mechanical skills. The workers were paid salaries instead of hourly wages.

To minimize start-up costs, the USW agreed to salaries and benefits which, if converted to hourly costs, probably totaled as much as $5 to $10 less than the terms of the industrywide agreement. But the pact provided for long-term job security, a "pay-for-knowledge" compensation system enabling workers to increase pay by learning new skills, and a gain-sharing plan. The expectation was that many of the employees would be earning $30,000 annually within two years.

Unlike almost all industrial labor contracts in the United States, the LSE agreement contains no grievance procedure. Management and labor settled disputes informally, avoiding legalistic rules and paper work. All decisions affecting the workplace were made by consensus.

I was impressed by the workers' enthusiasm and sense of working in a familylike atmosphere. Although most of them had worked in a traditional steel plant, they had no regrets about the chance to work under a new system. "Back in the old mill," said twenty-eight-year-old Rodney Clay, "nobody was open to new ideas. I didn't feel I was allowed to put out like I could. Now there's something new every day, and I feel like I'm a part of it."

"The worker today is much more sophisticated than I was when I started work," LSE General Manager Donald Vernon told me. "They're in a better position to accept business responsibility, and they really want it. Making decisions by consensus takes longer than the old way, but gets everybody committed. Most of the time the teams come to the same decision on an issue as I would as a manager. When you think about it, why shouldn't a guy who is a union member not come to the same conclusion as a management man?"[4]

By 1987 LSE was operating flat out and had added twenty employees. The new work system provided a model for efficient, highly competitive businesses in which labor became the initiator of competitive behavior, not a barrier to it. With experiences like LSE and LTV's LMPT program as a base, Williams and his fellow officers and top staffers believed that if wages had to be bargained still lower in the steel industry, LTV would be likely to negotiate innovatively on union concerns.

Before starting talks, however, Williams wanted to develop an industrywide strategy. "If the companies weren't going to coordinate bargaining, we were going to coordinate bargaining," he said.

Developing a Strategy for 1986

In the summer of 1985, the Steelworkers commissioned Mike Locker to prepare a comprehensive study of the steel economy, identifying its major problems and suggesting solutions. The study was to cover prices, costs, and company performance and to project what would happen in the steel industry through the 1980s.[5]

By this time, USW officers and staffers had worked with Locker on several different issues and felt comfortable with his economic and political approach. Locker grew up in a union family in New York City; his mother and father belonged to the New York Federation of Teachers. Active in the radical student movement of the 1960s, he graduated in 1966 from the University of Michigan with a master's degree in

sociology. Locker conducted research on corporate ownership and control and eventually formed a consultancy with
another researcher, Stephen Abrecht (who left the firm in late
1985). They specialized in helping union clients develop corporate campaigns.

As his background indicates, Locker had a left-of-center
political philosophy. He acknowledged the strengths of a free
market economy but believed that a society should engage in
planning to manage the impacts of large structural change,
such as the decline of the steel industry. Smith and Williams
were certain that Locker would see the problem generally as
they did, that the USW and its members should not be expected to bear the full cost of making the companies more
competitive through wage cuts. But the union officials did not
direct Locker to produce any specific conclusion.

In late November, Locker delivered a preliminary draft of
the study, and Jim Smith flew to New York to discuss it with
him. It was then the union economist discovered that "Mike's
view that it [the steel industry decline] was fundamentally a
government policy problem was even more forceful than our
own. We quite possibly might not have put as much emphasis
on the government policy aspect of what we were doing if he
hadn't been so totally convinced" that this was the best way
to deal with the problems.

Locker's major findings were these:

1. The Reagan administration's Voluntary Restraint Agreements (VRAs) were not being implemented in a forceful manner. Instead of holding the penetration level of imports to the
promised 20.3 percent of domestic consumption, the program
had allowed foreign producers to take about 25 percent. In
1985, the study said, imports would "rob the domestic producers of shipments equal to almost 4 million tons."

2. Contrary to conventional belief, U.S. steel consumption
had not declined significantly. The traditional means of measuring steel demand did not include indirect imports, or the
amount of steel contained in imported manufactured products. In 1985, 13.7 million tons of steel came into the country
in this form, a 136 percent increase since 1977. Total steel
usage had dropped only 2 percent, still maintaining a market
big enough to sustain the domestic industry at its 1985 size.

3. The major operating losses incurred by steelmakers
since 1982 were caused by "a massive surge in direct and
indirect imports" which drove down prices and shipments.

For the first time, the domestic steelmakers no longer had the market power to recoup higher costs by raising prices. Average prices had dropped nearly 10 percent between 1982 and 1985. A fall of this magnitude prevented the producers from realizing profits, despite substantial cost-cutting in many areas, especially wages.

Two trends contributed to the price decline. Policies of the Reagan administration and the Federal Reserve produced a rapid rise in the value of the dollar against foreign currencies in 1980–84, lowering the price of imports by 40 percent and boosting the price of exports by 70 percent. This effect was exacerbated by deep price-cutting on the part of U.S. steelmakers, especially U.S. Steel, from 1983 to 1985.

4. From 1982 through 1985, the steel industry reduced employment costs by 35 percent. This included cuts in hourly and salaried employees, reductions in compensation, and productivity improvements. As a result, the companies slashed the number of man-hours required to produce and ship a ton of steel from 8.3 in 1980 to 6.1 in 1984, or 27 percent.

Locker could see no letup in the decline of the industry unless the federal government stepped in with more help. For example, the study identified three programs the government could undertake: proper enforcement of the VRAs, a 20 percent restriction on indirect steel imports, and investment in public projects to increase the use of steel. If these programs had been instituted in 1985, they would have enabled U.S. companies to ship 10 million more tons, raise operating rates to 79 percent of capacity, increase prices by 5 percent, boost per-ton profits by $23, and employ twenty-five thousand more workers.

But Locker did not hold out hope that continuing shrinkage of the industry could be avoided, given the excess production capacity in the world. Even in his most optimistic scenario of trends between 1985 and 1989, thirty-one thousand fewer steelworkers would be employed at the end of the period. "The union must prepare for the downside by considering concessions for desperate companies; it also must make sure its members will benefit from possible improvements," the study said.

The analysis pointed out that many of the problems facing the industry lay outside the sphere of traditional collective bargaining. These included government trade policy, interest rates, the value of the dollar, government expenditures, and bank policies. To address these problems, Locker concluded, the union "should try to *get the parties that control these forces to the 'bargaining table.'*"

The Locker study was authoritative, well-documented, and contained a large amount of useful information. Based on the premise that the United States should retain as much steel capacity as possible, the study called for government solutions. A study based on the opposite premise but using the same facts could have urged against government help. Even Bruce Johnston, while denigrating the study as being "about five years behind the learning curve," conceded that it showed the steel industry as being "in deep trouble."[6]

After studying the Locker report and discussing it with his staff, Williams proposed a two-pronged strategy that would fulfill the union's publicity needs and bargaining needs at the same time. The USW should use the study as the basis for contending that the union alone couldn't solve the problems of the industry and that the government, especially, had to be involved. Even if the chances of the latter were nil, the expression of need would create an image in the public mind. The union, meanwhile, would consider wage concessions at financially troubled companies if they joined the union in publicizing the plight of the industry. The two-pronged approach was simple, clearcut, and easy to present in the media.

On December 16, 1985, Williams released copies of the Locker report to about five hundred local presidents meeting in Chicago. No action was taken, but Williams endorsed Locker's conclusions about the need for government intervention in steel. "Our members are very much of the view that they have sacrificed and made an enormous contribution," Williams said at a news conference. "It's time for other actors to make their contribution, principally of course the administration and the government." In typical Williams fashion, he added: "I have a lot of hope. I just can't believe that America is so blind as to [allow] this kind of destruction [in the steel industry] to continue."

There existed practically no hope, however, that the Reagan White House would respond positively to the union's plea. Nevertheless, the release of the Locker report and Williams's comments on it pushed the issue of government aid into the headlines. At the same time, the tactic carried the important political message to local union officials that the USW president was thinking in strategic terms and that he might even have a plan, however vague, for imposing order on the chaos in the steel industry. It was a far different message from McBride's "face up to reality" lectures in 1982.

On the same day as the BSIC meeting, however, an announcement by U.S. Steel indicated that this company had no intention of waiting to have order imposed on it by the union. Big Steel disclosed an agreement with Pohang Iron & Steel Company (Posco) of South Korea for joint ownership of U.S. Steel's finishing plant in Pittsburg, California, which produced sheet and tin products. Posco would supply about four hundred thousand tons of hot-rolled, semifinished steel coils annually to Pittsburg, displacing hot bands presently supplied by USS's Geneva Works in Provo, Utah. The Korean firm also would invest about $300 million in a modernization program at Pittsburg. U.S. Steel envisioned this deal as enabling it to become a low-cost producer on the West Coast and eventually to shut down the Utah plant.

The Geneva Works still produced steel in open hearths, and a modernization program would cost $1 billion. Posco, however, was a modern, efficient plant and steel labor costs in Korea totaled only $2.30 per hour. USS could save $50 per ton by importing the Korean coils. If the U.S. government allowed import restraints on Korean steel to expire in 1989, Posco could supply Pittsburg's entire annual needs of more than a million tons. Obviously, U.S. Steel—once the leader of industry efforts to obtain import quotas—would lobby against an extension of the limitations on steel from Korea. This would set it against the USW and perhaps other steel companies.

USW leaders didn't like the Posco arrangement, so reminiscent of the aborted British Steel deal, but there was little they could do about it. Meanwhile, USS said it regarded the Pittsburg plant as a new entity and would demand a separate labor contract there to help the plant become competitive with California Steel Industries (CSI), which operated the old Kaiser Steel plant at Fontana. Half-owned by Japanese and Brazilian interests, the nonunion CSI imported semifinished slabs from Brazil and had relatively low labor costs.

A "competitive" contract would mean wage cuts and work-rule changes. "It won't be done without a lot of bloodletting," vowed a local USW official at Pittsburg. The Posco news also dismayed unionists at Geneva, which shipped 70 percent of its hot bands to Pittsburg. "Geneva is gone," said a bitter Kay B. Mitani, vice-president of Local 2701 at Geneva. The Posco deal, he said, would unite Geneva workers to fight against any concessions in 1986 negotiations. "Why should we concede anything?" he asked. "What does it matter, shutting down now or four years from now?"

Many other local presidents at U.S. Steel plants made the same point during the Chicago meeting. Yet a pragmatic resignation to further concessions also was apparent. Don Thomas, president of the Braddock local, said older members eligible for shutdown pensions would be willing to vote against concessions. But he sided with younger members who wanted to save the plant. "A job is better than nothing," he said. "We need the company, and the company needs us."[7]

The timing of the Posco announcement, coming on the very day that the USW began posturing in public on 1986 negotiations, seemed suspicious to many unionists. In fact, it might have been coincidental. Bruce Johnston had asked for a meeting with Williams the previous week to inform him of a new development, but Williams was busy with other matters. Whatever the case, the Posco deal could only strain further the relations between U.S. Steel and the union. By this time, many people in the industry and the union were predicting a strike at USS the following August. Williams made no such forecast, but he admitted frankly that without the no-strike guarantee of an ENA a strike somewhere in the industry was "a real possibility."

Creating a "Level Playing Field"

Williams reconvened the BSIC in Washington on January 15, 1986. By then, he and his staff had worked out their strategy for 1986 in more detail, and the local presidents adopted it as a "statement of objectives." The USW, the statement said, would open early negotiations on requests for further concessions only if a company met two conditions.

First, the firm must agree to participate in a "steel crisis action" program in which the two parties would undertake lobbying efforts on steel imports and other economic issues. For example, the union wanted company commitment to the idea that the steel industry's crisis was "a national disaster which cuts across all industries, threatens the entire manufacturing sector and must be solved if our nation is to remain a first-class industrial power." To address this problem, the campaign would demand government action to reduce the value of the dollar, lower interest rates, start a national program to rebuild the nation's infrastructure, retrain displaced workers, and reform the tax code to stimulate capital formation.

Second, the company must "make all of its books and records available to our financial experts." The USW would consider concessions only if an examination of the books showed the firm was in "dire economic and business circumstances." Further, the BSIC statement said the union would not enter into two-tiered wage agreements in which employees are paid different rates for the same work and would insist on contracting out restrictions, among other things.

The conference also decreed that agreements reached in 1986 with the major steelmakers must be approved by a rank-and-file vote at the company involved. After years of internal disputes over this issue, the USW's half-century old method of ratification by committee was tossed aside with little difficulty. Williams and other USW leaders proposed the change for two reasons. Now that the union had to negotiate on a company-by-company basis, summoning the BSIC to vote on each company's contract would be a cumbersome and costly procedure. It also would raise the question of whether union representatives at one company should have a voice in determining the terms under which employees at another firm must work.[8]

In a deeper sense, rank-and-file ratification would be another step in Williams's efforts to involve ordinary steelworkers in union affairs. Williams had a greater faith than past USW leaders, such as Abel and perhaps McBride, in the ability of the rank and file to make reasonable decisions. Under the new procedure, a tentative labor agreement must first be approved by a negotiating committee, usually consisting of all local presidents in the company, and, second, by the executive board. Only then could it be submitted to a membership vote, and Williams would have discretion in deciding whether to mail ballots to the members' homes or conduct secret-ballot votes in meetings.

The steel crisis action program and offer to negotiate early, with conditions, were the visible parts of the USW's overall negotiating posture in 1986. There also existed a secret strategy developed by Williams and his advisers for the actual bargaining. It addressed what would be the two main issues in the talks, the amount of wage reductions to be granted each company and a method of dealing with contracting out. This strategy had been stitched together at high-level union sessions, starting in late November, involving Williams, primary staff people, and chairmen and secretaries of USW committees at the six companies.

These leaders realized that the USW could no longer maintain uniform wages and benefits across six companies with increasingly diverse businesses. But the union must try to negotiate some kind of standard and enforce it from the top. One thing already had become apparent from labor cost data submitted to the union by LTV and National Steel: A large gap existed between the total hourly employment costs at each company. LTV's costs were about $3 per hour higher than National's.

Information received later from other companies disclosed that the costs ranged from about $22 at Inland Steel to more than $26 at LTV. During the thirty years of industrywide bargaining, the USW had seen only average costs for the involved companies, computed by the American Iron & Steel Institute. The union negotiators had guessed that companies with a higher ratio of pensioners to active workers would have higher hourly costs, but they hadn't suspected how large the variation was. "We were all amazed when the companies opened the books and we saw how much different they were in employment costs," said Buddy Davis.

Even before the full range of costs was known, Williams decided how to deal with the spread in negotiations. The USW should create a "level playing field," as he called it, by narrowing the employment-cost gap. In negotiating with each of the six companies, the USW first would determine approximately how much relief the firm needed to cope with its financial problems. The union would adjust this amount up or down to make each settlement come out at the same, or nearly the same, hourly employment cost. This strategy entailed changing the firms' positions relative to each other on the existing scale of costs, making the least competitive in labor terms more competitive and the most competitive less so.

As the union studied the cost gap, Kleiman later testified, "it began to make sense to us that we could probably . . . make competition more fair, and at the same time address the problems." In other words, the union as the supplier of labor would adjust its price on a company-by-company basis to try to keep all of the players in the game.

To stop the hemorrhaging of jobs from the bargaining units to outside contractors, the USW leaders decided to obtain an airtight ban on the practice and insist that each company accept identical contract language. At each firm the union also would seek: (1) a termination date of July 31, 1989; (2) the restoration of wage cuts through profit-sharing and stock issuance, or a combination of both; (3) seats on the board of directors (this goal was considered expendable if the company accepted the rest of the program).[9]

A U.S. Steel Initiative

On the day after the BSIC approved the strategy laid out by Williams, the union sent letters to the six steelmakers, offering to open early talks, subject to the conditions approved by the conference. Within twenty-four hours, it received affirmative answers by telegram from five of the six—LTV, Bethlehem, National, Armco, and Inland. U.S. Steel didn't respond. David Roderick later told reporters that he saw no need "to adopt [the union's] political philosophy in order to negotiate a labor agreement."

In fact, however, USS's reason for not replying was more complicated than that. It involved a meeting on December 30, 1985, between Bruce Johnston and Lynn Williams. Neither the meeting nor its background was disclosed then or, as far as I know, at any other time until the publication of this book. I first learned about both in a 1987 interview, when I pressed Johnston on the issue of cooperation. If U.S. Steel wanted to make fundamental changes in the labor agreement and work practices—and the fact that it did was demonstrated by its demands in 1986 bargaining—why hadn't the corporation made an effort to obtain labor's commitment through partnership arrangements rather than through confrontation?

Responding to the challenge, Johnston became animated. He related the following story.

In August 1985, he had begun thinking about how to avoid a confrontation with the USW in 1986 and, at the same time, ensure that U.S. Steel would have the lead role in negotiations. Although he did not admit it publicly, Johnston realized that the contracting out issue had driven deep wedges into the corporation's relations with the union. Could the two sides "find any shared solutions" to their problems? Johnston concluded that "we needed to reform our labor agreement totally."

He decided to form a task force to put together a new

package of benefits and work-practice concepts. Johnston invited his old bargaining partner, George Moore, who had retired early from Bethlehem in 1984 and joined a New York law firm, to work with him. Moore readily accepted the chance to get back into steel and joined USS as a full-time consultant after Labor Day, 1985. For chairman of the group, Johnston picked Thomas Sterling, vice-president of employees relations in USS's steel operations.

Johnston directed the group to suggest ways of reformulating benefit plans and contract provisions "to meet some of the Steelworkers' concerns . . . but meet them in a way that allows us to be competitive and to remain a viable player." He encouraged the task force to feel "free to do all the innovating you can." Johnston himself could see possibilities of converting hourly workers to salary, eliminating time clocks, introducing salary guarantees and profit-sharing. To deal with the USW's concerns about contracting out, the company came up with a gain-sharing plan that would reward USW members—those that were left on the job—for suggesting work that could be subcontracted.

In return for "a more appealing total package" for employees, Johnston said the company "might be able to gain better and more effective use of our labor." Very likely, the proposed agreement would wipe out the 2B clause in the contract and give management the unrestricted right to determine crew size and work practices.

Moore worked full time on the project, receiving aid as needed from other task force members, and finished the report by Johnston's December 15 deadline. The recommendations weren't couched in contract language, but each had been deeply explored and was accompanied by cost assumptions at varying gradations of implementation. Johnston pored over the study and decided that it provided a good basis for negotiations. Inviting Williams to his office on December 30, Johnston told him about the Moore study—without disclosing any specifics—and proposed that the parties open negotiations as soon as possible after New Year's Day.

"I remember saying to Lynn that it made a lot of sense to have U.S. Steel go first in the negotiations," Johnston later testified. "We'd been the pattern setter since the union was recognized by U.S. Steel in 1937. We had the staff. We had the geographic dispersion. We were the largest company in terms of employees and in terms of product lines." The two sides could negotiate in a crisis-free atmosphere and take six months to craft a new type of agreement that would become the model for the industry. Johnston also wanted assurances by the union that it wouldn't negotiate lower-cost agreements with the other producers.

What was Williams's response? "Well, he got up and he nodded his head and said, 'I will ponder this, and I'll get back to you.'" But the USW never did respond to this specific proposal, Johnston said. He didn't hear from Williams again until the mimeographed "steel crisis" letter, bearing the USW president's stamped signature, arrived from Washington.

Johnston didn't consider the union's form letter a real response to his proposal and so he didn't answer it. He reiterated Roderick's statement that "we were not interested in pursuing a campaign with the union to change various governmental and business policies," he said. "We think it's a massive failure of communication to tell our employees that the answer to their problems lies in Washington, D.C." The company suspected, he went on, that the USW wanted "to take the focus off the high employment costs, which were the biggest problem that the industry had with the union."

Williams later said the "steel crisis" form letter was the union's answer to U.S. Steel's offer to open talks. He should have realized that Johnston would take the lack of personal response as a studied insult. But the USW had replied to offers from other companies with the same letter. By not answering the union's letter, Williams said, the corporation missed a chance to "make us sweat." USS could have accepted part of the program but not all and offered to bargain early, for example. "This might have put the onus of failing to negotiate early squarely on our backs. But in their arrogance they never answered the letter. I interpreted Roderick's remark about our 'political program' as a deliberately hostile response."

Indeed, the union's program contained nothing that would stigmatize a cooperating steel company with the "political philosophy" of organized labor. Faster depreciation, tax loopholes for the steel industry, tighter enforcement of import restraints, job retraining—these goals hardly indicated a turn to the left. The companies had supported all of them before and had joined the union in ad hoc campaigns to achieve them. As for blaming government, Johnston and Roderick each previously had declared that steel's predicament resulted from labor, management *and government* policies. The 1983 steel labor agreement contained a long section committing the union and the industry "to join mutually in the task of petitioning the federal government to respond" to steel's problems.

The U.S. Steel leaders felt, undoubtedly correctly, that the United States lacked a national consensus "to protect a high wage industry from competition in any greater degree than the president had already agreed to do," as Johnston said. Williams, however, had said the publicity campaign was only one part of the USW's approach; the other part was to negotiate concessions.

All of the above suggests that U.S. Steel wanted to avoid the implication that it accepted, or even acknowledged, the USW's attempt to exert leadership on any matter except cutting wages. First, there was no room in USS's corporate ideology for the notion that a union should tell management what to do. Second, Roderick, Johnston, and Graham believed that USS, being the largest and healthiest company, should take the lead in "rationalizing" the industry—and should be at the head of the survivor's list at the end of that process. The USW, on the other hand, was determined to prevent U.S. Steel from "winning the unrestricted right to decide the future of this industry," as Jim Smith once put it.

In any case, receiving no response from USS, Williams and his aides made up their minds about the order of bargaining toward the end of January. The union opened negotiations with LTV, Bethlehem, and National, but quickly decided to focus on LTV because of its level of distress.

A New Kind of Bargaining at LTV

Since the Republic acquisition, the parent LTV Corporation had lost nearly $4 billion and debt had climbed from $1.6 billion to $2.6 billion. The steel segment was dragging down the parent, which also had aerospace and energy subsidiaries. The company had an unfunded pension liability of $850 million. Some LTV directors had talked informally about a bankruptcy filing.[10]

The LTV negotiations began on January 17, 1986, at the Pittsburgh Hilton. Tremain spoke for the company. Tony Rainaldi headed the union's top team, assisted by Joe Coyle and key staffers, Bernie Kleiman, Jim Smith, and Sam Camens. At the first session, a "sound off" meeting in which both sides aired their problems, Tremain laid the full burden on the table:

The company needed a $5 reduction in hourly labor costs. To avoid adding new costs, he asked the union to defer 55¢ per hour in wage and COLA increases due on February 1.

Before beginning that bargaining, the two sides negotiated a "steel crisis action agreement," which committed the company to a joint political campaign. This wasn't an easy process for either party. Although the USW occasionally had aligned with the industry on tax issues, it never had pledged in a written agreement to lobby for such a capitalistic measure as a provision to aid capital formation. It took several days to draft an agreement, which both sides finally signed on January 29.

They immediately began working on wage deferral and two days later emerged with an "interim progress agreement." The union agreed to postpone the February 1 increases, which, because of compounding, totaled 78¢ per hour in labor costs. However, LTV agreed to keep a 25¢-per-hour SUB contribution due to expire, because layoffs remained high at the company. In the end, LTV netted hourly savings of 53¢, and the two sides set a March 15 deadline for negotiating a new pact to replace the one scheduled to run out at midnight, July 31.

But the USW forced the company to pay a stiff price. First, LTV agreed that in the main bargaining it would leave untouched one of the primary accelerators of labor costs in the industry, pension benefits. LTV had wanted to propose changes in rules for early shutdown benefits to reduce costs, but the union took an adamant position. The rank and file at LTV, and most other steel companies, had seen that Wheeling-Pittsburgh terminated its pension plans and feared that their firms would do the same. Tremain finally gave in on this issue.

The USW also extracted several other company concessions, including limitations on contracting out and scheduling of overtime during the interim period. But another item drew the most incredulous attention of labor relations executives at other steel firms, although it received almost no notice in the press.

LTV agreed to pay USW grievance committeemen for time spent representing workers. This completely reversed a USW policy practically as old as the union itself. Under the old practice, the local union reimbursed grievance representatives for time spent off the job in filing and discussing worker complaints. In the auto industry, UAW grievance officials had been paid full-time by companies since the 1950s. But the

Steelworkers had stuck to the principle of paying its own people to represent the members—until now.

Executives at LTV's competitors questioned why a company that claimed to be in such dire straits would agree to a provision that would increase costs. It would provide an incentive for union officials eager to escape boring or dirty jobs to seek out complaints and file even more grievances. Tremain made this point in bargaining. But the USW negotiators, under heavy pressure from the locals, insisted on the change, claiming that the loss of dues income had drained local union treasuries. One conclusion to be drawn is that the USW had switched to the view that companies should help maintain the union apparatus. Tremain's staff estimated the provision would increase costs by $250,000 a year.

All in all, wrote *Industry Week*'s veteran steel reporter, Donald Thompson, LTV's competitors were "outraged" by the provisions of the interim agreement. "They already left everything at the table but their socks and shorts," he quoted one company negotiator as saying.[11]

Alone among the steelmakers, however, LTV won deferral of the February 1 rises in labor costs, estimated at $27 million on an annual basis. The action also went a long way toward assuring LTV's customers that they would have an uninterrupted flow of steel from LTV.

Tremain also believed an important political benefit flowed from the steel crisis and interim agreement negotiations. The top union people consulted daily with LTV's forty local union presidents, who voted on each agreement. This process helped gain the commitment of the local union officials to LTV's plight and the eventual outcome of bargaining. The local unionists were even more involved in the subsequent negotiations for a new contract. Each president served on one of ten committees that actually negotiated new provisions. This rarely had happened in industrywide bargaining.

With the interim agreement in place, the USW and LTV went to work on a long-term contract. Wage relief and contracting out were the crucial issues.

LTV figured that it had contracted out about 7 percent of the current man-hours worked in the plants, the equivalent of sixteen hundred jobs. By imposing a virtual prohibition on most subcontracting, the USW hoped to force the company to pull much of that work back in-house. In economic terms, this strategy seemed at odds with what the union wanted to accomplish at LTV. In a letter to all members at LTV, the USW noted that in the past it had gone into negotiations to improve wages, benefits, and working conditions. "This year," the letter continued, "our fight is to *save the Company!* We don't like it, but that's the way it really is." To save the firm, the USW had to cut labor costs. Forcing the company to pay USW wages for work that could be done more cheaply on the outside would—on the face of it—increase the cost of doing business. How did the union explain this paradox?

In the first place, said Sam Camens, who headed the USW's contracting out committee, subcontracting work didn't always save money for the companies. "Even if it saved them a little," he said in a 1986 interview, "if we're going to save LTV as a company, we're going to save it for our members. We're not going to save the company so they can contract out jobs."

The sides dickered over contracting out for several weeks and essentially wrote an entirely new provision. The USW won significant prohibitions against the practice:

1. No work inside the plant could be subcontracted out unless it involved new construction or unless the company could meet two conditions: (a) It must pass a "consistency" test by showing that it consistently subcontracted the work over a period of years; (b) if the firm passed the first test, eleven factors had to be considered to show that it would be "more reasonable" to hire a contractor. For example, if the company could show that restrictive work rules inside the mill prevented efficient performance of the job, it could hire outside employees. But it couldn't subcontract work merely by showing that it would be cheaper to do so. A "cost" factor was not included among the reasonable standards.

2. No maintenance and repair work could be performed outside the plant unless the company passed the reasonableness test. This meant that the union could recapture much of the repair and machine shop work. LTV also agreed to measures enabling local unions to closely monitor contracting out practices and file complaints under a new expedited arbitration procedure. The union could obtain an arbitrator's ruling *before* the work actually was contracted out.

Although the USW wanted to solve contracting out problems at LTV this provision really was aimed at U.S. Steel. USS's refusal to bargain early under the union conditions helped the USW determine its contracting out strategy. After LTV, the USW intended to win an identical provision at

National, Bethlehem, Inland, and Armco, and demand that
USS—at the end of the bargaining queue—accept what would
then be an industry "pattern."

Tremain contended that the subcontracting agreement was
not especially onerous for LTV. More than a year after the
agreement took effect, he estimated that it created jobs for
three hundred to five hundred USW members. Yet the company actually saved money by recalling people from layoff to do
this work, Tremain said, because of the high cost of jobless
benefits and health insurance premiums for laid-off workers.[12]

After the negotiators got over the contracting out hurdle
and other issues, they attacked the wage reduction problem.
Up to the last seventy-two hours of bargaining, Tremain stuck
to his demand for a $5 cut in wages and benefits. Lynn
Williams, who had delegated bargaining authority to Rainaldi
and Coyle, now appeared at the bargaining table for the first
time to make the critical final decision.

The accountants from Arthur Young had determined how
much money LTV was losing in steel operations and why.
Using this data and analyzing LTV's future prospects, the
three advisers from Lazard Frères—Keilin, Gotbaum and
Bloom—recommended several actions LTV should take, such
as eliminating salaried jobs, refinancing debt, and securing an
Internal Revenue Service waiver of a $185 million pension
payment due in September. On the basis of the company's
agreement to do these things, Lazard Frères estimated how
much the union should give up to enable the company to
survive, given certain assumptions. If steel prices rose modestly and volume stayed about the same, and if the USW
agreed to reduce labor costs by $3 to $4 per hour, LTV could
get through 1986, the banking firm said. The Lazard men gave
three different figures in that range, with Keilin indicating he
would go a bit higher than $4.

Tremain finally lowered his asking price to $4, but Williams
said this was too much and bargained the number down to
$3.15 worth of wages and benefits. Because of compounding,
this would enable LTV to shave $3.60 per hour from a total
employment cost of $26.40 and wind up at $22.80. This
number would become the union's target for the rest of the
industry.

Two of the Lazard analysts had suggested a number higher
than $3.60, but Williams decided to keep the give-back as low
as possible within the suggested range. "I don't know of any-

thing we could have done, short of taking a strike, that would have changed that number materially," Tremain said a month after the negotiations. "Lynn said, 'Cole, it's a question of what can we sell, and what should we ask our people to take? We don't think we can sell $5. But we're not going to hide behind that. We don't think steelworkers alone should have to bear that big burden.'"

The reductions in the LTV package included an hourly wage cut of $1.14, suspension of COLA for the duration of the forty-month contract, and cuts in various benefits. The USW termed the reductions as "investments," because all of the give-ups were to be returned over a period of time through a profit-sharing and stock ownership plan. The USW also asked for two seats on the board of LTV Corporation, the parent firm. Tremain and Hoag had no voice in this decision; LTV management in Dallas rejected the demand. The USW leaders accepted the decision, feeling that workers wouldn't be willing to strike over this issue.[13]

LTV took on substantial new pension costs by granting early retirement to eligible steelworkers affected by plant and production-unit closedowns at ten separate locations. It also set up a $20 million fund to finance severance pay for workers not eligible for pensions.

The pension and severance provisions probably provided the margin of victory in a ratification vote. When the two sides reached agreement on March 15, LTV's local presidents voted approval, 31 to 6. But the contract was ratified by only a 61 percent majority, 13,162 to 8,474, in a mail-ballot referendum concluded on April 4. The combined yes vote at the Aliquippa and Pittsburgh Works, where most workers had been laid off, totaled 4,141, compared with 297 no votes. At more active plants, such as the Indiana Harbor and Cleveland Works, employees defeated the contract by two to one margins.

The LTV ratification demonstrated once again that workers in a multiplant company have a greater interest in the success or failure of their own plant than in the company as a whole. The no votes by active workers did not result from a lack of effort on the part of USW leaders. They put themselves on the line in preratification meetings. A letter signed by Williams, Rainaldi, and Coyle and sent to all employees urged a yes vote "to save your company and your job." Many workers still felt that give-backs would not guarantee their jobs. They were right (see below).

Other steelmakers almost uniformly panned the LTV contract. *Industry Week* reported that "in the eyes of competing executives the LTV Corp. subsidiary went for the quick fix at the high price." Criticizing LTV's plan to pay back workers in preferred stock, David Roderick called it "a stock-dilution machine." Another industry official complained that the plan put steelworkers "right at the head of the line, ahead of every other shareholder."[14]

Rightly or wrongly, the Steelworkers believed their members deserved that position.

An Innovative Agreement at National Steel

The USW had started negotiations with Bethlehem and National in late January but switched them to a slower track than LTV. This was partly because the union didn't have enough staffers with top-level bargaining skills to negotiate more than one contract at a time. Primarily, however, union negotiators at these companies had to postpone the hard bargaining until the LTV settlement gave them an employment cost target, $22.80 an hour, to shoot at.

After the LTV ratification, Williams gave the go-ahead to bargainers at National Steel, where the union expected to gain some highly innovative provisions. It only took a few days. On April 9, National and the USW reached agreement on a pact that knocked down labor costs by $1.51 per hour, including 99¢ worth of actual wage and benefit cuts. The agreement reduced National's employment costs from $23.72 per hour to $22.21. By the end of the first year, the compensation level would be at least within 10¢ per hour of LTV's because National agreed to a profit-sharing plan with a guaranteed annual yield of 50¢ per hour.

The National bargaining became an important test for Williams's "level playing field" strategy. The company asked for a $3 per hour cut. Buddy Davis, the chief union negotiator, thought this was too much, but he would have granted more than $1.51 if Williams had allowed it. Although he supported Williams's strategy, Davis thought the USW should have deviated slightly from it in National's case because the firm agreed to a comprehensive employment security program. "I didn't really have autonomy on the wage and contracting out issues," Davis said in a 1987 interview. "The bottom line had to be within a few cents of the LTV settlement. As it was, I had to spend a couple of hours arguing with Lynn that the 50¢ profit-sharing payment would bring us up to LTV."

Williams, in fact, realized that the good relationship forged at National since 1984 might tempt Davis to ignore both the wage reduction and contracting out pattern. To enforce the employment cost standard, Williams ordered Jim Smith to participate in the last few days of talks at National. The subcontracting restrictions annoyed the half owner of National Steel, Nippon Kokan, which engaged in extensive contracting out in its own plants. Bernie Kleiman entered bargaining in the late stages to assure National's top negotiator, Stan Ellspermann, that the USW had to have contracting out language identical to LTV's. "Frankly," said Ellspermann, "I resented that I had to eat that contracting out language because other companies had been abusing the issue. That language was not developed for LTV or National Steel. It was developed for the big guys down the street [U.S. Steel]." Ellspermann added: "The Steelworkers were saying basically, 'We're not going to give you concessions and help you eliminate jobs, then have you give the jobs to the contractors.' When you think about it, that's not a totally unreasonable point of view for them."

Aside from the arguments on wages and contracting out, Ellspermann and Davis produced a novel agreement with tightly linked security and incentive pay concepts aimed at increasing productivity. The USW even agreed to incorporate in the contract a commitment that "to ensure job security, substantial productivity improvements must be made to reduce costs and to enable the company to operate with increased effectiveness."

Unions rarely commit themselves so specifically to raise productivity because workers believe this can only lead to job loss. At National, this fear would be considerably lessened because the company guaranteed that no USW member would be laid off for the full term of the agreement, except in the case of a "disaster" such as a permanent plant shutdown or a bankruptcy court order. If business conditions forced a production cutback, affected workers would be assigned to other tasks or various training programs and participation activities like problem-solving groups.

Although a number of nonunion companies like IBM followed "no-layoff" policies, few if any unionized companies had granted such strong employment security promises as National. In return, Davis agreed that National's local unions could negotiate work-rule changes to eliminate unneeded jobs. Although the national union refused to commit to a specified number of jobs to be cut, Davis agreed to push the locals as hard as he could. Through a policy of attrition, National hoped to reduce its work force of seventy-two hundred hourly and five thousand salaried employees by a substantial number within five years. The bottom-line goal was to reduce man-hours per ton of steel shipped from six to three. At the same time, National pledged to invest $1.2 billion in the company by 1989.

As at LTV, the restrictions on contracting out would go counter to eliminating jobs. National figured that work done by contractors inside and outside the plant added up to about one thousand jobs.

Another incentive to cut jobs was contained in a gain-sharing plan designed by National and accepted by the union. Quarterly bonuses would be based on two weighted criteria and would differ from plant to plant: (1) increases in "prime" (high quality) tons of steel shipped per employee (30 percent); and (2) the level of manning reductions at each plant and the home office (20 percent), as well as on a companywide basis (50 percent).

The USW and National also created a multilayered Cooperative Partnership Program intended to give the union a much larger voice in decision making from the shop floor to the companywide level. If the concept worked, this would be the most intricately linked participative structure in the steel industry. It was built on three vital principles: (1) that the company should become "the highest quality supplier of competitively priced steel products"; (2) that workers should be treated as "responsible and trustworthy"; and (3) that management must adopt "a nonautocratic" style of managing people.

Davis was able to exert some autonomy on the issue of board seats. Feeling that having a union official sitting on the board could be a conflict of interest, he didn't make such a demand. But under the partnership program, he and Detroit Director Harry Lester would meet quarterly with top officials of National Steel to talk about capital investment, business plans, technological changes, and other issues that companies normally don't discuss with unions.

One new element in the partnership program was that all employees were required to be involved. The two sides agreed that they "can no longer afford the luxury of allowing such an undertaking to be a solely voluntary process." Workers wouldn't be fired if they refused to attend a participative group meeting, Ellspermann said, but the idea was to put peer pressure on them to become involved.

On April 28, National employees ratified the contract by a 60 percent margin, 3,412 votes to 2,247. The pact received high praise for its innovative features in many newspaper and magazine articles. But some referred misleadingly to the agreement as a "Japanese-style" contract, largely because of its employment security provision.

However, "lifetime job security" is not written into labor contracts in Japan. Actually, Ellspermann and Davis began talking about job security concepts long before NKK bought 50 percent of National. Moreover, National's gain-sharing and profit-sharing plans and its form of participation differed markedly from versions of those concepts used in Japan. "The Japanese influence was positive," Ellspermann said, "but we were on this track before NKK came into the picture."[15]

Settlements at Bethlehem, Inland, and Armco

The USW next set its sights on Bethlehem, which had lost $1.9 billion in the previous four years. Through plant shutdowns and cuts in the salaried force, Bethlehem had reduced its employee ranks from 83,000 in 1981 to 43,900. Anxious to make an agreement, the company opened its books to Lazard Frères and proposed a $5 per hour reduction in labor costs. A new, inexperienced management negotiating team led by Curtis H. Barnette, a senior vice-president, ran into difficulties early in the talks with local presidents who complained about the company stalling on settling grievances. The union even broke off talks in April upon learning that the Bethlehem board had granted a golden parachute to departing Chairman Donald Trautlein.

After clearing these barriers, the two sides began serious negotiations and reached agreement on May 26. Led by Paul McHale, the USW's district director in Bethlehem, the union granted wage and benefit cuts of $1.96 per hour. With the effect on administrative and tax costs, the pact reduced Bethlehem's hourly labor costs by $2.35 to about $22.50, within about 30¢ of the LTV and National agreements.

To repay the concessions, Bethlehem agreed to a profit-sharing plan and stock-issuing arrangement similar to LTV's. If not enough profits were generated to make full reimbursements, the company would issue preferred stock convertible to Bethlehem common shares on a one-for-one basis. The workers lost about 99¢, or 8 percent, in hourly wages and three holidays, among other benefits. But Bethlehem agreed to improve regular pensions, and it also confirmed the "pattern" agreement on contracting out. The company and the union agreed to develop gain-sharing plans on a plant-by-plant basis.

Employees at the company's newest and most profitable plant, Burns Harbor, rejected the agreement by a three to one margin on June 16. But the possibility of saving jobs, along with the pension increase and a company pledge to use the cost savings only to maintain steel operations, drew an overwhelming favorable vote at four other plants. The pact was approved by a 58 percent majority, 11,600 votes to 8,368.

Six days after the Bethlehem ratification, the USW and Inland announced agreement on a contract freezing wages and cutting benefits by 40¢ an hour. Inland had reported a net loss of $178.4 million for 1985, but it returned to profitability in 1986. One of the healthiest steelmakers, Inland entered bargaining with labor costs nearly $2 an hour lower than the industry average. It asked for a $2 cut but pulled back on this request and escaped opening its books.

Inland's fourteen thousand hourly employees ratified the accord by a wide margin on July 8. The company emerged from the talks with a total compensation cost of about $21.60, considerably lower than the other firms. However, because of its financial strength Inland almost certainly would have to pay annual bonuses under a profit-sharing plan earmarking 10 percent of pretax profits for employees. The union also won an additional week of vacation, an offer of early retirement to six hundred workers, company agreement to set up a gain-sharing plan, and the "pattern" contracting out provision.

The shortage of USW staff members twice forced postponement of the Armco talks. The union extended the contract beyond the expiration date and finally settled in November 1986. The USW moved even further away from a uniform wage policy at Armco, negotiating different rates at three plants for a total of sixty-eight hundred employees. The union reduced hourly labor costs by $2.25 at the Kansas City (Kansas) plant, which produces bars and rods in competition with minimills. The Baltimore plant received a $3.25 per hour reduction to make it competitive with other specialty steel plants. The USW agreed only to freeze labor rates at Ashland, Kentucky, to bring this basic steel plant—which had a successful LMPT process—into line with LTV, Bethlehem, and USX.[16]

Epilogue: LTV and the Steel Pension Crisis

On July 17, 1986, three months after LTV's new contract took effect, the corporation filed for chapter 11 protection. The $3.60 per hour in labor-cost reductions could generate enough cash to offset expenses only if steel shipments and prices rose. A 2 percent price rise, at a time when the average price of steel was $500 a ton, would have yielded $25 million to LTV in three months. But the opposite happened. Within six weeks after the labor settlement, the demand for steel subsided and prices dropped. Wall Street analysts issued pessimistic reports on steel in general and LTV in particular. The company's worried trade creditors demanded cash payments and called in short-term loans amounting to hundreds of millions of dollars. Drained of cash, LTV couldn't continue operating day-to-day, much less make a scheduled $185 million payment to its pension funds in September.

As it turned out, the annual value of the concessions granted by the USW was $126 million, or about $60.5 million for the five-month period before the pension payment fell due. If the USW had given the full $6 per hour in concessions asked by LTV, the savings would have totaled only $100.8 million in five months. Not even a $10 per hour cut would have saved the company from bankruptcy without volume and price improvements.

In a sense, the USW and LTV had been forced to do alone what in most industrialized nations would be accepted as largely a government responsibility. The union and the company had to cope with the massive social and economic consequences of an industrial decline caused in part by external factors over which they had no control, such as the rise of foreign steelmaking, the lure of the United States as a steel market, overseas lending by American banks, and the stimulative effect of the strong dollar on imports.

LTV Steel in reality consisted of three old-line steelmakers joined by merger: J&L, Republic, and Youngstown Sheet & Tube (acquired by LTV in 1978). In 1970 the three companies

separately employed 108,000 people, supported 27,000 retirees, and shipped 15.1 million tons of steel. By the end of
1986, LTV employed about 25,000 active workers, paid pensions to 71,000 retirees, and shipped only 8.7 million tons.
From a ratio of 4.0 actives to each retiree, the company swung
around to a ratio of 3.5 retirees to each active. The addition of
dependents of both groups raised the total number of people
dependent on the company and the union to 250,000.[17]

LTV, of course, didn't acquire Republic and Youngstown
out of a sense of social obligation. If mergers had not occurred,
however, the two firms would have failed, and the government would have been forced to deal with some of the consequences. Indeed, the government ultimately was drawn ever
deeper into the crisis in the steel industry as a cascading series
of events followed LTV's bankruptcy filing.

Upon entering chapter 11, LTV Corporation terminated
health and life-insurance benefits for seventy-six thousand
retirees, including sixty-one thousand in the steel subsidiary.
Corporate management in Dallas ordered the cancellation,
acting on the recommendation of outside lawyers who believed a bankruptcy judge would not approve continued payment of $120 million a year for the coverage. Hoag and
Tremain had no voice in the decision, which threatened to
undermine their good relations with the USW. The union
protested that the action violated contract provisions providing for the benefits. It was not assuaged by LTV's offer to
arrange for insurance coverage under which retirees would
pay the full cost.

It was another major dilemma for the USW. Few of its
negotiating victories over the years was more popular with
members than continued health care coverage after retirement. On humanitarian and institutional grounds, the USW
could not allow the LTV action to go unchallenged. Yet it did
not want to drive LTV into liquidation by striking the entire
company. Williams and his aides decided to risk a strike at
one plant where the union had good leverage. They called out
the forty-four hundred workers at the Indiana Harbor Works
in East Chicago, Indiana, LTV's most modern and profitable
mill. Williams was prepared to apply more pressure by shutting down LTV's big Cleveland Works.

Citing the effect of the walkout on operations, the company
applied to federal bankruptcy court in New York City for
permission to reinstate the plans. Despite the opposition of

LTV's bank creditors, Judge Burton R. Lifland ruled on July 30 that LTV could reinstate the original insurance coverage. Williams called off the six-day strike. Once again the USW had "won" a strike at a bankrupt company and had given a strong signal that similar action would greet other efforts to reduce employee benefits in bankruptcy.

Another flammable issue arose in January 1987, this time involving LTV's pension benefits which are provided by plans separate from the insurance coverage. The plans covered fifty-nine thousand current beneficiaries, as well as some forty-four thousand active employees with vested rights. Although the plans contained about $1.5 billion in assets, they were underfunded by $2.2 billion. The federal Pension Benefit Guaranty Corporation (PBGC), believing that bankrupt LTV could not continue providing pensions to all recipients without further reducing the assets, stepped in and took over the obligation of paying the pensions. It had the statutory authority to do this.

Under its rules, however, the PBGC refused to honor the plans' commitments to provide certain kinds of pension benefits. Among other deletions, the rules eliminated a $400 monthly supplement paid to some eight thousand early retirees as the result of plant shutdowns or work-rule bargaining (see chapter 16). This prompted demonstrations by LTV retirees and workers at several plants. The pensioners who gathered outside the gates at Aliquippa were no longer the old men who had nothing else to do, but angry retirees in their fifties and early sixties whose pension payments had been cut by as much as half.

The USW had objected to the PBGC taking over the pension plans in the first place and demanded that the company address the pension shortfall at the bargaining table. Realizing that it could face another strike, LTV resumed negotiations and granted a "follow-on" pension plan restoring $300 of the $400 supplement and making up part of the other benefits cut by the PBGC. This meant that retirees would receive a portion of their pension benefits from the PBGC and the remainder from LTV. On average, the pensioners would receive 92 percent of the benefits originally negotiated by the union.

The USW, in return, endorsed a plan enabling LTV to eliminate about five hundred jobs by offering severance pay. The new agreement also contained two unique provisions which further solidified the parnership between the USW and LTV.

The parties agreed to install in six more places radical reforms of the kind that had worked so well in the Cleveland electrogalvanizing plant. And the USW won the right to jointly administer the new pension plan by naming three of six trustees who, among other things, would make investment decisions for the fund. It was the first instance of this in the steel industry. Nevertheless, the contract initially met opposition from LTV local presidents. On a second go-round they accepted it, as did the rank and file in a June 1987 vote.[18]

The agreement solved one problem but created another. The PBGC heatedly objected to the makeup plan. "If you eliminate all the pain, you destroy the pension insurance system," said Kathleen P. Utgoff, the agency's executive director. She contended that LTV's action would encourage other steel companies to enter bankruptcy, shed their pension obligations, and gain a comparative advantage (lower costs) over their competitors. Lynn Williams argued the reverse, saying that "a steel company will be *less* likely to opt for bankruptcy and pension-plan termination if it knows that it will then be faced with a costly new pension plan." In fact, according to Tremain, the "follow-on" plan increased LTV's labor costs by $2 per hour, raising its hourly compensation to about the level of the other major companies.

In any case, the PBGC in September 1987 "gave back" to LTV three pension plans it had taken over eight months before. LTV contested this action in court but, pending a decision, resumed monthly payments under the plans. However, LTV said it would pay benefits out of the plans' assets without contributing more money to the funds. If the company attempted to fund the plans at a cost of more than $300 million a year, the resulting cash squeeze would push it into liquidation. The irony was that if the plans were drained of assets, the PBGC would be legally required to take them over again—but $1.5 billion poorer than they were when the agency gave them back.[19]

By now, the PBGC itself was in deep trouble, largely because of the steel-industry crash. In 1987 it faced a deficit of $4 billion, 75 percent of which was attributable to underfunded steel pension liabilities (including LTV's). The agency also had taken over Wheeling-Pittsburgh's plans, underfunded by $498 million, as well as those of eight smaller steel firms, including McLouth and Mesta Machine.

The PBGC had been established in 1974 with passage of the

Employee Retirement Income Security Act (ERISA), which contained many provisions aimed at preventing employer misuse of retirement benefits. It was conceived as the administrator of an insurance system financed through premium payments made by all companies with pension plans. The agency's original mission was to pay the pensions of employees whose companies went out of business. But in 1974 no one anticipated that a large portion of an entire industry would face liquidation; nor did the experts foresee that firms would be able to slough off their pension obligations by entering chapter 11.

The havoc caused by the steel pension takeovers affected the thousands of companies involved in the insurance system. To stave off insolvency at the agency, Congress in late 1987 raised the premiums from $8.50 per covered employee to a minimum of $16. Employers with underfunded plans would pay more, depending on the amount of underfunding. Congress also tightened minimum funding standards. Companies with healthy pension plans, of course, objected to these increases. USX and Inland Steel, which alone among the major steelmakers had fully-funded pension plans, also opposed other pending bills aimed at giving additional help to steel firms with underfunded plans.

Meanwhile, the USW had reason for complaint. It also objected to the dumping of pension plans by bankrupt employers, and to the reduction of benefits by the PBGC when it assumed obligations. Actually, the Steelworkers could rightly claim large credit for the very existence of the PBGC. The USW had lobbied hard for ERISA and more than any other union—or any other institution, for that matter—was responsible for its passage. In the 1980s, the union correctly warned that the damage caused by imports eventually would jeopardize steel pension plans and force government intervention at a cost that would partially offset whatever savings consumers realized from the importation of low-cost steel. No one listened.

By late 1987 the steel pension problem had become a crisis. Analysts said that the steel industry needed to reduce steel-making capacity by an additional 20 million tons. The USW disputed this assessment, but some additional plant shutdowns seemed inevitable. The problem was that few companies could afford the cost of paying shutdown benefits averaging $75,000 per employee (and reaching $150,000 in some instances). Bethlehem's pension plans, for example, already were underfunded by more than $2 billion. One study estimated it would cost roughly $4 billion to retire 20 million tons of capacity by 1992. The high closure costs, coupled with the need to spend $8 to $10 billion for capital improvements, meant that more steel producers could go into chapter 11 forcing the PBGC to take over their pension plans. The steel financial crisis also endangered $10 to $15 billion in retiree health insurance plans, which were not insured by the federal government. But the failure of these plans almost certainly would lead to Congressional action to maintain benefits.

One way or another, the federal government could wind up paying billions of dollars to help the steel industry restructure itself. The government had handled the steel crisis on a case-by-case basis through PBGC takeovers of pension plans, thereby maintaining a fiction of nonintervention in the marketplace. By default, a tiny, obscure agency entirely lacking in qualifications to set economic policy was forced to make decisions that would determine the fate of a basic American industry. Indeed, court testimony showed that the PBGC's decision to return the pension plans to LTV was not based on any written policy. The agency's governing board of three cabinet members had not even met for eight years. This presented a powerful argument for government, management, and labor to resume the tripartite process and formulate the most efficient way to help the industry restructure. But Reagan administration officials in 1987 considered and rejected proposals that the White House take the lead in developing a restructuring plan for the steel industry. The prospect was for continuing turmoil, more bankruptcies, more battles between the PBGC and steel companies, the possibility of further strikes if benefits are terminated, and a never-ending worry for retired steelworkers about income security. In late 1987, a total of three hundred thousand steel employees were vested in underfunded plans at thirteen companies.[20]

Chapter 19

The "Crisis in Steel" Campaign

In January 1986, full page ads began appearing in steel-town and national newspapers under attention-catching headlines. "Steel Yourself," warned a headline in the *New York Times*: "America's most basic industry is dying and nobody but us seems to care. Are we to depend on the Japanese, the Koreans, the Brazilians, the Europeans to defend our country?" The *Pittsburgh Press* carried another ad titled "The Steel Domino," which declared that "America's economy is like a row of dominoes. If the steel domino falls, others will also fall."

These and other ads were part of a "Crisis in Steel" program sponsored by the Steelworkers and the five companies that signed the USW's steel crisis action agreement. In addition to the publication of ads and pamphlets, the program consisted of rallies in steel communities, "marches" on Washington by busloads of workers and retirees, and lobbying by union and company executives at state and federal offices. The purpose was to impress upon legislators and government officials the need to help steel and other basic industries by enacting legislation on imports, tax reform, and job retraining. A "Communities in Distress" campaign publicized the problems faced by nineteen steel towns in the Northeast and Midwest.

The eleven-week communities campaign led up to a "Save American Industry and Jobs Day," partly sponsored by the AFL-CIO, on June 21. Rallies held at fifty-five sites were linked by satellite television, which carried speeches from each site by governors, senators, labor leaders, and chief executive officers, as well as music and entertainment. The rallies received little attention outside regions with concentrations of heavy industry, and less than overwhelming support in the affected areas. Only about six hundred people showed up at a rally in Aliquippa, one of the most distressed communities, where thousands had been expected. "It's certainly not the weather. It's not the publicity," said Rev. Jay Geiser of St. Titus Church in Aliquippa. "These people are just whooped spiritually. A lot of them feel nothing can be done."

The companies and the union poured a total of $3 million into these activities, $1.8 million of which came from the union. The companies essentially tagged along on a program directed by the USW. The very existence of a political campaign financed jointly by labor and management indicated how drastically the fortunes of the steel industry had changed. Forty years ago, most steel firm chairmen wouldn't be seen publicly with labor leaders and rarely met with them in private on anything but labor problems. In 1986, Williams summoned the five chief executive officers to meetings on at least two occasions to coordinate the crisis campaign, and they came.

The campaign did not produce major new legislation to help the industry. It kept pressure on Washington to deal with "unfair" import competition and probably contributed to an atmosphere—critics called it a "protectionist" climate—in which Congress began debating an omnibus trade bill aimed at restraining imports. For the most part, however, the campaign was a public relations maneuver, and a fairly successful one. Without it, the news media might well have covered the steel labor negotiations of 1986 as if the main issue were a comparison of USW wages with those in Japan, Korea, and Brazil. The campaign forced other questions to the fore, such as whether banks should make financial sacrifices, as workers did, to help steel firms survive. On the whole, the media did not picture the USW as a recalcitrant union holding the weak steel companies captive to high wages and featherbedding practices. Public opinion doesn't determine bargaining outcomes, but a wave of negative opinion in the steel situation could have put the Steelworkers under great pressure.

As the rallies continued through the spring, and the USW reached settlements at LTV, National, Bethlehem, and Inland, management at U.S. Steel observed the doings with grim faces and caustic words. "So Williams called the five CEOs in on steel crisis and said, 'You kneel down and do as we say,'" commented one management man. "The only time the love-in works is when the union gets everything. Williams said, 'We got to have contracting out, and that's the way it is.'"

Ever since the companies rejected Phil Murray's attempts to participate in planning during World War II, the USW had functioned in a reactive role in the steel industry. Now it was trying to exert leadership over a divided, demoralized industry. Williams generally shied away from making large claims about the union's intentions. But Jim Smith probably reflected Williams's thinking in early 1986 when he described the union's new role to me. "If anyone is going to stabilize this industry and make anything out of it, it's going to be the union. Somehow, some way, we're going to have to take the same kind of control over this industry that John L. Lewis took over the coal industry. We haven't had a leader since Philip Murray that had the guts to recognize that, and now we do."

As the Mine Workers president in the 1950s, Lewis pulled a fragmented coal industry together to deal with problems caused by coal's decreasing use as a fuel. Among other things, Lewis kept a restraining hand on the wage throttle and allowed, even encouraged, mechanization of the coal mines. Unfortunately, he neglected to provide retraining for three hundred thousand Appalachian miners who lost their jobs. The Clothing & Textile Workers Union and the Ladies' Garment Workers also have played leading roles in their respective industries. In these cases, however, the union was—and still is, in the case of the garment industries—the single, most powerful institution in the industry.

The steel industry had two equally powerful institutions, the Steelworkers and U.S. Steel. As spring turned into summer in 1986, they squared off for the labor battle of the decade. Before it began, however, the USW's thin staff had to contend with many other trouble spots.[1]

Trouble in the Metal Industries

Despite the peaceful steel settlements in the first half of 1986, the USW got itself bogged down in several labor disputes that exposed its growing weakness in industries where it traditionally had been strong. Inexperience on the union side of the bargaining table played a role. In February and March, the union conducted a sixteen-day strike against four container manufacturers: National Can, American Can, Continental Can, and Crown Cork and Seal. Instead of winning wage increases, the union had to settle for lump-sum "bonuses" of

$400 in the first year of the agreement and $300 in each of the last two years. General Motors and Ford had introduced the bonus concept in 1984 to avoid the compounding effect of increasing the wage base.

Leon Lynch served as chief negotiator in containers, a position that the USW vice-president hadn't occupied before in a major industry. In June and July, Vice-President George Becker—also serving a first stint as top bargainer at a national firm—led a five-week strike at fifteen Aluminum Company of America (Alcoa) plants. One complicating factor was the USW's previous granting of a 20 percent wage and benefit cut to Alcoa's competitor, Kaiser Aluminum. Using supervisors as production workers, Alcoa continued to produce aluminum during the walkout and wound up with a labor-cost reduction of 95¢ an hour. Alcoa refused to open its books or grant stock or profit-sharing in return for concessions—all goals that the union had set out in its policy statement for 1986 bargaining. The union, however, resisted Alcoa's demands for other cutbacks, worrying about the impact this would have on its negotiations at U.S. steel.

To informed observers, the USW's stand at Alcoa indicated that the union still thought in terms of pattern bargaining throughout the four large metals industries—steel, aluminum, containers, and copper—despite trends in international and domestic economies which had affected the four industries in significantly different ways. Commented one management observer: "They've had this practice of pattern bargaining in place for thirty or forty years, and dealing with situations on an individual or company basis is a massive change for the union. I think they are having great difficulty in changing from inflexibility to flexibility."

However, the USW demonstrated flexibility in the copper industry, where it had lost the disastrous Phelps Dodge strike in 1983–84. Under Secretary Ed Ball, the USW and its union coalition partners negotiated wage and benefit cuts of about 20 percent at Asarco, Newmont, Kennecott, and other companies. The workers realized, Ball said, "that in order to keep their industry alive and to regain or keep their jobs they would have to take this painful step."

In western Pennsylvania, workers at Latrobe Steel ratified a 20¢-per-hour wage cut in May 1986. Sharon Steel employees approved pay cuts in January 1986 and again thirteen months later to try to save the company. Eventually, however, the Sharon company filed for chapter II protection. In late 1987, a USW attempt to negotiate a worker buyout with the company's owner and chairman, industrialist Victor Posner, collapsed.

One of the uglier disputes of 1986–87 occurred at the three plants of Babcock & Wilcox west and north of Pittsburgh. The union rejected demands for reductions of $9.20 per hour, offering $3.50 instead. This turned into a bitter strike, culminating in a June 1987 decision by McDermott International, B&W's parent, to close the three tube plants. These were, except for Aliquippa, the last steel operations left in Beaver County, once a large steelmaking and fabricating center. The union initiated an employee ownership bid in late 1987.

In midst of these activities, U.S. Steel and the USW quietly traded sixty-day notices that their contract would terminate at midnight July 31, 1986, and that each party wished to negotiate a new agreement. It wasn't as though either side had forgotten the expiration date. Such notices are required under federal law as one means of promoting peaceful collective bargaining. But procedures, of course, can't change attitudes.[2]

U.S. Steel: Round One

On June 12, about two hundred local unionists, plant managers, and negotiators for U.S. Steel and the Steelworkers gathered at the David L. Lawrence Convention Center in Pittsburgh. The first two days of talks consisted of sound-off sessions. Speaking for management, Bruce Johnston said the company would make "a total effort" to reach agreement by July 4 so that U.S. Steel's customers could be assured of a continuing supply of steel after the strike deadline. Both sides spoke of finding peaceful solutions to their problems. No one who attended the meetings—indeed, no one who worked at the company or lived in a town with a U.S. Steel plant or knew anything at all about the company and the union— could have believed that a peaceful solution was even remotely possible.

Discussions between top-level teams headed by Johnston and USW Treasurer James McGeehan started on June 16 at U.S. Steel headquarters. As if in recognition of the USW's fall from industrywide bargaining power, the negotiating site was downgraded from the sixty-first floor to a conference room on the twenty-fifth floor. Assisting Johnston were Thomas Sterling, a vice-president who headed employee relations in the corporation's steel division, and George Moore. Still working for U.S. Steel as a consultant, Moore took notes and didn't speak much in the negotiations. Sterling, a Georgia native who joined USS at the Fairfield Works in 1969, was both an engineer and a lawyer. An experienced negotiator, he made a good impression on the union bargainers. Johnston, however, did most of the talking for U.S. Steel. The company team, as Sam Camens saw it, consisted of "two note-takers and Bruce ... and Bruce can talk for hours."

McGeehan's committee included Bernie Kleiman, Jim Smith, Camens, and two directors with extensive U.S. Steel operations in their districts, Lefty Palm of the Mon Valley, and Thermon Phillips of Alabama. It was the first time in the fifty-year relationship between the two parties that a USW president hadn't led bargaining with U.S. Steel. Williams had decided against personally heading the union committees in steel talks because daily meetings over a period of many months would force him to ignore other union business and curtail his traveling. Williams liked to move about the United States and Canada, visiting locals, talking about the union cause, and creating a climate for political and legislative change.

Williams's absence irritated the company men. "I like Lynn personally," Johnston once told me, in his way of leading up to a critical observation about a person. "But Lynn came up the seminar side of the union, attending conferences. He didn't come from the shops. You raise living standards only by raising productivity. I want a guy I can bargain with."[3]

McGeehan, however, did not lack personal experience. A white-haired, conscientious man in his late fifties, he dealt with people on a straightforward basis, eye to eye, without a hint of deviousness. Company negotiators referred to him, with only mild sarcasm, as "Jimmy, the guy who loves everybody." A long-time staff representative and director of District 7 in Philadelphia, McGeehan had served as chairman of the U.S. Steel committee in the last years of industrywide bargaining. In the 1986 talks, the union staff men deferred to McGeehan as chairman, but Smith did much of the talking on economic issues, Camens on contracting out, and Kleiman on all issues. After each session, the team briefed Williams, if he was in town.

A major problem arose before serious bargaining began. U.S.

Steel had refused to open its books to the same degree as had LTV, Bethlehem, and National Steel. Declaring that it would not grant access to information concerning nonsteel businesses, the company gave the union an audited report on the performance of a hypothetical "steel and iron ore division" for six years ending December 31, 1985. It showed that expenditures on property, plant, and equipment exceeded "internal cash generated" in each year, resulting in pretax losses of $2.4 billion for the six years.

The union assigned Mike Locker to analyze the information. He criticized it on several counts. The corporation had not included in "Artificial Steel Company," as Locker dubbed it, the assets of businesses that are intimately related to steel—coal mines, inter- and intraplant railroads, Canadian iron ore facilities, and Great Lakes shipping—all of which formed part of LTV's and Bethlehem's steel business. U.S. Steel excluded these assets on grounds that they derived revenues from nonsteel customers as well as the company's own steel operations. But this had the effect of "skimming off the profitable parts of what is normally included in the steel business and leaving in the unprofitable pieces," Locker said. The union refused to accept the figures as valid, and U.S. Steel refused to supply the kind of data that Locker considered necessary for a comparison with other steelmakers. Mired in this impasse, the "open the books" demand disappeared as a debating point, but it remained a source of friction.

U.S. Steel, however, published its data in a pamphlet, *USS Today*, which was mailed to all employees. Printed on glossy paper with colorful charts and graphs, the pamphlet also listed concessions that the USW had granted to competitors. It showed U.S. Steel's expenditures to aid communities, retirees, and displaced workers, including $750 million in 1985 alone to fund unemployment and retiree benefits. Signed by Johnston, the negotiator, the pamphlet concluded on a note that bore the unmistakable rhetorical style of Johnston, the pamphleteer: "no union, no local government, no state government, no competitor, no academic critic, no political critic, no clergy group and no federal agency, even remotely approaches what USS had done and paid to cushion the people" affected by the industry's decline.[4]

USW negotiators weren't moved by the propaganda. By late June, the bargainers had made no progress on what both sides called the "linch pin" issues, contracting out for the union and total employment costs for the company. Both USS and the USW had committed themselves in advance of the talks to publicly stated goals and now were locked in classical position-bargaining. They had set up impenetrable defenses and were shadow-boxing behind the sand bags.

The crucial bargain would involve company acceptance of some degree of restriction on contracting out and union acceptance of some degree of reduction in wages and benefits. But the union demanded satisfaction on contracting out before it would make a move on the company's economic demands. USW bargainers had taken an intractable stand on this issue: Unless the company agreed to the "pattern" provision signed by the other steelmakers, there would be no agreement. The company, meanwhile, wanted to obtain a substantial wage cut without bending much on contracting out. Chairman Roderick had said publicly that he would settle for nothing less than the Wheeling-Pittsburgh deal of late 1985—that is, a reduction of total employment costs to $18 an hour.

In fact, Bruce Johnston's first proposal, on June 27, was for $18. At this level, Johnston said, the company would grant the pattern contracting out provision. The union refused to take the offer seriously, noting that Johnston hadn't even put it on paper. It would have meant a cut of about $7 per hour in wages and benefits and put USS far below all other major steelmakers except Wheeling-Pittsburgh.

Meanwhile, customers were defecting from USS in droves, a movement that had begun as early as April. By mid-June production at the Gary Works had dropped to one-third the normal level. On July 1 General Motors cancelled its August orders. Nevertheless, USS boosted output in mid-July and shipped the product to outside storage facilities to build a strike hedge. By August 1 it had stockpiled about 1 million tons.

The company clearly did not expect a peaceful settlement. Neither side had an incentive to make a big move in bargaining, and Johnston's July 4 target date passed without progress. So desultory were the talks that the two sides agreed to send home nearly one hundred out-of-town union and company bargainers who sat on twelve subcommittees. The top negotiators kept only two subcommittees, one on benefits and one on nonwage issues.[5]

The negotiations were recessed for a long Fourth of July holiday. Before the teams resumed meetings, U.S. Steel sprang an announcement that pointed up the Steelworkers' declining importance in the corporation's future. On July 8, Chairman Roderick announced a restructuring that divided the company into four "stand alone" operating units. Steel plants and domestic iron ore facilities were grouped in one division called USS, with Tom Graham as president. Marathon, Texas Oil & Gas, and U.S. Diversified Group made up the other three units. The last three accounted for 75 percent of the corporation's revenues, compared to steel's 25 percent share. To reflect this changing mix of businesses, the company adopted the name USX Corporation, using the X which had long been its stock exchange symbol.

Henceforth, Roderick said, each division would have to generate its own capital resources instead of depending on other parts of the corporation—a commandment that cast doubt on the future of the ailing steel unit. Roderick told reporters that the steel division probably would be converted to a wholly owned subsidiary (in early 1987, the corporation moved further in this direction). As a subsidiary, USS could be sold or cast into bankruptcy in the event that the steel business continued to decline.

Comments and jokes about the new name raced around Pittsburgh. The *Post-Gazette* remarked editorially that the decision to rename the company "was designed to X out memories of when the company—and this region—thrived on steel." The most telling commentary, however, was provided by a cartoon in the same newspaper by Tim Menees. It shows "USX" in big red letters hovering over dark-suited executives celebrating with champagne. In the foreground, a burly, hard-hatted steelworker says, "I think we're the 'EX'!"

If nothing else, the timing of the name change, and the cartoon, fit into the USW's propaganda campaign. The union had prepared a pamphlet to respond to Johnston's *USS Today*. Using Locker's analysis of "Artificial Steel," the publication derided the company's financial statement, though with nothing approaching the showy editorial display of *USS Today*. Seizing the moment, the USW's public relations department reprinted the *Post-Gazette* cartoon on the pamphlet's cover under a questioning headline, "Is Steel the EX in USX?" and mailed it to forty-five thousand USX employees.[6]

The propaganda skirmish produced some excitement. But the company's rhetoric and the union's derision contributed nothing to the solution of the bargaining problems.

Johnston lowered his economic demands on July 9. This proposal produced conflicting calculations by the company and union which are almost impossible to reconcile. Apparently it was an incomplete offer, calling for "give-ups" ranging from $2.75 per hour to more than $3. The union rejected it and countered on July 15 with an offer merely to freeze wage and benefit levels in a three-year agreement. After USX eased its stance somewhat on contracting out, the union offered on July 27 to cut benefits by 75¢ an hour.

These numbers mean little, however, because the two sides were caught up in a fundamental disagreement over "costing." At issue was the method to be used in determining USX's employment cost base. Under the company's method, hourly employment costs totaled $25.35 in July 1986. The union figured it at $24.21. Jim Smith insisted that USX overstated its costs by $1.14 in relation to the way LTV and Bethlehem computed theirs. With a gap of that size between the base calculations, the two sides could not come close to an agreement on how to put USX on "the level playing field."

Each side could cite some justification for its method of costing (see notes for explanation). Going beyond accounting technicalities, the dispute was less attributable to a difference in arithmetic abilities than to a complete absence of trust between the parties. The bitter clashes since 1983 over plant shutdowns, local negotiations, contracting out, proposed employee buyouts, the British Steel negotiations, the South Works imbroglio—all of these confrontations had convinced leaders on both sides that the other was trying to get away with something.

Since USX figured its costs differently from the other five companies, the USW concluded that the corporation insisted on the higher cost base to get a larger relative concession than its competitors. "What they said to us in essence was, 'Let us proceed to rationalize the American steel industry by bankrupting our major competitors and driving them out of business.' That's really what they believe and what they want," Smith told me.

Roderick and Johnston, however, believed that the USW wanted to punish USX by saddling it with relatively higher costs than the other majors. The USW's "attempt to hold us to a less advantageous settlement than our predecessors is not acceptable," Johnston said later.

To some degree, the perceptions of both sides were correct.

The union would not have let considerations of fairness in trude if USX agreed to accept higher cost levels than the other companies. USX, on the other hand, did try to get a lower cost agreement than its competitors. Wanting to retain most of its old freedom to contract out, USX made only a very slight movement in Round One toward the provision granted by four other companies. In addition, it demanded many concessions that the other firms did not receive, including: (1) the right to unilaterally change crew composition and eliminate jobs; (2) compression of the existing thirty-two job classifications into less than ten; (3) elimination of 1,105 jobs throughout the corporation in return for about 1,000 "mutual" pensions.

The last item alone was likely to arouse angry passions in the local unions, for *they* would have to do the cutting. The corporation specified how many jobs should be eliminated at each plant and insisted that the national union affirm the number in writing. McGeehan and his team insisted that they didn't have enough knowledge of plant conditions to do that and offered to bring in local negotiators. But USX rejected that idea, saying the company had tried this in local negotiations and failed. The only way it would work, Johnston said, was for the union negotiators to commit the locals in a binding agreement.

The severity of these proposed changes gave the union reason to conduct a strike authorization vote among the USX rank and file in mid-July. "The company has put on the table contract changes that would turn back the clock on workers' rights more than fifty years," McGeehan said. Predictably, the workers voted "unanimously" to shut down USS if necessary.[7]

The Impact of the LTV Bankruptcy

On July 18, the twentieth day of negotiations, "all hell broke loose," as a company man put it. LTV had filed for bankruptcy protection on the seventeenth. This posed a major competitive threat to USX. In the chapter 11 reorganization process, LTV would be exempted from paying certain bills and likely would ask for further union concessions. How could USX agree to a new price for labor when its biggest competitor might gain an even lower price?

As soon as the union negotiators sat down at the table, the company men had at them. Johnston later paraphrased some of what he told the union people. "You told us, 'What a wonderful relationship we have at LTV, trust and faith and LMPTs.' All they did was wait until they got their labor agreement in place and then turn around and file for bankruptcy. This is the outfit you trust? Certainly, Cole Tremain dealt with you in good faith, but while he was running a concession negotiation with you, somebody else at LTV Corp. was running a bankruptcy strategy." Johnston suspected that LTV intended to file for bankruptcy after USX locked itself into a long-term labor agreement. LTV lawyers must have been gathering the bankruptcy documents even as Tremain negotiated the agreement, Johnston said. How else explain that they were able to assemble ten thousand pages of legal documents and thirty separate petitions involving twenty thousand creditors in such a short time?

The same suspicions had occurred to Smith and Kleiman, who participated in the LTV negotiations. Indeed, when the bankruptcy news broke on July 17, they postponed that day's bargaining session at USX to look into the LTV situation. But the USW officials concluded that neither Tremain nor his bosses had "conned" the union. Smith pointed out that steel prices hadn't turned up by the forecasted 5 percent, volume slid drastically after the LTV ratification, and suppliers refused to extend credit. The company had run out of money. It was unsettling, nonetheless, to be accused of naiveté at the USX bargaining table.

There is some evidence, if not proof, that LTV acted in good faith. Gene Keilin and his associates at Lazard Frères engaged in a weeks-long examination of LTV's books in early 1986, a process that included exhaustive interviews with the company's financial executives. They saw no sign of preparations for a bankruptcy filing. "I am personally convinced that they had not determined to do it [file for bankruptcy] in advance of negotiations, and that they resisted doing it," Keilin told me.

The most compelling piece of evidence against a company plan to dupe the USW is that LTV sold its specialty steel division, the most profitable part of steel operations, while labor negotiations were going on. The management of J&L Steel Specialty Products bought the division, consisting of three plants, for $150 million. LTV used the money to pay off debts. If the company had planned to file for bankruptcy, it would have kept the division because of its asset value and avoided paying the $150 million to creditors while in chapter 11 protection.

Both Keilin and Cole Tremain also said that bankruptcy lawyers have the ability to assemble documents quickly. "I can tell you that the claim that all those documents couldn't have been gotten together in a short time is a lot of nonsense," Tremain said. "I know for a fact that they were produced in a very short period of time. I'm personally convinced my leaders did not have that decision in their head while I was negotiating."[8]

The LTV bankruptcy added another tough issue to the already complicated bargaining at USX. The company asked for a provision enabling it to reopen the new contract if LTV achieved further cuts. The two sides did not seriously bargain over this issue in July, because they were far apart on all other issues.

As the negotiations wore down to the last week, three issues dominated the talks: overall cost, job eliminations and manning, and contracting out. The union had little opportunity to elaborate on other issues, such as restrictions on overtime. Although twenty thousand people were laid off, USS refused to call them back and recharge the rights to insurance benefits. Instead, it worked the existing force overtime; 11 percent of all working hours were overtime, according to the union. The USW also wanted to make changes in the grievance procedure and strengthen the LMPT process, as it did at other firms.

At USX, Sam Camens later told me, the word "participation" was never mentioned. "Every other company," he said, "negotiated in an atmosphere of, the only way we get out of this terrible steel crisis is by developing a joint approach. We work jointly with the government, and we try to develop worker participation and input to maximize the efficiency of the plants by joint efforts, not by fighting each other. U.S. Steel is off on another tangent. They didn't develop that joint attitude and get that type of discussion going."

Johnston's strategy, the union people perceived, was to control the bargaining agenda and "get us down to the last night before the deadline and then do something [make a final offer]," Camens said. "And he got us down to the last day and was able to keep us off our agenda and make us talk his talk. It was always talk on his agenda. He thought he could get us down to the crisis, and we wouldn't want a strike. I'm sure he thought that."

USX management had an equally critical view of USW

officers and negotiators. Johnston accused them of being un
willing to take the political risk of recommending large wage
concessions and job eliminations to members. "They wait for
the threat of extinction and then go to the members and say,
'We've got to make concessions. You want in the life boat,
here's the price,'" Johnston said. "There's no political risk in
that. The International fears political challenge."

On Tuesday, July 29, Johnston told the union negotiators
that the company was in the final phases of shutting down
equipment. Therefore, he made what he termed the company's "bottom-line" proposal to avoid a strike. It included a
$1.50 per hour wage cut and benefit reductions. The two sides
disagreed on the amount of concessions involved, USX valu-
ing it at $2.70 per hour the first year, dropping to $1.99 by the
end of the contract because of escalation in health care costs.
The USW calculated the reductions at $3.34 per hour. The
dispute involved another disagreement over costing methods
(see explanation in notes).9

McGeehan indicated disapproval of the proposal but didn't
give a final no. The USW team spent the next day caucusing
at union headquarters. It was apparent that USX, after weeks
of moving very little on contracting out, would not take the
giant step required to accept the pattern provision by midnight, July 31. Late Wednesday afternoon, Kleiman later said,
"we began to realize...that we were very far apart, and we'd
better start thinking about what happens if we don't reach
agreement."

The idea of trying to maneuver USX into a position of
helping to pay for the stoppage through unemployment compensation—as Wheeling-Pittsburgh Steel did in 1985—had
been lingering in the background. During negotiations, the
USW had taken special care to avoid making strike threats in
the event it had to make a case for lockout before state UC
agencies.

McGeehan and his team, along with lawyers Carl Frankel
and Jim English, discussed the pros and cons of offering to
continue working under the terms of the old contract. English
and Frankel thought it likely that Pennsylvania, Ohio, and
Utah would rule in the union's favor, because of court interpretations of the law in those states. They worried, however,
about Alabama, Texas, Illinois, and especially Indiana, where
USX employed the largest number of steelworkers. Another
concern was that USX might accept the proposal on a day-by-

day basis, continue to produce steel, and suddenly cancel the agreement after strengthening its position. The union tactic, however, had an obvious public relations advantage: If USX turned down the offer, the union could picture itself truthfully as having been shut out of the plants.

Averting a strike also would meet another concern. The district director in California, Robert A. Guadiana, had disregarded orders and signed a six-month contract extension with the joint-venture firm formed by USX and the South Korean Pohang Iron & Steel Company at Pittsburg, California. This threatened USX's Geneva plant, which supplied most of Pittsburg's semifinished steel. If Geneva went on strike, Pittsburg would find other suppliers and possibly decide to stay with them permanently. USW leaders were furious with Guadiana but could not overturn his decision. Concerned about their future, three Geneva locals had asked the top negotiators to consider extending the contract for them.

For these reasons, McGeehan decided to go ahead. Union lawyers drafted a letter to the company, and McGeehan read it on Thursday morning, July 31 to some fifty presidents of USX locals gathered in the Pittsburgh Hilton. Receiving their approval, McGeehan dispatched the letter by messenger at 10:30 A.M. and telephoned Johnston to say it was on the way. It stated that the USW "unconditionally offers to continue working on and after July 31, 1986, under the terms and conditions of the 1983–86 agreement." If the USW should later decide to halt work, the letter said, it would give the company forty-eight hours' notice and use that period to help arrange an orderly and safe shutdown.

Johnston later said the offer didn't surprise him. He showed it to Roderick and said he would call McGeehan and try once more to bargain an agreement. Roderick shook his head. "Well," he said, according to Johnston, "I think you're bidding against yourself. You have already made three offers, and I don't think you're going to get very far by holding another meeting."

Johnston replied that he wanted to be sure he had done everything possible to avert a walkout. Indeed, for legal reasons he now had to demonstrate that the company tried to settle the dispute. But he was at a tactical disadvantage. If he explicitly rejected the offer to work, many state agencies would term the stoppage a lockout. By accepting the offer, USX would be producing steel at a substantially higher labor cost than its major competitors. There was another problem. The process of shutting down equipment, which started on July 24, cost the company $10 million. Idling blast furnaces, open hearths, and coke ovens also shortens their campaign life before they must be relined. It would cost USX about as much to restart the equipment, and if the USW decided later to strike, the corporation would bear a second round of shut-down costs and further decrease the useful life of the furnaces and ovens.

Yet if the corporation really wanted to avert a shutdown it would have absorbed these costs. In the past, there had been contract extensions at the last moment, after the company had begun shutdowns. It appears, however, that Johnston, Graham, and Roderick had been planning for months to "take" a strike and did not want to back down now. The demands made by the company, especially concerning man-ning and work practices, indicated that it hadn't expected to settle the dispute at this stage.

The company later tried to make the case in unemployment compensation hearings that the union had played unfairly by not offering to work until after the equipment shutdown had started. This point had some merit, but the USW replied that it would have undermined its own position if it asked for a contract extension several days or a week before the strike deadline.

After the two sides got to this point, there was no saving the situation. Johnston sent a letter to McGeehan in mid-after-noon on July 31, neither accepting nor rejecting the offer to work but declaring that the union was attempting "to convert the coming strike...into a legal fiction of a lockout." John-ston asked for one more meeting, which was taken up partly by legal sparring between Johnston and Kleiman over the strike-lockout issue.

With twenty-two thousand workers set to go out on strike in about six hours, the Kleiman-Johnston debate was not especially edifying behavior. It symbolized how deeply the two sides had fallen into a legalistic snakepit. Johnston con-demned the offer to work, asking how the company could attract customers if the union could "pull the trapdoor in forty-eight hours." McGeehan suggested that the union could specify a longer period, but Johnston dismissed the idea.

The final session got nowhere. The bargaining teams re-viewed the bidding and agreed that the USW had offered benefit reductions worth 82¢ per hour. USX stood pat on its

July 29 proposal, calling for cuts estimated by the company as $1.99 for the contract term and by the union as $3.34. According to Kleiman, Johnston probed with the idea that he might be willing to subtract 32¢ from the concession demand. That wasn't enough to make a difference, the USW replied. After two hours of this inconclusive talk, the meeting ended. At about 9 P.M., I later learned, Johnston telephoned Kleiman, and more inconclusive talk ensued.

That was the last contact. Whatever it would be called, strike or lockout, the cessation of work at twenty-five USX plants, ore mines, and other facilities was now virtually certain.[10]

Going on Strike . . . or Lockout

The evening of July 31 was a mild one for the middle of summer, with temperatures in the seventies. Thousands of people strolled in Point State Park and lined up on Liberty Avenue outside the Hilton Hotel, awaiting a parade associated with Pittsburgh's annual regatta. The scene reminded me that almost exactly four years before, on July 30, 1982, during an earlier regatta, the Steelworkers had rejected the steel industry's first attempt to bring the union to heel.

I went to the Hilton mezzanine, where McGeehan and his negotiators reported the results of the final bargaining session to the USX local presidents. As they met in the Brigade Room, the regatta's masquerade ball got under way in nearby Ballroom 2. Women in stringy straw skirts and skimpy tops, barechested men in loin cloths, masked Lone Rangers, and Hawaiian-shirted drunks wandered about the mezzanine, drinking from paper cups.

Hours passed. The band in the ballroom became tinnier and louder, and the revelers began shedding masks and other portions of their costumes. Shortly after midnight, the local presidents poured out of their meeting, announcing that the "lockout" was on. Most of them looked serious and worried. Nobody doubted that it would be a long ordeal, and it was questionable whether there would be jobs to go back to when the work stoppage had ended. Ron Weisen, however, was proud of the union. "The only way we're going to rebuild this union is to take on U.S. Steel, take 'em on out in the street, and we intend to beat 'em." Weisen, however, had nobody to lead in this fight. The last department in the Homestead Works had been closed earlier in 1986.

A few minutes later Jim McGeehan announced to reporters:

"We have not been able to reach a mutually acceptable agreement. We have proposed to extend the agreement. Our proposal was rejected by the company. We are locked out."

Out at the plants, USW members on the 11 P.M. to 7 A.M. shift reported to work but were refused admittance to the plants. Each was handed a piece of paper which said that because the union rejected the company's offer of a new contract and "has refused to allow its members to work here under the terms of USS's last offer, USW has struck USS, and for these reasons operations have ceased." Meanwhile, USW pickets showed up at the gates bearing strike placards with "on strike" replaced by "locked out."

I visited a few picket lines at the Mon Valley plants in the first two days of the work stoppage and found, not surprisingly, strong support for the union's position. The men on the picket lines did not see the strike, or lockout, as resulting from technical differences over costing methods or the company's demands for wage concessions. In their view, they were engaged in an all-out fight to prevent USX management from whittling down the work force by bringing in outsiders to do steelworkers' work. It was a matter of job ownership.

In Clairton I found the local USW president, Charlie Grese, on the picket line at the main gate. Seven men sat in metal chairs spaced across a wide dirt street leading to the gate of the USX plant. Just inside the high, metal-link gate, two men in white shirts and ties stood with their hands behind their backs, looking out. Some of the pickets wore blue T-shirts with "Local 1557, Clairton Works" on the front, and on the back, the local's bargaining position on concessions: "Enough is enough."

"We're talking about self-preservation," Grese said. "We're trying to salvage the work that we have in these plants. The company wants the absolute right to contract out our work. It's been getting progressively worse. The company is taking us on in more and more areas. They're even expanding their contracting out violations to the production end. It's starting to flow into the coke batteries, the patchers' jobs [workers who maintain coke-oven doors to minimize the escape of heat and fumes]. They have brought people in to spray and patch the ovens. If we let them get away with it, they'll bring people in to operate our machinery. We can't let them do that."

A couple of miles down the river and up the hill on the left, a polite captain of pickets at the Irvin Works suggested that I talk to Local President Don Conn if I wanted information about the . . . lockout. (In those first days, all of us—reporters, union officials, and people on the picket lines—had to pause and think before choosing the right word.) "I'll find him for you," he said. He whipped out a walkie-talkie and said, "Hello, South Gate, this is North Gate. Come in, South Gate." Static poured out of the instrument. "Base, Base, come in. Base," he said. More static. The captain frowned and hefted the walkie-talkie as if he wanted to throw it across the river. He apologized and said I would probably find Conn up at Base, the union hall, and would I please tell them up there that this blinking machine was on the blink?

When I got to Base, I reported the communications problem to a woman in the front office who directed me to the president's office. Conn, a large man with curly hair, gave an impression of being in a constant state of agitation. He also expressed anger about contracting out but was most bitter about his perception of the company's dealings with the Mon Valley plants. "U.S. Steel used to be a steel-producing company," he said. "We made the profits for them to buy Marathon and Texas Oil & Gas, real estate. They didn't put the money in the valley. They took it all out of the valley and left us with nothing."

With rising vehemence, Conn added: "Give us the ability to compete with them [foreign steelmakers], and we'll compete. With the right equipment, we'll outproduce them and out quality-control them. But we're not going back fifty years to please Dave Roderick."

During my tour of the picket lines I couldn't help noticing all the closed plants, abandoned stores and buildings, the sagging and closed bridge across the Mon at Clairton, and areas that had been cleared for redevelopment but never redeveloped. It was a strange environment for a strike. One usually thinks of strikes as occurring in the midst of plenty, when the workers want to get a fair share—or sometimes more than a fair share—of the employer's profits. But the USX stoppage was a defensive strike occurring in the midst of scarcity and decline, very much like the decade of coal-mine disputes in the 1920s.

By 1986 no one expected that a strike against one steel company, even the largest, would have much impact on the national economy. More interesting was the effect in the Pittsburgh region. When the steelworkers struck the twelve largest steelmakers in 1959, a total of one hundred thirty thousand people were employed in the primary metals industries in the Pittsburgh area. Of that number, one hundred thousand went on strike. In 1986, metals employment had shrunk to thirty-three thousand of which three thousand were on strike, according to Norman Robertson, Mellon Bank's chief economist. The 1959 strike idled 12 percent of the region's work force, calculated Joseph Lang of the Pittsburgh National Bank, compared with 0.4 percent in 1986.[11]

The UC Decisions

An argument can be made that strikes are no longer won on the picket line but in the courts and government agencies that oversee labor relations. The USW's strike-to-lockout gambit proved highly successful, even more so than the union had expected. Indeed, the political and tactical moves involved in this episode added up to one of the more fascinating aspects of the entire work stoppage.

On August 21 the Pennsylvania Department of Labor and Industry ruled that USX employees had been locked out by the employer. Some sixty-two hundred workers were eligible to receive unemployment compensation of $192 per week for a single steelworker and up to $220, with dependents, for twenty-six weeks. Minnesota already had issued a similar decision for thirteen hundred iron ore miners.

The issue was never much in question in Pennsylvania. The state supreme court had ruled in 1960—and reaffirmed the ruling on several occasions—that the "sole test" of which side initiates a stoppage is whether the workers offer to continue working "for a reasonable time" under the old terms. If the employer refuses to extend the contract and "maintain the status quo," he has locked out the employees. USX appealed the decision partly on grounds that the stoppage occurred because the union took an unreasonable bargaining position.

But a referee for the Unemployment Compensation Board of Review upheld the ruling. Statements made and positions taken during negotiations did not matter, he said, adding that "entitlement must turn on the actual conduct of the respective sides and not upon the rhetoric of the negotiations." In effect, all that mattered was which side issued the order that caused the plants to be closed.

USX damaged its own cause by labeling the stoppage a lockout in at least one instance. This happened in Duquesne, where the USX plant that had been closed in 1984 still employed security guards who belonged to the USW. When a guard reported for work on August 1, he was handed a typed notice by a USX manager which said that the guards "will be locked out of the involved facilities indefinitely."[12]

Ohio, Texas, and Alabama also issued decisions favorable in whole or in part to the union. Although Utah at the outset denied benefits to USX employees at the Geneva Works, the USW eventually won at the second appeals step on a ruling which stated the same principles that governed in Pennsylvania.

The biggest case for both sides occurred in Indiana, where the largest number of USW members—seventy-four hundred—were still employed in USX plants, primarily the Gary Works. Indiana's law differed significantly from Pennsylvania's and Utah's. A finding of lockout required evidence that a bargaining impasse did *not* exist. Realizing the value of public and political pressure, USW attorneys persuaded the Indiana Employment Security Division to hold hearings on the issue in a huge arena, called the Genesis Center, in Gary and open them to the public. Every day for eight days, over two thousand steelworkers and their supporters sat in the stands, listening to the testimony.

Initially, the crowd booed and hissed when USX executives testified. When union officials appealed for decorum, the spectators stopped the anticompany behavior. After state and company officials left the arena at the end of each session, either Kleiman or Frankel explained the significance of the day's testimony to the audience. The open hearing was a brilliant idea, which union staffers credited to Kleiman. It enabled the International to demonstrate its ability to fight for the rank and file. At various times during and after the work stoppage, Roderick and Johnston made statements questioning whether ordinary, nonactivist steelworkers supported the union in the six-month dispute. Addressing this question, Frankel described the atmosphere in the Gary arena, where thousands of steelworkers gathered. "When you walk into the Boston Gardens, is there any doubt who those people are cheering for and who they dislike? That's the way it was in Gary," he said.

The Indiana agency awarded UC to the locked out workers,

and both a referee and a review board affirmed the decision. To show that an impasse existed in the talks, the company argued that the two sides moved further apart toward the end of negotiations. Kleiman appeared for the union. In Utah, he had testified that the USW and the company were "very, very far apart" on all major issues. He later admitted to me that "we wanted to make the difference appear as large as possible," In Indiana, he made those same differences appear to be bridgeable.

Company men later grumbled that Kleiman "shaded" his testimony from state to state. But the Indiana referee also believed that "fluidity" existed in bargaining because the USW did not take three steps necessary under its rules to call a strike: (1) a strike vote by the concerned local union presidents; (2) a vote by the executive board to set a strike date; (3) a strike authorization by the USW president. The first and third of these steps wouldn't have taken much time. But the board either would have had to be polled by telephone, which would have taken some hours, or summoned to Pittsburgh. Neither of these procedures had been started by the time of the last negotiating session on July 31.

The Indiana review board found that, because of a drop-off in orders, USX began a "total shutdown" of the Gary Works as early as July 12 and completed it the morning of July 31. This meant that virtually all workers would be unemployed regardless of what happened in negotiations. Nevertheless, employees reported to work at 11 P.M. on July 31. The company turned them away, the board concluded, not because an impasse had been reached in negotiations "but because USX concluded that with no production scheduled at the Gary Works in any event, it was in its economic and strategic interest not to allow these employees to work."

Illinois was the only state that denied benefits to USW members. In this state, a work stoppage resulting from any labor dispute disqualifies affected workers. About five hundred to seven hundred people at USX's South Works had to last out the six-month dispute on the union's $60 a week strike benefit. Yet these members, as Frankel pointed out, were no less militant throughout the stoppage than nearby Gary workers who received UC. "Would the people have stayed out and been as militant as they were without 'comp?'" Frankel asked rhetorically. "On the whole, I think the answer is yes in this situation because the depth of feeling was there."

Perhaps strikes and lockouts still are won on the picket lines, especially if the employer acts like USX. "USX did enough in the last three years to prepare us for this event," Lefty Palm observed after the plants had been shut for four months and the members remained united. "They said, 'Screw you, go file a grievance.' They prepared us mentally and physically."[13]

Nevertheless, "lockout" doesn't describe adequately what took place at USX from August 1986 to February 1987. It wasn't as if David Roderick, like Henry Frick in 1892 at Homestead, locked the workers out after rejecting collective bargaining. Roderick refused to allow steelworkers to work at their offered price, but the union refused to work at the company's offered price. Although the law may not recognize such a term, I think of the USX shutdown as a lockout/strike. It was a mutual rejection.

Lasting It Out

The USW's Strike and Defense Fund contained $210 million at the start of the lockout/strike, enough to finance a very long strike of twenty-two thousand workers if necessary. The union announced it would pay a lockout benefit of only $60 per week, an amount that would have made it difficult for steelworkers to remain idle for a protracted shutdown unless they also received UC benefits. The $60 figure, however, had been the standard strike benefit allocated by the union for some years, and the leaders felt it would be unfair to previous strikers to raise the stipend at the outset of the USX closure. If need be, the executive board could change the allocation.

The USW had always followed a conservative policy in handing out strike benefits. Unlike the UAW, for example, which granted benefits to each striker as a matter of "right," the USW made payments on the basis of "need" only. Instead of sending checks to individual workers, the International made payments to each local based on its count of dues-paying members affected by the lockout. The locals distributed the money according to their own rules for determining who was needy. Some handed out vouchers for free or discounted food and some passed out checks. This system is obviously vulnerable to corruption. Some union members complained that local officers discriminated in favor of friends and followers on doling out benefits.

The 1986 bargaining round drained a large amount of money

out of the strike fund. By the end of the stoppage, the union had distributed $23 million to USX employees in benefits and spent another $8 million to buy Blue Cross and Blue Shield coverage when USX discontinued health insurance for those locked out. The strike fund also paid about $3.1 million to Lazard Frères and other consultants.[14]

In the 116-day strike of 1959, steelworkers received neither UC nor union strike benefits and yet they endured nearly four months of strike, taking part-time employment, odd jobs, and generally scrounging around to make ends meet. The strikes of the 1940s and 1950s had taught them how to do this. In 1986, not only were job opportunities extremely limited in the depressed steel towns, a large part of the steel work force had had no experience with strikes. If the USX workers had had to exist on the USW's meager strike benefits, could they have endured for six months without being tempted to go back to work? Certainly, if any movement back to work got started, there would have been large-scale violence in the communities of the Mon Valley, Gary, and South Chicago. The city of Gary even passed an ordinance forbidding scabbing. While it is of doubtful constitutionality, it demonstrated that some of the mill towns still lived by the trade union ethic.

Recognizing this, the company—to its credit—made no attempt to solicit strikebreakers. But it did not forswear all efforts to defeat the strike by continuing to supply the demand for steel. In some mills, supervisors apparently managed to produce a small amount of finished product. The company tried to ship finished steel out of independent warehouses, where an estimated 1 million tons of steel had been stored before the shutdown, and even out of the plants themselves. Williams named George Becker, a man known within the union as an exceedingly tough administrator, to be "coordinator of lockout activities." The union strategy was to hold up shipments, force USX to go to court for restraining orders, and do everything necessary to cause problems for the company. "We fought them every step of the way," Becker said. "It was very costly and time-consuming for the company to go through the court procedures." Pickets were arrested for this kind of activity in Birmingham, Fairless Hills, Gary, and Geneva.

One of the more serious flare-ups occurred in Lorain, Ohio. When the company tried to ship pipe out on a fifteen-car train on November 26, about 150 union members blocked the tracks. As a large force of local police moved in to clear the tracks, scuffling broke out. Fourteen unionists were arrested and four treated at a hospital, including District Director Frank Valenta who suffered a broken nose.

The Steelworkers received a considerable amount of help from the AFL-CIO and many affiliated unions. The federation established a task force chaired by Murray Finley of the Amalgamated Clothing & Textile Workers Union to provide support for the USX strike. Locals of other unions were particularly helpful in organizing an "informational" campaign to discourage motorists from buying Marathon products. Marathon is sold under seven brand names, but the unionists eventually identified more than three thousand service stations that handled its products. On Saturday mornings the union people appeared at the Marathon stations and distributed leaflets informing drivers that USX had bought Marathon with profits produced by steelworkers, a statement that was true only in a very broad sense. "USX will use your gas $$$ to starve out its steel plant employees," the leaflets said.

Very likely, the Marathon campaign hurt USX no more than would a needle prick in the arm. The company's sales to retail outlets increased substantially in 1986. The union halted far short of mounting an all-out corporate campaign to harass USX, though it considered a range of other things it could do. For example, the union might have filed objections with the Securities & Exchange Commission when USX spun off a chemical division to gain cash. Williams held back on a number of activities that could have created regulation difficulties for Marathon and Texas Oil & Gas (TXO).

For its part, USX took many steps to arm itself for combat, despite protestations of peaceful intentions. The acquisition of TXO in early 1986 helped USX escape its dependence on the currently depressed and permanently cyclical steel business. As a secondary benefit, TXO undergirded the corporation's ability to withstand a steel strike. During the first sixty to ninety days of the shutdown, the inflow of cash from TXO, Marathon, other nonsteel businesses, and the sale of inventoried steel may well have exceeded the overhead costs of operating USS, the steel segment, without production.

USX managed to ship 2 million tons of steel from July through September according to analysts. Being shut down, however, hurt USX deeply in the marketplace and helped its competitors. LTV fired up a blast furnace at Cleveland that had been idle for two years and recalled 575 workers. By the end of September, it and other steelmakers had posted two 3 percent price increases on flatrolled products. Steel-watchers estimated that USX would lose $3 million to $5 million a day during the stoppage. The dispute was destined to become an old-fashioned test of strength between the company and the union. But long before it reached a decisive stage, a far different kind of adversary appeared out of nowhere and threatened the very existence of USX.[15]

Chapter 20

The Icahn Takeover Bid

In the fall of 1986, USX's steel division—the traditional old U.S. Steel, minus oil, gas, chemicals, and real estate—almost became a worker-owned company. This is more than a "might have been" tale. It is about the grim determination of three negotiating parties, each fighting for a different concept of how corporations ought to be managed and for what purpose. The battle hung in the balance for a few critical weeks and ended on a deeply ironic note.

The story actually started in August 1982, when more than one hundred twenty-five thousand steelworkers were unemployed and the Mon Valley mill towns began their final plunge to industrial obscurity. In that month, a bull market got started on Wall Street. Cheered by falling inflation, falling oil prices, falling interest rates, and—in the later stages of the trend—the falling dollar, people and institutions pumped vast amounts of money into company stocks. Stocks that lagged behind attracted the eye of the new financial buccaneers known as "corporate raiders." In 1985 and 1986, the raiders and investment bankers drove the market ever higher with feigned and real takeover bids, forcing the targeted companies to spin off, recapitalize, and buy themselves out.

It didn't seem to matter that the economy grew at an anemic rate in 1985–86 or that company earnings remained mediocre. The Dow Jones Industrial Average soared for five years and broke the 2,000 barrier in early 1987. The raiders insisted that corporate America was failing to produce earnings for the shareholder commensurate with the real value of company assets. Only the raiders' heroic risk-taking would force management to redeploy assets and restructure the business to be more efficient and competitive. Critics warned that all this

539

cutting up and reassembling would ruin U.S. competitiveness. How could managers do the necessary spending for new equipment, research and development, and job training—all important for long-term competitiveness—if they had to produce short-term big dividends in order to survive?[1]

The market ignored this crucial question. The surge of buying, selling, and merging produced thirty-eight transactions of more than $1 billion each in 1986 alone. On August 13, 1986, the fourth anniversary of the bull market, the Dow Jones opened at 1844.49, a 137 percent increase over the same date in 1982. A week later, USX corporation went on the buying block.

In 1986 no raider with a hunger for assets convertible to cash would even glance at a steel company. USX, however, no longer *was* a steel company. It now owned Marathon, with its valuable oil reserves, and Texas Oil & Gas. As early as May, Wall street began speculating that big investors were accumulating shares of U.S. Steel (before it was renamed). The company looked like an inviting target. The stock was selling at about $21 per share, with a total value of about $5 billion, but traders said the company's energy assets put the real value closer to $20 billion. In mid-August, when USX's price dipped to the $14 to $15 range, the big investors began acquiring names and faces. They were led by the Australian Robert Holmes à Court, who disclosed on August 21 that he would purchase up to 15 percent of USX common stock.

In September, rumors spread that large blocs of stock were being accumulated by other raiders, including Carl C. Icahn, who in 1985 bought controlling ownership of Trans World Airlines; T. Boone Pickens, the Texas oilman; and a Minneapolis entrepreneur, Irwin L. Jacobs. Adding to the appeal of the undervalued stock were speculations that USX had overfunded its pension plans by up to $2.5 billion. Though wildly erroneous, the talk of overfunding helped bid up the price.

To keep the raiders at bay, USX announced it would conduct a thirty-day study of restructuring possibilities to raise shareholder value. The company's stock price climbed to about $25 in the next two weeks. On October 6, Icahn leaped out of the pack, disclosing that he held 9.8 percent of USX shares (later increased to 11.4 percent). He made a "friendly" proposal to pay $31 per share, about $8 billion, for the remainder, threatening to make a tender offer to shareholders unless management accepted his proposal by October 22 or increased the value of the shares above $31. Icahn said he was not seeking "greenmail," or the purchase of his stock by USX at a higher price than other shareholders would receive.

The issue remained in doubt for the next several weeks. USX seemed highly vulnerable to a takeover. It had lost $235 million in the first six months of the year; it owed $6 billion in long-term debt; oil prices were stuck below $20 a barrel; and the labor dispute was draining $60 million to $100 million a month out of the company.

Chairman David Roderick searched desperately for a solution. He discussed restructuring plans with outside advisers and talked constantly with Icahn. "One thing I'm certain of," said a financial man who knew Icahn and Roderick, "for Roderick it was very difficult dealing with Carl in a hostile takeover situation. It was unpleasant, wearing, tiring. Roderick is very autocratic and imperial. He just offers himself up to Carl, who is persistent and tough and handles his own negotiations." The chairman could not have been pleased when Icahn himself described USX management as "arrogant."

USX's top managers, meanwhile, also took steps to protect themselves in the event of a hostile takeover. The USX board, it was later disclosed, approved employment contracts, or "golden parachutes," for ten executives, calling for a cash payment of up to three times current salary, plus other benefits. These terms were not excessive by the standards of other companies, and there was some economic justification for measures designed to keep valued executives in the corporation. Moreover, USX's top management had taken salary cuts along with all other salaried people when the lockout/strike started. For example, Roderick's cash compensation for 1986 dropped from the $1 million he earned in 1985 to $735,382. But if Icahn should win, Roderick and his colleagues would not be cut adrift without severance pay like the hourly workers at National Tube (see chapter 18).

Icahn presented an enigmatic profile to Lynn Williams, who had a keen interest in the Roderick-Icahn battle. On labor issues, Icahn, a fifty-year-old former Wall street arbitrager, seemed to send mixed signals. ACF Industries, where Icahn scored his first large takeover and served as chairman, was a union company. When he bid for TWA in late 1985, the Air Line Pilots' Association helped him. Eager to escape the clutches of another bidder, Frank A. Lorenzo, the antiunion chief of Texas Air, the pilots granted Icahn wage concessions in return for stock and profit-sharing. But a few months later, Icahn hired permanent replacements to break a strike of flight attendants. Although the attendants offered to take a 15 percent pay cut, they resisted the 22 percent demanded by Icahn. Nearly five thousand attendants lost their jobs as the strike petered out.[2]

Speculation arose that Icahn would try to enlist the aid of the Steelworkers in his fight with Roderick. Indeed, though it was not known at the time, Icahn had sent out feelers to the USW as early as July. He was purchasing USX stock then but had not accumulated enough to make a disclosure under government regulations. Icahn's people made their first contact with Mike Locker. From the beginning, Locker emphasized that the union would be more interested in negotiating a buyout of USX's steel segment with Icahn than in merely giving him concessions to aid a takeover.

Locker told Gene Keilin of Lazard Frères about Icahn's interest. As financial adviser to the pilots' union in the TWA situation, Keilin had negotiated personally with Icahn and knew him well. Now, representing the Steelworkers, Keilin had some background talks with Icahn, who said he would like to meet Lynn Williams. The banker arranged a meeting in mid-October.

The USW president was equally interested in meeting Icahn. As the takeover battle unfolded, Williams decided that the union wasn't "just going to sit back and take whatever happened," he later said. "We were thinking about some initiatives of our own."

Williams and Jim Smith, accompanied by Keilin and Locker, met with Icahn in New York on Thursday, October 16. The twenty-seventh-floor offices of Carl Icahn & Company reminded Keilin of a warship. Three-quarters of the space was taken up with the equipment of a trading operation—tickers, computer terminals, and shirt-sleeved traders working at the machines—rather like an engine room. Icahn's office, overlooking Sixth Avenue, was like the captain's quarters, outfitted with oriental rugs and antiques. On a couch sat a pillow with Happiness Is a Positive Cash Flow embroidered on it.[3]

Icahn did most of the talking, pleasant but very businesslike. He gave his standard speech—Keilin had heard it before—about the ineptness of American management in all industries, not just steel. Most managers, Icahn said, placed their own interests above those of shareholders and workers. He spoke frankly about his own chief interest: cash flow. He did not analyze companies in terms of profits and losses but in how much cash could be raised by various means. Icahn needed cash to do more takeovers.

"He didn't know anything about the steel business," Smith later said. "All he knew about were the numbers." He thought the pension funds were overfunded by $1 to $2 billion. (USX later revealed that the excess amounted to less than $300,000.) Icahn also mentioned that USX's steel segment generated about $400 million annually in depreciation charges. He would reinvest some of that to maintain equipment but extract about $300 million in cash.

Icahn pressed the USW men on what kind of labor agreement they would be interested in if he acquired control. He indicated generally that he would be willing to give profit-sharing and stock ownership in return for wage concessions. Williams and Smith did not want to talk specifics at this meeting but agreed to resume discussions early the following week.

"I made a minimal effort to explain some of the governmental policy problems on the future of the steel business," Smith said, "and he was totally uninterested in that. He didn't indicate any interest in being a steel tycoon. He was coveting the cash flow from the steel business. We left there with our mind made up that we didn't want him in our industry."

After the meeting, Williams called Pittsburgh and asked Jim McGeehan to come to New York the next day, Friday, with other members of the USX negotiating committee. Meeting in Keilin's office, they had little trouble agreeing that "it would not be good for the workers for Icahn to own or operate USX," Smith said. If he managed to buy control of USX, the union's response should be—as Mike Locker originally suggested—to buy USS, the steel division, from Icahn and run it as an employee-owned company, though likely with some Icahn money in it.

Williams went back to Pittsburgh Friday night, but some of the others stayed for sessions the next week with Icahn. On Saturday, the USW president received an unexpected call at his home from Roderick. The chairman suggested that the two meet "eyeball to eyeball." The timing left little doubt that Roderick had found out about his talk with Icahn and had decided to intervene before the USW made a deal. Now it would become a three-way negotiation.

When Williams and Roderick met that afternoon, it was the

first session between the two sides since July 31. Roderick wanted to resume bargaining, saying that there would be "a window of opportunity" for the corporation to obtain steel orders if it had a labor agreement by November 1. Williams agreed, telling Roderick that the union "was interested in settling the dispute" with USX. But the union leader left open the possibility that he would continue to talk to Icahn. As Roderick was going out the door, he turned and asked "with elaborate casualness" whether Williams and Icahn had met. Yes, Williams replied, without adding more.

Roderick did not appeal for the USW's help in routing Icahn, Williams said later. "But he certainly disparaged Icahn. I wouldn't describe either of them as great listeners. Roderick argued essentially that 'nobody can run this company better than I can.' If Icahn took control, it would mean going into a great deal more debt. He said the best alternative for the employees is to keep Icahn out."

After the Saturday session with Roderick, Williams flew back to New York and met for several hours Sunday with Keilin, Locker, Smith and other members of the USX bargaining team. Roderick's offer to negotiate had presented them with a crucial choice. Should the union align itself with Roderick or Icahn?

If Williams decided that the USW's primary interest lay in defeating Roderick, he conceivably could strike a deal with Icahn like the pilots' union did at TWA. By agreeing to a specified wage cut if the financier won control of USX, the USW could strengthen his ability to raise takeover funds. But that would mean escalating the USW's "cold war" with USX into a "hot war," as Locker put it. Up to this time, the company and the union had waged their battle under "cold war" rules. Roderick hadn't tried to operate the plants with scabs, and the USW hadn't mounted a tough corporate campaign against USX.

Roderick surely would consider a USW decision to bargain with Icahn as a "hot war" tactic. He might respond in kind, opening the plants and bidding for the loyalty of the employees. Once the two sides started down this road, there would be no stopping short of total victory or total defeat. The union strategists talked about other considerations. Icahn was hardly a pushover, and the union couldn't be certain it would get a reasonable agreement with him. Williams also worried about creating conflicts within the union if he rushed into employee ownership at USX. Should the union have decision-making control of the industry's largest company, one with enough market power to damage—even break—weaker firms whose employees also were union members? It was a difficult question that no union so far had had to answer in practice.

Williams concluded that the union could win its battle with USX without starting a hot war. After considering all the factors, he decided against negotiating with Icahn for a quick agreement to knock Roderick out of the box. "My judgment was that pursuing the Icahn approach didn't hold anything positive for our people," the president said, unless an Icahn takeover forced the union to consider buying out USS.

The USW decided to move forward on a two-track negotiating scheme, bargaining in Pittsburgh with USX while Keilin talked in New York with Icahn. "We instructed Keilin to tell Icahn we were *not* interested in negotiating a labor agreement with him," Smith said. "But we *were* interested in talking about a buyout of USS if he got hold of USX." In this event, the USW would grapple with the issue of union ownership rather than leave control of USS in the hands of the man to whom happiness was a positive cash flow.

Williams, of course, didn't announce his decision to Roderick. Keilin continued to talk with Icahn about the buyback possibility. "The idea of selling steel was appealing to Carl," Keilin said. "He was interested in oil. We never came together on a price, although we narrowed the gap." They suspended their talks when Round Two of the USW-USX negotiations began on Tuesday, October 21. Icahn did not want to be used as a lever by the union at USX.

Given the circumstances in late October, Icahn looked like a winner in his USX bid. The financier planned to supply $1 billion for the venture from his own capital firms. The investment banking firm of Drexel Burnham Lambert said it would raise the remaining $7 billion by issuing high-risk, high-yield "junk bonds." Drexel Burnham had been highly successful in financing other takeovers, and many analysts believed that Icahn eventually would take over USX.

This prospect shocked Pittsburgh's political and corporate leaders. Mayor Richard S. Caliguiri reflected the virulent reaction to Icahn when he described the investor as "a mercenary, a quick-buck artist. He takes millions of dollars out of corporations and destroys them." The corporate leaders were not much concerned about the steel plants in the Mon Valley;

their primary fear was that Pittsburgh would lose USX's corporate headquarters, just as it lost Gulf Oil when Chevron acquired Gulf in 1984. Nor would there have been much cheering if the USW had revealed its buyout strategy. Employee ownership was not a popular concept in Pittsburgh business circles.

Nevertheless, the possibility that workers would own and control the old U.S. Steel remained very high in late October. It would be dashed by other events. But the USW's fateful decision against helping Icahn was a major factor. "Our purpose was not saving or destroying Roderick," Williams said in 1987. Yet the irony was that, in choosing the cold war course, the Steelworkers helped Roderick—a bitter enemy on labor issues—save USX.[4]

USX: Round Two

The USX labor talks resumed. On October 29, Roderick told a news conference that the company needed a settlement within ten days or it would lose orders for the first quarter of 1987. But each time major progress seemed possible in bargaining, one side or the other backed away. Gene Keilin, looking on from New York, labeled the talks as "manic-depressive, now up, now down."

Toward the middle of November, union negotiators sensed a diminution of the urgency which USX had announced in October. By this time, Roderick realized that there would be no Williams-Icahn deal, and the chairman seemed to be edging out Icahn in persistence. He kept putting off an announcement of major restructuring moves supposedly recommended by outside advisers and denied Icahn full access to USX books. The corporation also raised $3.7 billion in cash by drawing against credit lines and transferring the company's profitable chemicals division into a new public company, later purchased by the division's managers. Although Roderick failed to raise USX's stock to $31, Icahn hesitated to make his threatened tender offer to shareholders. "I think Carl underestimated Roderick," Locker said.

On November 14, the disclosure of a Wall Street scandal of enormous proportions took the remaining potency out of Icahn's takeover bid. Federal investigators charged Ivan F. Boesky, "America's richest and best-known arbitrager," as the *Wall Street Journal* described him, with violating insider trading laws. Acting on tips supplied by cohorts at investment banks, Boesky had made huge profits by buying into companies about to be attacked in takeover bids and selling the stock after a runup in the price. He pleaded guilty to one felony count of engaging in insider trading, paid a $100 million fine, and was barred from the securities industry for life. The investigation explored, among other things, Boesky's connections with Drexel Burnham. This news had a "chilling effect" on Drexel's ability to raise money for clients like Icahn. The experts began dismissing him as an imminent conqueror of USX, though he remained a long-term threat.

In Pittsburgh, the labor talks had made a little progress because of the personal involvement of Roderick and Williams. Although the two leaders didn't negotiate over specific issues in their half dozen meetings, they helped the bargaining teams move off dead center on some items. It was unusual for a chief executive to perform such a role. "Roderick wasn't interested in getting into depth on issues," Bernie Kleiman said. "But he was willing to play around with ideas. He was certainly more open than I expected him to be."

It can be helpful in some situations for a chief executive to hear union arguments first-hand instead of relying on the interpretations of company negotiators. In the USX talks, Roderick helped resolve a major problem this way. The company had refused to give stock to employees in return for concessions, arguing that steelworkers should not be rewarded with ownership in USX's nonsteel assets for giving up wages in steel. The LTV and Bethlehem contracts, however, enabled workers to recoup at least part of their sacrifices through profit-sharing and stock ownership plans.

This issue came up at a meeting involving Williams, McGeehan, Kleiman, and Smith for the union and Roderick, Johnston, and USX financial executives. "We explained to Roderick that we couldn't have sold our members at LTV and Bethlehem the kind of cuts they took unless they had a program to get their money back," Smith said. "If the sacrifices were regarded as an investment, maybe we could sell the idea." As Smith perceived it, Roderick "began to realize what the problem was for the first time." He suggested that the negotiators design a new plan with a minimum payout that would match the value of the stock recovered under the LTV and Bethlehem programs. By the end of Round Two, the issue was largely resolved.

Roderick also came up with an interesting procedural idea. Up to now, union and company negotiators had followed traditional bargaining practice: One side would make a proposal and the other would respond with a counterproposal. The chairman suggested that they exchange proposals containing their best efforts to reach agreement. Williams thought the approach worth trying. It was something like "final-offer arbitration" in which two parties have an incentive to submit their best offers to an arbitrator, knowing that he must select one instead of halving the difference. They exchanged offers on November 15, but without the threat of arbitration, neither side had moved very far. Neither was ready to make significant compromises to halt the dispute. McGeehan summoned the USX local presidents to Pittsburgh on November 21 and handed out copies of the two proposals. If it was true, as the company believed, that large numbers of USX workers were critical of the union position in the dispute, the attitude of the local officers did not reflect this sentiment.

The officeholders at USX locals had changed substantially since 1982–83. Anticoncession presidents like Weisen, Bonn, and Grace and moderates like Bilcsik, Brown, and Cyprian no longer were part of the group, having lost office through election defeat, plant closing, or failing to run for reelection in 1985. In some ways, however, the new presidents were more militant, especially on the issues of contracting out and eliminating jobs. Larry Regan, president of the largest local, 1014 in Gary, staunchly opposed company efforts to reduce the size of mill crews.

Although the USW negotiators recommended rejection of the company offer, Regan and other presidents chastised the bargaining team for *even presenting* the company's proposal to eliminate more than one thousand jobs through a "remanning" plan. One didn't have to press an ear to the door of the meeting room in the Pittsburgh Hilton to hear the loud, angry voices inside. "We're not here to eliminate jobs," one president shouted. "We're here to get the jobs back." Jim McGeehan replied that the union negotiators weren't recommending the provision but merely reporting it. "We have not been cozy with the company," he said, obviously annoyed. "We have never been cozy with the company. Don't get up with a negative, sarcastic attitude and attack us." Things calmed down, and the presidents later that day voted unanimously to accept the bargainers' recommendation against the agreement.

The proposal to eliminate jobs had become a major issue. The company increased the number of jobs involved to 1,500, up from 1,105 in July. To make the demand more palatable, it granted a union request to give two early pensions for each job eliminated. However, the union negotiators still refused to commit the locals, in writing, to the elimination of a specified number of positions at each plant. They realized that many mill departments probably were overstaffed, but hesitated to intervene in what had always been a local matter. Although this was partly a political consideration, the leaders continued to argue that they didn't know enough about local conditions to make such a commitment. This was a valid point but one that competitive conditions in the world steel market made weaker each day.

To some degree, the national officials tried to persuade the locals to change their minds on this issue. For example, Jim Smith discussed the problem during a briefing of union employees at USX's Geneva Works in Utah after the Round Two bargaining failure. Smith provided me with a tape recording of his talk. He referred to the manning issue as an "idle time" problem, meaning that the natural bent of workers was to maintain large crews so that each member would have substantial relief time during the shift. As a former local president himself, Smith said, "I know that one of the things we did was to negotiate for more idle time. We're not a damn bit different from politicians or business executives. A little more time on coffee breaks suits us, too. There's nothing wrong with that in a society that can afford it.... I'm not ashamed of it when the industry could afford it. But it doesn't make sense to preserve every luxury of idle time at a time when we're asked to take reductions in our standard of living in order to keep our mills open."

An agreement on the jobs issue, however, would not have settled the USX labor dispute in November 1986. Many other items remained unresolved at the end of Round Two bargaining.

On economic matters, the two sides closed the gap only slightly. The union raised its concession offer from 82¢ per hour to $1.29, including a 31¢ wage reduction. The corporation valued its concession demand at $2.36 an hour, down from the $2.70 of July. But the union claimed that the $2.36 actually rose to $3.38 because the company insisted on charging employees for costs not counted at LTV and Bethlehem (see notes for explanations). The disagreement over base costing continued, the USW accusing USX of wanting an advantage of $1.50 to $2 an hour over its competitors and the company complaining that the union wanted to force it to pay up to $1.50 more than its competitors.

On contracting out, the USW held its ground, refusing to retreat an inch from the model provision negotiated elsewhere. Having forced the weaker firms to accept stringent restrictions, the union strategists felt any compromise at USX would give it a further competitive advantage, especially since the company had achieved low man-hours per ton by contracting out. The company claimed to have offered the union 85 percent of what it wanted in contracting out. This was a public relations ploy: USX based the claim on a word count of its provision, but the other 15 percent would have allowed the company much greater flexibility than the other firms.

USX unveiled one major surprise. If the union signed a four-year agreement, the company said, it would build the first continuous caster in the Mon Valley, at the Edgar Thomson plant. Without a caster, the Mon Valley probably would gradually be phased out as a steelmaking center for USX. The company also said it would erect a new hot strip mill at the Irvin Works and install a new caster at Fairfield. For years, the union had been asking for investment guarantees and now it could have them. The price, however, was forbidding. The union would have to accept a less-than-pattern agreement for four years, until 1990, although pacts at the other steelmakers would expire in 1989.

The company was not unaware of the embarrassment that might accrue to the union if it turned down a plan to save steelmaking in the Mon Valley. USX hadn't held a news conference when the first round of bargaining failed in July, but on November 21 it called in reporters and camera crews for a session with Bruce Johnston, in time to make the evening news programs. Over the next few days, however, there was a singular lack of editorial comment about the union's rejection of the offer. Did the Pittsburgh region really care whether USX remained in the Mon Valley? Management began to wonder.

Meanwhile, the union at its own news conference announcing rejection of the company proposal suggested that a panel of mediators be named to help solve the dispute. McGeehan listed the names of several former secretaries of labor and a few academics who might serve on the panel. The proposal was sincere but not without its own publicity value. Without completely squashing the idea, Johnston and Roderick reacted negatively to it.[5]

On November 21, the work stoppage was three days shy of the record 116-day steel strike of 1959.

USX: Round Three

USX and the Steelworkers had halted their propaganda campaigns during a negotiations blackout in November. After the bargaining ended, each side returned to the attack. Bruce Johnston wrote his third letter to union workers, charging that their leaders' bargaining position "has already cost every employee far more money than our proposed concessions ever would have." Jim McGeehan branded the letter "an amateurish effort from the era of high-button shoes" to divide and conquer the union.

The November bargaining failure especially irritated Williams. He had muted his public criticism of USX during most of 1986, not wanting to strain relations any more than necessary. He had backed away from an all-out war against the company and allowed himself to be lured into the Round Two talks, only to see them collapse.

"With all the new ideas out there in labor relations, this is as confrontational as any negotiations I've ever been in," he told me on December 9. "It's the kind of bargaining I used to do when I broke in. We'd try to show each other how smart we are. Johnston is a brilliant man. I wish we could direct his talents to problem solving instead of leading a debating society. The company displayed total intransigence about accepting what everybody in the industry has accepted on contracting out."

The old debate over the union's granting of larger wage relief to failing and bankrupt firms than to healthy companies also cropped up again. In his letter to employees, Johnston said: "When competitors are given selective union wage cuts and government pension bailouts, those companies continue contributing to one of the industry's biggest problems—oversupply. Excess steel capacity has to leave the market before the industry can recover; the biggest question is whose jobs will go—yours or theirs?"

Johnston may have converted some employees with this argument, but not the USW leaders. They faced the same problem that confronted Lloyd McBride in 1982 in dealing with employees in failing plants. "The real world is that people at those plants and in those communities don't go away when we refuse concessions," Williams told me a few days after Johnston's letter was made public. "The productive capacity at those plants doesn't go away." Whether the union consents or not, he said, reorganized companies usually emerge from chapter 11 with lower labor costs.

The concession agreements at some of the weaker companies undoubtedly hurt USX competitively, although the degree of damage was unclear. But what would Johnston have the union do? I presented the argument to him in an interview on December 11: If the union takes healthy companies down to the employment-cost level of the bankrupts, the competition at that level would force more bankruptcies, resulting in yet lower labor costs. Such a situation could degenerate into a long downward spiral. How far should the union be expected to go? Johnston shrugged and answered: "At whatever wages employees are willing to work, the union has to be there." It was an interesting theoretical argument.

In this interview, Johnston also contended that the union's method of putting a value on hourly employment was not only erroneous but also the principal reason that such a disparity existed between the union and company proposals. If the union's costing method was so patently wrong, I asked, why not accept the USW's mediation offer? Experienced labor mediators would spot this defect in the union's position. "We're not paid to get a third party to make our decisions," Johnston answered.

Whatever the reason, USX shortly began having second thoughts about mediation. On December 16, Roderick called Williams and suggested they explore the idea of using one mediator, Sylvester Garrett, the former chairman of the U.S. Steel Board of Arbitration. The next day, Williams approved the plan. Union officials hadn't proffered Garrett's name in November only because they suspected USX would reject him. Garrett served as the chief umpire in arbitration cases between U.S. Steel and the USW from 1951 to 1979, the year that Roderick became U.S. Steel chairman. The union figured that Roderick, who disliked some of his predecessors' policies, would view Garrett as part of that past.

Yet the "liberal" academics and former secretaries of labor, such as Willard Wirtz and Ray Marshall, recommended by the union would be even less likely to receive USX's approval. They knew little about steel, while Garrett had participated in most of the arbitration decisions that established the basic interpretations of the USW labor agreement. At the age of seventy-five, Garrett still arbitrated labor grievances at a group of iron ore companies. (Despite Garrett's long eminence in the labor field, his wife, Molly Yard, was better known to the public. A women's rights activist for many years, she was elected president of the National Organization for Women in 1987.)

Beginning December 21, Garrett met with the two parties at length every day, except Christmas Eve and Christmas Day, through January 1. Essentially, he conducted informal hearings, having each side testify about its position on every disputed issue. On January 3, the mediator delivered his findings at a joint meeting with Williams, Roderick, and the top negotiators. Garrett made three recommendations to help the parties solve their major points of dispute:

1. Employment costs. Garrett noted that USX's "cash" method of computing hourly employment costs produced a base of $25.35 at the end of the old contract. Bethlehem Steel's cost, calculated the same way, was $25.40. Therefore, the mediator recommended the two sides use the Bethlehem package of wage and benefit reductions as a basis for negotiations. The USW had granted Bethlehem cuts of $2.35, ending up—under this costing method—at $23.05. The USW and USX should negotiate the amount of reduction at USX and deduct it from $25.35, instead of the $24.21 cost base used by the union up to this point.

2. Contracting out and manning. USX should accept the contracting out principles adopted by five other steelmakers, Garrett said. He also recommended that the USW accept a job elimination plan, if not the specific one proposed.

3. Steel bankruptcies. Garrett said the parties "might wish to deal" with the fact that the depression in the steel industry had driven LTV into bankruptcy and might force others into chapter 11. Although purposely vague, this recommendation indicated that Garrett thought USX made a strong argument for reopening its contract if a competitor achieved lower labor costs through chapter 11 proceedings.

The USW viewed the economic and manning recommendations as "distasteful," and the company undoubtedly thought the same about Garrett's method of resolving the subcontracting issue. But people on both sides recognized the need for a third party to break the deadlock. The fact that an outside mediator made the proposals enabled each side to rationalize the compromises it would have to make. So the union and the company accepted Garrett's recommendations. If they failed to accept the recommendations, the mediator said, he would make them public.

Other impending problems faced the union and the company. If the shutdown continued beyond mid-January, Roderick said, the company could lose about half its steel orders for the second quarter of 1987. While the Icahn threat had subsided, Roderick probably would feel more secure about that, too, if he had a labor agreement. (On January 8, 1987, a few days after the labor talks resumed, Icahn canceled his offer to buy USX but retained his 11.4 percent stock holdings.)

The union faced the imminent exhaustion of UC benefits; they would begin running out in February, forcing the workers to get along on the $60 a week USW stipend. Meanwhile, the USW's pension, SUB, and health and life insurance agreements with USX expired at the end of 1986. To protect these benefits in the event of a long strike, the USW years ago had won industry approval to have the plans expire five months after the basic labor agreement. A strike longer than five months had seemed out of the question in the days when the federal government moved quickly to prevent or halt a steel shutdown.

When the pension agreement expired, USX was obligated to pay pensions to existing and future retirees. But the company announced that future retirees would lose health insurance coverage, as well as the accrual of pension benefits after December 31. With many older USX employees on the verge of retirement, this action created a potentially explosive situation. Steelworkers were willing to sacrifice practically anything except pension and health insurance coverage.

Roderick also had warned publicly that "the longer the strike goes, the greater the chances are" that some USX mills would not reopen. He had issued a similar warning in the summer of 1986. Although USW negotiators said they received assurances from Johnston in Round Two talks that all plants would reopen, the chairman now said that the union "had no commitment whatsoever from USX."[6]

Following Garrett's recommendations, the USW drew up a proposal based on the Bethlehem agreement, with a wage and benefit reduction of $2.35 per hour and the same contracting out language. USX produced a counterproposal which contained few changes of significance in contracting out and still called for economic concessions of more than $3. When the union "raised hell" about this and threatened to break off, the company agreed to base further discussions on the pattern contracting out provision. The two sides began negotiating on January 7 and worked intensively under a news blackout for the next week.

Since the union had increased its wage give-ups, the most crucial issues became contracting out and manning. It took four or five days, including two all-night sessions at USW headquarters, to settle most of these issues. Thomas Sterling and Jared H. Meyer, an arbitration lawyer, negotiated for the company with a union team of Kleiman, Camens, and Frankel. Williams, who stayed overnight at the union office, made strategic decisions. He and Roderick conferred a number of times, mainly to keep the talks on track.

It became a matter of pride with Williams that the union forced USX to engage in what Sylvester Garrett called "as difficult a negotiation as any I have experienced in my involvement in collective bargaining since 1937." Williams ordered the union bargainers to grant only minor variations on the major issues, especially contracting out. "We made them [USX] work at every issue," Williams said. "We ground and ground and ground right up to the end."

Negotiators in subcommittees resolved most of the difficult questions on contracting out and manning, but both sides "saved" some issues on each item to be resolved in a final trade-off at the top level. Roderick and Williams assigned this task to Johnston and Kleiman, who worked one-on-one during the last two days of bargaining. Some union and management negotiators who deal with each other for many years develop a warm, professional relationship, sometimes becoming good friends. For Johnston and Kleiman, it had always been a cool, often strained, relationship. Johnston regarded Kleiman as "the typical adversary, arbitration lawyer, always presenting long lists of things." And Kleiman thought of the USX executive as a "debater rather than a problem solver." Nevertheless, the two lawyers settled many major and minor issues in the last two days, checking with Roderick and Williams on significant questions. Both sides made some important concessions to put together a final package.

The discussions seemed endless. On the night of January 16, 1987, Williams and about ten staffers and directors were sitting at the table in his conference room, eating pizza. Each had a list of local presidents to call for a ratification meeting. As Dick Fontana entered to get approval of a news release, the phone rang. Williams picked it up and listened for a moment. It was McGeehan calling from the USX headquarters. He told Williams the last issue had been resolved.

Fontana particularly remembered the understated behavior that accompanied this important moment. Williams put the phone down. "Let's do it!" he said, meaning to carry out the telephone assignments. There was some shaking of hands, but no cheering or clapping. Everybody was too tired.

Two days later, on Sunday, January 18, the USX local presidents met at the Pittsburgh Hilton and approved the contract, 38 to 4. Don Conn of the Irvin local and Al Pena, president of the Lorain local, had come with mandates to vote against the contract. Two other presidents, Larry Regan and Bob Bratulich, head of an iron ore local, voted against it because they thought the concessions too stiff.

Eighty-eight percent of the 27,088 USW members eligible to vote cast mail ballots, approving the new agreement, 19,621 to 4,045. The lockout/strike ended on January 31. It had lasted for 184 days.[7]

The 1987 USX Settlement

As befits a battle of behemoths, USX and the USW emerged staggering from their six-month ordeal but still on their feet. Neither had landed a knockout blow. Although both claimed victory on various issues, an objective assessment would have to await the test of time.

The union had started Round Three by offering the Bethlehem package of $2.35 per hour in wage and benefit cuts. By the end of bargaining USW negotiators had increased the reductions at USX to $2.52 an hour for the first year by granting temporary additional relief in the areas of vacations, holidays, and shift premiums. This was part of a last-minute trade-off on several items. The company reluctantly retracted its demand for a reopener clause in the event other steel companies filed for bankruptcy and also agreed to provide shutdown pensions and severance pay to some five thousand workers idled in the past by plant closings.

The cost reductions would total $2.20 per hour in the second year and $2.07 in the third and fourth years. USX's hourly employment cost in the first year would go down to $22.83, without figuring in the cost of the shutdown pensions. Therefore, USX gained a cost advantage over Bethlehem in the first year and improved its position relative to National and Inland but did not achieve its objective of winning the same decrease as LTV.[8]

Both sides compromised on the costing controversy. The union accepted USX's method of calculating base costs, and the company dropped the attempt to charge to employees certain costs associated with matters like combining craft jobs. Had the work stoppage ensued because of these technical disputes alone, twenty-two thousand employees would have lost six months' pay for no justifiable reason. That the two sides used the costing differences to claw for negotiating advantage demonstrated a less than mature relationship. But the lockout/strike really resulted from more fundamental disagreements.

In the contracting out dispute, USX did not intend to take a six-month work stoppage without being able to *claim* a triumph. In the end, the union compromised in four areas and agreed to an off-pattern provision. The company accepted the basic prohibition against subcontracting, with some exceptions. On inside-the-plant jobs, it won variations involving:

1. work contracted out before March 1, 1983. If the company could show a "consistent" practice of hiring contractors for this work, it could continue to do so.

2. work contracted out between March 1, 1983, and April 1, 1985. USX could continue contracting out this work if management could prove, first, past consistency and, second, that it was cheaper to import contractors than assign the work to USW members. In other words, cost could be applied as a "reasonableness" test.

The USW also agreed to allow USX to subcontract finishing work to outside shops in order to meet a customer's demand relating to quality and delivery. Without this exception, USX might lose the customer and therefore have less work for USW members. So the union saw this change as beneficial.

On work going outside the plant, the company gained the

right to continue contracting out work if a local earlier had allowed the company to hire contractors in exchange for some benefit, such as pensions for affected workers. The company thus had "paid" for the right to take the jobs out of USW hands. This became a matter of principle with Roderick, who had told Williams that he didn't want to "pay twice" for the work by restoring it to the USW.

However, the union refused to give the company a major exception involving "mixed practices," or work that sometimes had been done by USX workers and sometimes by contractors.

In the first few months after the new agreement took effect in 1987, Roderick and Graham boasted to Wall Street analysts that the company had avoided the restrictive "boilerplate language" on contracting out that its competitors agreed to. These claims overstated the case. A comparison of the contracts shows that, apart from the exceptions, the USX provision is identical to the others. This includes the right of workers to obtain expedited arbitration on disputed work *before* the company can give it to contractors.

The union claims that as many as two thousand jobs are affected by contracting out. To decide whether to grant the exceptions, USW negotiators checked with local presidents on when management established the consistent practices and concluded that the changes will not result in a significant amount of contracting out. The company concluded otherwise. The importance of the exceptions will depend on how arbitrators interpret them, especially the term "consistent," which is not defined in the contract. It may take years to make the final assessment of which side "won" or "lost" on this provision, although early arbitration decisions favored the USW (see below).

The top union negotiators relented on the manning issue, agreeing that the locals must negotiate the elimination of 1,346 jobs. In return, USX would grant two pensions, with $400 monthly supplements, for each worker displaced by combining and eliminating jobs and also recall one person from layoff for each two pensions. In other words, the union would lose permanently 1,346 jobs, but 1,346 laid-off workers would get back on the payroll as steel orders increased. And nearly 2,700 would receive pensions.[9]

USX granted a profit-sharing plan that would enable employees to recoup a portion of their give-ups and receive at least as much as the value of stock received by Bethlehem and LTV workers. But the company escaped what it considered to be onerous provisions accepted by those firms on the grievance procedure. In November, USW negotiators had disparaged the company's request for a four-year agreement which would give it a longer period of labor stability than the other major producers. Now, the union accepted the four-year term ending on January 31, 1991.

In return, the company pledged to build casters at the Edgar Thomson and Fairfield plants, modernize the 84-inch strip mill at the Irvin Works, and keep the open hearth furnaces at Fairless open during the contract term. To ensure that USX would keep its word, the union insisted that these promises be spelled out in side letters, in language that USW lawyers thought would be legally binding.

The agreement contained other provisions that would be positive for both sides, including a training and development program for laid-off steelworkers funded by the company at $600,000 per year. Roderick, according to his public remarks on the subject, felt that USX had a social obligation to help displaced employees get back into the economic mainstream. It signaled an encouragingly progressive attitude about community affairs. USX also proposed, and the union agreed, to create a new job of "team leader," an hourly worker who would perform many of the functions of the old foreman both in craft and production units. Conceivably, this could lead to substantially more workplace autonomy for rank-and-file workers.

During the later stages of negotiations, Roderick and Williams also had discussed the need to improve management-union relations, another positive sign. The feeling began to grow that perhaps the long struggle had changed stripes on both sides. Then came the announcement of February 4, 1987.[10]

The Final Blow

A society that bases its economic decisions on statistics describing business activity necessarily lives in the past. By the time the experts develop the aggregated data into a snapshot of the economy, the scene has changed. We make vital decisions on the basis of what was, adjusting the data according to models of past experience. This produces only a shadow of economic truth, not the real thing. We have accepted unreal economic time as a fact of life, and what we know always lags behind what is.

On February 4, 1987, thirty-seven hundred USX employees learned a painful truth about lag time. They were told that their jobs had ceased to exist at midnight, December 31, 1986, when they were picketing to save those jobs. Not only that. Their jobs had not existed on January 27 when the USX board of directors approved a $1 billion writeoff of assets, in the very week that they voted to save their jobs by ratifying the new contract. Only after the ratification ballots had been counted and the lockout/strike declared over did these workers learn that USX had decided to close *their* plants. Using the lag-time concept as a weapon, USX had struck the final blow in the steel labor war of 1986–87.

On January 27, 1987, eleven days after USX and the union reached a tentative agreement and while the ratification vote was going on, the company released its fourth quarter results for 1986, disclosing that it had lost $1.83 billion. Of that amount, $1.03 billion was charged against the steel division for writing off the value of plants and production units that would be, or had been, closed. Theoretically, they had been shut as of December 31. USX, however, declined to identify the affected mills, even though the board of directors, meeting on January 27, formally approved the write-offs. Wall Street analysts speculated that USX was waiting for results of the ratification vote before naming the closed plants. These speculations were printed in national publications like the *Wall Street Journal* and *New York Times*, as well as the Pittsburgh papers.

Four days later, Lynn Williams announced that the contract had been ratified and issued a statement on the results: "We have reaffirmed the right of workers in America to have some say about the shape of their industries and their workplaces in the years ahead in the most fundamental way possible, which is, 'Am I going to be there?'"[11]

The USW president spoke too soon for the thirty-seven hundred members who were *not* going to be there. On February 4, Roderick held a news conference in Pittsburgh and disclosed which facilities were involved in the $1.03 billion writeoff. They included miscellaneous production units at several plants that had not operated in some cases for many years. This was not surprising. But Roderick also announced the closing of three entire plants (what was left of them)—

McKeesport's National plant, a small sinter plant at Saxon-burg north of Pittsburgh, and the Geneva Works in Utah—as well as most of the plant in Baytown, Texas. These plants were put on "indefinite idled status." This new term limned a sort of industrial limbo, as one reporter noted, and Roderick agreed with the description.

The company would remove the supervisory forces and padlock the gates but still not declare the plants permanently closed. They stood a chance—"although somewhat remote," Roderick said—of reopening if market demand called for it. In the meantime, the workers at these plants who still remained on the employment rolls would not be able to claim shutdown benefits. On the other hand, although they had helped ratify the contract, they would not be working under it. Limbo.

It was as if Roderick had poked a burning stick in the union's eye, almost relishing the idea of punishing people because they hadn't accepted the company's terms to begin with. In the long run, the effects of this incident would pass, but it was no way to make a start on reforming a relationship.

The USW could also be faulted for lack of diligence in questioning the company. There was no doubt that the union had been warned. In the summer of 1986, Roderick told reporters that a strike "would raise a great risk to a number of them [USS plants] as to whether they will ever open again." Although Jim McGeehan considered Roderick's warning in July as a "bargaining ploy," the union nevertheless asked USX negotiators in November what would happen to the plants if the dispute were settled. According to union bargainers, John-ston said the company planned to resume operations at each plant "when work became available for them." The company first would start up Gary, its most efficient plant, and reopen the others as the orders flowed in. The term "indefinite idled status" was not used in these discussions.

McGeehan interpreted Johnston's reply as meaning that work would not be available immediately at every plant but that all eventually would reopen. This is what McGeehan told the local presidents when they asked about the status of their plants. "I must have said it twenty times before the presidents, so imagine how I felt on February 4," McGeehan said. Jim Smith also understood Johnston's reply this way, al-though he caught intimations that some plants might not open for a long time. Indeed, he told company negotiators that he did not want to hear more details about how long it would take to reopen some plants. "I didn't want to listen to Johnston's forecast about when some plant might open because I would have been obligated to report that, and I didn't think Johnston's opinion on matters like that deserved much consideration," Smith recalled.

In the third round of bargaining, the union negotiators didn't pursue the issue, thinking that the company men would tell them if circumstances had changed. Even when the two sides drafted an agreement to finance a feasibility study of the Geneva Works, management did not reveal that the plant would be closed. Moreover, the managers of the National, Saxonburg, and Geneva plants sent letters to their employees in late December 1986, trying to persuade them that the union should have accepted the company offer to settle the dispute in November. The clear implication of the letters was that these plants would resume operations.[12]

Perhaps the union bargainers should have inferred from Johnston's comments that the four plants would remain closed for a long time. But they were not told that USX would write them off as assets and put them on "indefinite idled status." Nor were the plant managers told. On Monday, February 2, after the USW announced the contract had been ratified, the National plant called in about twenty-seven workers to prepare equipment for startup. Even then, Williams remained nervous about what Roderick might announce on the fourth. Somebody on the union staff had suggested withholding the ratification results until after Roderick's news conference, but Williams didn't think that an appropriate way to end a six-month work stoppage. He announce the results on Saturday, January 31, and made a "courtesy call" to Roderick on Monday to tell him about the vote. "I said, 'I hope you don't have any surprises for us at the news conference,'" Williams said. "He didn't indicate that he did."

At the February 4 conference, Roderick announced the plant closings and reminded reporters that he had raised the possibility the previous July. Asked if the work stoppage caused the shutdowns, Roderick said it was "a factor" and added: "I don't feel that anybody has been cheated or anybody was not alerted to the fact that if you shut down these plants, the more marginal plants would not reopen. I mean they were warned. We said it to the International, we said it locally, we said it to you all here. And now the economic hammer has dropped. And it's unfortunate, but that's the way it is."

When Williams heard this news, he was furious. He interrupted a meeting in his office to take an urgent call. "When he came back, he was absolutely livid," a visitor said. "He could hardly talk."

I called Robert Cross, a millwright at the National Works whom I had interviewed the day the lockout started. "I never dreamed this company would be this way with their people," Cross said. "I put in thirty-seven good years in the plant and for them to treat us like that...it just isn't right. I always thought real good of the company until the last few years. Now they don't have any heart or compassion. I didn't sleep for two nights, wondering what I'm going to do. Everybody here voted for the contract. If they knew this was to happen four days later, they would have reversed the vote."

A few months later, USX heeded appeals by the local unions involved and declared the National and Saxonburg plants permanently closed so the workers could receive shutdown pensions and severance pay. In Utah, about seven hundred workers were represented as plaintiffs in a $1.4 billion class-action lawsuit filed against USX, alleging that the company reneged on a promise to keep the plant in operation until 1989. USX later sold the Geneva plant to a group of local investors. The new company lost the business of supplying USX's California plant with semifinished steel. But it found other markets, kept the existing work force, and even began to hire new workers in late 1987.[13]

A Concluding Note

The USW spent $32 million from its strike fund on the USX shutdown, and its members lost six month's pay. Was the cost worth it? The union managed to level out employment costs among the five big firms, an achievement that puts the companies on a more equal competitive footing on this one measure of costs. The more important measure, cost per ton, would depend on many other inputs, including technology and—not least—whether a company could gain the commitment of employees through its management methods.

The union also succeeded in restricting the ability of the corporation to give union jobs to outsiders and therefore maintained a rough standard among the integrated producers on contracting out. There can be no doubt that USX tried to defeat the union on this issue, with a good deal less than total success. The contracting out provision took effect May 1, 1987. During the next eight months, the USS Board of Arbitration handed down forty expedited decisions on company attempts to move work out of the plant or bring contractors in. According to a USW tabulation, the union scored twenty-two clear victories, three "substantial" wins (it was awarded most of the work at issue), and split one decision with USS. The union lost thirteen cases on the merits and one because of a late filing.

Most important, in only one of the lost cases did the company cite an exception obtained in negotiations, but several of the prounion decisions involved exceptions. The other cases arose under parts of the provision identical to those in all USW contracts. Although the pattern of wins and losses conceivably could change, it appeared that USS's effort to carve significant caveats from the contracting out pattern accepted by other companies did not succeed.

Perhaps the Steelworkers' biggest gain in forcing the USX lockout was intangible. There is a point at which the employees of any company, unionized or not, will rebel against the working conditions established by management. That point was reached at USX between 1983 and 1986. To prevent a further deterioration in conditions, the union had no choice but to strike. While it may sound stilted, there is a certain dignity involved in standing up for rights and standards in the workplace, and the union enabled its members to do that.

It is not clear what price USX paid, in terms of lost revenues and other costs, for locking out its workers. Roderick contended the new contract would result in savings of $400 million over four years. But the company lost $3 million to $5 million per day during the stoppage, according to analysts, in addition to about $100 million to shut down and restart its plants and an unknown amount because of the loss of key managerial employees. One analyst contends that USX recovered its losses after the shutdown because its absence from the market for almost eleven months (counting the long closing and starting-up periods) restricted supply and resulted in price rises that would not have been available otherwise and that USX enjoyed when it resumed production. Of course, its competitors reaped the benefits of higher prices during most of the time USX was closed down.

The rising prices probably helped USX offset some of its losses; how much is uncertain. It is questionable whether the company's bargaining achievements justified a six-month work stoppage. Its major victories, the ability to eliminate 1,346 jobs and the four-year agreement, were significant. But the company probably would have been able to gain these advantages, as well as the minor modifications in the contracting out pattern that it achieved, by settling the dispute months earlier than it did.

However, USX went into negotiations with one of the lowest per-ton labor costs in the industry, according to analysts, the union, and the company itself. If it can produce as much or more steel than before with fewer workers, it will have achieved an important gain in productivity, despite the loss of whatever good will remained among its employees.

In July 1987, USX reported that output in the steel operations was nearing 3.5 tons per man-hour, making its plants among the most productive in the world. Most of the locals had cooperated with management in eliminating jobs as required by the new agreement, although the company had to go to arbitration in three cases, winning all of them. According to Bruce Johnston, the company planned to cut as many as two thousand jobs by the end of 1987. But this planned reduction could be largely offset by the number of employees USS must retain if it continues to lose arbitration decisions on contracting out.[14]

One move toward labor-management unity was USX's new profit-sharing plan. Under the novel idea suggested by Roderick himself, the plan produced an average bonus for 1987 of $1,700. Since USX's steel division did not make a profit, the amount to be shared was calculated by a formula based on Bethlehem and LTV stock prices. The price of Bethlehem common stock, in an unforeseen development, rose sharply in 1987, enabling hourly workers at USX to recoup 90¢ of the $2.52 per hour they gave up in the first contract year. Nonunion, salaried employees, however, had no chance to recover a 20 percent-plus cut in compensation since 1982, and their morale sank in 1987.

The company made other efforts to retrieve the loyalty of its employees. Management of the Mon Valley Works sponsored a get-together in Pittsburgh for twenty-two hundred steelworkers and their spouses. Roderick told a group of *Business Week* editors that USX hoped to initiate some cooperative labor programs, including LMPTs, at its plants. Ten months later, however, the corporation still hadn't extended the LMPT concept beyond the Braddock plant. There were some attempts by local managers to work more closely with the rank and file, though not in a joint, union-company process. USX's top-level management was not interested. Roderick was a first-rate financial manager who brought the corporation through hard times without violating certain standards; pension funds, for example, were not to be tampered with because they, in a sense, belonged to the employees. But he provided much less than visionary leadership in the field of human and industrial relations.

The bargaining relationship between USX and the Steelworkers did not improve after the six-month shutdown. Still based on overly legalistic descriptions of how labor and management must behave, it remained rigid and confrontational. One measure of the health of a relationship is the grievance and arbitration backlog. As of September 30, 1987, eight hundred forty-nine cases were pending in arbitration at USS, compared to the normal load of two hundred to four hundred in the 1970s, before relations turned sour.

Yet the union-management atmosphere differs from plant to plant. At the Fairless Works, Local 4889—which already had won ten of eleven expedited arbitration cases—docketed two hundred twenty additional cases in the third quarter of 1987. "USX is still USX," said the local's grievance chairman, Carmen Rettzo. "They don't want any interference from the union. They don't want suggestions. They keep talking to Wall Street about how everything is fine. But they haven't changed one bit." At Clairton, however, relations had improved between the local and a new plant manager. The team leader concept was being implemented. "We are actually running the plant," said Richard Pastore, grievance chairman at Local 1557. "They [managers] are finally saying, 'It's time to ask the employees for their input and listen to what they say.' They are starting to realize they can't make it unless the employees are involved." [15]

Chapter 21

The Once and Future Valley

During the six years covered by the major part of this book, manufacturing in the Monongahela Valley exploded into death the way it had exploded into life. The extent of economic devastation in the valley is difficult to comprehend, the more so when one realizes that it happened in little more than one generation. Steelmaking and other manufacturing reached a peak of activity during World War II, began a long, slow descent in the 1950s, and plunged to the rocks of near destruction in the 1980s. If the concentration of heavy industry in the Mon Valley was once an awesome sight, the trail of industrial wreckage was just as awful in 1987. One hardly knows where to begin a physical description of it. I shall start at Monessen, thirty-nine river miles up the valley from Pittsburgh.

The dreams of rivaling the Ruhr Valley's Essen are dreamed no more in Monessen. The Wheeling-Pittsburgh plant is permanently closed. The new rail mill may some day operate, but that is uncertain. Scratch one steel plant.

A few miles to the north sits old Donora, with small industrial plants situated on the riverbank in an industrial park once occupied by smoky steel furnaces and a zinc plant spewing deadly fumes. Donora partially recovered from its shutdown disaster of two decades ago. Twenty-three hundred jobs existed in the industrial park at its peak of activity, but these have dwindled to nine hundred in recent years. White houses on the Donora hillside sparkle in the sunlight, paying homage to a scattering of church steeples. North of Donora a half-dozen old river villages slumber in decline, not much bothered by the wail of towboats' air horns. The Mon carried 23 percent less freight tonnage in 1985 than in 1975.

On the east bank, a few shabby brick factory buildings are all that remain of a Combustion Engineering plant that until a few years ago made coal crushers and other equipment. Across the river, I see a couple of sticky black gob piles where they used to mine coal, and the Elrama powerplant with its tall, red and white stack. But mine equipment and glass factories have halted operations. The sixteen-mile stretch from Donora to Clairton once thrived on the income from Monessen and Donora steelworkers, river boatmen, hinterland farmers, coal miners, and small businessmen. Today, the setting is pleasant, but its economy is crippled.

Clairton, along with U.S. Steel's Clairton Works, has shrunk physically. The entire hot end of the plant, closed in the 1960s, has been torn down, and fields of rubble mark the southern boundary of the city's advance. But the coke works and the interconnecting Tarben Chemical plant, which converts coking wastes into carbon derivatives, will survive. Silver and green coal conveyors soar about the plant in crisscrossing patterns, partly obscuring long batteries of coke ovens and quenching towers that still send up plumes of steam. Cleansed of impurities, the steam is much less lethal these days, and vegetation has made a timid return to the long-suffering cliff across the river.

A bit down the valley, the pale blue Irvin Works sits in its sylvan setting near the top of the west bluff. It, too, will survive, along with its next-door customer, GM's Fisher Body plant. Across the river from the Irvin Works, however, the wrecks of two old plants lie side by side on Glassport's riverbank: Pittsburgh Steel Foundry and Copperweld. Several small firms now operate in parts of the old Copperweld plant, possible seedlings that may sprout into something larger. A United States Glass plant, for which the town was named ninety years ago, was destroyed by a tornado in 1963.

The going gets grimmer from here to Pittsburgh. Industrial graveyards line the Mon and extend up its narrow tributaries. The Youghiogheny flows into the Mon a few miles below Glassport, and on its south bank workers recently have demolished the long black buildings of McKeesport Tin Plate, years ago abandoned, where thousands of women workers called "tin floppers" used to pack tin sheets into boxes. Farther up the Yough, the Fort Pitt Steel Foundry shops stand silent; an attempt at employee ownership couldn't save them. The hangarlike structures of the venerable Kelsey-Hayes Wheel Company, which left years ago, still occupy a mile or so of

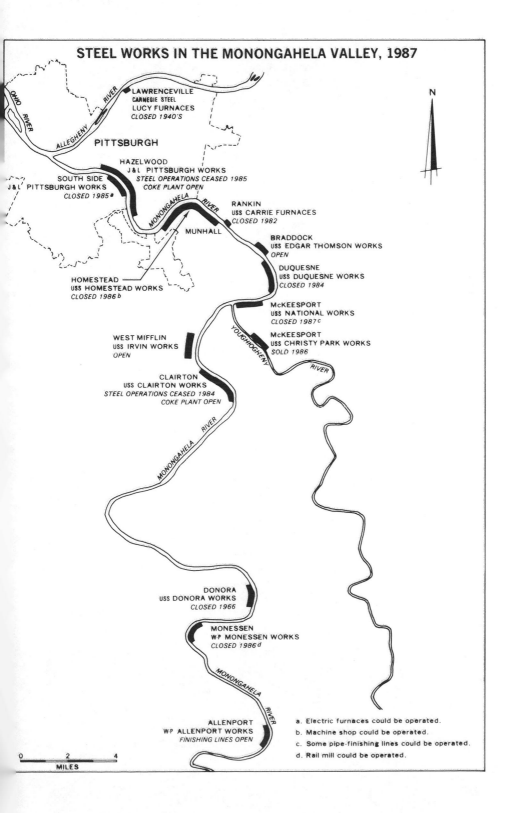

STEEL WORKS IN THE MONONGAHELA VALLEY, 1987

OHIO RIVER

ALLEGHENY RIVER

PITTSBURGH

LAWRENCEVILLE
CARNEGIE STEEL
LUCY FURNACES
CLOSED 1940'S

HAZELWOOD
J&L PITTSBURGH WORKS
STEEL OPERATIONS CEASED 1985
COKE PLANT OPEN

SOUTH SIDE
J&L PITTSBURGH WORKS
CLOSED 1985 ª

MONONGAHELA RIVER

RANKIN
USS CARRIE FURNACES
CLOSED 1982

MUNHALL

BRADDOCK
USS EDGAR THOMSON WORKS
OPEN

DUQUESNE
USS DUQUESNE WORKS
CLOSED 1984

HOMESTEAD
USS HOMESTEAD WORKS
CLOSED 1986 ᵇ

McKEESPORT
USS NATIONAL WORKS
CLOSED 1987 ᶜ

McKEESPORT
USS CHRISTY PARK WORKS
SOLD 1986

WEST MIFFLIN
USS IRVIN WORKS
OPEN

YOUGHIOGHENY RIVER

CLAIRTON
USS CLAIRTON WORKS
STEEL OPERATIONS CEASED 1984
COKE PLANT OPEN

MONONGAHELA RIVER

DONORA
USS DONORA WORKS
CLOSED 1966

MONESSEN
W·P MONESSEN WORKS
CLOSED 1986 ᵈ

MONONGAHELA RIVER

ALLENPORT
W·P ALLENPORT WORKS
FINISHING LINES OPEN

a. Electric furnaces could be operated.
b. Machine shop could be operated.
c. Some pipe-finishing lines could be operated.
d. Rail mill could be operated.

N

0   2   4
MILES

space along the Yough. About fifty employees of a new firm, Steelmet Incorporated, clatter around in the immense space, processing stainless steel scrap. On the other side of the river, CPM Industries has taken over USX's Christy Park Works, the old bomb plant.

Steelmet, CPM, and the businesses ensconced in Glassport's Copperweld plant may be what wave of the future there is for manufacturing in the Mon Valley: small shops that produce for a special market.

On the western edge of McKeesport, along the Monongahela, the National Works of USX, for years the premier pipe manufacturer in the world, is slowly being knocked to earth. Perhaps a shop or two will survive in some sort of minimill operation, but generally speaking, one and a half miles of riverbank real estate will soon be available. Just downriver from National, the lifeless structures of Firth-Stirling and the Duquesne Works of U.S. Steel face each other across the river.

Turning up the Turtle Creek defile, I see Westinghouse Electric's huge East Pittsburgh plant, the working home of ten thousand people in the early 1970s. The last eight hundred jobs will be phased out by 1989. Around a bend in the creek, all but six hundred jobs have been eliminated in Wilmerding's ninety-year-old Westinghouse Air Brake plant, now owned by American Standard, which also has shut down its Union Switch & Signal plant at Swissvale, on a ridge overlooking the Mon. Total jobs lost: eighteen hundred.

At the mouth of Turtle Creek, in Braddock, the Edgar Thomson plant—Carnegie's original steel works—still produces steel and appears to have won an extended life with the promised installation of a continuous caster. But a few miles west of ET stand the decrepit remains of the Carrie Furnaces at Rankin and across the river the vast and empty Homestead Works (its machine shop may be bought and operated), and adjoining it on the downriver side, the large complex of buildings that once constituted Mesta Machine.

In Pittsburgh, only a coking operation still works in Jones & Laughlin's plant on the north side of the Mon. Aside from a refurbished building taken over by a small fabricating company, the plant has been leveled to make way for a high-technology center. Across the river, the other half of J&L's historic Pittsburgh Works, which dates from 1857, still looms over the picturesque ethnic enclave of Pittsburgh's Southside with its old frame houses. This plant also is closed, although a citizen organization is fighting to have a public authority take over and operate the electric steel furnaces. The combined Pittsburgh Works employed more than fifteen thousand people in the 1940s.

Down the Ohio River, the Aliquippa Works of LTV has gone from ten thousand employees in 1980 to about five hundred. American Bridge and Armco plants are closed in Ambridge. U.S. Steel has pulled out of McKees Rocks on the Ohio and Saxonburg north of Pittsburgh. Many other steel-related and metals plants have closed down since 1970. Among them are Alcoa at New Kensington in the Allegheny Valley and three Babcock & Wilcox plants in Beaver County which may reopen as employee-owned plants. The list goes on.

In the Mon Valley alone, some seventy-five thousand steel jobs have disappeared since 1950, and nearly thirty thousand since 1980. As of August 15, 1987, USX had a total of 3,481 hourly and salaried employees working at Clairton (coke plant), Braddock (iron and steel furnaces), and the Irvin Works (hot strip mills). Wheeling-Pittsburgh employed about 450 people in its remaining finishing plant at Allenport, and some 250 workers operated the LTV coke plant in Pittsburgh. Total Mon Valley steel employees in 1987: approximately 4,200.[1]

This accounting does not begin to tell the full story. Each 1,000 jobs lost in the primary metals industry forces the loss of 130 additional jobs at firms that supply the industry, according to one estimate. To put it another way, 2.7 additional jobs are dependent on each steelworker. The most specific count of job loss can be found in a study by Robert P. Strauss and Beverly Bunch of Carnegie-Mellon University. They computed the decline in revenues from the loss of occupational taxes levied on each worker in nine Mon Valley communities. From 1983 to 1987, these towns lost a total of 9,710 jobs of all kinds, a decrease of 60 percent. Thousands more jobs would be added to the total if the study had included all industrial communities in the Monongahela and Turtle Creek Valleys.[2]

Such widespread carnage may be unparalleled in U.S. industrial history, especially within such a short period of time. An entire industrial civilization lies in ruins here. The closings profoundly altered the lives of several hundred thousand residents of Pittsburgh's valleys; produced new kinds of sociopolitical activism; and permanently transformed the economic character of the Pittsburgh region.

The Human Price

Residents of the Mon Valley and surrounding areas paid an appalling price for the decline of manufacturing in their region. The unemployment rate in the four-county Pittsburgh Metropolitan Area peaked at 15.9 percent in January 1983, with 168,500 persons seeking work. The rate declined to 7.6 percent at the end of 1986 for the entire area, but stayed above 20 percent in many mill towns.

What was the extent of unemployment in the valley? Official figures published by the state do not reflect part-time employment or the number of discouraged job seekers who have given up looking for work. Nor do official statistics tell us about the psychological impact on people and the financial impact on their communities. However, some excellent studies by sociologists and political scientists at Carnegie-Mellon and the University of Pittsburgh help illustrate the scale of suffering.

In 1986, researchers from Pitt's School of Social Work assessed the real extent of job and income loss in Duquesne and an adjacent section of West Mifflin, an area with a combined population of about 19,335. The survey encompassed 401 households and 1,009 persons, of whom 39 percent were employed or looking for work and thus designated "labor force members." The results, published in a study titled *Steel People,* showed that about 21 percent of the labor force was unemployed in 1986 and another 8 percent worked part time but wanted full-time jobs. Forty-six percent of households with labor force members had at least one member laid off at some time during the past five years; in one out of three households, someone had been jobless for a year or more. Such volatile employment activity, the study said, could only be described as "job chaos." Moreover, Duquesne households suffered a loss in purchasing power of $4,000 per year between 1980 and 1986.

In addition to the mill towns, many surrounding communities had large numbers of families directly affected by layoffs. Indeed, unemployment during the steel depression appears to have been as pervasive in many areas of the Mon Valley as it was during the Great Depression. I cite, for example, one neighborhood in White Oak, a sprawling suburban borough outside McKeesport. Only sparsely populated in the 1930s, White Oak expanded rapidly during and after the 1940s as people moved out of the mill towns. On average, it is a middle-income community, but the residents include a healthy sprinkling of high-income business and professional people.

Larry and Maureen Delo live in a pleasant neighborhood of two-story brick and shingle homes on the side of what used to be known as 'Pill Hill.' Thirty years ago, only doctors and other professionals could afford expensive homes on the crown of the hill. But large tracts of middle-income houses had crawled up a once-wooded slope. By the early 1970s, the developments contained a high proportion of families whose breadwinners worked in factories throughout the Mon Valley, foremen as well as wage workers.

When I visited the Delos in February 1986, three of their four grown sons still lived at home, only one of whom had a full-time job. In eleven of the fourteen nearby homes, including the Delos', at least one person lost a full-time job at a steel mill. Ten had worked for USX as hourly workers. In the Delo household, Larry lost his managerial job and his son Tom, a recent college graduate who had been unable to find a business or professional job in the Mon Valley, was laid off as an hourly steelworker. Three of the former steelworkers retired early, three took part-time jobs, one was retrained as a baker, two found full-time work, and one left the area after being divorced from his wife because of difficulties arising from the extended layoff.

At a quick glance, White Oak did not look like a distressed community. But a closer inspection revealed For Sale signs in front of many homes. Real estate values had plummeted, but houses weren't selling. Although most churches in the area handed out free food two or three times a month and conducted special collections for the unemployed, few people would admit going to a food bank. The subject was seldom discussed even between the closest of neighbors.

One way or another, most people got along. The Delos—Maureen, Larry, and Tom—managed to make ends meet by buying and running a candy franchise. They learned how to make novelty chocolates which they boxed and delivered themselves. Their living room had become a barnyard of cellophane-wrapped animals—brown bunnies, yellow ducks, white geese—in various stages of preparation for delivery. Larry, who had spent his entire working life in steel mills, peddled their product at retail candy stores throughout the region. He had never sold a thing in his life but discovered that adversity can make the body and mind dance to unknown music.

Many others found ways to make do on their own. The Pitt study of Duquesne found that members of 8 percent of the households surveyed tried to start a homegrown business. While 46 percent of the families had been affected directly by layoffs, only 17 percent reported efforts to get help from a social agency, church, or other institution. James Cunningham, a professor of sociology at Pitt, concluded in *Steel People* that "residents of the valley are not sinking with their obsolete mills and factories. These western Pennsylvania people, knit into traditional families and stubborn communities, persist. The social fabric of the valley is reinforced by shared values of hard work and self-reliance, by churches to which they have commitment, and by loyalty to a place with a history rich for the nation as well as for their own households."[3]

A large number of people moved out of the valley. Fifteen percent of the households surveyed in Duquesne had a member who left the community to seek a job, 72 percent of them to another state. Most of the movers were under twenty-five, confirming the conventional belief that older industrial workers generally do not relocate. A 1986 survey by the *Pittsburgh Post Gazette* disclosed that 74 percent of the graduating seniors of Duquesne High School planned to leave the area, 62 percent felt there were no suitable jobs in Pittsburgh, and 57 percent planned to go on to other education, compared with only 40 percent in 1981.

Out-migration can relieve some of the pressure of too many people seeking too few jobs. Its disadvantage is that the loss of young people can sap the community's vitality and make it less attractive to new employers. Nevertheless, parents and older people advised this course of action. At the Delos' urging, their three unemployed sons eventually left White Oak and found jobs in Maryland. Tom, the college graduate and former steelworker, held out the longest. Angry about what happened in the Mon Valley, he wanted to stay and help bring the area back to life. But only menial jobs were available.

A large majority of the long-term jobless experienced "significant psychological distress," said another study of Duquesne residents. "After 16 to 18 months of unemployment, the large majority of these men and women are either frustrated, bitter, angry, resentful, bewildered, humiliated, or desperate," wrote

Ray M. Milke of the University of Pittsburgh. Alcohol and drug use increased substantially, especially among younger workers. Five percent of workers questioned said they had seriously considered suicide. A later study of the unemployed in McKeesport, Duquesne and Clairton reported seventeen suicides in the three towns from the beginning of 1984 to the middle of 1985. The study did not specifically relate the suicides to joblessness but said that the 1984 suicide rate in the three cities was more than double the national rate.[4]

Mon Valley Fragmentation Revisited

The mill towns were left with enormous problems of population loss, declining tax base, inadequate services, eroding infrastructure, and fiscal distress. The loss of real estate, occupation, and earnings tax revenues when steel moved out devastated the city budgets. USX property accounted for 41 percent of Homestead's total assessed valuation, 23 percent of Clairton's, and 22 percent of Duquesne's. As the company closed sections of plants, it requested lower valuations, which of course reduced tax revenues. If the towns raised taxes to make up the shortfall, they risked driving out more people and businesses. The population decline of the past decade had already led to an increase in tax delinquencies as people abandoned homes they no longer could pay for. A total of 500 abandoned homes littered McKeesport neighborhoods in 1986, resulting in an estimated $517,000 loss in tax income, or nearly 10 percent of tax revenues.

Another problem comes with high unemployment and an aging population. McKeesport's jobless rate hovered near 20 percent. The percentage of the population over sixty reached 25 percent in 1980, almost double the number of 1950, and was expected to rise to about 30 percent in 1990. The growing need for health and welfare services for this kind of population meant an increasing use of land and buildings by tax-exempt, nonprofit social agencies and hospitals. The more help that comes to town, the more property goes off the tax rolls.

When Christine Seltzer, McKeesport's finance officer, explained this problem to me in early 1986, a tax-paying business had donated its property to a food bank and moved out of town. It was a magnanimous gesture, Seltzer said, but added: "Shall I really welcome a food bank?"

Meanwhile, the towns were forced to cut services drastically. Clairton laid off its ten policemen in 1985 and asked the state police to patrol the city. It cut firefighters from sixteen to ten, street workers from twenty-eight to fourteen, and laid off all twenty-one ambulance employees. Duquesne dropped ten police officers from a force of twenty-three, and McKeesport eliminated twenty firefighter positions and eight police officers. With the loss of public works employees, the towns' streets and other infrastructure would go for years without adequate maintenance.

The study by Strauss and Bunch outlined a grim fiscal crisis for the nine communities that belonged to the Steel Valley Council of Government. The description could also stand for many other communities in the valley. Despite increased property taxes in eight of the municipalities, expenditures would exceed revenues by a total of $705,000 in 1987, rising to $2.4 million in 1989. Clairton alone would have a deficit of $631,000 in 1989, Duquesne $608,000. McKeesport, according to Mayor Louis Washowicz, would wind up 1987 with a deficit of $300,000 to $400,000. The towns could hold off paying bills only so long before the courts stepped in. Yet the alternatives to municipal bankruptcy—raising taxes and cutting services—were self-defeating.

The Monongahela mill towns faced the very real possibility of extinction as government entities. The extreme fragmentation that resulted from the early patterns of settlement and growth now came home to roost. The Mon Valley consists of more than the river towns. It is generally defined as communities along the valleys drained by the Monongahela and Youghiogheny rivers and Turtle Creek, an area of 140 square miles comprising more than 270,000 residents and 38 municipal governments. Strauss and Bunch suggested that, for some of the towns, there may be no way out of a tightening fiscal noose except consolidation of governments. Such combinations would best involve adjacent towns, such as Homestead, West Homestead, and Munhall. However, wide geographic gaps existed between some communities. For example, it might make sense for Duquesne, Clairton, and West Elizabeth (a small town south of Clairton) to consolidate. Unfortunately, Duquesne and Clairton are separated by many miles and two communities which are not members of the council.

The political and social fragmentation of the valley mill towns, so helpful in enabling the old U.S. Steel Corporation to organize and control its great assembly line, has turned into a life-threatening liability. The political and social barriers to consolidation are awesome: the undesirability of merging better-off communities with worse-off towns, the resistance of political office-holders, and latent racism in mainly white neighborhoods, to name a few. For old-time fans of high school football, the idea of making one town out of Clairton and Duquesne is tantamount to merging Iran and Iraq.

McKeesport, which in the past took the staunchest stand against "regionalism," has indicated a change in attitude. "The 'isolationist' theory of McKeesport is not the theory today," Mayor Washowicz told me in September 1987. "I'm convinced that if we don't merge through the political process, we'll be forced to do it by court action."

When I visited Washowicz in early 1986, he had reservations about the forty-five-acre "industrial park" that would replace the National Works. All of the old mill grounds were being converted to industrial parks, Washowicz noted, and so McKeesport was not optimistic that *its* park would attract a large high-tech employer. "How many jobs in high-tech can we create?" he asked. He had proposed replacing the steel mill with a dog track. Opponents had raised moral questions about the legalization of gambling and inhumane treatment of dogs. Washowicz, however, saw his moral duty as doing "anything possible to allow people who want to work to care for their families."

McKeesport had been overwhelmed, like all other Mon Valley communities, by external forces over which it had no control—trends in the international steel economy and decisions by corporate and political leaders in Pittsburgh to keep the valley basically a manufacturing plant as long as it served a market. When the market collapsed, the valley largely was abandoned by the industrial establishment and the nation. Every other steelmaking country in the West poured substantial help into regions affected by steel closings, including retraining and relocation programs for workers and job creation programs for the communities. The United States did little of the former and none of the latter.

For forty years, the bureaucratic structure of the steel industry precluded innovation in product entrepreneurialism or industrial relations. This inertia in industry was mirrored by the lack of political innovation in the mill towns. The people did little to arouse themselves from decades of political apathy. The steel industry would always be there, wouldn't it?

By early 1986, Barney Oursler, co-director of the Mon Valley Unemployed Committee (MVUC), had been in closer contact with ordinary people in the valley than most politicians. "People have no understanding of what's happening to them," he told me. "They played the game, did what they were supposed to do, and then they were thrown away, and they have no idea why. The whole political climate in the Mon Valley was, if you're making a decent income, you have no reason to care about political things."[5]

The trauma of the 1980s, however, gave birth to a social and political activism that hadn't been seen since the steel organizing drive of the 1930s. The activism occurred in three distinct forms, one stressing service to aid the unemployed, a second relying on protest to force change, and a third emphasizing political organizing and coalition-building to maintain steel jobs in the valley.

Activism: Service and Protest

When I first met the leaders of the Mon Valley Unemployed Committee in 1982, they had no headquarters but worked out of their homes, cars, and union halls. By early 1986, the committee had moved into a McKeesport storefront at the south end of Fifth Avenue not far from the Youghiogheny. A large banner hung in the window with words written in big red letters: IF YOU THINK THIS SYSTEM IS WORKING, ASK SOMEONE WHO ISN'T.

Three or four people sat at desks, making phone calls and doing paper work. A large part of the committee's work still consisted of counseling people on obtaining government benefits, retraining, health care, help in paying mortgages. Occasionally, the committee organized demonstrations and lobbying efforts to gain specific goals, such as Pennsylvania's mortgage assistance law, the first in the nation. In 1987, the MVUC also played a major role in pushing state agencies to resolve administrative and funding problems relating to retraining programs under the federal Trade Readjustment Act.

Later in 1986, the MVUC merged with similar groups in Beaver County, the mid-Monongahela Valley area, Turtle Creek, Pittsburgh, and the Allegheny-Kiski Valley to form the

Unemployed Council of Southwestern Pennsylvania. Oursler and his co-director, Paul Lodico, also became directors of the regional organization. The committees had started with no financing except for personal donations. But companies and foundations began to see the value of a nongovernment organization that could mobilize public support to force action on the part of government bureaucracies and politicians. In 1986–87, the regional council operated on a budget of $125,000 to $150,000, supplied by personal donations and grants from companies and foundations, some of which were represented on a board of directors.

In 1987, the valley had been through five years of depression, and the mill-town people had little to look forward to. Their jobs had disappeared and would not come back; their children had to leave to obtain decent jobs; their communities faced bankruptcy; and help was nowhere in sight. Why hadn't the people rebelled? I asked Paul Lodico.

Chronology was part of his explanation. When the mills first went down, he said, workers thought they would be called back. For one hundred years, steelmaking had been up and down in the Mon Valley, and collective instinct held that this was just another downturn. In the meantime, people survived well enough on jobless benefits. When the benefits ran out, they had to scratch for a living and had no time for revolution. Good pensions won by the union cushioned the pain for many thousands of older workers. Finally, after two or three years, the perception filtered in that the mills would never come back. Younger people, who would be most likely to revolt, began to relocate.

In addition, as Lodico said, "people here have a solid, conventional outlook" formed by three or four generations of fairly settled life in the valley. After the great waves of immigration began to recede in the 1910s, few external forces invaded the valley to produce sociopolitical volatility. The migration from the South of blacks and whites had bypassed the Pittsburgh region. A strong tradition of resignation to economic forces bottled up the natural anger and frustration that accompanied the great shocks of the 1980s.[6]

The frustration finally burst forth in 1983, when small groups of people began demonstrating at Mellon Bank branches in the Mon Valley. They were organized by the Denominational Ministry Strategy (DMS), a Protestant clergy group that advocated church activism to help the unemployed. The DMS became closely allied with a tiny, tightly knit group of local union officials and rank-and-file steelworkers called the Network to Save the Mon/Ohio Valley. The Network's principal leaders were Ron Weisen, Mike Bonn, and Darrell Becker, president of Local 61 of the Industrial Union of Marine Shipbuilding Workers of America. Local 61 represented workers at Dravo Corporation, a Pittsburgh firm engaged in bargebuilding and heavy construction.

The protesters held rallies outside bank offices, sprayed skunk oil to drive away customers, and sometimes stuffed dead fish into safety deposit boxes. They achieved national notoriety on Easter Sunday 1984 when about two dozen protesters, including Weisen and Bonn, marched into Shadyside Presbyterian Church in Pittsburgh and told the congregation that their lifestyle was "evil and against all biblical example" because they spent their money on selfish interests and ignored the unemployed steelworkers. The church was targeted because DMS/Network had discovered that Tom Graham and Thomas F. Faught, Jr., president of Dravo, as well as other prominent business leaders were among its members.

Mellon Bank had been singled out for several reasons. The DMS and their union allies contended that the bank loaned money abroad instead of in the valley, thereby "decapitalizing" steel and local industries. Further, they wanted Mellon to spearhead a campaign to obtain a government program for jobs and disaster aid for the valley. Mellon, of course, did loan money to several steelmaking countries, including Britain, Brazil, South Korea, West Germany, and Japan. However, it was not as though Mellon invested in these countries while rejecting loan applications from USX. By 1983, USX had pretty much determined to phase out most of the valley plants. There is no doubt, however, that Mellon would have considered any steel venture a high-risk proposition. Bank officials generally remained silent about the protests. But a senior vice-president complained about the clergymen's objectives to the *New York Times*. "I can't turn back the clock and bring back the steel industry," he said. "They are looking for a scapegoat."

Regardless of the merits of the investment argument, the DMS campaign created a powerful symbol of conflict between Pittsburgh corporate and banking interests, as represented by Mellon, and the blue-collar workers of the Mon Valley. The evidence is that many valley residents believed the DMS's message but spurned its tactics.

Mellon had invited this unwelcome attention in early 1983 by its behavior in a bankruptcy case involving Mesta Machine Company. The bank foreclosed on its loans to Mesta and froze the company's bank accounts, an action that denied payment of $430,000 in wages and salaries for work already performed by Mesta employees. Despite complaints by the Steelworkers, Mellon delayed in asking a bankruptcy court to approve the release of funds. Incensed by the delay, Lloyd McBride in late May sent letters to 180,000 members and 499 locals in western Pennsylvania, urging that they transfer Mellon accounts to other banks. The campaign broadened into community-wide efforts joined by the DMS and the Network. Thousands of people began a minor run on the bank, and Mellon quickly moved to have the Mesta employees paid. Its carefully cultivated image as "a neighbor you can count on" suffered severely.

Although the agendas of the DMS and the Steelworkers briefly came together in the Mesta campaign, the union stayed far away from other DMS activities. District 15 Director Lefty Palm said the USW also decried the lack of investment in the Mon Valley. "But our answer is a different one than violence in the streets and setting up picket lines and shutting down branch offices," he said. "They are saying they want Mellon to take their billions of dollars and give it to U.S. Steel and tell U.S. Steel to modernize their plants. If you believe that they are going to do that, you just fell off a Christmas tree."

After the Mellon boycott, other supporters of DMS and the Network, began to drop away as the two groups intensified their confrontational tactics. One man who left the Network at this point was Mike Stout, the grievance chairman of Homestead Local 1397, once a member of Ron Weisen's rank-and-file team, and an articulate young militant. "I was part of the Network up until June 1983 and was one of the major organizers of the boycott in the Mesta affair," Stout told me in 1985. "A lot of people were involved then, but when the Network shifted direction, I got out. When they sprayed skunk oil in the banks, the people who actually smelled it were wives of laid-off steelworkers. It didn't do any good to attack local businesses as stooges of the bank. We should have been mobilizing public opinion, not attacking it."

The reason for DMS's use of "shock" tactics became clearer when the *Post-Gazette* in early 1984 published a profile of the apparent top strategist of the DMS. He was Charles Honeywell, a professional urban organizer who had studied the teachings of Saul Alinsky, a well-known Chicago urban organizer in the 1940s. Alinsky learned confrontational methods by studying the CIO's sitdown strikes of the 1930s and John L. Lewis's organizing tactics in the coalfields. Honeywell had worked in Pittsburgh as a community organizer in the mid-1970s, left the area, and returned in 1979 to advise a group of pastors on matters involving church leadership in the community.

One of the more bizarre incidents in the DMS saga occurred at the Trinity Lutheran Church in Clairton. In October 1984, members of the congregation complained to the Lutheran hierarchy about the behavior of their pastor, Rev. D. Douglas Roth, a DMS militant. When church officials transferred Roth to another parish, he locked himself in the Clairton church and refused to appear in court to face civil charges brought by his bishop. Finally arrested on November 13, Roth was sentenced to ninety days in jail for contempt of court. A few months later, the Western Pennsylvania-West Virginia Synod voted to defrock him for violating the church constitution.

DMS/Network also turned against potential allies, particularly the Mon Valley Unemployed Committee and Tri-State Conference on Steel, an organization promoting the rejuvenation of steel mills. Even Local 1397, the origin of most of the Network support, became a target of the DMS. In May 1986, DMS ministers burst into a union meeting at the local and disrupted the proceedings. Weisen couldn't contain the anger of members who were there, including Stout who led the local in voting to deny use of the union hall by the ministers for a fundraising dinner.

Stout now broke publicly with the protest groups. The DMS, he told the *Pittsburgh Press*, "is giving the struggle to save jobs a bad name." Stout reported that former Homestead steelworkers had been rejected for jobs in other states because of Homestead's reputation as the center of DMS and Network activity. "They see 'steelworker' and 'Homestead' and that's it," Stout said. Some corporate executives and Chamber of Commerce officials in the Mon Valley also reported that companies changed their minds about relocating to the Pittsburgh area because of the DMS publicity.

If companies actually did write off Pittsburgh on grounds that DMS/Network characterized worker attitudes in the valley, such decisions were not based on factual evidence. While the two groups received a lot of publicity, at no time did they command the loyalty of more than a few dozen steelworkers and other unionists. They had allowed their religious and ideological zeal to absorb all their energies. Believing in absolutes, driven by a sense of dread that produced only negativism, DMS/Network went screaming in the night.[7]

Campaigns to Save Plants

It was a frosty, clear evening, January 28, 1985. Snow and ice crunched under the tires of my car as I drove north on River Road opposite McKeesport. In the last moments of sunset, silhouetted city lights stood still on the shiny surface of the Mon at the mouth of the Youghiogheny. The rivers looked like they had been cast solid into a timeless, inverted *T*. I longed to see shadowy skaters gliding about, hands behind backs. In the 1800s, according to histories of the area, people skated and raced sleighs on the thick ice of the shallower Yough in wintertime. Workmen also sawed blocks of ice out of the frozen river and packed them in sawdust for storage in ice houses and eventual sale to businesses and residents. In those days, people used the resources at hand for fun and profit.

Today, the burning issue in the valley was whether manmade resources, the closed mills and cold furnaces, should be saved and operated by people who needed jobs, or abandoned in the nation's headlong race away from steelmaking.

I drove on into Duquesne. The immediate focus of the save-the-mills campaign, the tall, lonely Dorothy No. Six blast furnace, loomed darkly against the snow-covered bluff across the river. When U.S. Steel shut her down the previous May, many employees of the Duquesne Works wore black arm bands and somebody played "Taps" over the plant's public address system. Now the entire plant was closed. But the furnace was only twenty-one years old, a youngster as blast furnaces go, and capable of operating if repaired and outfitted with new equipment, according to a feasibility study prepared by Mike Locker. The study would be discussed tonight at a town meeting in Duquesne.

As president of Duquesne Local 1256, Mike Bilcsik had talked of keeping the Duquesne plant open under employee ownership as early as 1983. But he had no means of developing a formal plan until the Tri-State Conference on Steel became involved a year later. Tri-State was a nonprofit, public interest group organized to preserve or reopen manufacturing plants in western Pennsylvania, northern West Virginia, and eastern Ohio. Started in 1979 by political and church activists, the conference first attempted, unsuccessfully, to prevent the demolition of steel plants in the Youngstown area. Among the origninal leaders were Rev. Charles Rawlings of the Episcopal Diocese of Ohio and Staughton Lynd, a lawyer and former university professor who specialized in labor history studies from a radical left perspective.

Tri-State moved into the Pittsburgh area in 1981, organized by Lynd and local union and religious leaders, including Mike Stout; Monsignor Charles Owen Rice, Pittsburgh's "labor priest"; Rev. Garrett Dorsey of St. Stephen Roman Catholic Church in Hazelwood, the chairman; and others. It adopted a strategy initiated by the city of Pittsburgh in its famous "Renaissance" of the 1940s and 1950s. The city created a public authority under state law with the power of eminent domain to acquire land for urban redevelopment. The same power, Tri-State lawyers contended, could be used by a new "Steel Valley Authority" to acquire the land, buildings, and machinery of existing plants for economic development. A plant rejuvenated by this method could be operated by an existing company, a newly created, employee-owned firm, or a combination of both.

The conference gained little public attention until October 1984, when U.S. Steel announced it would tear down the hot end of the Duquesne plant in November. This provided the catalyst for action. "People began to realize that this stuff [steel mills] is really going down the tubes," Mike Stout explained. "This helped us pull the forces together."

Tri-State united with Bilcsik and gained the support, somewhat surprisingly, of USW leaders in Pittsburgh. The national USW usually had remained aloof from valley politics. Moreover, the union leadership harbored some resentment toward Lynd, a staunch critic of the USW's internal politics. But when Lynn Williams took command of the USW in 1984, he shifted the union toward increased activism to halt plant shutdowns.

The USW and Tri-State mounted a "Save Dorothy" campaign that swept through the Mon Valley like a tornado. Although it eventually failed and perhaps promised too much from the beginning, the campaign was an important turning point. For the first time, politicians, institutions, and ordinary people worked together to try to achieve something for their area.

The basic plan called for acquiring the Dorothy blast furnace, as well as Duquesne's basic oxygen shop and "primary" mill (a mill that rolls steel ingots into semifinished slabs, billets, and bars), and running them as a worker-owned enterprise. To determine whether the plan was feasible, the USW, the Allegheny County Board of Commissioners, and the Pittsburgh City Council commissioned a $150,000 study by Mike Locker.

In just four months, the coalition created intense excitement in the Mon Valley, winning the support of several state legislators, mayors, and council members, and national politicians such as Democratic Congressman Joseph Gaydos of McKeesport. It was a valley-bred, lift-yourself-by-the-bootstraps movement, not a project handed down by corporate leaders in Pittsburgh. It was also the first sign since the start of the big recession in 1981 that something might actually be done to arrest the fast erosion of steelmaking and jobs in the valley.

The Dorothy campaign irritated U.S. Steel officials from the outset, especially because outsiders had questioned the corporation's business judgment in shutting down a plant. At first they refused to allow the community groups to inspect the equipment in the plant. As the campaign grew, however, Roderick relented. He permitted a plant inspection by Locker, along with USW and county officials. The company made additional information available for the study, and twice delayed a deadline for destruction of the iron and steel furnaces.

Locker's study, completed in January 1985, concluded that the furnaces and primary mill could be viably operated as a producer of semifinished steel, with a potential market for 1.3 million tons annually within a 300-mile radius of Duquesne. Startup costs and operating capital for three years would total $90 million, assuming the plant were run on a low-cost basis. Locker identified several potential customers that consumed semifinished steel—usually imported from abroad—although he hadn't determined whether these specific companies actually would purchase from Duquesne.

The town meeting on January 28 showed indisputably that widespread support for the plan existed in the valley. About six hundred people packed the basement bingo room of the Sts. Peter and Paul Greek Orthodox Catholic Church, near the top of a hill overlooking the plant. This was no rump session of malcontents but an authentic cross section of citizens. In the crowd were many steelworkers, especially older men, a fair number of women, and owners of small businesses and officials from many surrounding communities.

The most stirring gatherings I have covered as a reporter always involved people who were trying to exert control over situations that had stifled them in one way or another—some antipoverty meetings in the 1960s, a few conferences where rank-and-file workers talked about self-determination in the workplace, and the 1972 founding convention of the Miners for Democracy, the reform movement in the UMW. The Duquesne meeting was of this genre. At first, the mood of the audience was controlled and cautious, but after a while I sensed a great feeling of hope welling up as speakers described various aspects of the Save Dorothy plan. Bilcsik, Stout, Locker, and Allan Sarver—a former Duquesne superintendent who had been chosen by the union to manage the worker-owned plant—were applauded and cheered. Ron Weisen, who attacked the plan, was booed.

Locker did not omit the difficulties of raising $90 million, and possibly an additional $150 million if a continuous caster were needed, to take over the plant. Sarver delineated his plan for operating with about six hundred workers, 30 percent less than the number employed for the same equipment by U.S. Steel. Wages and working conditions were not discussed. There could have been no doubt in anybody's mind, however, that the employees would have to work in smaller crews and for far less than the wages they had earned at U.S. Steel. But the attitude of the audience seemed to be that they would give anything to get back their steel mill. In later conversations, Bilcsik, who had refused to negotiate crew cuts with U.S. Steel, talked enthusiastically of "creating a new work system" geared to high productivity in a worker-owned plant. This was an interesting reflection on the failure of the bureaucratic industrial relations system that existed in the old Mon Valley plants.

The Save Dorothy campaign continued through 1985. Tri-State concentrated on political action at the local level, resulting in creation of a Steel Valley Authority by nine municipalities in the Mon and Turtle Creek valleys. Mike Bilcsik and Local 1256 volunteers maintained a round-the-clock watch on the Duquesne plant gates to make sure that U.S. Steel didn't sneak demolition crews into the plant in the dead of the night. Because of old animosities, they didn't believe Roderick's promise that USS would leave the plant standing until Lazard Frères completed a final feasibility study.

The Lazard study, released in early 1986, killed the plan. While the market for semifinished steel existed as described by Locker, Lazard found that the potential customers wanted the kind of high quality steel that could be produced only with a continuous caster. The amount of debt that would be incurred by a new company to build a caster—if it could even find banks willing to loan the required amount—made the project prohibitively costly, the study showed. Tri-State, the USW, and their supporters reluctantly abandoned the effort. Dorothy was demolished in late 1986.

The Tri-State Conference turned to other projects, principally an effort to stop the closure of two American Standard plants, Union Switch & Signal in Swissvale and Westinghouse Air Brake in Wilmerding. This, too, was unavailing, primarily because the new Steel Valley Authority lacked funds to compensate American Standard for its property. Once again, however, Tri-State had broad public support for the effort, including the backing of Pennsylvania's two Republican senators, Arlen Specter and John Heinz.

In 1987, Tri-State and the Authority took up yet another cause, that of saving electric steel furnaces at LTV's old Southside plant in Pittsburgh. In this case, LTV said it would sell the furnaces to any organization that made a fair and realistic offer.

In the end, the coalition of the Tri-State Conference, the USW, and Mon Valley communities challenged traditional corporate practices in a more fundamental way than had DMS/ Network. In a community that for decades had accepted the domination of corporate thinking and decision making— demonstrably wrong in too many cases—a new interest grouping had spoken for affected people. This did not go without notice by perceptive people in Pittsburgh. When the Dorothy Six battle was lost, the *Post-Gazette* commended USW officials "for their tireless efforts and broad imagination" in trying to keep the plant alive. These comments were on target, although the newspaper should have extended the commendation to the Tri-State Conference and many local political leaders.

However, some public policy issues still remain unresolved about Tri-State's campaigns. If the Steel Valley Authority succeeds in taking over a closed plant and installing a "new owner," the power of the state will have been used to modify the free interplay of commerce. Public monies in one way or another will be involved in the new enterprise, either because of government backing of bonds or loans or outright grants from state or local governments. It would be difficult to demonstrate that subsidization represents the most efficient use of public resources. Saving or creating jobs may be judged the higher good, but a publicly established enterprise should be required to exist on its own after an incubation period. There is also a very difficult social question involved in organizing broad public movements to save jobs. Inevitably, the polemics of the situation distort the reality, so that people are led to *expect* far more than can be delivered.

David Roderick raised both of these issues in an extended comment on the subject at a news conference in 1986. The chairman said he thought such campaigns were "not very productive." The idea that a community-based group can gain ownership of a steel plant "that U.S. Steel, with its professional management and the expertise of Mr. Graham and his people ... can't operate profitably, that somehow some people who don't know a lot about the business, all of a sudden can get a hold of it and with some money from heaven proceed to make it some viable entity, to me is a little bit of a travesty."

Roderick continued: "We never shut down anything that's profitable.... I just think it holds out so much hope for people that somehow this process is going to get them back working again in the mills and so forth, and with the overcapacity that's in the world steel market of 200 million tons, there's 20 to 25 million tons of excess capacity in the United States, that's false hope. And what we ought to be trying to do is convince the people and convince the community to be trying to direct our assets to get these people retrained so that they can be quickly employed in the other productive elements of our society and not sitting there waiting for yesterday to come back. Yesterday is gone."[8]

It must be said, however, that USX (née U.S. Steel) had a long history of ignoring or stalling action on public criticism of its internal operations. This was not villainy, simply the corporation's accustomed way of behaving, whether in reaction to attacks on the twelve-hour day in the early 1900s, criticism of the industry's pricing policies in the fifties and sixties, protests against its environmental policies in the seventies, or criticism of its autocratic decision-making processes in the 1980s. Just as the Steelworkers found it difficult to stop protecting unneeded jobs in the mills, so USX had not yet understood how to deal with increasing employee and community involvement.

Chapter **22**

The Post-Manufacturing Era in Pittsburgh

Forty-five years ago, Clinton Golden and Harold Ruttenberg— the intellectuals of the young steel union movement—hypothesized that "community disciplines" forced upon two generations of immigrant steelworkers in the Monongahela Valley had created labor stability (see chapter 7). In the late 1980s, an opposite image of labor *instability* hovers over the empty mills in Pittsburgh's industrial valleys, retarding the economic recovery of the region. The name given to this image is "bad labor climate," which—true or not—turns away prospective employers. This is the last bequest, the final irony, of a century of steelmaking in the Mon Valley: The alienation and antagonism produced by steel's industrial relations system is now making it difficult for Pittsburgh to adjust to the loss of its steel and manufacturing base.

Exploring this idea requires a brief economic profile of the region. From 1979 to the middle of 1987, Pittsburgh and the surrounding metropolitan area lost 127,500 manufacturing jobs, including 63,100 jobs in basic steel. Although retail trade, finance, and services continued to grow in the 1980s, only 75,800 new jobs were created in these sectors. In 1987, steel accounted for 2.7 percent of total employment, down from 8.8 percent in 1979 and 16 percent in 1960. While non-steel manufacturing employment may grow somewhat, the loss of steel jobs almost certainly is permanent.

Unemployment in the region would have been much higher than the 7.1 percent recorded in June 1987 if a decreasing population had not reduced the region's labor force. Population of the five counties fell to 2,337,400 in 1985, a drop of 3.5 percent from 1980. Per capita income averaged $12,456 in 1985, only 8 percent above 1979 despite a 49 percent rise in inflation. Without the population decline, the per capita increase would be even lower. The reason for the flattening effect is clear. In 1979, some 63,000 steelworkers averaged $10.26 per hour; but most of these well-paid jobs had disappeared by 1986 when steel pay averaged $14.28.[1]

The importance of relatively high manufacturing wages cannot be overstated. When Henry Ford instituted the Five Dollar Day at his Detroit auto plants in 1914, it was not because of a philanthropic urge. Ford first recognized what became the accepted economic philosophy of American industrialists: that they could profit from mass production only if the working masses had enough money to buy Tin Lizzies and all the other goods that mechanization produced in such large quantities. Wage earners in manufacturing gradually came to form the core of the American middle class whose purchasing power fueled U.S. economic growth.

It was precisely this sector of the population that declined significantly in the Pittsburgh area in the 1980s. The phenomenon was cited as a serious problem in Pittsburgh by several corporate and university executives in a report by two professors at the University of Pittsburgh, Roger S. Ahlbrandt, Jr., and Morton Coleman, published in early 1987. One corporate leader said: "There's no question in my mind that western Pennsylvania has seen the elimination of a material slice of what I call the middle class, which is what makes a geographic area really work. The people who are making $25–35,000 a year, they're the people that in large measure support the charities, that support a lot of the community activities; and they are the people that participate, which to me is a very important thing. Now, why has America been so successful? One of the reasons, I think, is that we've had a very strong middle class. . . . I don't think that the middle class is growing any more; it certainly isn't growing in the Pittsburgh area."[2]

It does not appear that this trend will be reversed soon. Manufacturing jobs constituted 27.9 percent of the region's total employment in 1975 but only 15 percent in 1987. Service jobs, meanwhile, surged to 29.6 percent of the total, up from 19.3 percent in 1975. Although service jobs generally are lower paid than manufacturing, Pittsburgh's growth as a service center has some positive implications. The fastest growth has occurred in occupations characterized as "private services," which includes health care, education, and business support services. Pittsburgh has particular strengths in these areas. It is a regional center for health services, a leading financial center, and has a research base for the growth of advanced technology industries. Carnegie-Mellon University and Pitt are both strong in materials research relating to metallurgy, and Pitt is a center of biomedical research. Pittsburgh is home to a large number of engineers, scientists, and technicians engaged in research in corporate and university laboratories. In the past, the corporate research has been confined to existing lines of business. But this research, it is said, produces ideas which could be the source of new technology.

Although advanced technology is still a small part of the region's economy, it is growing. Penn's Southwest Association, a nonprofit group that seeks new businesses for the Pittsburgh area, focuses mainly on high-tech companies. As a result of its efforts, two dozen firms with 1,200 employees located there in 1986. Another promotion group, the Pittsburgh High Technology Council, says 640 high-tech firms— mainly small—are located in nine counties. The city also is building a high-tech industrial park on the site of the old Jones & Laughlin plant on the north side of the Monongahela. The park should benefit from the presence of Carnegie-Mellon, a leader in robotics, artificial intelligence, and computer software research.[3]

Given the existence of this research capability, Pittsburgh's hope of becoming a high-tech center is not entirely forlorn. However, southwestern Pennsylvania lacks a critical component of a high-tech economy: a computer manufacturing plant. Although a further concentration of high-tech firms in the Pittsburgh area would be a valuable business generator, the industry is not expected to create large numbers of jobs. If the region is to experience healthy growth, it must also attract well-paid manufacturing jobs to replace the loss of steel mills and other heavy industry.

The decline of heavy industry hurt Pittsburgh in terms of prestige, as well as income. For many years, the city was ranked as the nation's third largest center of corporate headquarters, behind New York City and Chicago. In 1986, however, Dallas caught up to and tied Pittsburgh, each being home to fourteen of the largest industrial companies in terms of annual sales. Pittsburgh's ranking, which is considered a selling point in attracting business, may drop to fourth in 1987. Wheeling-Pittsburgh moved its headquarters to Wheeling. Joy Manufacturing, converted from a public to a privately owned company, will no longer qualify for the *"Fortune 500"* list.

Not even Mellon Bank escaped the trouble that befell Pittsburgh in the 1980s. Managed by the Mellon family until 1967, the bank was a 118-year-old symbol of the financial strength that undergirded Pittsburgh's industrial growth. In trying to become a national bank, however, Mellon made a number of poor decisions, including an effort to cash in on the energy boom in the early 1980s. In 1982, just as oil prices headed down, Mellon established a Dallas loan production office, which produced an excess of bad loans. In the fourth quarter of 1986, Mellon wrote off $55 million in bad loans and set aside another $98 million for future losses. After the bank posted a $59.8 million loss in the first three months of 1987, the first quarterly deficit in its history, Chairman J. David Barnes and several other executives resigned. Frank Cahouet, president of the Federal National Mortgage Association, became chairman.

Mellon had provided venture capital for many of the great industrial enterprises in the Pittsburgh region, including Alcoa and Gulf Oil. In the mid-1980s, critics said, the bank no longer understood entrepreneurialism that starts on a small scale. The relative lack of risk-taking (for whatever reason) to start new businesses presented a major problem. In 1986, according to Dun & Bradstreet, Pittsburgh fared poorly among cities of like size in new business startups. A total of 821 new ventures in the city employed 4,879 people, an increase from 1985, but well below cities like Denver, Seattle, Phoenix, and St. Louis in numbers of new firms and employment.[4]

All of these statistics and projections do not mark Pittsburgh as a hopeless case by any means. It is a vigorous city with a history of adapting to change. But the mediocre record of business startups and the inability to attract a major new manufacturer has disturbing implications. They can be put in better perspective by referring to a new theory of regional growth and decline, which also provides the link that I have suggested between industrial relations and the Pittsburgh region's difficult adjustment to the loss of manufacturing.

How Mature Industries Impede New Businesses

The author of the theory, Douglas E. Booth, an economist at Marquette University, contends that the dominance of large, mature industries stifles the rise of new companies. Early in its life, a future mature industry is entrepreneurial in nature. It is characterized by the constant efforts of the founder-owners to make production more efficient and profitable, a description that fits Andrew Carnegie and Henry Frick. As the industry matures, entrepreneurs are replaced by managers who focus on large-scale production of existing products rather than the creation of new ones, Booth argues. Stability rather than innovation becomes the goal in all phases of the business.

When an industry's orientation shifts from entrepreneurship to management, companies lose their skills in operating "flexible enterprises capable of carrying out rapid innovation," Booth writes. Suppliers of business services cater to the larger enterprises and ignore the small orders placed by new businesses. The entire economic infrastructure of a region is shaped to serve the mature industry. And the greater the dominance of an industry in a region, the longer it takes for new industries to rise. The applicability of Booth's analysis to the steel industry in the Monongahela and Ohio Valleys is self-evident. By the early 1900s, most businesses, governments, and social institutions depended so thoroughly on steel that the ability to innovate had atrophied.

Booth also argues that the presence of large, mature industries limits the amount of venture capital for new businesses. The Pittsburgh region's physical assets were downgraded in value by continuous usage. In the Mon Valley, steel and related industries monopolized prime riverside space, exhausted the most accessible coal deposits, defaced the mill towns, polluted the air, and contributed—along with the communities—to the fouling of the rivers. Investors became reluctant to gamble on new enterprises in this environment. More than twenty years ago, an economics professor at Pitt, Edgar M. Hoover, pointed out the great irony of the situation. The people who had benefited most from the growth and profitability of the mature industries became the least likely to reinvest in the region. "After private fortunes have been built up on the basis of the industrial pioneering of previous generations," Hoover wrote, "their heirs understandably are not always so eager as were the forebears to share the risks of staking untried local innovators."

Stability and predictability also replaced innovation in industrial relations and the management of large organizations.

The steel entrepreneurs formed massive organizations under bureaucratic management systems to exploit the valleys' resources. These industrial bureaucracies stifled initiative and competitiveness on the plant floor. This factor is of enormous importance in helping to explain steel's decline and the inability of the region to attract new manufacturing industries. It has not received much attention in the past. With the exception of Booth, regional economists seem not to have been aware of analyses of work organization and bureaucratic management published in the last twenty years by insightful students like Warren Bennis, Michel Crozier, Richard Edwards, Larry Hirschhorn, and Michael Maccoby.[5]

Methods of managing people that companies deemed appropriate to the organization of work and quality of the labor force in the early twentieth century—management by control—continued essentially unchanged, though modified by the coming of unions, into the 1980s. Mature corporations organize work bureaucratically in order to produce on a large scale and minimize costs, Booth says, adding: "It is precisely this bureaucratic form of production organization that prevents corporations from pioneering in new lines of production." The management bureaucracy in the Mon Valley alienated workers, dulling the impetus to change practices at the bottom of the organization where productivity is most quickly affected. From the early days of the USW, steel management continually repulsed union efforts to take an active role in making the workplace more productive. In the face of this opposition, the union abandoned its efforts toward worker involvement and concentrated on achieving some control over the size of the work force through contractually protected work practices.

At the same time, a curious form of job security lulled the employees and mill towns into somnolence. By the end of World War II, steelworkers in the Mon Valley assumed that their industry always would provide a source of more or less steady work at good wages. There were ups and downs, layoffs and recalls, but these cycles occurred in a lifelike rhythm that embedded itself in the valley psyche like the inevitability of seasonal change. So strong and deep was this feeling of security that many unemployed steelworkers in the 1980s would not accept the idea that their jobs were gone for good. Job security can be a very important element in a competitive labor agreement, one which guarantees jobs for current members of the work force in return for productivity growth. The Mon Valley type of security, based solely on historical precedent and a cycle of life and work established by generations of steelmaking, in the long run was debilitating.[6]

Taking into account all of the above factors, it would be difficult to imagine a human relations formula better calculated to ignore economic reality. When the time came to adapt to new levels of world competition, the Mon Valley labor-management establishment could not break out of its cocoon of inertia. Given the great structural shifts in the world steel industry, Mon Valley production would have shrunk drastically in any case. But the tragedy is that the old bureaucratic industrial relations system now overshadows the future for the Mon and Ohio Valleys.

Now it is said that a "bad labor climate" in the valleys stymies further economic development. This is one conclusion of a study by the influential Allegheny Conference on Community Development (ACCD), a volunteer organization of Pittsburgh's most important chief executives, founded in the 1940s by Richard King Mellon to provide corporate leadership in the city. At that time its strength lay in the ability to "get things done" by marshaling corporate staff and financial resources. Because of interlocking board directorships, social ties between corporate leaders, Mellon Bank's financial stake in the corporations and the city, and Richard Mellon's undisputed personal leadership, the Conference was a remarkably cohesive group in its first decade or two. The ACCD's alliance with the Democratic administration of Mayor David L. Lawrence in the 1940s and 1950s resulted in Pittsburgh's justly acclaimed urban Renaissance.

After Mellon died in 1970, the Conference gradually lost the ability to shape events. It continued to be a catalyst for action where broad consensus existed among the chief executives, such as neighborhood revival and the improvement of housing and education. Although the ACCD had supported regional planning, the Conference had minimal involvement in economic issues until 1984 when it published an economic development strategy for southwestern Pennsylvania. The report made special mention of the labor climate, noting that "Pittsburgh is regarded as an area where *labor-management relations* are more often adversarial than in other communities." But the discussion that followed contributed little to a better understanding of Pittsburgh's poor image. The ACCD recommended creation of a top-level labor-management committee to deal with the problem. Although such a committee was named, the Conference in the succeeding three years did little else to grapple with the issue.[7]

Two possible reasons suggest themselves to account for this inactivity. One is that some of the CEO's preferred to let the problem solve itself through the increasing de-unionization of the region. In late 1986, Morton Coleman estimated that only 17 percent of eligible employees in the area belonged to unions, down from well over 20 percent a few years ago, and the ratio continued to decline.

A second possible reason has to do with the structure of the Allegheny Conference and Pittsburgh's corporate ethic. The Conference apparently acts only by consensus. Just as union leaders shy away from criticizing one another, so do the Conference's CEOs. To undertake a positive program based on changing the bureaucratic management system and forming partnerships with unions would imply criticism of the largest corporate member of the body, USX.

Richard M. Cyert, president of Carnegie-Mellon University and an activist in recruiting business to the region, probably summed up the situation accurately in an article by Charles C. Robb, a Pittsburgh journalist. There is no question, Cyert said, that the region "is perceived as having a labor force that thinks only in terms of unions and that the unions are tough and that nobody in his right mind would build a manufacturing plant in the area." Moreover, an official of the nation's largest consulting firm on business locations described "the whole western half of Pennsylvania" as if it were posted with bad labor climate signs.

The region's poor image consisted of several constituent parts, some illusion and some fact, according to Robb. One was the steel industry's reputation for high labor costs. (Even with two rounds of wage cuts, USX's average wage rate still approached $11 per hour, and average earnings for incentive workers go well beyond that. In contrast, companies can open plants in South Carolina and pay $6 to $7 per hour.) But USX employees represent only a small portion of the region's manufacturing force. Other facets of the image were that Pittsburgh is strike-prone and heavily unionized. Neither perception is accurate in 1987, although both seemed to comport with what corporate officials outside—and inside—the area *want* to think about Pittsburgh.[8]

In the Ahlbrandt-Coleman report, seven chief executives (quoted anonymously) make negative comments about western Pennsylvania and lay much of the blame on unions. "We have an image that if you come to Pittsburgh you're automatically going to be unionized and, of course, there are dozens of states that do not have that image, and we're competing with them to attract companies to Pittsburgh," remarks one executive. Another speaks of "the extraordinary truculence of the unionized labor environment in the Pittsburgh area." The authors do not endorse these comments but nevertheless conclude that "western Pennsylvania is unlikely to have any large manufacturing company move in or open a new plant."

My own conversations with executives in newer growth industries have revealed an even broader perception of inefficiency in the old industrial bureaucracies like steel. They see Pittsburgh as a sooty, dark place of giant steelmaking machines, vast company bureaucracies involved in dreadful mismanagement and demanding protectionist relief, Luddite groups of preachers and workers trying to smash the region's capital-generating technology, and adversary labor relations of the worst kind.

Although it is *possible* for an anticompany ideology to pervade an area, there is no evidence that this is true of the Mon Valley. Radical ideologies are not a staple of valley life. Observers for generations have commented on the area's conservatism. One would be hard-pressed to find many people who would cross a picket line, but refusing to take someone else's job hardly implies anticapitalistic feelings. As for unions dominating business, this was never true of the steel industry in the valley. If anything, steel companies dominated the USW locals.

Indeed, only 1 of 173 Mon Valley companies surveyed in 1986 by the Mon-Yough Chamber of Commerce made negative comments about its employees. The survey was conducted, in part, during the USX work stoppage, and USX managers apparently were not available for questioning. But the owners and managers of other businesses, primarily small firms, disputed the idea of a bad labor climate. "The bad labor image is because of the steel industries," said Art Parker, executive vice-president of the chamber, "but when you talk to other employers, they pay tribute to their workers."[9]

This suggests to me that if Pittsburgh does have a bad labor climate, the source is *not* poor worker attitudes rising from a bottomless pit of union "truculence." Rather, the real source is the top-heavy, management-by-control industrial relations system put in place by the steel corporations. This is what first alienated workers and unions. This is what prevents workers from better understanding the competitive challenge. And this created the climate in which USX employees could engage in a work stoppage for 184 days in 1986–87. I believe that a large part of the truculence that Pittsburgh executives profess to see in Steelworkers locals can be traced directly to a management determination to treat employees like animals to be herded into the workplace and corralled there with minimum opportunities for involvement.

Yet, as Robb points out, enlightened management practices can be found in successful, smaller firms in the Pittsburgh area. Many high-tech companies give employees a high degree of autonomy in their work and, possibly as a result, avoid unions. Weirton Steel and Wheeling-Pittsburgh both have excellent worker participation programs. And in the city itself, several old-line Pittsburgh companies have changed radically the way they manage people and deal with their unions. National Steel, Allegheny Industries (formerly Allegheny-Ludlum), Alcoa, PPG, and Westinghouse all have established programs to involve employees on a broad scale throughout their companies, in efforts that started at the top and have continuing commitment from high-level management. Among major companies, the lone holdout is USX.

Ben Fischer, who directs Carnegie-Mellon's Center for Labor Studies, is promoting the idea that Pittsburgh, and Pennsylvania generally, could be the centers of positive labor-management relations. "Just as Pennsylvania was a birthplace of the modern labor movement, it should also be a central focus of a new labor relations system," he said. "We should not concentrate on depicting ourselves as a more peaceful place than people think but on developing new, collaborative labor-management systems." The USW supports the idea and, in fact, uses its labor education center at Dawson, Pennsylvania, to train union and management officials in collaborative techniques.

The Mon Valley is the main victim of the bad labor image and corporate Pittsburgh's reluctance to deal with it. "The labor climate is probably a significant issue because few people are going to think hard about doing much of anything in the Mon Valley because of a perceived labor militancy," says one chief executive in the Ahlbrandt-Coleman report. "You could say that it is unfair to a lot of conscientious, hard-working people, but I think that's probably the truth."

One result of this attitude is that the institutions trying to promote the region have displayed a "Greater Pittsburgh mentality" in their work, says one chief executive. They concentrate on developing Pittsburgh and its immediate environs to the detriment of the Mon Valley and other outlying areas. "The business community has little experience in dealing with problems outside of Pittsburgh's central business district, and many of the large corporations have made business decisions which have exacerbated the problems of the mill towns through plant closings and work-force reductions," conclude Ahlbrandt and Coleman.

This lack of attention to mill-town problems is resented by Mon Valley residents, former steel supervisors and workers alike. Although management methods and union attitudes separated the two groups of employees in the mills, the conflicts largely disappeared outside the plants. Supervisors up to and including department superintendents lived side by side with union members in suburban communities like White Oak. The mill shutdowns of the 1980s drew the two classes of employees much closer together in a recognition that they were all part of the same labor pool, at the disposal of corporate leaders in Pittsburgh. The brutal treatment of management people when USX reduced its work force reinforced this feeling (see chapter 16).

One day in early 1986, Larry Delo and I were talking about the plant closings and the absence of any noticeable effort on the part of corporate Pittsburgh to revive the valley. Delo, a former loyal management man, summarized it this way in his quiet manner: "It seems it's all been orchestrated by the people in power in Pittsburgh. Somebody made the decision that Pittsburgh would be a white collar town. The undertone that I get is that white collar is what is wanted here. We hear all this stuff about how Pittsburgh is Number One. They forget that the steelworkers spent a lot of money. They didn't buy Volvos and BMWs, the Yuppie cars, but they bought Chevies and Pontiacs, pickup trucks and boats."[10]

A year later, some hope for the valley grew out of a report filed by the Mon Valley Commission, a body created by the Allegheny County Board of Commissioners. The commission suggested ideas for developing former mill properties in the valley and initiated a planning process. USX agreed to sell its former plant sites at McKeesport and Duquesne to a county redevelopment agency. The commission and the new governor of Pennsylvania, Robert P. Casey, also announced their support for the construction of a north-south expressway transversing the valley to open it up to truck traffic and possible new industry. Planning and building the expressway, however, would take most of a decade—if it got built. Twenty years ago, a similar project was announced with much fanfare, only to be quietly shelved.

There was much shaking of heads then, too, about the Mon Valley's bad labor climate. Those who think America's industrial relations system—the manner in which we organize and carry on work—has no impact beyond the plant should look into the history of the Mon Valley.

Edgar Hoover, the Pitt professor who directed the regional study of the early 1960s, pointed out at the time that the Pittsburgh area's old advantages—the rivers, the riverbank mill sites, and the proximity of the coalfields—had been devalued by market shifts and new technologies. "People rather than physical geography will play the leading part in shaping the region's future," he wrote. This lesson has yet to sink in.[11]

The Mon Valley in Retrospect

In winter, the wooded hills and bluffs march purple-hued through the valley, unchanged, too steep for habitation and unavailable—one hopes forever—to billboards and commercial messages. Cold and silent, the hills brood over gray, green, and white frame homes dotting the lower reaches, and creeks still run down to the rivers as they have for many thousands of years. Springtime brings a light green, fuzzy look to the hills, with a brilliant pink splash here and there of a flowering crab apple tree. Summer sets in with the deep green density of beeches, oaks, maples, poplars, and sumacs merging in undifferentiated variety on the hillsides. The valley is a beautiful, verdant place.

Once considered ungainly barriers to industrialization, the hills of the Mon Valley in the end may be its salvation. The valley has lost most of its industry, but it has retained its great natural resources, minus several thousand acres of underground coal. Indeed, it doesn't take nature long to reclaim what man abandons. When I drive through the mill towns, I notice large empty lots where houses or mill structures once stood, now reverting to tree-filled hillsides and ravines or fields of crabby weeds. So rapid is this retreat to nature that a first-time visitor to the valley might mistake some abandoned areas for original, unused land. I know of one such place in White Oak, several acres of thickening vegetation where once stood four restaurants, a bar, a drive-in theater, picnic groves, amusement rides and galleries, and a huge public swimming pool: Rainbow Gardens, where for forty years steelworkers and their families spent happy summer days. It has all disappeared as if it never existed.

In this case and many others, peoples' grasp exceeded their reach. Tearing down in order to build anew, they destroyed what existed before realizing that the mill towns' days already were numbered. This effect was especially noticeable in McKeesport.

In the late 1960s, McKeesport tried to redevelop and renew some fifty acres of the business district to compete with outlying shopping centers. The McKeesport Redevelopment Authority tore out the mercantile guts of the city and filled some of the holes with massive parking garages, leaving others unfilled. The plan entailed the forced removal of hundreds of small businesses that lined Fifth Avenue, many of which did not want to leave. Louis Washowicz, the mayor of McKeesport, worked for the Redevelopment Authority at the time. "I moved out over three hundred businesses in one year," he said ruefully in 1986. "It wasn't a wise plan. We lost all the 'Mom and Pop' stores that are healthy for a community. We moved them out en masse, even the ones that didn't want to go."

The idea was to attract a major retailer or two and turn downtown McKeesport into a shopping mall with fancy ancillaries such as restaurants and cocktail lounges. But the planners tore down buildings before they had commitments for replacements. As a result, vacant lots are scattered along Fifth, particularly north of Sinclair Street. The McKeesport National Bank holds down the corner of Fifth and Sinclair, but one can look north and see a half-mile-long vacant space on the east side of Fifth. Part of the space is metered for parking, but most of it sits unused.

The biggest disaster of the redevelopment, however, is Midtown Plaza. The city ripped out businesses on both sides of

Fifth Avenue between Sinclair and Locust. In their place squats an elephantine, public-worksy-looking structure with ground floor shops and an overhead parking lot that, unaccountably, spans Fifth Avenue and blocks out the sun. If anyplace needs sunlight, it is a narrow street in a Mon Valley milltown. Yet the planners turned most of one block into a tunnel. When the plaza opened in the mid-1970s, people were afraid to walk through the area at night.

On the east side of the plaza is an enclosed mall, a wide corridor stretching the length of an old city block under the parking garage. In the original plan the corridor would be lined with high-class shops. In February 1986, when I visited the plaza, a small discount store and a coin shop were the only for-profit ventures in the place. The other stores were either vacant or given over to charitable and welfare services for elderly people. About thirty or forty men, obviously retired steelworkers, stood along the corridor in small groups, some of them leaning on canes, talking, gazing about. A sign in the window of the Steelworkers Oldtimers Foundation offered to help on "transportation, nutrition, socialization, recreation." I walked up and down the corridor on each side and observed dozens of old people engaged in various "senior citizen" activities. Ten older women sat in an art class earnestly sketching and painting still-life arrangements. Back in the corner, about thirty women played a desultory game of bingo. Nearby, a group of men, mostly in their seventies, sat and stood around a small pool room, watching the slow-moving action at two tables. One of the players, a thin black man with a white cap and white beard sat motionless, awaiting his turn, his cue stick standing straight up between his legs. A couple of men dozed in their chairs.

In a city where 25 percent of the people are over sixty there *should* be social centers for the retired and the elderly. It appeared as if this one, however, hadn't been planned but that the old people, needing someplace to go, just began loafing there, the way gangs of kids hang out in vacant lots. I was glad they had something to do. But these people were not really part of the general community; they were isolated from it in a downtown business district that catered to the very young, the poor, and the old. Just as American communities have little idea how to cope with an excess of young people, they don't know what to do with a lot of old people, either.

Back on the sidewalk, I took a stroll down Fifth Avenue to

see how it had changed. Up to the age of seventeen, I had never seen a more vibrant place than this narrow street packed with businesses and pulsing with activity. Now it seemed weak, anemic, barely alive. The pedestrian traffic was so thin that I could actually count the number of people on both sides of the street in a two-block area. Some twenty to thirty people, compared with the hundreds that would have filled the same area forty years ago, or even twenty years ago. By the mid-1980s, many thousands of people who once routinely shopped in downtown McKeesport had switched to outlying malls. Old friends told me that they hadn't even driven through the city for years. For all the banality and materialism that goes hand in hand with American consumerism, economic activity that draws people together in one place is a vital part of community life.

When I was growing up, the retail anchor of the whole street was Cox's. Started in 1875 as a dress shop, Cox's grew into a high-quality clothing store for both sexes and attracted people from up and down the valley for decades. It had expanded over most of one block. Closed now, the store was just a lot of plate glass windows shrouded with plastic sheets. Robert Cox, the last entrepreneur, had extended the Cox empire to several shopping malls but it had all faded, and now even the flagship was dead.

Once upon a generation, Bob Cox had seemed eternally youthful; he played softball with us when we were teenagers in White Oak. But McKeesport's decline turned him into a recluse. He lived alone on an upper floor of his empty building. One morning at 7 A.M., people observed him shadow-boxing in front of a gazebo in a small park near his store. In 1985 Cox announced he would auction off the store and start the bidding at $325,000. Nobody showed up. Now a large sign in a second-floor window pleaded: "Make an Offer."

Everything I had seen overwhelmed me. "Oh, man," I said to my tape recorder, "is this depressing!"

A statue of John F. Kennedy stands in John F. Kennedy Memorial Park, a fifty-foot square with a paved walk and some shrubbery next to the McKeesport Municipal Building on Lysle Boulevard. Kennedy brought national attention to McKeesport on that one day in 1962 when he campaigned here during congressional elections, and the Democrats of McKeesport never forgot that favor. Up went a statue of Kennedy in the city's most valued public place. John McKee, the founder of

McKeesport, can be seen in statuary form in the city's main cemetery.

But there are no statues here of John Flagler, the man who brought the National Tube Company to town in 1872 and put McKeesport on the industrial map for a century, or of Phil Murray, who opened the steel organizing drive in 1936 at a rally on the bank of the Youghiogheny next to the city dump.

Chapter 23

A Great Industrial Failure

In the 184-day shutdown of USX's steel plants, the USW proved that it still could mobilize workers to engage in old-fashioned conflict. But throwing up picket lines recedes as a viable strategy as the union loses monopoly control of the supply of labor. Nor does relying on import restraints ensure the long-term health of an industry in a world of increasing economic interdependence. The challenge for the USW in the 1980s was to formulate innovative strategies for keeping its industries competitive in cost and quality. This called for a different kind of leadership, one willing to accept new ideas, to leave behind the old bureaucratic union structures that paralleled management organization, and to take the political risk of leading change in the plants.

Lynn Williams proved his willingness to change many of the old ways. The long USX conflict and the major bankruptcies in steel obscured many joint ventures between the USW and employers. These efforts would not have been possible under a more traditional president. Most important, Williams adopted the concept that the union must share responsibility for making a business more successful. He could have been more forceful in pushing the locals to engage in productivity bargaining. And the USW has not resolved the problem of maintaining general industrywide labor standards while encouraging labor-management entrepreneurialism at the local level.

Williams also strengthened the union internally. In late 1987, the executive board adopted a "mission statement" which stated, as one objective, strengthening USW locals and allowing them more autonomy. Recognizing that a new form of union organization is necessary, the USW began experimenting with what is known as "associational unionism," providing services to associate members for whom the union does not bargain. This includes retirees and workers who supported the union in organizing efforts but were out-voted by opponents.

In 1987, four years after the U.S. economy began to rebound from recession, the domestic steel business finally enjoyed an upturn. Even with USX back in production, steelmakers managed to raise prices further and make the increases stick. The six largest producers were expected to earn $1 billion by the end of the year. But the permanence of the steel recovery continued in doubt and, in any case, had occurred only after a disastrous shrinkage. The industry eliminated 43 million tons of raw steelmaking capacity, or 28 percent, from 1982 to 1987, and analysts said yet another 20 million tons of capacity would have to be shut down. While the 1987 upswing was encouraging, the larger story remained that of the industry's decline and why it occurred.

The collapse of the American steel industry represents one of the great industrial failures of modern times. It is not the sole instance of a once prominent industry caught in a technological revolution or a sudden shift in comparative advantage from one region to another. Had the industry been blessed with the most astute of leaders—and it was not— even that management could not have maneuvered the industry fully out of the path of unstoppable economic forces of the kind that killed the horse-drawn carriage or moved the textile industry from New England to the South. However, the toppling of American steel as the premier metal-maker for the West produced consequences far vaster than most industrial fade-outs, in terms of plants shut down, workers dislocated, families uprooted, communities virtually destroyed, and entire regions thrown into economic decline.

Only historians will be able to assess the extent of damage to the American economy from the reduction in steelmaking ability and the loss of 337,552 steel jobs from 1974 to 1986. This book has focused on the labor and industrial relations issues that contributed to the downfall. But many other factors were involved. To reiterate: the steel companies failed to invest adequately in new technologies at critical junctures, failed to engage in competitive pricing when imports first penetrated the U.S. market, and failed to streamline bureaucratic management systems. Large steelmakers clung too long to the concept of vertical integration when market and technological shifts favored a concentration on fewer products and the purchase of raw materials from outside sources. In some periods, government policies forced the industry into unfair competition with foreign producers.

The overwhelming external problem, of course, was the explosive growth of steel production in developing countries in the 1970s and 1980s. At the same time, demand for steel flattened out as car makers and other manufacturers increasingly used substitute materials. These trends brought on a worldwide restructuring of steel which will continue for many years. Not just the United States, but all of the traditional steelmaking countries suffered—the United Kingdom, West Germany, France, other European nations, and Japan. All faced the common necessity of reducing capacity. Steel output in the West fell from 494 million metric tons in 1974, when the crisis started, to 368 million tons in 1986.[1]

There is one more reason for the American industry's decline: the inability of the industrial relations system to adapt, soon enough, to changing economic conditions. More than any other factor, this one resulted from a human failure of tragic proportions, a failure to unite people in a common endeavor.

Relating Wages to Productivity

The steel companies and the Steelworkers must be faulted for negotiating excessive wage increases in the 1970s, when wages rose for several years at a faster rate than productivity. The problem began in the immediate aftermath of World War II when both sides adopted the concept, although never admitting so, of engaging in wage-price bargaining. The USW directed its negotiating efforts at forcing government approval of steel price increases to cover wage boosts. The steel companies focused on the same goal, while pretending to hold the line on wage inflation. In a very real sense, the union played a role in setting prices but had no role in creating conditions that would allow the companies to grant wage increases without raising prices; in short, by increasing productivity.

In later years, steel labor and management failed to respond to ample warnings of the industry's declining competitiveness; they did not reach an understanding on a long-term, wage-productivity relationship, despite many opportunities to confront the issue, starting in the late 1950s. I do not under-estimate the difficulty of arriving at such a formula. Indeed, given the many variables that influence productivity, a per-manent formula, such as the 3 percent guaranteed under the ENA, is not supportable economically. What labor and man-agement need is a long-term flexible arrangement with peri-odic reassessments of wage, price, and productivity trends. A combination of minimum fixed wage increases and gain-shar-ing or profit-sharing bonuses seems the best way of assuring flexibility.

The move to industrywide bargaining in the mid-1950s fur-ther institutionalized the tendency to engage in uncom-petitive, administered pricing. It is understandable that the USW wished to take wages out of competition; workers should not be treated like commodities. Industrywide wage deals, however, can be inefficient for a national economy and self-defeating for workers and their employers if a substantial amount of production remains outside union control. Al-though USW members may have produced more than 90 percent of the steel sold in the United States in the 1950s, the union's coverage slipped to little better than 50 percent by the late 1970s, when imports and nonunion minimills accounted for 40 percent of shipments to the domestic market.

In the last 10 to 20 years of its life, industrywide bargaining produced large short-term gains for workers and strike-free stability for employers. The long-run consequences, however, were damaging—for employees, companies, shareholders, and the union—and not just because of uncompetitive wage bar-gains. The flow of information from management to workers in individual companies suffered from the necessity of dealing with industrywide averages. Given the exclusion of local union officers from economic negotiations, the steel rank and file remained uneducated about their companies' specific competitive situations. The ever present suspicion of a "sell out" by top union leaders forced the latter in some cases to go beyond what they knew to be wise. Perhaps most important, negotiating at the industrywide level tended to produce a false sense of well-being in the industry because the outcomes seldom reflected the growing anger and discontent on the shop floor.

To have acted on these incipient problems twenty years ago would have required a wisdom and clairvoyance rarely given to ordinary mortals. Still, the essence of leadership in indus-

trial relations ought to be the ability to foresee the adjustments in group relations necessitated by economic changes
and shifts in social values. Some people on both sides of the
steel bargaining table did look ahead but could not capture the
attention of a satisfied officialdom. One hopes that the searing
experience of the 1980s will lead to better receptivity of farsighted ideas.

In any case, industrywide bargaining has come to an end in
the steel industry. In negotiating individually with the steel
companies in 1986, the USW tried to maintain a semblance of
standards across the industry on wages, benefits, and work
practices. In reality, steel wages and hourly employment costs
are likely to diverge widely in the next few years. The union
may be able to demonstrate a rough equality between employment costs at the half dozen major companies, but because of
different profit-sharing and gain-sharing yields, true labor
costs will vary significantly in the years ahead.

For all practical purposes, pattern bargaining also has ceased
to exist among the USW's four major industries: steel, aluminum, containers, and nonferrous mining and smelting. At
most, bargains made in one industry will have some influence
on those contemplated in another. Similarly, the USW appears to have broken its habit of rivaling the UAW at the
bargaining table. At a time when industries have different
growth rates, different levels of international competition,
and different regional concerns, pattern bargaining cannot be
defended in economic terms.

Although pattern bargaining has ended for the forseeable
future, Americans have a natural tendency to compare wages.
Inevitably, steelworkers in one company that pays comparatively low wages will demand parity with those in a
higher-paid firm, leading to political problems for union leaders. Particularly troubling will be comparisons between gain-
sharing and profit-sharing bonuses. In early 1987, Ford workers
received 1986 profit-sharing bonuses averaging $2,100 while
General Motors made no profit-sharing payments, forcing the
UAW to demand a change in the GM formula. There is no
reason why the USW should escape having to deal with similar
disparities, particularly since four of the big six steel companies in 1986 not only granted profit-sharing but also won the
right to install gain-sharing plans to increase productivity.[2]

There is only one way to deflect this potential problem in
companies that cannot afford to grant parity: a constant shar-

ing of information with local union officials on financial conditions and prospects. One advantage of the new, company-by-company negotiating format is that it gives the management of each firm an incentive to build a closer relationship with its local unions. It must be noted that LTV, Bethlehem, and USX rejected the USW's request for board seats, despite deep wage cuts accepted by the union in 1986– 87. Nor have the steel companies emulated Ford and the UAW which, in lieu of union board representation, created an array of joint bodies for a periodic exchange of information and ideas. Most of the steel firms provided an extraordinary amount of information to the union during 1986 negotiations to justify wage cuts. The question now is whether they will continue this practice if the business improves.

"The Road Not Taken"

Steel's industrial relations system contained an even more basic flaw than the method of negotiating wages. How a society organizes itself to be productive is a fundamental building block of civilized life, and the organization of work is at the core of an industrial relations system. In steel, the Tayloristic structure of jobs and control management alienated workers and caused a deterioration of group relations. The companies clung to these methods long after they had ceased to satisfy production needs and the psychology of modern workers—if, in fact, the methods ever were supportable on economic and sociological grounds. Even after iron and steel production moved away from dependence on the old craftsmen of the nineteenth century, most mill operations were run by key wage workers, such as rollers and first helpers. Management, however, stripped them of authority to make decisions in the name of standardizing mill work under bureaucratic control systems. Only now are producers belatedly trying to reinvest authority in hourly workers by creating the position of team leader.

The steel union's largest failure—and it is a failure of most unions—has been its reluctance to tackle issues concerning the organization of the workplace. As I said near the beginning of this book, the USW in essence abandoned the workplace to management. Of course, it maintained a strong presence on safety matters (though less so on occupational health issues until the 1970s), the grievance procedure, and the control of crew size and idle time. Local unions policed the workplace to ensure that management lived up to detailed job descriptions and various "rights" spelled out in national and local agreements. While this classical "job-control unionism" guaranteed due process to workers, it did not really challenge management on operational decisions and ways of improving quality and output. Once workers are protected against arbitrary discipline and assignment, these issues are by far the most important that arise in any work situation *because ultimately they determine the competitiveness of the firm and the long-range job prospects of the workers.* It is questionable whether a union that does not engage management on these issues is giving good representation to its members. Moreover, the local unions' negative power to maintain certain work-force levels, regardless of the effect on efficiency, could only endanger future job prospects.

This retreat from the workplace is all the more curious in steel because of the USW's unique history of experimentation with worker participation in the 1930s and 1940s. Phil Murray, however, did not see much future in this approach, given his distrust of employers. For Murray, industrial democracy had much less to do with worker autonomy on the job than with the power of organized labor to share in decision making at the industry level and in the national economy. When it became clear that American society was not prepared to move in that direction, Murray gave up the effort and concentrated on wage negotiations.

It must be repeated that industrial management generally fought a desperate rearguard battle to prevent unions from encroaching on managerial prerogatives. Managers believed that they were saving industry from a socialistic breakthrough that eventually would lead to nationalization. The militancy of CIO unions may have given them some cause to believe in this possibility, but the history of organized labor in the United States might have shown its improbability. One piece of evidence was that not even the more militant and visionary union leaders like Walter Reuther deluded themselves into thinking that a labor party could succeed in the United States. The great majority of workers, following American tradition, were too independent to believe that they belonged to a large, permanent working class whose interests varied significantly on political and economic issues from all other classes.[3]

This evidence did not penetrate the shield that owners and managers had erected around themselves, one that permitted managers to believe that they had a monopoly on thinking and creativity. The myth of managerial omniscience has been a very destructive force in American life.

In the auto industry, management resistance deflected Reuther from his early goal of winning a voice in determining the wage-productivity-price relationship. Instead, in 1948 he assented to a traditional role for the union when GM offered a long-term contract with virtually automatic annual wage increases protected by a COLA provision. The UAW, and all unions that followed its lead, thus gained what seemed to be their permanent acceptance by large industrial management, leading to the presumption that the labor movement inevitably would expand. This wage buyout worked in that it raised workers' living standards. But it institutionalized the notion that the union member was merely a contract worker who sold his labor by the hour and had no interest in the success of the employer. And management's acceptance of unions turned out to be far less permanent than labor leaders believed.

In abandoning the CIO's prewar goal of sharing power with employers, the key industrial unions thus took a different road from the one on which they had started out. The latter, suggested Ronald Schatz, a labor historian at Wesleyan University, might be termed "the road not taken," borrowing Robert Frost's expressive phrase. If management blocked this road, it also must be said that union members did not spontaneously storm the barrier. They *seemed* little interested in collaborating with supervisors and having a representative on the company board. Of course, it is difficult to express an active interest in an unknown process, the rewards of which could only be speculated on, in comparison to the prospect of more money in the pay envelope. Both labor and management failed to realize forty years ago the benefits of creating the kinds of partnerships now exemplified in the most advanced union-management relationships.

The road that *was* taken led to reasonable stability and a procedural kind of labor relations that stressed dispute resolution, mediation, fact finding, and crisis-averting and crisis-solving mechanisms such as the Taft-Hartley emergency disputes injunction. The most comprehensive work so far published on recent trends in workplace relations and collective bargaining, *The Transformation of American Industrial Relations,* by Thomas A. Kochan, Harry C. Katz, and Robert B. McKersie, shows how the postwar system evolved into a dead end for the labor movement in a time of worldwide economic change. The New Deal industrial relations system, as these writers see it, derived from the national policy goal of labor peace established by the Wagner Act of 1935. The differing interests of labor and management were to be balanced in a Darwinian system of competing economic strengths. Workers had the right to organize, bargain, and strike, but managers ran the business. The issues that unions could bargain over were limited, a formulation consistent with the business unionism philosophy ingrained in the American labor movement since the days of Samuel Gompers. With the exception of the UAW, the USW, and a few other CIO unions, American unions wanted no part of decision making at the "strategic level" (the corporate management level), because this might compromise their efforts to represent their members. At the workplace level, workers' rights were linked to specific, minutely defined jobs in the Tayloristic tradition, and these rights were spelled out in highly formalized contracts.[4]

Unions and managements focused on the middle, or collective bargaining, level of their relationship to the virtual exclusion of the workplace and long-range strategies to meet competition. In nonunion firms also, little attention was paid in the first two decades of the New Deal to workplace issues that affected competitive ability. When a newer, more competitive model began to emerge in the nonunion sector, American unions were left behind.

Beginnings of a New Industrial Relations System

In the late 1950s and early 1960s, the more perceptive corporations began to see the value of integrating human-resource policies with investment and product decisions. They began experimenting, usually in new plants, with job enrichment, flexible work hours, semiautonomous work teams, and other practices that increased job satisfaction. Companies discovered that these reforms reduced the probability that the new plants would be organized. Firms that were once highly unionized, such as General Electric and Westinghouse Electric, began adopting this approach by building new plants in areas with little active unionism.

Some union leaders condemned the new practices as anti-union. The companies were by no means unhappy at avoiding unions through such reforms. But if operating nonunion plants in the South resulted only in 10 to 20 percent lower labor costs, this may not have been sufficient to meet the heightened competition from abroad and at home. The possibility of improving output and quality also motivated companies to reform the industrial relations system. Perhaps the best evidence of the importance of employee involvement comes from companies that voluntarily reformed their work systems to give workers more voice.

It took some unions years to understand this crucial change. Kochan and his co-authors conclude: "It was management that initiated the new nonunion personnel practices at the workplace and accelerated their expansion through strategic investment decisions. When management adjusted to economic pressures by broadening its activities to the workplace and strategic levels, it benefited from a significant power advantage and operated as the driving force in introducing change in industrial relations. The slow response of unions to management's broadened scope of activity has meant that their ability to influence critical events, and therefore, to effectively serve the needs of American workers has declined."[5]

The reforms range across a broad spectrum of personnel policies, but the most basic involve the design of work itself. The goal is to unite tasks and skills so that workers manage a larger array of tasks and, therefore, can use their own discretion and creativity in solving production problems. The most advanced work systems use semiautonomous teams with multiskilled workers who perform a series of related tasks to produce an entire product or a significant subassembly. Like the system used at LTV's electrogalvanizing plant in Cleveland (see chapter 18), these teams largely manage themselves: They determine their own work pace; schedule their own hours of work; have a voice in hiring, firing, and promotion; redesign their workplace to suit changing production needs; and often make business decisions regarding their product. This "sociotechnical" approach integrates the social and psychological needs of workers with technological requirements.

Semiautonomous teams call for meticulous planning, the right kind of employee (those who *want* to work in an uncontrolled, though self-disciplined, atmosphere), and participative managers. Only companies willing to risk a large investment in planning, training, and startup costs can succeed. But the potential for increased productivity is extraordinary. Many plants designed and operated with sociotechnical methods are 25 to 50 percent more productive than their conventional counterparts. Numbers in this range have been confirmed by companies such as Procter & Gamble, Cummins Engine, Tektronix, Gaines Foods, and Sweden's Volvo. Other companies that have adopted teamwork include General Motors, Ford, Xerox, Honeywell, Digital Equipment, Hewlett-Packard, GE, Westinghouse, TRW, Best Food, and Shell Canada.

In the mid-1970s, fewer than two dozen manufacturing plants in the United States organized work on a team basis. Ten years later, several hundred factories and offices used teamwork or variants of it, and the concept was beginning to spread rapidly. The companies listed above are successful enterprises operated by hard-headed managers who continue to crank out quarterly dividends for shareholders. They have revolutionized the way work is organized and managed, not primarily because it makes employees feel good but because it is profitable. Despite this evidence, only a limited number of companies have converted to the new ways, primarily because it is exceedingly difficult to tear down management bureaucracies and institute real change. Many companies blanch at the prospect of such deep reform.

As American manufacturing companies increasingly adopt computer-based technology, the new model of work organization becomes increasingly important. This is because many functions in the workplace—materials handling, parts manufacture, assembly, inventory control, and testing—will be integrated by computer. Instead of working at narrow tasks, employees must monitor and oversee the computer system, engage in troubleshooting, and change programs to move quickly from one product model to another. Richard E. Walton of Harvard, a pioneer theorist and installer of sociotechnical systems, points out that job-control unionism becomes irrelevant with the use of computers. "The integration no longer makes it possible to define jobs individually or measure individual performance," he said. "It requires a collection of people to manage a segment of technology and perform as a team."

Teamwork and other innovations that expand worker responsibilities and discretion also enable a company to redesign its products more frequently. In *The Second Industrial Divide,*

Michael Piore and Charles Sabel make a good case for shifting from the old mass-production model to a craft-based "flexible specialization" concept in which workers play an important role in innovation. A wholesale conversion to flexible specialization, as urged by Piore and Sabel, seems questionable, but clearly this model will spread. Where it is instituted, the new industrial relations system must be used.[6]

This brief discussion has only touched some of the main issues involved in the nation's budding changeover to a new industrial relations system. I believe that the old system gradually is being replaced, but organized labor has yet to find its place in the new system.

Issues for Unions

No one knows how far worker participation has spread in the United States, but some type of participative activity very likely has been tried in a large majority of manufacturing plants. The most common form is the problem-solving committee, variously known as LMPT, QWL Group, or Quality Circle. But this is only the beginning of a real participation that cuts across the entire company with mechanisms linking shop-floor teams with committees of department superintendents and union stewards, plantwide steering committees, and companywide policy committees. In perhaps a few dozen situations, unions also are represented on the corporate board, a strategic-level involvement which will grow.

Participation and cooperation are delicate things in a country where the adversary principle has meant so much, and many union-management efforts have collapsed after brief efforts or even after some years. However, by the mid-1980s an impressive number of collaborative relationships had lasted for several years—impressive, that is, in comparison with the predictions of skeptics who said that cooperation would fail in the United States. For example, the International Union of Electronic Workers (IUE) and the Packard Electric Division of GM turned toward cooperation in 1978 at Packard Electric's large complex of plants near Warren, Ohio. The relationship has survived a number of calamitous events, sometimes waning but never dying out, and today is one of the most advanced systems in the nation. The IUE local in Warren has a voice in most of the crucial business decisions made by the division, because management has learned that business strategies can live or die on the basis of human-resource policies.

The union and company jointly manage several functions formerly the exclusive province of management, including plant design and location, job redesign for new model production, training and retraining. Problem-solving committees pervade the plants, and the division began to implement team-based production in the early 1980s. Plant managers consult with union stewards on practically every decision, including the promotion of people to managerial posts. Indeed, the Packard Electric plants have come closer than any that I have visited to what must be a goal of collaboration: a continuous, natural process of rank-and-file participation in the managing of work. An employee does not have to attend a problem-solving group session to advance ideas or suggest changes that would improve quality or efficiency.

GM, in conjunction with the UAW, is also planning to build a small car, Saturn, in a Tennessee plant with a new work system that may become the most sophisticated in the United States. If anything should have opened management's eyes to the value of allowing greater autonomy in the workplace it is the GM-Toyota joint venture at Fremont, California, in a refurbished plant which for years assembled cars under GM management. It was, as Doug Fraser puts it, "the worst car plant in the GM system" because of quality problems and a combative union-management relationship. Absenteeism averaged 21 percent. GM closed the plant. When it reopened in 1985 under Toyota management, the work force consisted entirely of former Fremont workers who had been trained under the Japanese management system. They work in teams and have a great deal of discretion in managing their own production areas. Even though Toyota did not install new technology, this plant now turns out a higher quality car than any other GM plant. Absenteeism averages only 2.5 percent, including absence for illness and days off. It is not that the Japanese discovered something new about human nature, or have a patent on the most efficient organization of work. It is simply that they put to work what American work reformers and behavioral science theorists have been trying to tell management for decades: that people generally *want* to work, *want* to produce a good product, and will do so if given a chance to exercise responsibility.

One such theorist is Warren Bennis, former president of the

University of Cincinnati and now a professor at the University of Southern California. In the early 1960s, Bennis explained the need for worker autonomy in a basic, yet striking, way that is difficult to forget. Creating a competitive workplace, he wrote, calls for the use of the *scientific method*—not *scientific management*—to solve constantly changing technical problems. "The institution of science is the only institution based on, and geared for, change." But the spirit of inquiry can't flourish in a militaristic environment ruled by mechanization and blind obedience. This is why "democracy in industry is not an idealistic conception but a hard necessity." Bennis added: "For democracy is the only system of organization which is compatible with perpetual change."[7]

The most advanced systems have as their goal the establishment of a partnership to run the enterprise. Once again, I must repeat that this does not mean managing a plant or company by committee. It does mean that management consults with labor on important decisions, especially those directly affecting workers. This kind of consultation pervades a new, three-year agreement negotiated by the UAW and Ford in September 1987. The contract guarantees job security for all present workers, as well as a set number of jobs, throughout the company (because demand is volatile in the auto industry, employees may be laid off during business slowdowns, though they will receive SUB benefits for up to two years). The agreement also contains an array of employment programs jointly administered by union and management. The UAW does not have a representative on the board of directors, a consequence of a Justice Department opinion that antitrust laws *could* be violated if the UAW had board representatives at more than one auto producer (it already has a representative on the Chrysler board). This is one of many areas of potential conflict with the law that should be treated by amendments to the National Labor Relations Act to reflect the new competitive environment.

Ford's new environment began in 1980 when Peter Pestillo of Ford and Don Ephlin of the UAW began to put in place the company's extensive Employee Involvement program. In 1982 Ephlin and Fraser, then UAW president, made a breakthrough on job security at Ford. It was in 1982 also that the UAW signaled it could forego the traditional 3 percent annual wage increase and Ford granted its first profit-sharing plan. These were major changes in Ford's industrial relations system, but skeptics said they were only temporary adaptations to an extremely adverse business climate. Five years later, the two sides not only had expanded job security and the Employee Involvement program, they also negotiated a contract with only one 3 percent annual wage boost in three years. In 1988 and 1989, employees will receive bonus payments averaging $1,150 and $1,200 respectively. These payments will *not* become part of the wage base.

Looking back on what has happened at Ford in the 1980s, Doug Fraser, now retired, summed it up this way for a group of business executives in 1987: "We're never going to go back to the old days. From this day forward, we're going to continue to change to meet the greater international competition."[8]

LTV Steel and National Steel, along with the USW, are now creating equally advanced systems of extensive worker and union participation. Indeed, the National agreement of 1986 goes further than any other labor agreement in *requiring* the participation of *all* workers in activities related to employee involvement. All present workers at National also are guaranteed their jobs for the term of the contract. Layoffs cannot take place even during business turndowns. In return, workers are expected to help make the plant more efficient by suggesting ways of eliminating jobs, knowing that they will not wipe out employment for any employee currently on the payroll.

When I visited National's Great Lakes Division plant in mid-1987, the new agreement had produced many striking changes in plant operation, brought on by employee ideas. The ideas went far beyond more efficient ways of wielding a wrench or reducing production wastage. In one department, a team of rank-and-file workers wrestled for months with the problem of making their cold mill more efficient. The team analyzed the working operation of the mill, noting the tasks performed by each worker, and finally recommended a solution. Eight positions could be eliminated by shifting tasks to other workers without unduly burdening the remaining production employees, the team concluded.

These were rank and filers defying tradition by telling management how to eliminate jobs. No union member can be very happy about reducing employment opportunities for future union members. But the competitive situation is such that unless the steel companies streamline their operations, there will be *no* jobs in the future.

The National Steel–USW Cooperative Partnership Program does not stop on the plant floor. At the Great Lakes plant, James Howell, division general manager and general superintendent, not only shares information with Local 1292 President Ray Bonds; he invited Bonds and another union official to be involved in developing the 1987 business plan for the division. The unionists sat in on the business meetings and contributed ideas to the discussion. "I still make the decisions," Howell told me, "but I consult all the time with the union people. I think it's a great way to run the business." The plant has sharply reduced production costs.

Giving workers autonomy over their jobs is *not* merely a prescription for treating a vague malaise of worklife to make employees "happy." Nor should the significance of how society organizes work and relations between managers and employees be confined to a single plant, company, or industry. Participation should be the foundation of a free, competitive economy. That this has not been the case in the American steel industry, I suggest, helps explain the decline of the industry and also the difficult readjustment of the Pittsburgh region to the loss of its manufacturing base.[9]

Autocratic management will survive in many places in the United States. But participaion will continue to spread. With or without the labor movement, the American industrial relations system must be reformed if this country is to compete in the new world environment. I believe, however, that the presence of a union strengthens the participative process and, indeed, makes it possible in many situations.

The survival of the labor movement as a large and vital democratic force in this country may depend on its ability to accept innovation. Moreover, it is impossible to conceive of a free, democratic society without an independent labor movement.

A Note on Sources

Part of this book is about contemporary events, part about the history of steel unionism, and obviously I have had to rely on different sets of sources for the two parts. Much of the material on contemporary happenings comes from my own reporting. By this I mean that I was on the scene as an observer when many of the events occurred, from 1965 through 1987, and developed large files of notes, documents, and news clippings. I would not pretend, however, that a reporter's immediate *sense* of events accurately portrays the total *reality* of complex interactions that arise out of social, political, economic, and psychological forces. For the purposes of this book, therefore, I conducted numerous interviews of people who participated one way or another in these events. The interviews, mostly recorded on tape, expanded and to some degree reshaped my knowledge of what occurred, particularly with regard to decisions made behind the scenes. Although I have tried to avoid citing anonymous sources, in some cases I have had to honor requests for anonymity.

My detailed accounts of labor negotiations in the steel industry in 1982–83 and 1986–87 are based on interviews with the direct participants in bargaining, as well as many other officials and aides on both sides. Only five men participated in the top-level negotiations in 1982–83, J. Bruce Johnston and George A. Moore, Jr., for the industry and Lloyd McBride, Joseph Odorcich and Bernard Kleiman for the Steelworkers. Two of the negotiators, Moore and Kleiman, took personal notes. In reconstructing the most important meetings, I have pieced together interviews with each of the five bargainers. Some interviews took place shortly after the meetings. While it has been impossible to check each factual assertion with each of the negotiators, I believe that I have sufficient corroboration to make the statements that appear in this book. The direct quotations attributed to people in the closed sessions are *not* fictional creations but rather are based on the memories and notes of the participants. I wish to make it clear, however, that I alone am responsible for the version of events presented in this book. The fact that I have interviewed the participants does *not* commit them to my reconstructions.

Because of the death of Lloyd McBride on November 6, 1983, some of the events of 1982–83 remain clouded. McBride, as I make clear in the book, carried out the union's bargaining policy to a large degree by himself. Only he could tell the full story of those years from the union's point of view. Although I interviewed McBride for this book before he died, I was unable later to check back with him on his comments that were placed in question by subsequent interviews with other people. His comments and those of local union officials at meetings of the Basic Steel Industry Conference in 1982 and 1983 are drawn from transcripts of the meetings, to which the United Steelworkers gave me access.

Contract negotiations in 1986–87 were conducted at individual companies and involved many more people than the earlier round. My detailed descriptions of the talks at LTV Steel and National Steel are based on interviews conducted before and after the negotiations with company and union officials. My abbreviated accounts of bargaining at Bethlehem, Inland, and Armco relied on summaries and statements issued by both sides, as well as brief interviews.

The USX negotiations and work stoppage in 1986–87 presented formidable difficulties. Since these events form the dramatic climax of the book, I have recounted them in some detail. I was able to conduct separate interviews on several occasions with most of the six union negotiators (including Kleiman, who kept notes), as well as USW President Lynn R. Williams. Corporations operate in a more hierarchical fashion than unions, typically allowing only one executive to speak for the company on a given issue. Of the three USX bargainers—Johnston, Moore, and Thomas W. Sterling III—I could interview only Johnston, whose time was limited. My two sessions with him on the 1986–87 negotiations only partially covered the material.

Fortunately, another record of facts exists. After the USX work stoppage began on August 1, 1986, hearings were held in the seven states in which USX employed USW members to determine whether the dispute should be classified as a strike or lockout. The transcripts of those hearings, along with findings of fact and decisions by referees and boards of review, form an important factual record. Hearings conducted in Salt Lake City by the Industrial Commission of Utah provided an especially comprehensive record, since Johnston, Kleiman, and the union's chief bargainer, James N.

McGeehan, testified at length. I have relied heavily on a transcript of these hearings, supplemented by interviews, in writing about the bargaining that preceded the USX shutdown.

For accounts of USX activity outside negotiations, transcripts of quarterly news conferences held by Chairman David M. Roderick are important sources of information. However, my attempts to interview Thomas C. Graham, president of the steel segment of USX, were unavailing. Twice, by letter, I asked for an interview with Graham. The first request was rejected. USX responded to the second request by offering only to answer written questions. However, I received no answer to a letter posing questions.

The history portions of the book (chapters 4, 7, 11, and parts of 12) were developed mainly through research in various archives. My discussion of labor-management committees and labor's role in World War II government agencies, as well as the Murray Plan and the Reuther Plan, is based primarily on the voluminous records of the War Production Board in the National Archives in Washington. Other manuscript collections are cited. The Pennsylvania State University Labor Archives at State College, Pennsylvania, has the Clinton S. Golden Papers and the David J. McDonald Papers. The Morris L. Cooke Papers are at the Franklin D. Roosevelt Library at Hyde Park, New York. The Columbia University Oral History Collection holds the Lee Pressman Papers. The Philip Murray Papers are at Catholic University of America in Washington, D.C. Unfortunately, this file consists largely of administrative trivia, as many scholars have pointed out and as I discovered for myself. I have also read with some profit a number of interviews in the extensive series of interviews of former steelworkers and USW officials conducted by the Penn State Oral History Projects.

The U.S. Labor Department has a six-volume set containing transcripts and statements from the President's Labor-Management Conference of 1945. Files of *Steel Labor*, dating from the first issue in 1936, are held in the Labor Department library in Washington and the USW office in Pittsburgh. The USW maintains its own archives containing minutes of executive board meetings, reports, studies, correspondence, and other documents. With a few exceptions, however, the material is not organized for research. In an attempt to do this, the USW several years ago put some documents on microfilm, for example, the records of the union's participation in the Human Relations Committee in the early 1960s. The union was helpful in granting me access to these files. Unfortunately, they are incomplete, containing only the first page of each document. The minutes of board meetings are still off limits to researchers, although portions of one such meeting in 1945 were read to me by a union lawyer. In the Monongahela Valley, the McKeesport Heritage Center is an important source of local history; it contains microfilm records of the McKeesport *Daily News* dating from 1884, as well as other documents.

Regarding notes, I have tried to follow a consistent style for citing interviews. If the date and place of the interview are not stated in the text, they can be found in the notes. To keep the notes as brief as possible, I simply list the name of the source, place of the interview (or "phone" if it was conducted by telephone), and the date.

Following is a list of abbreviations used in the notes:

AISI	American Iron & Steel Institute
BW	*Business Week*
MDN	*Daily News,* McKeesport, Pa.
NYT	*New York Times*
PPG	*Pittsburgh Post-Gazette*
PP	*Pittsburgh Press*
PSULA	Pennsylvania State University Labor Archives
SL	*Steel Labor* (USW newspaper)
WSJ	*Wall Street Journal*
USS	United States Steel Corporation
USW	United Steelworkers of America

Notes

Chapter 1

1. Pension expenditures: Otto A. Davis and Edward Montgomery, "Study of Income Security in Basic Steel Industry," School of Urban and Public Affairs, Carnegie-Mellon University, as reprinted in *Daily Labor Report*, May 16, 1986, pp. D1–D10.

2. Employment statistics: Michael A. Acquaviva, Pennsylvania Department of Labor and Industry, Office of Employment Security, Pittsburgh, August 1987. The five counties included in the Pittsburgh Consolidated Metropolitan Statistical Area are Allegheny, Beaver, Fayette, Washington, and Westmoreland. The steelmaking area of the Mon Valley lay mostly in Allegheny County, though the southern portion, including Donora and Monessen, is in Washington County. Beaver County, located west and north of Pittsburgh, also once was a heavy steelmaking area.

3. It is difficult to estimate the actual amount of excess steelmaking capacity in the world. Many countries do not report accurate capacity statistics. An analyst in the iron and steel office of the U.S. Commerce Department's Basic Industries Section estimated that the capacity to produce raw steel exceeded apparent consumption by about 150 million tons (Ralph Thompson, phone, Feb. 7, 1987). AISI estimated 200 million tons (Janet Nash, phone, Oct. 30, 1987). Whatever the true amount, it was far and away sufficient to create widespread disruptions in steel markets.

4. Because the terms in the industrywide agreement were incorporated in each company's contract, it is technically correct to say that the USW negotiated eight contracts. For the sake of simplicity, however, when I use the term "the contract" or "the agreement," I will be referring to the industrywide pact unless otherwise noted.

5. My account of the BSIC meeting is from personal reportage and interviews with many participants. Capacity utilization: AISI, *Annual Statistical Report* (Washington, D.C., 1984). Steel wage and benefit costs: AISI, *Wage Trends in the Iron & Steel Industry* (Washington, D.C., 1985). This report said that hourly employment costs averaged $23.78 per hour in 1982 among the sixty-three member companies of the AISI, which represented 90 percent of U.S. raw steel production. However, these costs averaged $22 to $23 for eight companies in the bargaining group as of July 1982, before wage and COLA increases later in the year had taken effect (George A. Moore, Jr., New York City, Mar. 21, 1984). Average costs at the eight companies generally were slightly higher than those for the total AISI membership. Layoffs: AISI, "The American Steel Industry's Problems: Severe But Solvable," (Washington, D.C., February 1983). This position paper reported monthly unemployment totals for

1982. Jobless numbers for 1982 used later in this and other chapters come from this paper. In 1981 the AISI companies employed an average of 286,219 hourly workers.

6. In 1986 the hotel was acquired by the Westin chain and renamed the Westin William Penn.

7. For more on the strike avoidance emphasis in industrial relations, see Thomas A. Kochan, Harry C. Katz, and Robert B. McKersie, *The Transformation of American Industrial Relations* (New York: Basic Books, 1986), pp. 40–45.

8. Lawrence Delo, White Oak, Pa., Oct. 26, 1983.

9. Joseph Odorcich, Pittsburgh, Oct. 24, 1983.

Chapter 2

1. Statistics on union membership: Leo Troy and Neil Sheflin, *Union Sourcebook* (West Orange, N.J.: IRDIS, 1985), pp. 3–14. Impact in the Pittsburgh region: Morton Coleman, "Decline of the Mon Valley Viewed in a Global Context," in *Steel People: Survival and Resilience in Pittsburgh's Mon Valley,* ed. Jim Cunningham and Pamela Martz, a report prepared by the River Communities Project, School of Social Work, University of Pittsburgh, 1986, pp. 4–5.

2. Analysis of wage concessions: Daniel J. B. Mitchell, "Alternative Explanations of Union Wage Concessions: Deregulation and Foreign Trade or Prior Wage Trends," Institute of Industrial Relations, University of California, March 1986. Lane Kirkland quoted in John Hoerr, "A Host of Strikebreakers Is Tipping the Scale Against Labor," *BW,* July 15, 1985, p. 32.

3. Sumner H. Slichter, "Are We Becoming a Laboristic State?" *New York Times Magazine,* May 16, 1948, reprinted in *Potentials of the American Economy,* ed. John T. Dunlop (Cambridge: Harvard University Press, 1961), pp. 255–270. Slichter pointed out that 14 million of the 35 million nonsupervisory and nontechnical employees in private industry were union members. As a result, he said, "the American economy is a laboristic economy, or at least is rapidly becoming one."

4. According to one opinion poll, 44 percent of women workers said they would "definitely" or "probably" vote for a union, "if an election were held tomorrow to decide whether your workplace would be unionized or not." Only 24 percent of surveyed men answered likewise (the BW/Harris Poll, June 3, 1985). Data on employment changes in goods- and service-producing industries: Ronald E. Kutscher and Valerie A. Personick, "Deindustrialization and the Shift to Services," *Monthly Labor Review,* June 1986. pp. 3–13.

5. Organizing expenditures: Paula Voos, quoted in Richard B. Freeman and James L. Medoff, *What Do Unions Do?* (New York: Basic Books, 1984), p. 229. Membership declining because unions have fallen from public favor: Seymour Martin Lipset, "North American Labor Movements: A Comparative Perspective," in *Unions in Transition,* ed. Seymour Martin Lipset (San Francisco: ICS Press, 1986), chap. 17. The concept of the nonunion industrial relations system is developed by Thomas A. Kochan, Harry C. Katz, and Robert B. McKersie, *The Transformation of American Industrial Relations* (New York: Basic Books, 1986), chap. 3. For evidence of illegal actions by employers, see Paul C. Weiler, "Promises to Keep: Securing Workers' Rights to Self-Organization under the NLRA," *Harvard Law Review* 96, no. 8 (June 1983): 1780–81.

Attempts by economists to calculate the impact of each of the major factors involved in declining membership go beyond the scope of this book. But see Henry S. Farber, "The Extent of Unionization in the United States," and Richard B. Freeman, "Why Are Unions Faring Poorly in NLRB Representation Elections?" in Thomas A. Kochan, ed., *Challenges and Choices Facing American Labor* (Cambridge: MIT

Press, 1985). Farber and Freeman generally argue that structural factors such as employment shifts and the increase in white-collar and female workers account for only 40 percent of the declining success rate in union representation elections. Reduced organizing efforts are a large part of the problem. In addition, Freeman gives a 40 percent weight to increased management opposition, including illegal activities.

6. Declining membership noted in early 1970s: see Daniel Bell, *The Coming of Post-Industrial Society* (New York: Basic Books, 1973), p. 139. Bell observed in 1973 that union membership as a percentage of workers in nonagricultural jobs was less than in 1947. In other words, union membership was falling behind labor-force growth. When confronted with similar data at news conferences, George Meany, president of the AFL-CIO from 1955 to 1979, usually dismissed the evidence with a tart reply. His position was that union density (percentage) comparisons were unimportant as long as total membership continued to grow. And it did grow. According to Troy and Sheflin, *Union Sourcebook*, the AFL-CIO's membership peaked at 16.2 million in 1981. But employment grew faster. Membership figures, 1953–87, from two series: Troy and Sheflin, *Union Sourcebook*, pp. 3–10, and Bureau of Labor Statistics, news release, "Union Membership in 1987," Washington, D.C., Jan. 22, 1988. The two series differ somewhat, but density ratios are roughly the same. Analysis of membership growth, decline, and future projections: Freeman and Medoff, *What Do Unions Do?*, pp. 268–83. If all other factors bearing on the rate of unionization remain the same, organized labor's share of the work force will drop to 10 to 12 percent by the year 2000, but not fall much lower than that.

7. The most noted example of a false proclamation of union demise comes from early in the Depression. By 1932 organized labor's membership had dropped to a fifteen-year low of 3 million, and the shrinkage continued. George E. Barnett, a noted economist of the time, commented on the trend in a speech to the American Economic Association on Dec. 29, 1932. "I see no reason to believe that American trade unionism will so revolutionize itself within a short period as to become in the next decade a more potent social influence than it has been in the past decade." Within six months, the New Deal was attacking the Depression with a package of new laws, including legislation which encouraged and protected unions. By 1938 the labor movement had doubled in size to more than 6 million members. Quoted in David Brody, "The Expansion of the American Labor Movement: Institutional Sources of Stimulus and Restraint," in *Institutions in Modern America*, ed. Stephen Ambrose (Baltimore: Johns Hopkins University Press, 1967), pp. 11–12.

8. American management lost its way: For a classic statement of this thesis see Robert H. Hayes and William J. Abernathy, "Managing Our Way to Economic Decline," *Harvard Business Review*, July–August 1980, pp. 67–77. Criticism of steel management and comparison with the Japanese: Richard Bolling and John Bowles, *America's Competitive Edge: How to Get Our Country Moving Again* (New York: McGraw-Hill, 1982), chap. 8.

9. Manufacturing employment, 1979–86: Bureau of Labor Statistics, *Employment and Earnings* (Washington, D.C., February 1987), p. 45. Manufacturing employment declined from 21,040,000 in 1979 to 19,186,000 in 1986.

10. This brief historical outline is my interpretation of events. For a comprehensive treatment of the evolution of postwar bargaining, see Kochan, *Transformation*, chap. 2.

11. Wage increases in the 1970s: Freeman and Medoff, *What Do Unions Do?*, pp. 53–54.

12. Ben Fischer, phone, Feb. 25, 1984.

13. For the AFL-CIO's changed position on participation, see "The Changing Situation of Workers and Their Unions," a report by the AFL-CIO Committee on the

Evolution of Work, February, 1985. Williams quote: John Hoerr, "Worker Participation Then and Now," in *Participative Systems at Work: Creating Quality and Employment Security*, ed. Sidney P. Rubinstein (New York: Human Sciences Press, 1986), pp. 142–43. Survey of involvement programs: ibid., pp. 170–71.

14. This characterization of the controllers is based largely on Arthur B. Shostak and David Skocik, *The Air Controllers' Controversy: Lessons from the PATCO Strike* (New York: Human Sciences Press, 1986), pp. 21–41. This book, which has excellent background material on the controllers, the FAA, and the strike, views the controllers in a sympathetic light. My account of the strike also is drawn from my own reporting, news articles of the time, and a critical analysis of PATCO's strategy and tactics in Herbert R. Northrup, "The Rise and Demise of PATCO," *Industrial and Labor Relations Review* 37, no. 2 (January 1984): 167–84.

15. Kirkland's comments at news conferences: My notes, supplemented by AFL-CIO transcripts.

16. According to Northrup, "PATCO," pp. 174–75, PATCO leaders whipped up strike sentiment in regional meetings prior to negotiations to set up the following strategy. By accepting the FAA's proposal, Poli would seem reasonable. But rejection by the rank and file would make it appear that he couldn't control his members. This would convince Congress that it should grant PATCO's demands.

The updated contingency plan confounded PATCO: Shostak and Skocik, *Air Controllers*, p. 106. Picket quoted: "Strikers Are Confident They'll Get Their Jobs Back," *Chicago Tribune*, Aug. 4, 1981.

17. Winpisinger's offer to strike: Shostak and Skocik, *Air Controllers*, p. 107.

18. "The $10,000 demand killed us:" ibid., p. 101 (by coauthor Skocik). Amateurish performance: Northrup, "PATCO," p. 184.

19. Strike cost: "Controller Complaints Hang over the FAA," *BW*, Aug. 24, 1981, p. 35. Increases in air traffic and near collisions: Arthur B. Shostak, "Let the Air Controllers Return," *NYT*, Sept. 1, 1986. Lewis and Dole censures and congressional report: Shostak and Skocik, *Air Controllers*, p. 180. New controllers' union: "Air Traffic Controllers Choose Union Representation By More Than 2–1 Margin," *Daily Labor Report*, June 12, 1987, p. A9. The controllers cast 7,494 votes for a new union, the National Air Traffic Controllers Association, and 3,225 against. Traumatically cut off: Shostak and Skocik, *Air Controllers*, pp. 136–42.

20. Stanley Aronowitz, *Working Class Hero: A New Strategy for Labor* (New York: Adama Books, 1983), p. 74.

21. Pullman strike: Henry Pelling, *American Labor* (Chicago: University of Chicago Press, 1960), pp. 92–95. Price paid by controllers: Shostak, "Let the Air Controllers Return." Effect of PATCO firings on bargaining: "TVA's Hard Line Ends Years of Labor Peace," *BW*, Sept. 21, 1981, p. 31; William Serrin, "Unionists Anxious Over PATCO Strike," *NYT*, Oct. 21, 1981.

22. I have relied on three sources for accounts of the 1901 strike at U.S. Steel: David Brody, *Steelworkers in America: The Nonunion Era* (Cambridge: Harvard University Press, 1960), pp. 50–73; Gerald G. Eggert, *Steelmasters and Labor Reform, 1886–1923* (Pittsburgh: University of Pittsburgh Press, 1981), pp. 34–40; Selig Perlman and Philip Taft, *History of Labor in the United States, 1896–1932*, vol. 4 (New York: Macmillan, 1935), pp. 101–08.

23. Cutthroat competition in the 1890s: David Brody, *Workers in Industrial America: Essays on the Twentieth Century Struggle* (New York: Oxford University Press, 1980), p. 8. Formation of U.S. Steel: Eggert, *Steelmasters*, p. 28. "If a workman sticks up his head": ibid., p. 37.

24. Assuming indispensability: Brody, *Steelworkers*, p. 67.

25. Perlman and Taft, *History of Labor*, p. 109.

26. Working conditions, hours, and wages in the steel industry: Eggert, *Steelmasters*, pp. 86, 135, 150–60. The vice-president was William B. Dickson, who served as assistant to Charles M. Schwab, the first president of U.S. Steel, and rose to first vice-president. Dickson resigned in 1911, partly because his efforts to reform working conditions were opposed by other corporate officers.

Chapter 3

1. The term "official family" usually has been ascribed to I. W. Abel, the USW's third president, who mentioned it while testifying at a court hearing in the early 1970s. Critics of the union's political processes cite Abel's use of the term as evidence that a self-serving clique ruled the union. Although Abel did use the term, it is doubtful that he coined it. John Herling, who wrote a definitive history of USW politics from the union's founding to 1965, quotes a staff member who mentions "official family" while referring to an event in the early 1950s. See Herling's *Right to Challenge: People and Power in the Steelworkers Union* (New York: Harper & Row, 1972), p. 17.

2. Details on McBride's background: interviews with McBride, Gary Hubbard (McBride's public relations aide), and George Becker and Buddy Davis (cited below). Exciting job: Lloyd McBride, Las Vegas, Aug. 29, 1976; on employers, Pittsburgh, Aug. 19, 1983. Concerning McBride's Catholicism, his son Larry became a lay brother. Quotes about McBride: Bernard Kleiman, phone, Jan. 27, 1983; George Becker, Pittsburgh, Oct. 28, 1983.

3. Politics leading up to USW's 1977 election: personal reportage, 1975–77. McBride's concern about retirement: Buddy Davis, phone, Nov. 21, 1983. Davis, the director of District 34 in 1983, had been a staff member under McBride and became one of his closest friends. In a eulogy delivered at McBride's funeral in St. Louis on Nov. 10, 1983, Davis said that McBride did not expect to have a "normal retirement." Although he had announced his candidacy in the fall of 1975, McBride was having second thoughts about seeking the presidency. He suffered from cataracts, which had left him partially blind in one eye, and he wanted to spend more time with his family after many years of travel. Since one of his reasons for seeking election was to keep Sadlowski out of office, McBride intended to serve two terms if elected. "That meant he would be seventy years old at the end of the second term," Davis said. "He made the comment that not many men lived beyond seventy. Just being practical, he could see that, with the stress of that high office, he might be forgoing any chance to retire." Election results, 1977: USW, International Tellers, *Report on International Election* (Pittsburgh, Apr. 28, 1977).

4. Beginning of concession trend: personal reportage, 1979–82. Perhaps the earliest published article on the beginning of a wage-concession trend was by Zachary Schiller, "Unions Bend with the Recession," *BW*, Sept. 8, 1980, p. 110. Fraser memo: "Pleas for Wage Relief Flood into the UAW," *BW*, Feb. 16, 1981, p. 27.

5. Union membership data: Leo Troy and Neil Sheflin, *Union Sourcebook* (West Orange, N.J.: IRDIS, 1985). USW membership and industries: USW, Office of the President; Research Department; Public Relations Department, reports and news releases, various dates.

6. Unemployment rates, list of troubled companies: "Colt Is the Latest to Seek Union Help," *BW*, Feb. 15, 1982, p. 48. James Smith, phone, Jan. 27, 1982.

7. James Smith, Pittsburgh, Apr. 17, 1986 (except as noted, this citation covers further quotes by Smith in this chapter).

8. "Colt Is the Latest to Seek Union Help." Lack of policy: Lloyd McBride, phone, Mar. 31, 1982; Edmund L. Ayoub, Mt. Lebanon, Pa., July 26, 1985 (citation covers

further quotes by Ayoub in this chapter). Description of steel market: James Smith, Pittsburgh, May 4, 1982.

9. "Between a rock and a hard place": Lloyd McBride, Pittsburgh, Aug. 19, 1983. Silver mining case: Erik Larson, "Steelworkers in Idaho Bitter at Union Stand," *WSJ*, March 1982; USW, *Officers' Report*, Constitutional Convention, Sept. 20, 1982, p. 105; Order of Harold L. Ryan, U.S. District Judge, District of Idaho, Mar. 1, 1983. This case involved Bunker Hill Company, a subsidiary of Gulf Resources & Chemical Corporation, in Kellogg, Idaho. After Gulf Resources shut the mining and smelting operations in late 1981, an investment consortium offered to buy them and proposed a labor contract with a 25 percent cut in wages and other changes. Some members of the USW and other unions voted in January 1982 to accept the pact. Union officials refused to sign the agreement on grounds that it would "make a mockery of fundamental seniority principles." A class-action suit filed against the USW was decertified by a federal judge in 1983. The suit was settled out of court for less than $1,000. But the situation received national publicity.

10. Origin of Ayoub memos: Edmund Ayoub, Mt. Lebanon, July 26, 1985. Citation for memos: Edmund Ayoub, "Steel in the Seventies," a series of seven memorandums distributed to the International officers and district directors of the United Steelworkers. Figures quoted: Memorandum II, "Wages, Earnings, Employment Costs, and Productivity," Dec. 10, 1980; Memorandum III, "The Wage-Price-Productivity Relationship," Jan. 2, 1981.

11. First contract: *Agreement Between Carnegie-Illinois Steel Corporation and the Steel Workers Organizing Committee* (Pittsburgh, Mar. 17, 1937). 1980 documents (all published in Pittsburgh): *Agreement Between United States Steel Corporation and the United Steelworkers of America*, Aug. 1, 1980; *Pension Agreement* effective July 31, 1980; *PIB, Program of Insurance Benefits*, effective Jan. 1, 1981; *Program of Hospital-Medical Benefits for Eligible Pensioners and Surviving Spouses*, effective Jan. 1, 1981; *SUB, Supplemental Unemployment Benefit Plan*, effective Aug. 1, 1977 (1980 booklet not available); *SVP, Savings and Vacation Plan*, effective Dec. 31, 1978; *An Explanation of the Employment and Income Security Program*, 1980; and *Steel/ 1980: Summary of the Basic Steel Settlement*.

12. Definition of wages: *Webster's Third New International Dictionary* (Springfield, Mass.: G. & C. Merriam, 1968), p. 2568. According to this source, "wages" can refer to "bonuses, commissions, and amounts paid by the employer for insurance, pensions, hospitalization, and other benefits."

13. For this insight, I am indebted to Sidney P. Rubinstein, a pioneer consultant on worker participation and president of Participative Systems Inc., Princeton, N.J.

14. The description of wage and benefit terms in this section was based on steel and auto labor agreements, various years.

15. Origin of lifetime job security in steel industry and early anticipated cost: Ben Fischer, Pittsburgh, July 1, 1987; Bernard Kleiman, phone, Nov. 6, 1987. Description of benefits: personal reportage, 1977–87; USW, *An Explanation of the Employment and Income Security Program* (Pittsburgh, 1980). An exceptionally clear exposition of the income security program can be found in K. L. Hassan and T. Geoffrey, "Income Security in the U.S. Steel Industry," School of Urban and Public Affairs, prepared for the Center for Labor Studies, Carnegie-Mellon University, July 29, 1983. U.S. Steel payments: Otto A. Davis and Edward Montgomery, "Study of Income Security in Basic Steel Industry," School of Urban and Public Affairs, Carnegie-Mellon University, as reprinted in *Daily Labor Report*, May 16, 1986, pp. D1–D10. This study contains many figures on payments and numbers of recipients.

16. Total compensation in steel: see chap. 1, n. 5.

Chapter 4

1. First blast furnace built at National Works in 1889: USS news release, "National's 100th Anniversary, Plant Historical Background," Sept. 13, 1972. Construction on the furnace started in 1889, although it probably did not operate until 1890.

2. Thomas Bell, *Out of This Furnace* (1941; rpt. Pittsburgh: University of Pittsburgh Press, 1976), p. 3.

3. This historical sketch of the iron and steel industry in the Pittsburgh region is based on the following sources: J. H. Bridge, *The Inside Story of the Carnegie Steel Company* (New York: 1903); David Brody, *Steelworkers in America: The Nonunion Era* (Cambridge: Harvard University Press, 1960); Victor S. Clark, *History of Manufactures in the United States*, vol. 2 (New York: McGraw-Hill, 1929); William T. Hogan, *Economic History of the Iron and Steel Industry in the United States*, vols. 1, 2 (Lexington, Mass.: D.C. Heath, 1971); Joseph Frazier Wall, *Andrew Carnegie* (New York: Oxford University Press, 1970); *MDN*, 50th Anniversary Edition, June 30, 1934.

4. Clark, *History of Manufactures*, p. 205.

5. Brody, *Steelworkers*, p. 2.

6. Bridge, *Inside Story*, p. 173. The beehive coke ovens presented a colorful sight at night. Sitting side by side and glowing from their open tops, hundreds of ovens formed long, fiery lines on the horizon. As late as the 1940s, when some beehives still operated, people would drive to Connellsville from miles away to see the nightly show.

7. Bridge's description: ibid., p. 169. Carnegie financial results: Gerald G. Eggert, *Steelmasters and Labor Reform, 1886–1923* (Pittsburgh: University of Pittsburgh Press, 1981), p. 28.

8. The following are sources for the origin of plants that became part of U.S. Steel. McKeesport: Walter L. Riggs, "The Early History of McKeesport," undated ms, McKeesport Public Library; *A McKeesport Commemorative* (McKeesport Bicentennial Committee, 1976). Clairton: Hogan, *Economic History*, vol. 2, p. 523. Donora: ibid., p. 485. Christy Park: "Seamless Works in Christy Park Erected in 1897," *MDN*, June 30, 1934, and "Christy Park Kept Nation, City Booming," *MDN*, June 30, 1984.

9. Distances on the Monongahela: obtained from Jim Mershimer, U.S. Corps of Engineers, Waterways Management Branch, Pittsburgh, Sept. 9, 1987.

10. Bell was a good reporter as well as a novelist. Steelworkers and managers really did use the term "City Office," according to a letter written in 1936 by Clinton Golden, than an official of the National Labor Relations Board. Golden to Heber Blankenhorn, Apr. 20, 1936, Golden Papers, PSULA, box 1, file #13, personal file, 1935–36.

11. Total employees in Mon Valley: derived from USW, "List of Companies Having Contracts with Wage Openings During the Month of July, 1949," David J. McDonald Papers, PSULA, box 154, file #10. This memorandum gives membership figures for each plant involved in the July 1949 negotiations. The number of members in U.S. Steel's Mon Valley plants totaled 47,910. Pittsburgh Steel employed 7,770 in Monessen (figures for this company's Allenport plant are missing), and J&L employed 8,000 in its two Pittsburgh plants. The total is 64,258. To this must be added supervisors and managers who worked in the plants. In the late 1940s, according to the AISI, exempt managerial employees comprised about 20 percent of steel employment. Total employment in the Mon Valley plants must have been close to eighty thousand. The addition of employees at the Pittsburgh headquarters of the steel companies would bring the number close to one hundred thousand in the late 1940s. In 1987, the USS division of USX employed only 3,481 salaried and hourly workers

at three plants in the valley (Braddock, West Mifflin, and Clairton); Wheeling-Pittsburgh employed about 450 at Allenport; and LTV a few hundred at one Pittsburgh plant (1987 information from public relations departments of USS, W-P, and LTV). All other plants were closed.

12. Michel Crozier, *The Bureaucratic Phenomenon* (Chicago: University of Chicago Press, 1964), pp. 197–98 (emphasis in original).

13. The iron ore strategy is examined more fully in Donald F. Barnett and Louis Schorsch, *Steel: Upheaval in a Basic Industry* (Cambridge: Ballinger, 1983), pp. 299–307; quote on p. 305.

14. Figures on U.S. share of world production: Robert W. Crandall, "Trade Protection and the 'Revitalization' of the Steel Industry," a paper presented at the annual meeting of the American Economics Association, New York City, Dec. 28–30, 1985. Other data in this section on U.S. steel production and shipments: AISI, *Annual Statistical Report* (Washington, D.C., various years).

15. A cyclical business: George A. Moore, Jr., Easton, Pa., Feb. 11, 1984. "The crest was ahead": Eugene J. Keilin, New York City, Sept. 25, 1986 (this citation covers further mentions in this chapter).

16. Gary's pricing policy: Eggert, *Steelmasters*, chap. 3. No price competition: Barnett and Schorsch, *Steel*, p. 79. Barnett served as vice-president and chief economist of the AISI, and Schorsch was senior economist. Much of my brief overview presented here is based on their expert analysis.

17. "Budding competitiveness": ibid., p. 34. Domestic consumption grew 1.8 percent: ibid., p. 50. Investment and research in 1970s: ibid., p. 60; Japan's steel output: Charles Craypo, *The Economics of Collective Bargaining* (Washington, D.C.: Bureau of National Affairs, 1986), p. 168.

18. Growth of excess capacity: ibid., p. 47; see also chap. 1, n. 3.

19. Steel consumption and production: Crandall, "Trade Protection"; AISI, *Steel at the Crossroads* (Washington, D.C., January, 1980); William T. Hogan, *The 1970s: Critical Years for Steel* (Lexington, Mass.: D.C. Heath, 1972), p. 1; Office of Technology Assessment, *Technology and Steel Industry Competitiveness* (Washington, D.C., 1980), p. 155.

20. Rival materials: "Steelmakers Find New Battleground in Electrogalvanizing Plants," *PP*, June 10. 1986; "GM Plastic Cars Rattle Big Steel," *PP*, June 13, 1986. Growth of minimills: Barnett and Schorsch, *Steel*, p. 88. Prediction for 1990s by F. Kenneth Iverson, chairman of Nucor Corporation, quoted in Jonathan P. Hicks, "Big Steel Is Getting Smaller," *NYT*, July 31, 1986.

21. Steel strikes: E. Robert Livernash, ed., *Collective Bargaining in the Basic Steel Industry* (Washington, D.C.: U.S. Department of Labor, 1961), pp. 246–307. This study, commissioned by the Labor Department after the 1959 steel strike, remains the best history of steel-industry bargaining from 1937 through 1959. Steel strikes in the 1930s were waged mainly for union recognition. Strikes were prohibited during the war years. After the war, steel strikes were waged in support of collective bargaining. A list of the strikes and their duration: 1946 (26 days), 1949 (45 days), 1952 (59 days), 1956 (36 days), 1959 (116 days). In 1955, most plants were shut down for one shift. 1951 was the only year between 1946 and 1956 in which bargaining was not conducted.

22. For more on the wage-price relationship in steel negotiations, see Livernash, *Collective Bargaining*, pp. 3–18, 167, 246–307. Murray's dependence on government help is developed in Ronald W. Schatz, "Philip Murray and the Subordination of the Industrial Unions to the United States Government," in *Labor Leaders in America*, ed. Melvyn Dubofsky and Warren Van Tine (Champaign: University of Illinois Press, 1986), chap. 10. U.S. Steel's 1937 price hike was estimated at double the wage outlay

by economists at the Bureau of Labor Statistics and Central Statistical Board, quoted in Irving Bernstein, *Turbulent Years: A History of the American Worker, 1933–1941* (Boston: Houghton Mifflin, 1970), p. 471.

USW involvement in price hikes: Ben Fischer, Pittsburgh, July 1, 1987; Helmut J. Golatz and Alice Hoffman, "An Interview with David J. McDonald," Oral History Interview, PSULA, Feb. 20, 1970, pp. 31–32.

23. Livernash, *Collective Bargaining,* p. 16. It was in 1956 also that the steel industry and the USW began the practice of negotiating three-year contracts, usually without intervening wage reopeners.

24. Steel import figures cited here and below: AISI, *Annual Statistical Report* (Washington, D.C., various years).

Chapter 5

1. Abel's background, 1967 no-strike proposal and other events relating to negotiation of the ENA are based on my notes from personal reportage and interviews with Abel and others in the 1960s; USW, *Proceedings of the 13th Constitutional Convention,* Atlantic City, Sept. 19–23, 1966. The origination of the no-strike idea by Conrad Cooper was verified by Heath Larry, retired vice-chairman of U.S. Steel, in a phone interview, Nov. 25, 1986. Results of 1969 USW election: USW, International Tellers, *Report on International Election* (Pittsburgh: Apr. 18, 1969). Abel received 257,651 votes and Narick 181,122. Emil Narick later left the union and became an Allegheny County judge in Pittsburgh.

2. Little yield from modernization program: Michael K. Drapkin, "Steel Industry Money Woes Boost Pressure on Firms to Reach Accord Without Strike," *WSJ,* July 21, 1971. Concern about steel's survival: Hendrik S. Houthakker, Washington, D.C., May 20, 1971. In this interview, Houthakker, a member of the president's Council of Economic Advisers, criticized both the union and the steel companies and warned that the industry could go into a severe decline. Few people heeded his words.

3. Steelworkers' earnings: USW Research Department, "The Wage Status of the Steelworker Today" (Pittsburgh, 1971). Negotiations of 1971 and aftermath: personal reportage; see also John Hoerr, "Labor: The Steel Experiment," *Atlantic Monthly,* December 1973, p. 18.

4. I. W. Abel, Pittsburgh, Apr. 6, 1973. Unless otherwise noted, other comments by Abel in remainder of this chapter came from this taped interview.

5. In his election campaign, Sadlowski attacked the union's ratification procedure and raised questions about the ENA, but did not call for its repeal. Regarding rank-and-file ratification of the ENA, when Abel rejected the industry's first no-strike proposal in 1967, the USW's chief spokesman, Raymond Pasnick, declared that no such plan would be accepted without a rank-and-file vote. ENA opponents cited this statement in 1973, but the union denied it had stated such a position. The opponents were correct. I mentioned the statement in a 1967 story: "Steel Prospects Harden," *BW,* Dec. 9, 1967, p. 136.

6. In addition to these considerations for giving up the right to strike, each steelworker received a $150 bonus in the year that bargaining took place. Some published accounts have erred in stating the reason for the bonus. One example: "Anticipating dissent from the rank and file, the steel companies even agreed to give each steelworker a $150 cash bonus" (John Strohmeyer, *Crisis in Bethlehem* [Bethesda: Adler & Adler, 1986], p. 76). Actually, the bonus was a negotiated matter that resulted from the union's contention that workers should share the "savings" that would accrue to the companies from not having to open old facilities to meet hedge-buying demand and later shut them down. The total cost of this bonus to the ten CCSC companies in

1974 was $50 million. This was deemed to be the union's share of $80 million that the companies would save by *not* going through the boom-bust cycle. See Hoerr, "Labor: The Steel Experiment," p. 18.

7. Wage increases under ENA: "An Issue Imperiling Steel's Future," *BW*, Oct. 18, 1982, p. 94. "Industry believed its propaganda": Donald F. Barnett and Louis Schorsch, *Steel: Upheaval in a Basic Industry* (Cambridge: Ballinger, 1983), p. 70.

8. Steel output per manhour: Bureau of Labor Statistics, Office of Productivity and Technology, "Output per Employee-hour for SIC Code 331," (Washington, D.C., May 19, 1981). Nobody that I can recall made an issue of the 3 percent figure in 1973. The justification for it was not really questioned until later in the decade, when steel wages rose rapidly because of inflation. In a 1980 interview with Bernard Kleiman, the USW general counsel who helped negotiate the ENA, I confirmed the interpretation presented here of the union's justification for using 3 percent.

9. Barnett and Schorsch, *Steel*, p. 70. A selection of headlines indicates the flavor of pro-ENA editorials: "A New Era in Steel," *PPG*, Apr. 2, 1973; "The Better Way," *PP*, Apr. 1, 1973; "Wising Up in Pittsburgh," *WSJ*, Apr. 2, 1973. Government officials and labor experts also commented favorably: Ted Knap, "USW No-Strike Pact a Model, Nixon Says," *PP*, Apr. 5, 1973; Philip Shabecoff, "Growth of Arbitration Appears to Point to Era of Labor Peace," *NYT*, Apr. 22, 1973.

Negative comment came from one likely source: "A Battle for All Unions," *Daily World*, undated reprint. Among other things, this editorial said: "The Abel-Steel Trust plot is part of the Nixon-Monopoly plan for strangling the labor movement by bringing the top union leaders into a profits-first partnership with Big Business." Labor leaders outside the steel industry either complimented the USW or took a neutral position on the ENA. One exception was James R. Hoffa, the former Teamsters president who had recently been released from prison. Abel, he said in a television interview, "has softened up to the point in his old age that he would sacrifice the most precious thing workers have and that is the right to strike." "Hoffa Blasts Media, Watergate Probers," *PPG*, May 21, 1973.

10. Background information on Bruce Johnston and how he was viewed by union and company officials is based on my notes from many interviews over the years. Personal data was verified in interviews with Johnston, Pittsburgh, Feb. 27, Mar. 3, and May 1, 1986. Further quotes by Johnston are covered by this citation, unless otherwise noted.

11. "Real loner": Alvin L. Hillegass, phone, Sept. 22, 1986; R. Heath Larry, phone, Nov. 25, 1986.

12. Rise in labor costs, 1974–76: Council on Wage and Price Stability, *Collective Bargaining: Review of 1976, Outlook for 1977* (Washington, D.C., Feb. 10, 1977).

13. McBride's and Sadlowski's positions on the ENA: personal notes on debate between Lloyd McBride and Edward Sadlowski, WBBM-TV, Chicago, Nov. 16, 1976; J. Bruce Johnston, "A Steel Negotiator Looks at Election Issues," text of speech given before Pittsburgh Personnel Association, Dec. 16, 1976, pp. 9–10.

14. My statement that the steel industry allowed a larger than required wage increase in 1977 involves a USW rivalry with the UAW in collective bargaining (see chap. 8). In most negotiating years, the USW used the UAW's bargaining results of the previous year as a starting point and tried to win more. In 1976, the UAW had gained a first-year wage increase of 42¢ per hour. However, 10¢ of this amount represented a restoration of wages that the UAW previously had diverted to finance a dental plan, and 9¢ represented cost-of-living adjustments accumulated, but not paid, under the old contract. In other words, 19¢ of the UAW's 42¢ raise could not be considered "new money." The USW won a dental plan without diverting wages to finance it and received all COLA increases as cost-of-living adjustments. Nevertheless, the steel

industry granted a raise of 43¢ per hour in "new money" in 1977. See "Second Thoughts on Steel's No-Strike Pact," *BW*, Sept. 17, 1979, p. 80.

15. Sources for 1977 and 1980 internal company discussions and negotiations: George A. Moore, Jr., Easton, Pa., Feb. 11, 1984, and Philadelphia, July 7, 1987; Bruce Johnston, Pittsburgh, May 1, 1986; Anthony St. John, phone, Nov. 3, 1986; John Kirkwood, phone, Dec. 17, 1986.

16. AISI, *Annual Statistical Report* (Washington, D.C., 1984).

17. Ray Marshall, *Unheard Voices* (New York: Basic Books, 1987), chap. 7.

18. George Moore, Feb. 11, 1984; Lloyd McBride, Atlantic City, Sept. 17, 1982. One USW official who doubted that McBride made such a statement was Bernard Kleiman, Pittsburgh, Apr. 30, 1986.

19. This description of the industry's strategy to reopen the labor agreement is based on a number of interviews, including George Moore, Feb. 11, 1984, and Bruce Johnston, May 1, 1986.

20. Among the sources used as the basis for the derivation, development, and content of the steel study: Moore and Johnston, cited in n. 19; Edmund Ayoub, Mt. Lebanon, Pa., July 26, 1985. The Ayoub citation covers further quotes in this section by Ayoub.

21. "Why Are America's Steel Plants Closing?" *US Steel News*, July 1982. This issue of U.S. Steel's monthly news organ published much of the data contained in the steel study. Regarding the projection of hourly employment costs to $66.34 in 1990, Ayoub described this calculation as "nonsense." Such a number could be reached by projecting the 12 percent annual average rate of increase in the 1970s. "But it wasn't likely that would happen. Other economic factors would take effect. The industry would collapse before that happened."

San Francisco CEO meeting: George Moore, New York City, Aug. 2, 1984; Bruce Johnston, Pittsburgh, May 1, 1986.

22. George Moore, New York City, Mar. 21, 1984.

23. Lloyd McBride, Atlantic City, Sept. 17, 1982.

24. The history of Linden Hall is told in USW, *This Is Our Land* (Pittsburgh, n.d.).

25. Edmund Ayoub, "Economic Issues in the United States Steel Industry: The Implications for Government, Business, and Labor," prepared for International Executive Board, USW, April 1982. Ayoub's conclusion is stated separately in "Text to Accompany Economic Issues in the United States Steel Industry," pp. 21–23.

26. Linden Hall meeting: personal reportage and interviews with many of the participants, including Ayoub, Moore, McBride, Odorcich, and Johnston (cited above). Trautlein's large salary increase became an issue in Bethlehem. See Strohmeyer, *Crisis in Bethlehem*, p. 134. Marathon purchase: USS, *Annual Reports* (Pittsburgh, 1981, 1982).

Chapter 6

1. The story in this chapter about the National-Duquesne shutdown comes from several interviews and other sources. OPEC and the rise and fall of oil prices: Bruce Nussbaum, *The World After Oil* (New York: Simon and Schuster, 1983), pp. 61–75.

Primary interviews: John P. Ely, White Oak, Pa., Feb. 25, 1986 (Ely was general superintendent of the National Works from 1966 to 1969 and of the National-Duquesne Works from 1969 until his retirement in 1976); Albert L. Voss, McKeesport, Apr. 28, 1986 (general superintendent of National-Duquesne, 1977–84); Alvin L. Hillegass, phone, Feb. 26, 1987; Richard Grace, McKeesport, May 6, 1982; Manuel Stoupis, McKeesport, Feb. 17, 1986.

2. "Handwriting on the wall": Alvin Hillegass, phone, Feb. 26, 1987. "Strong demand": USS, *Annual Report* (Pittsburgh, 1981), pp. 6–7.

3. American Petroleum Association, "Well Completions and Footage Drilled in the U.S., 1970–82," (Washington, D.C., 1982). Rig count: Isaac Kerridge, economist, Hughes Tool Company, phone, Mar. 25, 1987. Drilling activity and pipe inventory: Richard Sparling, Independent Petroleum Association of America, phone, Mar. 10, 1987.

4. National-Duquesne layoffs: Voss and Grace, cited in n. 1.

5. Hillegass and Stoupis, cited in n. 1.

6. Unemployment statistic for April 1982: Michael Acquaviva, Pennsylvania Department of Labor and Industry, Office of Employment Security, Pittsburgh, phone, Nov. 4, 1987.

7. John A. Fitch, *The Steel Workers* (New York: Russell Sage Foundation, 1911). This was part of a several-volume study of the Pittsburgh area. Fitch, a writer from outside the area, conducted research in the Monongahela Valley for several months and wrote sympathetically about steelworkers and their families.

8. Saloons: ibid., p. 228.

9. Although these terms were well known, few steelworkers in my McKeesport days actually used them in ordering a shot and a beer. They hinted of affectation.

10. Lawrence Delo, White Oak, May 3, 1982.

11. LMPT provision: *Agreement Between United States Steel Corporation and the United Steelworkers of America*, Aug. 1, 1980, appendix I. This account of J&L's productivity focus and change in union relations is based on the following interviews: Thomas C. Graham, New York City, May 22, 1980; John H. Kirkwood, Pittsburgh, Oct. 27, 1983; A. Cole Tremain, Pittsburgh, Feb. 22, 1986 (citations cover all quotes by these sources in this chapter). Prussian authoritarianism: "Labor Relations for Tomorrow's Technology," text of a speech delivered by Thomas Graham at the Center for Labor Studies, Carnegie-Mellon University, Oct. 26, 1982. See also "Higher Output Via Workplace Reform," *BW*, Aug. 18, 1980, p. 98.

Chapter 7

1. "Menaungehilla": George P. Donehoo, *A History of the Indian Villages and Place Names in Pennsylvania* (Harrisburg: Telegraph Press, 1928), pp. 113–14.

The most reliable history of McKeesport's origins is Walter L. Riggs, "The Early History of McKeesport," undated ms, McKeesport Public Library. Washington's journal entry for Dec. 31, 1753, is quoted here. Riggs, a McKeesport lawyer who died in the 1960s, refused to credit undocumented stories published elsewhere about the city's founding. One such story was that David McKee settled in what became McKeesport as early as 1755 under the protection of Queen Aliquippa. Riggs could find no trace of McKee being in western Pennsylvania until 1768, by which time "that dusky sovereign was chasing the deer in the happy hunting ground." The advertisement for lots appeared in the *Pittsburgh Gazette*, Feb. 5, 1795, quoted in *A McKeesport Commemorative* (McKeesport: McKeesport Bicentennial Committee, 1976), p. 23.

Other sources for Mon Valley history in this section include: Walter S. Abbott and William E. Harrison, *The First 100 Years of McKeesport* (McKeesport: Press of McKeesport Times, 1894); Leland D. Baldwin, *Pittsburgh: The Story of a City, 1750–1865* (Pittsburgh: University of Pittsburgh Press, 1937; rpt., 1970); J. H. Bridge, *The Inside Story of Carnegie Steel Company* (New York: 1903); *MDN*, 50th Anniversary

Edition, June 30, 1934; William T. Hogan, *Economic History of the Iron and Steel Industry in the United States*, vol. 1 (Lexington, Mass.: D.C. Heath, 1971).

Snow blackened by coal dust: Zadok Cramer, *Navigator*, 1811, quoted in Richard Bissell, *The Monongahela* (New York: Rinehart, 1949), p. 107.

2. Boatbuilding on the Mon: ibid., pp. 52–89. First iron works: "W. Dewees Wood Plant Launched Iron Industry," *MDN*, June 30, 1934. Pittsburgh in the 1960s: Baldwin, *Pittsburgh*, pp. 218–47. Flagler and National Tube: Abbott and Harrison, *First 100 Years*; Hogan, *Economic History*, vol. 1; USS news release, "National's 100th Anniversary: Plant Historical Background," Sept. 13, 1972. "Quaint country seat": Bridge, *Inside Story*, p. 150. "Waving fields of grain": Robert Evans, "Duquesne Grows from Little Farm Center," *MDN*, June 30, 1934. Firth-Sterling: "Stainless Steel Introduced by Firth-Sterling," *MDN*, June 30, 1934.

3. Growth of industry: USS, "National's 100th Birthday," and Hogan, *Economic History*, vol. 1, pp. 275–81 (National Tube expansion); "Tin Plate Company Known World Over" (McKeesport Tin Plate) and "Fort Pitt Steel Casting Concern Is Progressive" (Fort Pitt Steel), *MDN*, June 30, 1934. Community growth, population density, and nationality characteristics of Mon Valley towns: Philip Klein et al., *A Social Study of Pittsburgh* (New York: Columbia University Press, 1938), pp. 51–53, 255. McKeesport population: Abbott and Harrison, *First 100 Years*, with updated census figures attached. "Unfit for urban living": Klein, *Social Study*, p. 215.

4. For the rural characteristics of mill towns, I rely partly on a study of Ellwood City, Pa., a onetime steel community north of Pittsburgh, by Robert H. Guest, in Charles R. Walker, *Steeltown* (New York: Harper & Brothers, 1950), supplement G, "Geographic Origins of Tube Mill Workers."

5. Sources for mill town poetry are as follows: Duquesne: quoted in Bridge, *Inside Story*, p. 163. Braddock: ibid., p. 85n. Steelton: John Bodnar, *Immigration and Industrialization: Ethnicity in an American Mill Town, 1870–1940* (Pittsburgh: University of Pittsburgh Press, 1977), p. 76. McKeesport: the claim of Almighty protection is in Abbott and Harrison, *The First 100 Years*, p. 177; cataloguing of assets in *MDN*, June 30, 1934; William Bingham Kay poem in same issue (no date is given for composition, but it must have been after 1900, since tin plate was first produced there in 1903). Aliquippa: quoted in Clarke Thomas, "United by Hard Times," *PPG*, July 24, 1986.

6. A personal note: My grandfather and grandmother on my father's side both came to McKeesport as a result of ads placed by U.S. Steel in European newspapers. Both emigrated as youngsters, my grandfather from Germany and my grandmother from England, brought over by older brothers who had been attracted by the ads and who already were working in the mills.

7. My compressed account of ethnic discrimination relies partly on personal knowledge and generally on Bodnar, *Immigration and Industrialization* (which has a rich store of documented evidence of segregation in a mill town), and Dennis C. Dickerson, *Out of the Crucible: Black Steelworkers in Western Pennsylvania, 1875–1980* (Albany: State University of New York Press, 1986). Importation of blacks to break 1919 strike: ibid., p. 92.

8. Bodnar, *Immigration and Industrialization*, pp. 102, 150–51. Diversity of nationalities: Klein, *Social Study*, pp. 55, 61; *History of McKeesport* (McKeesport: Chamber of Commerce, c. 1927), pp. 19–20.

9. Voting: John A. Fitch, *The Steel Workers* (New York: Russell Sage Foundation, 1911), p. 231.

10. Stories on union meetings prohibited in Duquesne and McKeesport appeared in *MDN*: "Police Break Up Meeting of Mill Workers," Sept. 8, 1919; full page "official statement" by Mayor Lysle, Sept. 20, 1919; editorial, Sept. 22, 1919. The newspaper

reported that 100,000 steelworkers were working in Allegheny County on Sept. 22, when a national steel strike began, but that only two plants, at Donora and Brackenridge, were closed by the walkout: "Each Side Claims an Advantage," Sept. 22, 1919. Crawford quote: David Brody, *Labor in Crisis: The Steel Strike of 1919* (Philadelphia: J.B. Lippincott, 1965), p. 94.

See also Thomas Bell, *Out of This Furnace* (1941; rpt. Pittsburgh: University of Pittsburgh Press, 1976). Bell, who worked in the Braddock mills as a young man, captured many of the themes of the present chapter in his novel: the hard life of Slovak immigrants in Homestead and Braddock, the dictatorial Irish bosses in the mills, U.S. Steel's efforts to control workers' votes, the power of immigrant political brokers like the saloonkeeper Perovsky, and finally the immigrants' breakthrough to freedom with the organization of the Steelworkers.

11. It is taken as common knowledge by many USW and town officials in the Mon Valley that the steel companies kept other industries out of the region. Documenting the allegation, however, is difficult. Some years ago, an official of Pittsburgh Steel (now Wheeling-Pittsburgh Steel) told me that the steel firms discouraged the placement of an aircraft manufacturing plant in the area during World War II. Andrew Jakomas, a former McKeesport mayor, said that U.S. Steel made it clear to public officials that the area's labor force "belonged to U.S. Steel" (see n. 20 below).

The friend who was discouraged from attending college is my brother-in-law, Ronald J. McKay, who today is a lawyer in McKeesport.

12. Discrimination against blacks: Dickerson, *Out of the Crucible*, pp. 154, 177, 188. I have used just a few statistics from this admirably researched study, which presents a picture of shameful discrimination in the steel mills and towns of western Pennsylvania.

13. For a good account of prewar steel earnings and living conditions, see Mark McColloch, "Consolidating Industrial Citizenship," in *Forging a Union of Steel*, ed. Paul F. Clark, Peter Gottlieb, and Donald Kennedy (Ithaca, N.Y.: ILR Press, 1987, pp. 45–87. Economist Walter Heller determined that in February 1942 an income of $2,409 per year was needed to sustain a family of four with modest expenditures.

14. Exuberance in McKeesport and Duquesne: "A Huey Street Disturbance," *MDN*, July 1, 1884, p. 1; Robert Evans, "Duquesne Grows from Little Farm Center." Prince Chow Chow: "An Operatic Crazy Quilt," *MDN*, July 1, 1884. Theaters in McKeesport: Charles C. Shaw, "Local Theaters Advance with City," *MDN*, June 30, 1934. Connelly reminisces about his boyhood in McKeesport in his autobiography, *Voices Offstage* (New York: Holt, Rinehart & Winston, 1968), pp. 1–29. Richie and Leaver: *McKeesport Commemorative*, pp. 51, 190–91. Miller and Dingledein: personal knowledge.

15. In 1985, when Murphy's was acquired by Ames Department Stores Inc., it had 383 stores in 19 states. William M. Bulkeley and Terence Roth, "Ames to Acquire G.C. Murphy for $48 a Share," *WSJ*, Apr. 24, 1985.

16. Bodnar, *Immigration and Industrialization*, p. 155.

17. Background on B&O problem: my reportage; Mon-Yough Conference on Community Development, news release, "McKeesport Grade Crossing Elimination," July 27, 1966.

18. Joseph Sabol, Jr., Duquesne, Dec. 20, 1965.

19. Kennedy-Nixon debate: John Vockley, "Junto Club Face-off Presaged Historic Debates," *MDN*, June 30, 1984.

20. Andrew J. Jakomas, McKeesport, Feb. 24, 1986.

21. Clinton S. Golden and Harold J. Ruttenberg, *The Dynamics of Industrial Democracy* (New York: Harper & Brothers, 1942), pp. 109–18. For a discussion of the

Golden-Ruttenberg thesis, see David Brody, "The Origins of Modern Steel Unionism: The SWOC Era," in Clark, *Forging a Union*, pp. 13–29.

22. Leaving town: number estimated in "McKeesport: A Detailed Technical Report," *Master Plan, City of McKeesport* (McKeesport: City Planning Commission, assisted by Pittsburgh Regional Planning Association, December, 1964).

Chapter 8

1. Douglas A. Fraser, Detroit, May 14, 1985.

2. GNP statistics: Council of Economic Advisers, *Economic Report of the President* (Washington, D.C., January 1987). The number of people employed in the United States grew from 100,907 million in 1980 to 111,306 million in 1986. High inflation in the late 1970s is one reason why annual wage increases rose about 10 percent.

3. The contract settlements mentioned here were reported in the following *BW* stories: "What the Teamsters' Pact Won't Do," Feb. 1, 1982, p. 17; "Detroit's New Balance of Power," Mar. 1, 1982, p. 90; "The Work-Rule Changes GM Is Counting on," Apr. 5, 1982, p. 30.

4. Percentage of steel output going to car manufacture: AISI, *Annual Statistical Report* (Washington, D.C., 1985). Steel and auto wages in 1946: "When Steel Wages Rise Faster Than Productivity," *BW*, Apr. 21, 1980, p. 144. Earnings in 1947: Jack Stieber, "Steel," in *Collective Bargaining: Contemporary American Experience*, ed. Gerald G. Somers (Madison: Industrial Relations Research Association, 1980), p. 197.

5. The story of the 1948 GM-UAW bargain has been told in many places. I have relied on the following: background knowledge derived from working as a journalist in Detroit, 1961–1964; personal interview with Nat Weinberg, a retired UAW economist who helped negotiate the bargain, 1979; John Barnard, *Walter Reuther and the Rise of the Auto Workers* (Boston: Little, Brown, 1983), chap. 6; Kathy Groehn El-Messidi, *The Bargain* (New York: Nellen, 1980); Alfred P. Sloan, Jr., *My Years with General Motors* (New York: Doubleday, 1964), pp. 395–403. Also see Michael J. Piore and Charles F. Sabel, *The Second Industrial Divide: Possibilities for Prosperity* (New York: Basic Books, 1984), pp. 73–84. Piore and Sabel state the case too strongly, given the diversity of American business, in declaring that the GM formula, combined with the minimum wage and other mechanisms, "tied wages throughout the economy into a single structure" (p. 82). That the GM deal was highly influential is not disputed.

6. The following background of the USW-UAW bargaining rivalry comes from my own reportage in both industries over many years; interview with Ben Fischer, Pittsburgh, Feb. 20, 1986, and July 1, 1987; Barnard, *Walter Reuther*, chap. 6. Output per man-hour in steel rose annually by 2.2 percent in the 1950s, 3 percent in the 1960s, and 2.4 percent in the 1970s, compared with 5.8, 3.6, and 3.5 in autos ("Why Are America's Steel Plants Closing?" *US Steel News*, July 1982).

7. Federal law requires that publicly traded companies publish the salaries of their top officers. Annual rankings of these salaries in *Business Week* lights in the eyes of chief executives a fierce competitive glow. Hourly employment costs in 1980: USW, Edmund L. Ayoub, "Economic Issues in the United States Steel Industry," Pittsburgh, April 1982, p. 42.

8. Ford's situation and settlement: "The Ford Pact Is No Panacea" and "Detroit's New Balance of Power," *BW*, Mar. 1, 1982, pp. 18, 90. Criticism of Ford deal: Bruce Johnston, Pittsburgh, May 1, 1986; Peter J. Pestillo, Washington, Sept. 9, 1982 (this citation covers later comments by Pestillo). Of course, Pestillo would be unlikely to

concede that he asked for, and received, too little from the UAW. It appears, however, that 1982 was the beginning of a stunning turnaround for Ford (see below).

9. The industrial relations debate: Dunlop and other observers were quoted in "Moderation's Chance to Survive," *BW*, Apr. 19, 1982, p. 123. Freedman's comment in Audrey Freedman, "A Fundamental Change in Wage Bargaining," *Challenge*, July–August 1982, p. 14; see also Audrey Freedman and William E. Fulmer, "Last Rites for Pattern Bargaining," *Harvard Business Review*, March–April 1982, p. 30.

In "The New Industrial Relations," *BW*, May 11, 1981, p. 84, I suggested that a new industrial relations model was emerging. I wrote that "work-level cooperation might well slip over into the bargaining process, particularly if unions—reflecting the concerns of their members—tie themselves more tightly to the success of the individual company, rather than try to keep up with national wage patterns and outdo other unions." Management, the story said, "must share information with workers, divide with them the gains resulting from increased participation, and work much harder to provide job security and to prevent the catastrophic blows of unexpected plant shutdowns." John Dunlop also took issue with this article, contending that there could be no "new industrial relations" as long as large numbers of employers remained hostile to unions. This showed up both in company resistance to organizing and in the 1978 employer-led success in defeating a labor attempt to reform the National Labor Relations Act. The industrial relations system, in Dunlop's view, encompasses not just the employment relationship within a company but also the national political atmosphere.

10. My account of bargaining at Ford in 1982 is based on interviews with Peter J. Pestillo, Washington, Sept. 9, 1982, and Donald F. Ephlin, New York City, Nov. 3, 1982. See also John Hoerr, "Smudging the Line Between Boss and Worker," *BW*, Mar. 1, 1982, p. 91. My Ford plant tour is reported in "What's Creating an 'Industrial Miracle' at Ford," *BW*, July 30, 1984, p. 80. Ford's success in 1986: John Holusha, "Ford Leads in U.S. Car Profits," *NYT*, Feb. 18, 1987; Jacob M. Schlesinger and Paul Ingrassia, "Ford Schedules Profit-Sharing Payments for '86, Averaging Over $2,100 a Worker," *WSJ*, Feb. 19, 1987.

11. The GM-UAW relationship is based on my personal observation as a reporter, interviews with many GM and UAW principals since the early 1960s, and accounts in many books, including Barnard, *Walter Reuther*, and Howell John Harris, *The Right to Manage: Industrial Relations Policies of American Business in the 1940s* (Madison: The University of Wisconsin Press, 1982).

12. Douglas Fraser, Detroit, May 14, 1985 (this citation covers further comments by Fraser in this chapter).

13. The story of Reuther's 1946 bargaining at GM also has been told in many places, most recently and succinctly in Barnard, *Walter Reuther*, pp. 101–09. Neither vision nor will to transcend traditional roles: ibid., p. 109.

14. Fraser conceded that the union may not have fully considered the implications of the pass-through proposal for Ford. Because Ford purchased about 35 percent of its car components from outside manufacturers, the actual UAW labor employed in auto production averaged 105 hours per car. GM contracted out only 25 percent of component production and required 150 hours of UAW labor to build each car. To achieve cost savings enabling it to cut car prices by the same amount as GM, Ford would have had to reduce wages and benefits by 40 percent more than its competitor. The UAW's Ford members probably would have opposed giving up that much more than GM workers. Whether the UAW could have solved this problem in bargaining is questionable.

15. Amanda Bennett, "GM's Bonus Flap: 'The Timing Was Wrong,'" *WSJ*, Apr. 30, 1982.

16. "Too little, too late": John Kirkwood, Pittsburgh, Oct. 27, 1983. "Very ada-

mant": Joseph Odorcich, phone, Mar. 23, 1982. "It's crazy to keep paying COLA": John Kirkwood, New York City, Mar. 24, 1982.

17. Wheeling-Pittsburgh agreement: "The Labor Givebacks Are Spreading to Steel," *BW*, Apr. 12, 1982, p. 40. It was reported at the time of the W-P negotiations that the USW's no-change position on COLA was primarily intended as a signal to the industrywide bargaining group that COLA was not a bargainable issue. In a phone interview on June 19, 1986, Jim Smith refuted that report and said that the decision to keep COLA was based on the circumstances at W-P.

18. Roger Smith's warning: Lloyd McBride, Atlantic City, Sept. 17, 1982; the verbatim comments are in a transcript of Basic Steel Industry Conference, Pittsburgh, June 18, 1982, USW files.

19. The USW used to save management: Ben Fischer, Pittsburgh, July 1, 1987.

20. My account of the drafting of the industry letter and its timing remains in dispute. In a taped interview on Sept. 17, 1982, McBride told me essentially that he had nothing to do with the letter. "Actually," he said, "I was pretty much in the posture of discouraging any kind of letter of that kind, although it seemed inevitable that it was coming. This letter was not something that was enthusiastically looked on by myself. It was their [the companies'] determination and their decision, and they were doing something that they had a right to do. So the timing of it was their decision. The phrasing of it . . . it was their letter in every respect. Now, in all probability I'm sure that they were aware of the local union elections, although that was not a topic of discussion between he [Johnston] and me."

I mention this discrepancy in versions because McBride expressed himself so strongly in the above interview. It was only after he died that I managed to interview George Moore (New York City, Mar. 21, 1984) and Bruce Johnston (Pittsburgh, May 1, 1986) about the same events. Separately, and each more than once, they recounted the version presented in this book. Furthermore, Joe Odorcich specifically recalled objecting to the use of "fungible" (phone, July 1, 1986). As for the timing of the letter, it is difficult to believe that McBride and Johnston had not discussed the local union elections. They were too important to be ignored.

Chapter 9

1. The economic climate in June 1982 is based on the following *BW* stories: "A Move Toward Global Reflation," June 21, 1982, p. 36; "Interest Rates Are Still the Villain," June 28, 1982, p. 40; "Bankruptcies Nudge Depression Levels," July 12, 1982, p. 14. The steel situation: "How Steel Is Pruning as the Chill Continues," *BW*, July 5, 1982, p. 25; Lydia Chavez, "The Year the Bottom Fell Out for Steel," *NYT*, June 20, 1982.

2. Letter from J. Bruce Johnston to Lloyd McBride, May 28, 1982. By citing labor, management, and government responsibility for the industry's decline, and by dwelling on what government must do to help the industry, the letter implied that collective bargaining could not solve all the problems. Four years later, however, U.S. Steel would refuse to sign a prenegotiation statement offered by the USW which said roughly the same thing and called for union-company cooperation in Washington as a precondition for bargaining an early agreement. The company labeled the union offer as "political."

3. Steel unemployment: AISI, "The American Steel Industry's Problems: Severe But Solvable" (Washington, D.C., February 1983). This position paper reported monthly unemployment totals for 1982. In June, 104,594 hourly workers and 6,859 salaried employees were laid off at AISI companies. Rank-and-file bravado: Chavez, "The Year the Bottom Fell Out for Steel."

4. Grievances: Harry Osborne, USW District 15 representative, phone, June 29, 1982; Lawrence Delo, phone, June 7, 1982. For an extended discussion of steel work rules, see chapter 12.

5. My report of what was said in the BSIC meeting is based on the transcript of the conference made by the USW. At a news conference following the meeting, McBride refused to identify the company that could face bankruptcy within a year. It became evident a year later that the firm was Republic Steel, which merged with LTV in 1983.

6. Henry Ford's quote from *Time*, Mar. 17, 1941, p. 19, quoted in Clinton S. Golden and Harold J. Ruttenberg, *The Dynamics of Industrial Democracy* (New York: Harper & Brothers, 1942), p. 22.

7. Elton Mayo, *The Social Problems of an Industrial Civilization* (Andover, Mass.: Andover Press, 1945), pp. 34–56.

8. Douglas McGregor, *The Human Side of Enterprise* (New York: McGraw-Hill, 1960), pp. 33–34.

9. John Kirkwood, Pittsburgh, Oct. 27, 1983.

10. I am indebted to Jack Barbash, professor of economics at the University of Wisconsin, for the idea that management has conditioned workers to think as if they were short-term employees but expects them to consider the long-term future of the company (interview, phone, Apr. 1, 1982).

By the early 1980s, an overwhelming majority of American companies still withheld large amounts of information from employees. See David Lewin, *Opening the Books: Corporate Information-Sharing with Employees* (New York: Conference Board, 1984). In a survey accompanying this report, only two of forty-nine major corporations reported that they shared financial and other information "fully" with employees. And these forty-nine companies, the report said, "are more information-sharing minded than is typically the case with U.S. business."

11. Slide show snafus: Lawrence Delo, phone, July 19, 1982; Charles Z. Molnar, phone, July 19, 1982.

12. The section on the origin of industrywide bargaining is based on years of personal reportage in the steel industry, as well as the following interviews: John Kirkwood, Feb. 14, 1984; Bruce Johnston, Feb. 27, 1986; George Moore, Easton, Pa., Feb. 11, 1984; Anthony St. John, phone, Nov. 3, 1986; A. Cole Tremain, Pittsburgh, Feb. 22, 1986; Ben Fischer; Stanley C. Ellspermann; William G. Caples (former chief negotiator, Inland Steel), phone, August 1982. Also see E. Robert Livernash, ed., *Collective Bargaining in the Basic Steel Industry* (Washington, D.C.: U.S. Department of Labor, 1961), pp. 85–91, 291–92.

13. Beginnings in 1956: ibid., pp. 87–88.

14. Incentive arbitration in 1968: personal reportage and files, supplemented by the recollections of Ben Fischer, Pittsburgh, Feb. 20, 1986, and July 1, 1987 (citation covers further comments in this chapter by Fischer).

15. Stanley Ellspermann, Pittsburgh, Feb. 20, 1986 (citation covers further comments in this chapter by Ellspermann).

16. J&L's proposal: John Kirkwood, New York City, Feb. 14, 1984.

17. Bruce Johnston, Pittsburgh, Feb. 27, 1986.

18. Wheeling-Pittsburgh's ouster: Paul D. Rusen (former USW negotiator at W-P), phone, July 22, 1987; Joseph L. Scalise (former vice-president, W-P), phone, July 29, 1987; Bruce Johnston, Pittsburgh, Feb. 27, 1986.

Chapter 10

1. Various aspects of my account of the July 1982 bargaining are based on the following interviews: Lloyd McBride, Pittsburgh, Aug. 2, 1982, and Atlantic City,

Sept. 17, 1982; Bruce Johnston, Pittsburgh, May 1, 1986; George Moore, New York City, Mar. 21, 1984; Joseph Odorcich, Pittsburgh, July 28, 30, 1982; Bernard Kleiman, Pittsburgh, Apr. 30, 1986; Edmund Ayoub, Mt. Lebanon, July 26, 1985; Carl Frankel, Pittsburgh, Feb. 18, 1986; John Kirkwood, phone, Aug. 2, 1982. This story is reconstructed from these principal interviews (including follow-up confirmation of facts), supplemented by personal reportage and notes compiled in 1982.

2. It is impossible to calculate precisely the dollar-and-cents gap between the final union and industry proposals. There are too many unknown cost impacts, as well as some confusion in the final position of the two sides on some items. Following is a very rough approximation of how the union proposal would have raised costs from Aug. 1, 1982 (before the wage and COLA increases of that date took effect) to Aug. 1, 1985:

COLA at 8 percent annual inflation (the assumption used by the negotiators), including its impact on the Social Security payroll tax and overtime pay = $3 per hour. Cost of providing major medical and hospital insurance = 90¢ per hour (the union did not ask for improved benefits, but the prices charged by hospitals and physicians were rising at a 15 percent annual rate). Increase in SUB funding = 50¢ an hour. Scheduled increase in pension rates on Aug. 1, 1982 = 10¢ to 20¢. From a total increase of about $4.50, deduct 26¢ per hour for reduction in Sunday premium pay from 1½ times regular pay to 1¼, giving a net rise of $4 to $5.

In his postnegotiation letter to employees, Johnston said the USW offer would have raised costs $4.54 per hour. But the union claimed that Johnston inflated the figure by including annual "roll-ins" of COLA to the wage base that the USW had agreed to cut.

Under the industry plan, the terminal cost would have been $2.20 to $2.25 per hour more than at the beginning. This was made up of the following increases: $1.10 for COLA (including 10¢ for its impact on overtime and taxes), 50¢ for SUB, 90¢ for health care, and 21¢ for various other benefits. The latter included pension and life insurance boosts already scheduled for Aug. 1 and a plan to give small amounts of stock to workers under a federal law that permitted the company to deduct the amount from income taxes. From this total, the industry would have subtracted 48¢ per hour by eliminating the 13-week sabbatical plan.

3. Inflation after 1982: COLA payments computed from CPI rise reported by Bureau of Labor Statistics, *Monthly Labor Review* (Washington, D.C.: U.S. Labor Department, various issues).

4. Walking away from $2 billion in savings: "A Costly Failure in Steel Bargaining," *BW*, Aug. 16, 1982, p. 22. As far as I know, the argument over the COLA-capacity proposal was reported only in this story.

5. Ayoub's and Smith's method of calculating savings: First, they added up the hourly savings that would have resulted from the Aug. 1 give-ups and the COLA deferrals (at 8 percent inflation), as well as from 3 percent wage hikes that would not be paid in 1983 and 1984 and the compounding that would not take place. This was like saying that one will save $10,000 next year by not buying a new car. Second, to translate hourly savings into total dollars, the union economists had to multiply the hourly amount by the number of hours that would be worked over three years. Because they did not have data on which to predict the number of hours that would be worked at the eight companies, they used an average worked in 1980–81 by all AISI companies. This meant that hours worked at dozens of firms were included.

Calculating the amount that would have been saved under the union plan depends on several assumptions, including the inflation rate and the level of operations. If inflation rose 8 percent and some 200,000 workers (two-thirds of the work force) worked full time in the first year of the contract, the eight companies would save $400 million to $500 million under the union's cash-flow offer.

6. Letter to "Each Steelworker-Represented Employee" at U.S. Steel, from J. Bruce Johnston, Aug. 6, 1982; letter to "All Members Employed by the Coordinating Committee Steel Companies," from Lloyd McBride, Aug. 12, 1982.

7. The industry view that McBride paid too much attention to opponents appeared in *What's New in Collective Bargaining Negotiations & Contracts*, Bureau of National Affairs, Oct. 14, 1982, p. 4. "Industry leaders," it said, "find it ridiculous that a union president with McBride's compelling record of negotiating very high settlements in the weak steel industry should feel impelled to display such sensitivity to dissident accusations that he is 'in bed with management.'" One trouble with this sentence is that, up to this point McBride himself did not have much of a record of negotiating industrywide agreements in steel. He helped I. W. Abel negotiate the 1977 agreement but was hospitalized during 1980 bargaining and did not make the final decisions in that round.

8. Lloyd McBride, Atlantic City, Sept. 17, 1982.

9. The historical analysis in this section is based partly on David Brody, "The Origins of Modern Steel Unionism: The SWOC Era," in *Forging a Union of Steel*, ed. Paul F. Clark, Peter Gottlieb, and Donald Kennedy (Ithaca, N.Y.: ILR Press, 1987), pp. 13–29; Melvyn Dubofsky and Warren Van Tine, *John L. Lewis* (New York: Quadrangle/New York Times Book Co.: 1977), p. 128; John Herling, *Right to Challenge: People and Power in the Steelworkers Union* (New York: Harper & Row, 1972); and personal interviews over many years. See also John Hoerr, "The Hot Issue of USW Democracy," *BW*, Sept. 27, 1976, p. 92.

10. Kempton on Murray: quoted in Thomas R. Brooks, *Clint: A Biography of a Labor Intellectual, Clinton S. Golden* (New York: Atheneum, 1978), p. 179.

11. Stifling of democracy, McDonald's rise to presidency, Dues Protest movement, 1957 election and convention beatings: Herling, *Right to Challenge*, pp. 5, 37, 11–22, and chaps. 5, 6.

12. This picture of Donald Rarick is based largely on my own observations and interviews with him and former supporters. Rarick had not liked paying dues from the beginning of his union career. Joe Odorcich particularly remembered Rarick from 1938 or 1939 shortly after U.S. Steel opened the Irvin Works. Odorcich volunteered to go to the plant as a "dues picketer" to force employees there to contribute dues before the checkoff system began. Rarick, Odorcich said, was one of those who complained about paying dues. My own conversations with Rarick convinced me that he was not in the mainstream of liberal unionism. In 1968, when he attempted to run a second time for president, Rarick had a hotel suite in Chicago during the USW convention there that summer. He introduced me, approvingly, to an advocate of the National Right-to-Work Committee, an anti-union organization.

13. One example of how the referendum process provides a forum for dissidents involves union publications. The Labor Management Reporting and Disclosure Act of 1959 (Landrum-Griffin) requires that unions conduct "fair" elections. Rulings of the Labor Department and federal courts under this statute require that union newspapers give equal space to all candidates.

14. Analysis of dissident movements, Ed Sadlowski, and 1960s conventions: personal reportage and notes. Black members argued that although blacks made up almost 20 percent of the union, they were the minority in every district. Even with support of the union leaders, a black could not be elected to a district or International office because of the white majority (a charge that proved true in the 1969 election). In four conventions, starting in 1968, the Abel administration opposed creating an office specifically for a black. The reasoning: This would extend special privileges to one minority and not to others. "I didn't hold office all these years as a Welshman but as a steelworker," Abel declared in 1968.

In 1976, however, Abel reversed course and proposed creation of a second vice-presidency to be filled by a black member who would be appointed by Abel. As an incumbent officer, the black would have a good chance of winning reelection. The motion passed and Abel appointed staff representative Leon Lynch, who has been reelected three times. As a result of this change blacks finally got representation in the top tiers of the union. But the expansion of posts also met a different political need of the union establishment, that of enabling more proadministration directors to rise to high office, thus securing their support.

15. Herling, *Right to Challenge*, p. 401, presents a statistical analysis of the 1968 convention, showing that 562 delegates were staffers. The votes controlled by international and district officers and paid staff employees totaled 13 percent of the weighted membership vote.

16. Mike Stout, Homestead, Feb. 17, 1986; Barney Oursler, McKeesport, Feb. 23, 1986.

Chapter 11

1. Mike Bilcsik told me his story and confirmed various details in several interviews dating from Feb. 28, 1983, when I first met him in Pittsburgh.

2. Description of rally: George Powers, *Cradle of Steel Unionism: Monongahela Valley, Pa.* (East Chicago: Figueroa Printers, 1972), pp. 89–90. See also "Steel Union Campaign Is Opened Here," *MDN*, June 22, 1936. In consigning this historic event to page 16, the McKeesport *Daily News* indicated it was less than enthusiastic about the prospect of unionism.

3. For an analysis of U.S. Steel's motives in recognizing SWOC, see Melvyn Dubofsky and Warren Van Tine, *John L. Lewis* (New York: Quadrangle/New York Times Book Co., 1977), pp. 272–77. The corporation had been outmaneuvered by employees who joined SWOC and took control of the Employee Representation Plans (company unions) that the company had set up. In addition, Taylor apparently wanted to avoid the kind of turmoil that engulfed the auto industry when its workers engaged in sitdown strikes. See also E. Robert Livernash ed., *Collective Bargaining in the Basic Steel Industry* (Washington, D.C.: U.S. Department of Labor, 1961), pp. 231–34. This study points out that U.S. Steel was earning substantial profits in 1937 for the first time since 1930, production exceeded 80 percent of capacity, and foreign armament orders were pending. USS might well have won a strike mounted by SWOC, but the union could have shut down some plants for an extended period of time. For more on SWOC's financial problems, see David J. McDonald, *Union Man* (New York: E. P. Dutton, 1969), pp. 121–28.

4. "Murray Urges Joint Parley to End Slump," *Steel Labor*, June 17, 1938, p. 1. In the late 1970s, *Steel Labor* was renamed *Steelabor*.

5. For more on Golden, see Thomas R. Brooks, *Clint: A Biography of a Labor Intellectual, Clinton S. Golden* (New York: Atheneum, 1978); influences bearing on Golden's views of participation, pp. 31–32, 100–07. Ruttenberg's background and views on participation: interview, Pittsburgh, Apr. 16, 1986.

6. Selig Perlman and Philip Taft, *History of Labor in the United States, 1896–1932*, vol. 4 (New York: Macmillan, 1935), p. 584. Gompers quote: Jean Trepp McKelvey, *AFL Attitudes Toward Production, 1900–1932* (Ithaca: Cornell University Press, 1952), p. 85. This book contains a comprehensive review of union-management cooperation in the 1920s. For accounts which relate the experiments of the 1920s to those of the 1980s, see Sanford M. Jacoby, "Union-Management Cooperation in the United States: Lessons from the 1920s," *Industrial and Labor Relations Review* 37, no. 1

(October 1983): 18–33, and John Hoerr, "Worker Participation Then and Now" in *Participative Systems at Work: Creating Equality and Employment Security,* ed. Sidney P. Rubinstein (New York: Human Sciences Press, 1987), pp. 138–76.

7. Origin of program at Empire Steel and Scanlon's role: Golden's remarks on Oct. 22, 1956, at dedication of plaque honoring Joseph Scanlon, Clinton S. Golden Papers, PSULA, box 6; Russell W. Davenport, "Enterprise for Everyman," *Fortune,* January 1950, p. 54. This article summarized the SWOC experiments, estimated the number of firms helped, and described Scanlon's later projects.

8. Cooke's background: Kenneth E. Trombley, *The Life and Times of a Happy Liberal: A Biography of Morris Llewellyn Cooke* (New York: Harper & Bros., 1954). Development of pamphlet: Cooke to Golden, Mar. 18, 1938, Golden to Cooke, July 6, 1938, Morris L. Cooke Papers, Franklin D. Roosevelt Library (hereafter called "FDRL"), box 142; SWOC, *Production Problems,* pp. 2–5. The engineering assistant who wrote the pamphlet was Francis Goodell. As of late 1938, SWOC had had 1,400 requests for the booklet, and eventually it went through five printings.

9. Curiously, no biography of Murray has been published. Details of his early years, intellectual development, and mature attitudes about some aspects of unionism are vague. For an uncritical biographical sketch, see Charles A. Madison, *American Labor Leaders* (New York: Frederick Ungar, 1950), pp. 311–34. For an excellent analytical essay on Murray, see Ronald W. Schatz, "Philip Murray and the Subordination of the Industrial Unions to the United States Government," in *Labor Leaders in America,* ed. Melvyn Dubofsky and Warren Van Tine (Champaign: University of Illinois Press, 1987), chap. 10. A recent compilation of new analyses about SWOC and Murray by historians Schatz, David Brody, Melvyn Dubofsky, Ronald Filippelli, and Mark McCulloch is *Forging a Union of Steel: Philip Murray, SWOC, and the United Steelworkers,* ed. Paul F. Clark, Peter Gottlieb, and Donald Kennedy (Ithaca: ILR Press, 1987).

10. Comments on Murray: Lee Pressman Papers, Columbia University Oral History Collection, p. 142; Harold J. Ruttenberg, Pittsburgh, Apr. 16, 1986 (citation covers subsequent quotes by Ruttenberg).

11. Morris Llewellyn Cooke and Philip Murray, *Organized Labor and Production: Next Steps in Industrial Democracy* (New York: Harper & Brothers, 1940; rev. ed., 1946). The story of the origin and development of this book is drawn from many sources, but most importantly: Brooks, *Clint,* pp. 197–200; Cooke Papers, FDRL, boxes 285–88, several files; Golden Papers, PSULA, box 3, several files.

Golden's quote about Murray: Golden to Arthur E. Suffern, Feb. 4, 1952 (emphasis in original), Golden Papers, PSULA, box 3, file 28. Murray's complaint about scientific management: Golden to Cooke, Nov. 29, 1939, Cooke Papers, FDRL, box 142, file 1. Quotes from *Organized Labor and Production:* pp. 192, 142, 81, 214. Knowledgeable observer: McKelvey, *AFL Attitudes,* p. 128. "In This Corner," *Steel Labor,* Feb. 28, 1940, p. 5. Another summary of SWOC's participation programs is in Sanford M. Jacoby, "Union-Management Cooperation in the United States During the Second World War," in *Technological Change and Workers' Movements,* ed. Melvyn Dubofsky (Beverly Hills: Sage Publications, 1985), pp. 100–29. This concise account goes somewhat astray, I believe, in characterizing Murray's attitude about plant-level cooperation and his involvement in writing *Organized Labor and Production.* Thus, Jacoby's conjecture that "Murray and Golden were enthusiastic about the SWOC plan" (p. 103) is correct only for Golden's attitude. Murray, I believe, was much more circumspect about the experiments, as my interview with Harold Ruttenberg and other evidence indicates. Murray's contribution to the book was minimal, as I show.

12. Clinton S. Golden and Harold J. Ruttenberg, *The Dynamics of Industrial Democracy* (New York: Harper & Brothers, 1942). Quotes from the book: efficiency,

p. 253; principles, pp. xxiii–xxvi; psychological underpinning, pp. 240–41; national planning, pp. 317–41. Bible for professionals: Brooks, *Clint*, p. 198. "Collective Bargaining Problems Told in Book Written by United Steelworkers Execs," *Steel Labor*, July 31, 1942, p. 6. Leftist criticism: "Bosses Praise Golden," *Weekly People* (organ of the Socialist Labor Party), July 27, 1946.

13. Labor's role and Reuther Plan: Brooks, *Clint*, p. 214; "Industrial and Labor Advisory Committees in the National Defense Advisory Commission and the Office of Production Management, May 1940 to January, 1942," *Special Report No. 24*, Historical Reports on War Administration: War Production Board (Washington, D.C., Dec. 9, 1946), pp. 1–64 (future references to War Production Board [WPB] Records should be presumed part of Record Group 179 at the National Archives). Murray to Franklin D. Roosevelt, Dec. 20, 1940, WPB Records, file 631.0423. Also in same file: Reuther explains his plan in a letter to *NYT*, Jan. 20, 1941; Bruce Catton to Robert W. Horton, Jan. 9, 1942. Victor G. Reuther, *The Brothers Reuther and the Story of the UAW* (Boston: Houghton Mifflin, 1976), pp. 225–230. "Speeding the Auto Conversion," *BW*, Jan. 31, 1942, p. 15.

14. Murray Plan: Golden and Ruttenberg, *Dynamics*, pp. 343–347; "How Steel Output Can Be Increased," *Steel Labor*, Jan. 31, 1941, p. 2; Jacoby, "Union-Management Cooperation During the Second World War," p. 104; Schatz, "Philip Murray and the Subordination of Industrial Unions."

David J. McDonald claimed credit for the industry council concept (*Union Man*, pp. 151–52). He said he got the idea during discussions with Catholic priests and proposed it to Murray. The latter rejected it but later surprised McDonald by formally announcing such a plan. People who knew McDonald well say his book contains many inaccuracies and must be treated with caution. McDonald repeated the claim to interviewers from PSULA's Oral History Project (Helmut J. Golatz and Alice Hoffman, "An Interview with David J. McDonald," Palm Springs, Feb. 20, 1970, pp. 17–19). What makes McDonald's account most questionable is that he says that John F. Kennedy was the first president "to act on the Industrial Council Plan" in forming the President's Labor-Management Committee at McDonald's suggestion. This group, made up of seven union presidents and seven executives from major companies, met to discuss economic and labor issues as a means of improving labor-management relations generally. This is a far different concept from the industry council plan, under which labor and management would jointly make production and pricing decisions for an industry.

15. Labor participation: *Special Report No. 24*, pp. 178, 67–95, 205–09; *Special Report No. 34*, Historical Reports on War Administration: War Production Board (Washington, D.C., Dec. 9, 1946), pp. 11, 161, WPB Records. Rejection of Murray Plan: Knudsen to Murray, Mar. 20, 1941, WPB Records, file 512.44. Although this letter contained no reference to industry councils, an early draft, dated Mar. 14, mentioned the concept, but said that Murray's "method of control" was not "practical or desirable at the present time." In early 1941, *Steel Labor* carried several stories and editorials on industry councils (issues dated Jan. 31, Feb. 28, Mar. 20, and June 25). Endorsement by the CIO convention was reported in "Industry Councils Will Unleash Nation's Energies and Achieve National Defense" and "Roosevelt's Letter to CIO—and Murray's Reply to the President," *Steel Labor*, Nov. 28, 1941, pp. 3, 7.

Nelson's comment on Reuther: interview with Bruce Catton, July 27, 1945, WPB Records, box 1030, file 245.13. Catton, who had served as director of information of the WPB under Nelson, also said that Nelson offered Reuther a high-level job in the agency. Reuther preferred to stay with the UAW so that he would not "lose his place in the political situation." Reuther, of course, was pursuing the UAW presidency.

16. Much of the information on labor-management committees was gleaned from a

manuscript by Dorothea de Schweinitz, "The History of the War Production Drive," December 1945, divided into Periods I to IV, WPB Records; "Period I," pp. 1–9 (formation of committees) and p. 27 (Wilson quote); "Period IV," pp. 8, 10 (number of committees and disbandment); "Period III," p. 87–90 (steel committees). Murray's speech: text of radio broadcast, Mar. 6, 1942, WPB Records, file 240.1C. Sovietizing industry: WPB memo, "The Struggle of Labor-Management Cooperation During War Time in the United States," 1943, file 245.1R. GM's hostility: W. Ellison Chalmers to Theodore Quinn, Dec. 13, 1943, file 245.1R.

Analysis of issues discussed: Jacoby, "Union-Management Cooperation During the Second World War," p. 112. Speedup wrapped in American flag: WPB memo, "The Struggle of Labor-Management Cooperation During War Time in the United States," 1943, file 245.1R. Committees at Ford, GM, USS: de Schweinitz quoted in Jacoby, "Union-Management Cooperation During the Second World War," p. 119. Steel committees: "Report on Labor-Management Production Committees Participated in by United Steelworkers of America, CIO," Dec. 11, 1944, WPB Records, file 512.446.

17. Wartime increase in steel earnings: Mark McColloch, "Consolidating Industrial Citizenship," in Forging a Union of Steel, ed. Paul F. Clark, Peter Gottlieb, and Donald Kennedy (Ithaca, N.Y.: ILR Press, 1987), pp 45–87. This paper attributes the 26 percent calculation to Nelson Lichtenstein, Labor's War at Home: The CIO in World War Two (New York: Cambridge University Press, 1982), p. 111. V-E Day earnings drop: Dennis C. Dickerson, Out of the Crucible: Black Steelworkers in Western Pennsylvania, 1875–1900 (Albany: State University of New York Press, 1986), p. 178. Wartime benefits for companies: Howell John Harris, The Right to Manage: Industrial Relations Policies of American Business in the 1940s (Madison: University of Wisconsin Press, 1982), p. 42. Wage demand at U.S. Steel and Murray quote: Livernash, Collective Bargaining, p. 252.

18. Material on Truman's conference was drawn generally from a six-volume set, U.S. Labor Department, The President's National Labor-Management Conference, Nov. 5–30, 1945. Murray's speech: Plenary Sessions, vol. 2, session of Nov. 7, 1945, document no. 3, p. 25. Positions on right to manage: Reports of Working Committees, vol. 3, session of Nov. 29, 1945, document no. 125, pp. 1–6 (management's position), and session of Nov. 28, document no. 120, pp. 1–4. (labor's position). Harris, Right to Manage, pp. 92, 97–98, 101, 85. Steel strike of 1946: Livernash, Collective Bargaining, pp. 249–54.

19. Brash, young intellectual: Brooks, Clint, pp. 180–81. Harold J. Ruttenberg, Self-Developing America (New York: Harper & Brothers, 1960), pp. 125–26.

20. Golden on Ruttenberg: Golden to F. J. Murphy, Feb. 9, 1947, Golden Papers, PSULA, box 6, file 11. Resignations: "Golden and Ruttenberg Quit Steel Workers Union," PPG, July 2, 1946. Golden's diary item: Brooks, Clint, p. 226. Comments at USW board meeting: from minutes of USW executive board meeting, Ashville, N.C., June 29, 1946, read to author by James English, USW associate general counsel, May 7, 1987. The transcript of the June 27, 1946, session, also held in Ashville, contained Murray's announcement of Ruttenberg's resignation and the president's appreciative comments about his work for the union.

"Peace at any price": Golden to Scanlon, Dec. 20, 1946, Golden Papers, PSULA, box 9, file 12. Bernstein's quote: taped interview with Meyer Bernstein, Oct. 19, 1976, Golden Papers, PSULA, box 2, file 24. Fischer on Golden: interview, Pittsburgh, July 1, 1987. Fischer also made the point that Murray frequently used rhetoric such as, "I do not want peace at any price" without necessarily aiming his comments at a specific person. In each convention between 1945 and 1951, the CIO called for the establishment of industry councils as a mechanism for national planning (Jacoby, "Union-Management Cooperation During the Second World War," p. 121).

21. "My heart in my work": minutes of USW executive board meeting, June 29, 1946 (see n. 20). "How to Deal with Labor Unions," *U.S. News & World Report*, Mar. 23, 1956, p. 76. In this long question-and-answer interview, Ruttenberg first said he resigned from the USW because he no longer felt challenged by his work as a union staffer and that he wanted a "line position of fundamental responsibility" in the steel industry. Halfway through the interview he introduced the idea that he left the union because he disagreed with Murray's collective bargaining philosophy.

Why Ruttenberg really resigned: Ben Fischer, phone, June 9, 1986. Meyer Bernstein also gives this reason in the 1976 interview cited in n. 20.

22. Frederick G. Lesieur, phone, May 11, 1986; Carl F. Frost, phone, May 14, 1986. In the 1960s Frost wrote several inquiries to the USW, asking if the union had taken an official position against the Scanlon Plan. He didn't receive a response. For more background, theory, and experience relating to Scanlon Plans, see Frederick G. Lesieur, ed. *The Scanlon Plan: A Frontier in Labor-Management Cooperation* (Cambridge: MIT Press, 1958; 11th printing, 1978), and Carl F. Frost, John H. Wakeley, and Robert A. Ruh, *The Scanlon Plan for Organization Development: Identity, Participation, and Equity* (Lansing: Michigan State University Press, 1974).

23. Otis Brubaker, phone, Nov. 6, 1986. Morreel story: Russell W. Davenport to Golden, Jan. 13, 1950; Golden to Davenport, Jan. 17, 1950; Golden to Davenport, Jan. 20, 1950; Ben Morreel to J. M. Scribner, Jan. 30, 1950; Davenport to Golden, Feb. 3, 1950; Golden to Davenport, Feb. 11, 1950, all in Golden Papers, PSULA, box 1, file 17. Davenport, a well-known journalist at *Fortune*, was approached by a Pittsburgh lawyer (Scribner) who asked Davenport to serve as an intermediary in bringing Phil Murray together with steel executives to discuss "a new *attitude* with regard to social responsibilities." Davenport asked Golden, although retired by then, to approach Murray with this idea. Golden did so and reported, somewhat surprised, that Murray would talk. Golden's correspondence with Davenport seems to have ended at this point. In 1986, when I asked the public relations department of LTV Steel (J&L's successor company) to check back on the Morreel suggestion, too much time had passed. The trail was cold, LTV was in bankruptcy proceedings, and further research by LTV personnel was out of the question.

24. CIO philosophy: Jacoby, "Union-Management Cooperation During the Second World War," p. 122. Fischer quotes: comments at a meeting of the Industrial Relations Research Association, New York City, Dec. 12, 1985. This theme, especially as it relates to the UAW, is best developed in David Brody, *Workers in Industrial America* (New York: Oxford University Press, 1980), chap. 5. See also Nelson Lichtenstein, "UAW Bargaining Strategy and Shop-Floor Conflict: 1946–1970," *Industrial Relations* 24, no. 3 (Fall 1985): 360–81. The focus on money demands in the 1940s and later is well documented in Ronald W. Schatz, *The Electrical Workers: A History of Labor at General Electric and Westinghouse, 1923–60* (Urbana: University of Illinois Press, 1983), pp. 151–60.

25. Rejection of tripartite committee: Livernash, *Collective Bargaining*, p. 304; Marvin Miller, New York City, May 18, 1987. History of Kaiser Plan: William Aussieker, "The Decline of Union-Management Cooperation: Kaiser Long Range Sharing Plan," *Proceedings of Thirty-Fifth Annual Meeting*, Industrial Relations Research Association, New York City, Dec. 28–30, 1982, pp. 403–09. See also George McManus, *The Inside Story of Steel Wages and Prices, 1957–1967* (Philadelphia: Chilton, 1967), pp. 86–89.

26. Human Relations Committee: Marvin Miller, New York City, May 18, 1987; Ben Fischer, Pittsburgh, July 1, 1987; Otis Brubaker, phone, Nov. 6, 1986; R. Heath Larry, phone, Nov. 25, 1986; various memos, USW Archives, microfilm roll no. 8400 (industrywide formula discussed in undated memo, "Kaiser Steel–United Steel-

workers Long Range Committee: Underlying Considerations in the Establishment of a Formula," and later confirmed by Miller). In 1965 the USW negotiated a cost-savings plan similar to Kaiser's at the small Alan Wood Steel Company in Conshohocken, Pa.

27. Kennedy vs. Blough: a contemporary account of this famous confrontation is in McManus, *Inside Story*, pp. 44–56. Bargaining under HRC in 1962: ibid., pp. 90–95. The HRC summing up: personal reportage and notes from the Abel-McDonald election fight of 1964–65; Miller and Fischer interviews cited above.

28. My account of the productivity committees is based on personal notes and interviews from the years when the committees were functioning. Wording of provision and USW conclusion that the committees represented a crew-cutting process: USW, "Programs in Steel," 1984, a report written by Donald Dalena, public relations, pp. 38–39. According to Fischer, the productivity effort also suffered because James P. Griffin, an assistant to Abel, headed the program for the union. Griffin, a former district director, wanted to run for USW president and did not aggressively push the committees.

29. Origin of LMPTs: John H. Kirkwood, Pittsburgh, Oct. 27, 1983; A. Cole Tremain, Pittsburgh, Feb. 22, 1986; Anthony St. John, phone, Nov. 3, 1986; Sam Camens, Pittsburgh, Apr. 29, 1986; Carl Frankel, Pittsburgh, Feb. 18, 1986; Edmund Ayoub, Mt. Lebanon, July 26, 1985. Camens, 1980: John Hoerr, "Beyond Bargaining: Unions and Bosses Try Trust," *BW*, May 5, 1980, p. 43. Background: San Camens, Pittsburgh, Apr. 29, 1986.

30. Mike Bilcsik, phone, Dec. 30, 1983.

Chapter 12

1. The background and material in this chapter was developed over twenty years in conversations with many scores of workers, supervisors, and company and union officials. In recent years I interviewed dozens of people specifically for this book. I list here only those I have quoted, with thanks to those not mentioned. William Behare, McKeesport, Apr. 29, 1986; Jack Bergman, West Mifflin, Feb. 24, 1986; William J. Daley, Duquesne, Apr. 22, 1986; Lawrence Delo, White Oak, many interviews, 1982–87; Edward Galka, phone, Dec. 28, 1986; Alvin L. Hillegass, phone, Sept. 22, 1986; Barney Joy, White Oak, May 1, 1986; Stephen A. Krivda, White Oak, Apr. 29, 1986; Joseph W. Lenart, McKeesport, Feb. 26, 1986; Rea McKay, phone, Dec. 31, 1986; Joseph Odorcich, Pittsburgh, Mar. 3, 1983, Feb. 18, 1986; Alan J. Sarver, Etna, Pa., Apr. 24, 1986; William Schoy, phone, Feb. 21, 1987; Manuel Stoupis, McKeesport, Feb. 16, 1986 (and many other interviews). Unless otherwise stated, these citations will cover all quotes from these sources.

2. Work hours and fatalities: Gerald G. Eggert, *Steelmasters and Labor Reform, 1886–1923* (Pittsburgh: University of Pittsburgh Press, 1981), pp. 44–51.

3. Richard E. Walton, "From Control to Commitment in the Workplace," *Harvard Business Review*, March–April 1985, pp. 77–84. Walton, a professor who has helped companies and unions design and implement participation processes since the early 1970s, has produced some of the best-informed articles and books on the subject.

4. Michael Maccoby, "The Problem of Work: Technology, Competition, Employment, Meaning," a paper delivered before the American Sociological Association, Washington, D.C., Aug. 26, 1985. My comments about Taylorism are based on Daniel Nelson, *Frederick W. Taylor and the Rise of Scientific Management* (Madison: University of Wisconsin Press, 1980), "pecuniary rewards," p. 102; Frederick W. Taylor, *The Principles of Scientific Management* (New York: Harper & Brothers, 1911), quoted, p. 7. Useful discussions of Taylorism can be found in Harry Braverman, *Labor*

and Monopoly Capital: The Degradation of Work in the Twentieth Century (New York: Monthly Review Press, 1974); Larry Hirschhorn, *Beyond Mechanization: Work and Technology in a Postindustrial Age* (Cambridge: MIT Press, 1984; rpt. 1986); Haruo Shimada and John Paul MacDuffie, "Industrial Relations and 'Humanware,'" Working Paper no. 1855–87, Alfred P. Sloan School of Management, MIT, December 1986.

5. Camens quoted: John Hoerr, "Worker Participation Then and Now," in *Participative Systems at Work: Creating Quality and Employment Security*, ed. Sidney P. Rubinstein (New York: Human Sciences Press, 1987), p. 146.

6. Development of job classifications: Jack Stieber, *The Steel Industry Wage Structure* (Cambridge: Harvard University Press, 1959), number of jobs classified, p. xvii; distrust of engineering methods, p. 10. Michael J. Piore and Charles F. Sabel, *The Second Industrial Divide: Possibilities for Prosperity* (New York: Basic Books, 1984), p. 115.

7. Retreatism: Michel Crozier, *The Bureaucratic Phenomenon* (Chicago: University of Chicago Press, 1964), p. 199. Knee-jerk reaction: John B. Myers, "Making Organizations Adaptive to Change: Eliminating Bureaucracy at Shenandoah Life," *National Productivity Review* 4, no. 2 (Spring 1985): 132–38.

8. My discussion of 2B is based on personal reportage and files, interviews with many union and management officials over the years (including those quoted in this chapter) and arbitration decisions listed below. Strict construction: Ben Fischer, Pittsburgh, July 1, 1987. 2B blocks new technology: John Strohmeyer, *Crisis in Bethlehem*, (Bethesda: Adler & Adler, 1986), p. 65. Negotiation mistake led to 2B: USS Board of Arbitration, *U.S. Steel, National Tube Division and USW, Local 1104*, case no. N-146, Jan. 31, 1953 (this is the first decision in which 2B was ruled to cover crew size). Past Practice Committee: Marvin Miller, New York City, May 18, 1987. The lack of cooperation of plant managers in studying the work rules problem is also noted in David J. McDonald, *Union Man* (New York: E.P. Dutton, 1969), p. 292.

Employment and shipments: AISI, *Annual Statistical Report* (Washington, D.C., various years). These statistics apply only to the AISI companies; nonmember minimills didn't have a 2B clause. For purposes of comparison, the AISI companies employed 291,483 hourly workers in 1980 and shipped 83.9 million tons.

9. My account of union corruption in District 15 is based on extensive personal reportage and notes from those years. Joseph Odorcich, Pittsburgh, Feb. 18, 1986.

Chapter 13

1. "Laid-off Steel Workers Angry About Benefits," *PPG*, July 31, 1982; Jean Bryant, "Steelworkers Demand Aid Extension," *PP*, Aug. 1, 1982. Steel production and employment decline: AISI, *Annual Statistical Report* (Washington, D.C., 1984); AISI, "The American Steel Industry's Problems: Severe But Solvable," (Washington, D.C., February 1983). The AISI figures are for all companies belonging to the association, including the eight coordinating committee firms. (In 1986 the AISI member-companies represented about 80 percent of raw steel produced in the United States.)

2. Reportage on the convention: my notes supplemented by USW, *Proceedings of the 21st Constitutional Convention*, Atlantic City, Sept. 20–25, 1982; delegates quoted: pp. 60–62, 80, 95–96. Complaints from members: according to George Moore (interview, New York City, Mar. 21, 1984), McBride told industry negotiators that his mail ran nine to one in favor of granting relief. The USW keeps files of such correspondence in its archives but in 1987 was not able to fulfill my request for a count of letters for and against. Richard Grace, McKeesport, Oct. 29, 1982. John Kirkwood, phone, Sept. 21, 1982; "The Pressure for New Steel Talks," *BW*, Oct. 4, 1982, p. 90.

3. Financial losses: McBride disclosed the company losses to local presidents at their ratification meeting. Transcript of Basic Steel Industry Conference meeting, Nov. 19, 1982, p. 59, in USW files, Pittsburgh. The account in this chapter of bargaining in October and November 1982 is reconstructed from interviews with Joseph Odorcich (Pittsburgh, Nov. 22, 1982, and Mar. 3, 1983), George Moore (New York City, Aug. 2 and 16, 1984), Bernard Kleiman (phone, Nov. 29, 1982), Edmund Ayoub (phone, Dec. 5, 1982; Mt. Lebanon, July 26, 1985), Thomas Duzak (Pittsburgh, Nov. 21, 1982), Carl Frankel (Pittsburgh, Feb. 18, 1986), James Smith (Pittsburgh, Apr. 17, 1986), James English (phone, Nov. 29, 1982, Dec. 30, 1986). Unless otherwise noted, all comments by these sources are covered by this citation.

4. Background on profit-sharing: John Hoerr, "Why Labor and Management Are Both Buying Profit-Sharing," *BW*, Jan. 10, 1983, p. 84; William C. Freund and Eugene Epstein, *People and Productivity: The New York Stock Exchange Guide to Financial Incentives and the Quality of Work Life* (Homewood, Ill.: Dow Jones-Irwin, 1984); "Did Chrysler Pay Too Much for Peace?" *BW*, Dec. 27, 1982, p. 33.

5. McBride's wage-cut proposal surprised Ayoub all the more because the president had approved the text of a speech made by the economist on Oct. 12. In this speech, Ayoub declared that the union's bargaining policy called only for "a deceleration of employment cost increases." He went on to say: "There are profound institutional reasons why a sharp, sudden curtailment in wages and benefits large enough to quickly resolve the competitive problem [is] simply out of the question." This is precisely what McBride offered to the industry less than two weeks later. "Steel Industry Problems and Prospects: Labor's View," text of speech delivered before Financial Analysts Federation Conference, Cleveland, Oct. 19, 1982.

Upon retiring at age fifty-five, Ayoub opened an art and framing store in Mt. Lebanon. He retained his connections with the USW and interest in labor affairs. In 1986 the USW chose Ayoub as its candidate for the board of directors at CF&I Industries, a right granted the union in return for wage concessions. Ayoub was elected to the board in 1987. Later that year, the union appointed him to be one of three union representatives, along with District Director Joe Coyle and a staff economist, Ray McDonald, on a union-management committee of trustees to oversee a new LTV Steel pension plan (see chapter 18).

6. The complexity of the COLA bonus plan is indicated in the following outline of steps required to compute profits:

a. Each of more than 10 categories of cost was weighted according to its ratio to total operating costs in the base year of 1977. The government's Producer Price Index (PPI) would be the basis for determining increases in each category of cost. PPI in plan year − PPI in base year = ratio of increase.

b. Steel shipments in plan year − shipments in base year = ratio of increase. PPI increase ratio × shipment increase ratio = cost increase adjusted by output. Cost increase × weight factor = cost escalation.

c. Sum of all cost increases = aggregate cost escalation index. Cost escalation index × actual cost total for 1977 = approximation of total dollars spent on operating expenses in plan year.

d. Actual hourly employment cost in plan year × number of hours worked = total employment costs. Employment costs + operating expenses = total costs. Revenues − costs = profits.

7. The extended vacation plan, far from being an example of largesse when first negotiated, enabled the industry to hold down labor cost rises in 1962. See George McManus, *The Inside Story of Steel Wages and Prices, 1957–1967* (Philadelphia: Chilton, 1967), p. 92. McManus covered the steel industry for *Iron Age*. Reporting on the results of 1962 negotiations, he writes: "The big benefit was the extended leave.

Dave McDonald estimated this would create 20,000 new jobs in basic steel. From the companies' standpoint, the beauty of the contract was its cost. The 15¢ package came to about 9¢ a year or 2 percent a year. This was the lowest cost increase in the postwar period."

Dissidents' meeting: the weekly newspaper *Labor Notes* convened a conference of anticoncession unionists from many industries in Detroit on Nov. 13, 1982. Dennis Shattuck, a grievance committeeman of USW Local 1010, who attended the conference, said one specific session dealt with the steel industry. Some fifty people showed up, Shattuck said, but only about five or six, including Ron Weisen and Mike Bonn, were local presidents. In addressing the BSIC meeting of Nov. 19, McBride charged that dissidents at the Detroit conference had developed a strategy to "disrupt" the ratification meeting. In a Dec. 16, 1986, phone interview, Shattuck described McBride's charge as "hyperbole," adding that "our meeting neither attempted to, nor was capable of, devising a grand strategy."

8. The wage and benefit cuts totaled $3.11 per hour, but the companies would not realize this amount of cost savings for the entire forty-four months of the new agreement because of wage restorations and a rise in health insurance premiums. The cuts and restorations are figured this way:

Hourly cost reductions: EV (50¢), one holiday (8¢), Sunday premium (28¢), wage cut ($1.50 + compounding = $2.25), wage cut for SAFE program (75¢) = $3.86. Because the companies would use the 75¢ wage reduction for the SAFE program, the total cost reduction at the outset would be $3.11 per hour. Cost additions: Restoration of the wage cuts, plus compounding, would add $2.25 in costs by the end of the contract. Health insurance costs would rise by an estimated $1.19, meaning that the terminal cost of the agreement would be 33¢ per hour above the starting cost of $23.35. The terminal cost could rise an additional amount at each company, depending on payments under the COLA bonus plan.

9. The story of the variations in COLA bonus payouts is based on previously cited interviews with Ayoub, Frankel, Odorcich, and Smith. The figures showing what the payments would have been from 1978 to 1981 are contained in a USW memo, Smith to Odorcich and Kleiman, Feb. 20, 1983. Allegheny-Ludlum is not included in this list because it had withdrawn from the coordinating committee by the time this memo was written. John Kirkwood, phone, Dec. 17, 1986; Stanley C. Ellspermann, Pittsburgh, Feb. 20, 1986.

10. Number of List 3 units at each company: listed in USW, *Summary of 1982 Basic Steel Settlement*, Nov. 19, 1982. For more on the demise of Bethlehem's construction subsidiary, see John Strohmeyer, *Crisis in Bethlehem* (Bethesda: Adler & Adler, 1986), chap. 10.

If the List 3 issue had been such a problem for the companies, for so long, why hadn't they forced the union to address it in previous years? One reason is that the federal government took a big interest in steel negotiations before the 1970s, and it is difficult to conceive of any president allowing a national steel strike to occur over the fate of nonsteel plants. Furthermore, as the USW bargainers realized, the CCSC couldn't obtain a consensus of its own members for tough action on the List 3 problem because four companies—National, Allegheny-Ludlum, J&L, and Armco—either had no List 3 plants or only a few. Although U.S. Steel owned thirty-three of the units, many were small and of less importance in the corporation's overall business objectives than those at Bethlehem, Republic, and Inland.

According to Ben Fischer, from the time McBride became president he was determined to get the List 3 plants out of the industrywide agreement. In 1977 he assigned Fischer to handle the problem in 1980 negotiations. But Fischer turned sixty-five, the union's mandatory retirement age, at the end of 1978, a few weeks before the effective date of a new federal law raising the mandatory retirement age to seventy. McBride decided if he allowed Fischer to stay on the staff, he also would have to keep two other staffers of the same age whom McBride really wanted to get rid of. Fischer retired, and the List 3 issue was not addressed.

11. Quotes from the BSIC meeting are taken from the transcript of the Basic Steel Industry Conference meeting, Nov. 19, 1982, p. 59, in USW files, Pittsburgh. Estimate of monthly earnings: based on average pay for hours worked in 1982 of $14.06 (AISI) and 179 hours of full-time work per month.

12. On only one other occasion, in 1977, had the BSIC rejected a tentative steel agreement. That vote probably reflected the anger of iron-ore delegates, who felt the agreement neglected their needs, as well as a nonspecific protest against the way negotiations were handled. Local presidents had had little involvement in the talks. Immediately after a majority of the presidents rejected the agreement in a voice vote, a roll-call vote was called for. Forced to record their vote and be held accountable by members back home, a majority of the delegates voted yes.

13. Worst contract: Jack Metzgar, "The Humbling of the Steelworkers," *Socialist Review*, no. 75/76, 1984, p. 57. BSIC composition: Gary Hubbard, USW public relations, phone, Jan. 18, 1983; Joseph Odorcich, phone, Dec. 27, 1982. Why the four presidents from Lackawanna voted against the pact, despite their certainty the plant would be closed, was explained to me on Jan. 3, 1983, a few days after the shutdown was announced. Local 2603 Treasurer Wiley Cole said the future of the plant became obvious two years before when Bethlehem announced a $750 million modernization program for other plants but none for Lackawanna. "If we'd given everything away, Bethlehem would simply have put the money in some other plant," he said. See "Steel Bargaining: The Last Chance," *BW*, Jan. 17, 1983, p. 94. $12,800 loss estimate: Midwest Center for Labor Research, *Concessions in Steel?* January 1983. This estimate includes not just the pay cut but also the amount of future COLA that would *not* be paid.

14. Problems with the SAFE program and payroll deductions to help the unemployed: Thomas Duzak, Pittsburgh, Nov. 21, 1982, and Mar. 1, 1983; Joseph Odorcich, Pittsburgh, Mar. 3, 1983. The low volunteer rate at Homestead and Duquesne was confirmed by the presidents of those locals. Meanwhile, according to Odorcich in the above interview, more than 90 percent of staffers in the USW headquarters contributed to an unemployed fund. Each of the five International officers contributed at least $100 per month, and all five turned down a raise scheduled in January 1983.

Two separate polls conducted later also indicated that aid to laid-off members, while important, was not the highest priority of most steelworkers. At Local 1066 in Gary, 53 percent of 1,750 members surveyed in January 1983 assigned high priority to the provision. Eleven other demands ranked higher. The Local 1066 survey was conducted by mailing a questionnaire in early January 1983 to each member (see chap. 14).

15. "A point of pride": Metzgar, "Humbling of the Steelworkers," p. 65. "Lloyd's not for window-dressing": Bernard Kleiman, phone, Nov. 29, 1982.

16. "Steel's Outlook: Mutual Misery," *BW*, Dec. 6, 1982, p. 95. Thomas F. O'Boyle and J. Ernest Beazley, "USW Chiefs, Backing Concessionary Pact, Slash Number Eligible to Ratify Contracts," *WSJ*, Jan. 13, 1983. Lloyd McBride, phone, Jan. 3, 1983.

Chapter 14

1. Pittsburgh area unemployment rates: Pennsylvania Department of Labor and Industry, *Labor Market Reports* (Pittsburgh: various issues).

2. Steel losses and closedowns reported in *NYT*: William Serrin, "Aftershock of

Layoffs Hits Lackawanna," Dec. 29, 1982; Raymond Bonner, "Bethlehem 4th-Quarter Loss Huge," Jan. 27, 1983; "Closing Imperils 8,000 Employed in Related Work," Dec. 29, 1982; "U.S. Steel Posts $363 Million Loss for 4th Quarter," Jan. 26, 1983; Raymond Bonner, "Armco Is Closing Some Units," Jan. 11, 1983.

From *WSJ:* "National Steel Posts Net Loss for 4th Period," Feb. 1, 1983; Ralph E. Winter and Gregory Stricharchuk, "Republic Steel, Workers Struggle to Cope with Declining Business," Jan. 31, 1983; "Mesta Machine Seeks Protection from Creditors," Feb. 10, 1983. Commerce Department prediction: "Big Steel's Winter of Woes," *Time,* Jan. 24, 1983, p. 58.

Aggregate statistics on steel financial, shipment, and production results: AISI, *Annual Statistical Report* (Washington, D.C., 1984). Marcus quote: Carol Hymowitz and J. Ernest Beazley, "Steel's Recovery Is Seen Set Back by Union's Vote," *WSJ,* Nov. 22, 1982. Forecasts: Thomas F. O'Boyle, "Steel Recovery Appears to Have Started But May Trail Earlier, Weak Forecasts," *WSJ,* Feb. 6, 1983.

3. Wheeling-Pittsburgh: "Deeper Sacrifices at Wheeling-Pittsburgh," *BW,* Jan. 17, 1983, p. 94. Salary cuts: "U.S. Steel Lowers Salaries 5 percent for 28,000, Will Post a 'Staggering' 4th Quarter Loss," *WSJ,* Jan. 17, 1983; "Bethlehem Steel Sets New Pay Reduction for 14,000 Workers," *WSJ,* Jan. 21, 1983.

4. Steel unemployment: AISI, "The American Steel Industry's Problems: Severe But Solvable," (Washington, D.C., February 1983). The 148,691 unemployed represents 38 percent of AISI average hourly and salaried employment of 390,914 in 1981. Criticism of local presidents: "Another Fumble by the USW," *BW,* Dec. 6, 1982. "Once again," this editorial said, "the local presidents have let their fellow workers down." I would have said that the presidents also had other reasons for rejecting the agreement, but reporters and editorial writers seldom agree.

5. Reporting on the groundswell comes from many sources: Jane Slaughter, "Opposition Grows to Steel Industry's Third Concessions Bid," *Labor Notes,* Jan. 27, 1983, p. 8; Phillip Cyprian, phone, Jan. 5, 1983; Bruce Bostick, "Labor Turns Away from Concessions," reprinted in *John Herling's Labor Letter,* Jan. 15, 1983; Mike Bilcsik, phone, Jan. 4, 1983, Jan. 7, 1987; Denney quoted: "Concessions Debate Sparks Renewed Interest in 'Right to Ratify' Contracts," *Labor Notes,* Jan. 27, 1983, p. 8.

6. Industry strategy: George Moore, New York City, Sept. 9, 1984.

7. Local 1066 poll results: Sidney P. Feldman, Indiana University Northwest, to James Brown, Feb. 11, 1983 (personal files, copy supplied by Brown). The strike finding came from a simple yes or no answer to the question: "Willing to strike to avoid concessions?" Contrary to conventional thinking about strike attitudes, the younger members of Local 1066 were less inclined to strike (64 percent of the under-thirty age group replied no to the strike question) than those in older age groups, probably because their jobs were at greater risk.

A sampling of the preference ranking of benefits: under columns headed "highest priority, improvement only" and "very important, maintain status," more than 90 percent of the members included job security, pension, Blue Cross, and sickness and accident; major medical, 87 percent; life insurance, 78 percent; wages, 77 percent; SUB, 69 percent; overtime premium, 73 percent; COLA, 63 percent; and aid to unemployed, 53 percent.

BSIC poll: USW files, Pittsburgh. Fear of strikebreakers: Joseph Odorcich, Pittsburgh, Feb. 18, 1986.

8. Jones & Laughlin had bought Colt Industries' Midland, Pa., plant, which produced the same kinds of specialty steel products as Allegheny-Ludlum. J&L wanted to operate some production units in the closed plant but told the USW it could do so only if labor costs were slashed. In April 1983, the union agreed to cut employment costs by $4 to $5 an hour. Since Colt had not been part of the CCSC, J&L contended

that it could operate the plant at lower than industrywide wage rates without violating CCSC rules. Apparently anticipating such an action and wanting the freedom to follow suit, Allegheny-Ludlum departed from the group and bargained individually with the USW. However, in April 1983, the company failed to win as much in concessions as the CCSC companies, let alone the Midland plant. The USW granted a wage reduction of 50¢ per hour, 75¢ less than the industrywide settlement. "Steelworkers Negotiate Smaller Concessions with Firm That Withdrew from Industry Group," *Daily Labor Report,* Apr. 4, 1983, p. A5.

9. Staff cuts and McBride's hospitalization: Harry Guenther (McBride aide), phone, Feb. 15, 1983; Edmund Ayoub, Mt. Lebanon, July 26, 1985. BSIC meeting: Joseph Odorcich, Pittsburgh, Feb. 18, 1986; James Brown, notes on BSIC meeting, Feb. 2, 1983 (copy supplied by Brown).

10. Odorcich's background: Joseph Odorcich, Pittsburgh, Feb. 3, 1983, Oct. 24, 1983, Feb. 18, 1986 (taped interviews), and many other conversations.

11. Election of 1969: previously cited Odorcich interviews, supplemented by author's notes taken in 1969 and conversations with Odorcich's supporters in Local 1408. District 15 vote tabulation and invalidation: USW, International Tellers, *Report on International Election* (Pittsburgh, Apr. 18, 1969). USW, news release, Apr. 30, 1969.

12. My account of Third Round bargaining is based on interviews with Joseph Odorcich and George Moore (previously cited); Bernard Kleiman, Pittsburgh, Apr. 30 and June 10, 1986; Thomas Duzak, Pittsburgh, Mar. 1, 1983; James Smith, Pittsburgh, Apr. 17, 1986; Edmund Ayoub (previously cited); many local union presidents.

13. Under the revised COLA bonus plan, the payments ranged from $2.50 for Armco to $0.23 for USS (USW memo, cited in chap. 13, n. 9).

14. The USW had never found a good solution to the problem of giving rank and filers a voice in steel bargaining. Many USW critics claimed that the UAW's system was more democratic. At each auto company, a rank-and-file committee of eight to ten members was elected to participate in top-level negotiations. Committees, however, are functionally incapable of closing a deal. The UAW rank and filers join in general discussions at the bargaining table, but the crucial decisions are made by a top negotiator who then seeks committee approval. The USW had not adopted this system for three reasons. (1) It would have been difficult to elect a representative committee for industrywide bargaining, given the number of companies involved and the distribution of members. (2) The USW's top-down tradition weighed heavily against diluting the president's authority. (3) U.S. Steel, the dominant industry force, strongly opposed expanding the bargaining teams on either side of the table.

15. Transcript of Basic Steel Industry Conference meeting, Feb. 28 and Mar. 1, 1983, in USW files, Pittsburgh.

16. McBride's approval on local bargaining: report of interview of Bruce Johnston by *BW* correspondent William C. Symonds, Jan. 16, 1984.

17. "The Use of Labor Cost Savings Under the 1983 Contract," memorandum dated Jan. 31, 1985, in USW Research Department files, Pittsburgh.

18. USW gave up more than it intended: decision, *In the Matter of Arbitration Between United Steelworkers of America and Coordinating Committee Steel Companies,* by Benjamin Aaron, Los Angeles, Nov. 28, 1983; John P. Moody, "USW Loses Arbitration on Benefits," *PPG,* Dec. 7, 1983; George Moore, New York City, Sept. 7, 1984; Joseph Odorcich, Pittsburgh, Feb. 18, 1986; Bernard Kleiman and James English, Pittsburgh, Mar. 4, 1986.

Chapter 15

1. Unemployment statistics: Michael A. Acquaviva, Pennsylvania Department of Labor and Industry, *Labor Market Report* (Pittsburgh: various issues up to August 1987). Paul Lodico, McKeesport, June 12, 1987. Jack Bergman, West Mifflin, Feb. 24, 1986 (covers further quotes in this chapter).

2. Steve Frazier, "Oil Recession Plunges Houston into a State of Mental Depression," *WSJ*, Mar. 25, 1986. Richard A. Pomponio, McKeesport, Mar. 2, 1986.

3. Value of mergers: Walter Adams and James W. Brock, "The Hidden Costs of Failed Mergers," *NYT*, June 21, 1987. Farm failures and Hagedon's quote: "What Five Families Did After Leaving the Farm" and "Big Operators Give Up," *NYT*, Feb. 4, 1987. Stright's story: Jane Blotzer and Eleanor Bergholz, "Moving on When Hope Runs Out," part of a special report, "When the Fire Dies," *PPG*, Dec. 30, 1985.

4. Wage increases and real wage decline: Bureau of Labor Statistics, *Current Wage Developments* (Washington, D.C., June 1987) and *Employment and Earnings* (Washington, D.C.: various issues). Strike statistics: news release, Bureau of Labor Statistics, "Major Work Stoppages: 1986," Washington, D.C., Mar. 6, 1987.

5. Although the active USW officers refrained from politicking before McBride's funeral, retired President I. W. Abel did not. According to two directors who talked to him in St. Louis on the eve of the funeral, Abel urged board members to vote for Frank McKee. This may have been on his presumption that McKee would run against Joe Odorcich, who had often been at odds with Abel. My sources were William J. Foley and Bruce Thrasher, the directors, respectively, of Districts 1 and 35. Abel later switched sides and campaigned for Lynn Williams in the special election of 1984.

6. My account of the election of an interim president is based on many interviews with USW staff and board members, including the following: Harry Guenther, phone, Oct. 17, 1983 (this citation covers future references). USW membership figures and Williams's comments on Odorcich: Lynn Williams, Pittsburgh, Nov. 11, 1983. Odorcich's comments on Williams: Pittsburgh, Nov. 18, 1983. McBride preferred Williams: Buddy Davis, phone, Nov. 21, 1983. Paul Rusen, phone, July 10, 1987. Odorcich pulling out: phone, Nov. 30, 1983.

7. While some Williams supporters felt that Odorcich was entitled to an interim presidency of a few months, they suspected that he would use the post to prepare the way for his own choice to succeed him, perhaps McKee or Frank Valenta of Cleveland. Could the two candidates have worked out a deal in which Williams would support Odorcich for an interim presidency in return for Odorcich's pledge to endorse Williams in the special election? Union politics rarely permits such a neat outcome. Williams told me he sent an emissary to Odorcich with such a message, but the latter said he didn't receive it. In any case, such a deal did not appeal to Odorcich. After Williams became president, Odorcich rejected other attempts at rapprochement.

8. Concession policy: personal reportage; USW news release, Washington, D.C., Dec. 13, 1983. The resolution condemned attempts by steel companies to entice USW locals to grant concessions in excess of those in the 1983 industrywide agreement. "If our union were to permit local level wage or benefit concessions, our wage and benefit programs would almost immediately become a shambles. Local after local would be whipsawed by threats of plant closing, diversion of orders from one plant to another, false promises, and many other pressure tactics," the resolution said.

9. McKee background: McKee campaign, "McKee for President," letter to all local unions, Dec. 8, 1983; Frank McKee, phone, Dec. 1, 1983, and Jan. 16, 1984. Phelps

Dodge: previously cited interviews with McKee, Odorcich, Williams, as well as with USW staff members; "Industrywide Wage Patterns Get a Test in Copper," *BW*, July 18, 1983, p. 58; "The Strikers Start to Crack at Phelps Dodge," *BW*, Feb. 13, 1984, p. 38.

10. McKee vs. Williams: previously cited McKee and Williams interviews. Williams on participation: phone, Feb. 15, 1984. Also see, "How Nationalism Is Splitting the USW," *BW*, Apr. 2, 1984, p. 32. Sambuchi quote: William Serrin, "Hard Times for Labor in Steel Industry," *NYT*, Apr. 1, 1984; Carol Hymowitz, "Canadian Claims Victory in USW Vote for President, But Foe Doesn't Concede," *WSJ*, Apr. 2, 1984. Vote results: USW, International Tellers, *Report on Special International Election* (Pittsburgh, May 18, 1984).

11. Happenings at 1984 convention: personal reportage; Bruce Thrasher, Cleveland, Sept. 23, 1984.

12. Williams background: several personal interviews with the author, including taped sessions in Pittsburgh, Nov. 18, 1983, and May 2, 1986. Williams as an organizer: William F. Scandlin, USW District 6 staff member, Cleveland, Sept. 27, 1984. I also relied on many interviews over several years with local union officers and Pittsburgh staff members, including Bernard Kleiman, Pittsburgh, Apr. 30, 1986. Journalistic profiles: "Lynn Williams of the USW Is Labor's Rising Star," *BW*, Oct. 15, 1984, p. 154; Robert Kuttner, "Lynn Williams and the Politics of Labour," *Saturday Night*, February 1985; Carol Hymowitz, "First Strike in 26 Years Puts Steelworker Chief in Midst of Latest Crisis," *WSJ*, July 23, 1985.

13. For more on the two-pronged approach, see Lynn Williams, "Decline of the Authoritarian Tradition," in *Participative Systems at Work: Creating Quality and Employment Security*, ed. Sidney P. Rubinstein (New York: Human Sciences Press, 1987), pp. 131–37.

Chapter 16

1. Roderick quoted: "Time Runs Out for Steel," *BW*, June 13, 1983, p. 84.

2. Worldwide restructuring: Peter F. Marcus and Karlis M. Kirsis, "Forces Creating Change More Powerful Than Business Cycle," World Steel Dynamics, PaineWebber Inc., presentation to Steel Survival Strategies Conference, New York, June 24, 1986.

3. Steel prices: Locker/Abrecht Associates Inc., "Confronting the Crisis: The Challenge for Labor," report prepared for USW, Dec. 16, 1985. Regarding U.S. Steel price-cutting, this report quotes Charles Bradford of Merrill Lynch, writing in the Oct. 9, 1983, issue of *Merrill Lynch Investment News*: "U.S. Steel has increased its market share, leading us to believe that they, not steel imports, are leading prices lower." Steel capacity and imports: AISI, *Annual Statistical Report* (Washington, D.C., various years); Chase Econometrics, "Crisis and Competition: The World Steel Industry, 1986–2000," vol. 2, 1986. Chase forecast a decline in U.S. capacity from 118 million tons in 1985 to 91 million tons in 2000. Steel employment declines: Bureau of Labor Statistics, Patricia Captiville, phone, Sept. 28, 1987.

4. Brazilian imports: AISI, *Annual Statistical Report* (Washington, D.C., various years). Import restraints: "Steel Giants Split Over Import Protection," *BW*, Nov. 28, 1983, p. 40; Art Pine, "Reagan Vows to Seek Voluntary Steel-Import Curbs; Restraints Held Unlikely to Stem Decline of Industry," and "Tentative Accord Is Set for Limit on Steel Imports," *WSJ*, Sept. 19, 1984, and Nov. 20, 1984. "201 filing": John J. Sheehan, USW legislative director, phone, Dec. 2, 1983. The ITC finding was flawed, according to a close anaylsis by Gene M. Grossman of Princeton University, in "Imports As a Cause of Injury: The Case of the U.S. Steel Industry," *Journal of International Economics* no. 20, 1986. The Trade Reform Act of 1974 requires that

the ITC recommend import restraints only if an industry has been damaged primarily by imports. Grossman found that a decline of 208,734 steel jobs from 1976 to 1983 was attributible to the substitution of other materials for steel, technological changes, and a general shift in employment away from manufacturing. Import competition wiped out 37,403 jobs (29,037 alone because of a more than 30 percent appreciation in the U.S. dollar). Excessive wages cost 5,047 jobs in the period.

5. New Strategies: USX, *Statistical Supplement to the Annual Report* (Pittsburgh, 1986); "National Intergroup: A New Name to Stress the Shift Away from Steel," *BW*, Sept. 26, 1983, p. 82; "National Tries to Move Even Further Away from Steel," *BW*, Oct. 22, 1984, p. 40.

6. My account of local bargaining at J&L is drawn mainly from these interviews: John Kirkwood, Pittsburgh, Apr. 17, 1986; A. Cole Tremain, Pittsburgh, Feb. 22, 1986; Peter J. Eritano, phone, May 1, 1983. J&L's 70/80 bargaining had an unfortunate sequel. After LTV entered chapter 11 in 1986, the Pension Benefit Guaranty Corporation (PBGC) took over the company's underfunded pension plans to ensure that retirees would receive their pensions. The federal agency eliminated the $400 monthly supplement for some 8,000 LTV retirees who had received early pensions for various reasons, including 1,200 who retired as a result of negotiations over manning (A. Cole Tremain, phone, Aug. 3, 1987). People who retired in their mid-50s had to swallow a drastic income reduction and would not be eligible for Social Security until age sixty-two (see chap. 18).

7. Roderick's background: USX, official biography; Ralph Nader and William Taylor, "David Roderick," in *The Big Boys: Power and Position in American Business* (New York: Pantheon Books, 1986), pp. 3–61. This is a critical profile of Roderick, based on taped interviews with him. "Economic hammer:" USX, transcript of Roderick news conference, Pittsburgh, Feb. 4, 1987. "U.S. Steel's Roderick: A Chairman Who Doesn't Flinch," *BW*, Feb. 25, 1985, p. 55.

8. British Steel: USX, text of a statement by David M. Roderick to the Congressional Steel Caucus, Washington, D.C., Apr. 19, 1983. "The Furor Over Steel's Buy-British Plan," *BW*, Apr. 11, 1983, p. 34. "Steelworkers Vow to Press Fight to Bar Possible U.S. Steel-British Steel Accord," *WSJ*, May 12, 1983. Importation of slabs: Steven Greenhouse, "Foreign Slabs Gain Hold in U.S. Market," *NYT*, Oct. 18, 1983; Thomas F. O'Boyle, "American Steelmakers Bring in Foreign Metal to Hold Down Costs," *WSJ*, Dec. 20, 1983.

9. South Works saga: Donald Stazak, Washington, D.C., Jan. 13, 1984. From the *Chicago Tribune:* U.S. Steel ad, Dec. 28, 1983; USW ad, Jan. 11, 1984; "Humbug from U.S. Scrooge," Jan. 3, 1984 (editorial); R. C. Longworth, "A Betrayal on South Works," Jan. 11, 1984. Also see, David Bensman and Roberta Lynch, *Rusted Dreams* (New York: McGraw-Hill, 1987), pp. 136–46.

In the South Works trial, the jury apparently was confused by the testimony of Illinois Attorney General Neil Hartigan on the nature of the state's commitment to defray the cost of the pollution control. Contending that the judge failed to properly instruct the jury on this issue, the USW's attorney, Stuart M. Israel, took the case to a federal appeals court. But the union dropped the appeal as part of the settlement of the 1986–87 work stoppage at USX. *United Steelworkers of America, et al., v. United States Steel Corp.*, case no. 86-2545, U.S. Court of Appeals for the Seventh Circuit.

10. Graham joins USS: "U.S. Steel Snags a Hotshot," *BW*, May 9, 1983, p. 38; "Graham Is Trying to Forge a Tougher U.S. Steel," *BW*, Oct. 10, 1983, p. 104; "The Toughest Job in Business," *BW*, Feb. 25, 1985, p. 50. Cuts in salaried employment: USS, *USS Today*, June 1986, p. 20; Lawrence Delo, phone, Oct. 29, 1984; Jack Bergman, West Mifflin, Feb. 24, 1986. J. Ernest Beazley, "USX's Graham Faces Pivotal Tests As He Seeks to Revive Its Steel Sector," *WSJ*, Feb. 12, 1987.

11. Steel compensation in Japan and Korea: Bureau of Labor Statistics, "Hourly Compensation Costs for Iron and Steel Production Workers, 21 Countries, 1975–1986" (Washington, D.C., February 1987 [unpublished data]). Contracting out: Jim McKay, "USW Contract Task: Slow Industry Erosion," *PPG*, Feb. 19, 1986. This article quoted Thomas Usher, senior vice-president of steel operations, as saying that two of the hours reduced could be attributed to contracting out. The "Graham revolution": three USW staffers told me separately that they had heard company men use this term during 1986–87 bargaining. Definition of employee: *Agreement Between U.S. Steel and the United Steelworkers*, Aug. 1, 1983, section 2A—Coverage. Management Rights, and section 3—Management. 1979 study: "Report of the Joint Steel Industry—Union Contracting Out Review Commission," Nov. 7, 1979, p. 55 (personal files). This report contains excellent historical and economic analyses of the contracting out issue.

12. Contracting out examples and awards: Charles Grese, Clairton, Apr. 23, 1986; Steve Kecman, Elizabeth, Pa., Dec. 14, 1986; James Brown, St. Louis, Nov. 10, 1983; Al Lupini, Washington, D.C., Jan. 17, 1986; Carl Frankel, Pittsburgh, June 11, 1987; Sam Camens, Pittsburgh, Apr. 29, 1986. Japanese contract workers: 150,000 estimate by Bureau of Labor Statistics, Patricia Captiville, phone, Sept. 28, 1987. Johnston quoted: John Hoerr, "Why Job Security Is More Important Than Income Security," *BW*, Nov. 21, 1983, p. 86. Clearwater Beach meeting: "U.S. Steel Shootout," *Local 1066 Banner*, Winter 1984–85, p. 1. Also, transcript of taped remarks by James Brown, Thomas Graham, and Bruce Johnston, provided by a participant in the closed meetings. John P. Moody, "USS, Steelworkers Trade Tough Talk at Meetings," *PPG*, Dec. 14, 1984. Holdaway quote: report by Gregory Miles, *BW* correspondent, Feb. 8, 1985.

13. My account of local negotiations at U.S. Steel in 1983–84 is based on many interviews with union and management officials, some of whom wished to remain anonymous. Specific sources include: Thermon Phillips, phone, Dec. 28, 1983; Charles Grese, Washington, D.C., Jan. 13, 1984; Louis Kelley, McKeesport, Apr. 22, 1986. Johnston denies threatening plants: phone interview, Dec. 5, 1983. Fairfield: USW, *Fairfield Works Agreement*, Dec. 24, 1983. Seventy nations exporting steel: Chase Econometrics, "Crisis and Competition." "Stonewalled": Johnston testimony in Industrial Commission of Utah, Department of Employment Security, *In re United Steelworkers*, transcript of hearing, Salt Lake City, Sept. 24, 1986, docket no. 86-A-4765-MC, p. 708.

14. Negotiations at Duquesne: Mike Bilcsik, many interviews from Dec. 30, 1983, to Jan. 6, 1987; *The Voice of 1256*, August 1983; USX, transcript of Roderick news conference, Dec. 27, 1983; "Big Steel vs. the USW: The Lines Harden," *BW*, Jan. 30, 1984, p. 84.

15. McBride moratorium on LMPTs: for example, McBride's action was cited by David Roderick at a meeting of *Business Week* editors (New York City, Apr. 13, 1987) when he was asked why U.S. Steel had not used LMPTs more widely. Johnston's attitude about LMPTs: phone interview, Dec. 5, 1983.

Chapter 17

1. Monessen history and statistics: Allen Feryok, Monessen Public Library, phone, July 30, 1987. Elliott quote: Lindsey Gruson, "As Steelworkers Strike, Mill Town Shows Decay," *NYT*, July 22, 1985. Loughhead quote: Linda Couts, "Employee Participation Is Key to Good Business, Weirton Steel Boss Says," *Alliance Review*, Oct. 1, 1986.

2. The USW–Lazard Fréres connection: Eugene J. Keilin, Washington, D.C., Jan. 13,

1984, and New York City, Nov. 5, 1984. For a profile of Keilin, see Irwin Ross, "Labor's Man on Wall Street," *Fortune*, Dec. 22, 1986, p. 123.

3. Background on employee stock ownership in the steel industry is based on personal reportage, 1980–87, including interviews with James Smith, John Kirkwood, and Bruce Johnston following the 1980 steel labor settlement. 1980 provision: Johnston to McBride, "Letter Agreement on Stock Ownership Plan," U.S. Steel–USW Agreement, Aug. 1, 1980, appendix N. The letter confirmed that each CCSC company "may, if it chooses, make necessary arrangements with the union to provide an employee stock ownership plan in addition to or in substitution for other provisions of this settlement agreement (except those relating to pensions)."

4. Failed buyout efforts: studies concluded that the South Works and Duquesne proposals were not economically viable. The Gadsden effort almost succeeded. When LTV acquired Republic, the Justice Department granted approval on the condition that LTV sell Gadsden to prevent LTV from cornering too much of the flatrolled steel market. After months of study, the union made a buyout offer through a 100 percent ESOP. But a federal district court ruled that an offer by the Breslin Group, an Ohio-based firm, should be accepted over the ESOP proposal.

See also Adam Blumenthal, "Interest and Equity: Uses of Employee Ownership and Participation in the Restructuring of the U.S. Steel Industry," unpublished paper, Harvard College, March 1986. List of steel companies with ESOPs: James Smith, phone, Aug. 28, 1987.

Participation and ownership: The National Center for Employee Ownership has conducted a number of studies of ownership and participation which draw this conclusion. For more information on these and other studies, as well as excellent analysis of the participation issue in ESOP companies, see Joseph R. Blasi, *Employee Ownership: Revolution or Ripoff?* (Cambridge: Ballinger, 1988). Board representation: personal reportage, 1986–87.

5. Lynn Williams, Pittsburgh, May 2, 1986.

6. The factual background of the W-P dispute is well laid out in two court decisions: *In re Wheeling-Pittsburgh Steel Corp.*, decision of the U.S. Bankruptcy Court for the Western District of Pennsylvania, July 17, 1985; *Wheeling-Pittsburgh Steel Corp. v. United Steelworkers*, U.S. Court of Appeals for the Third Circuit, May 28, 1986. Other documents: USW, letters to W-P members, various dates, 1985; USW, results of surveys, Fingerhut/Granados Opinion Research, Aug. 16 and Sept. 6, 1985.

Other sources include these primary interviews: Paul Rusen, phone, Apr. 17 and July 22, 1987; Joseph L. Scalise, phone, July 29, 1987; Richard Fontana, phone, July 25, 1987; Joshua Gotbaum, phone, July 10, 1987; James English, phone, Feb. 17, 1987. The following news articles were helpful: John P. Moody, "W-P Strike Ends Steel 'Truce,'" *PPG*, July 22, 1985; Gregory L. Miles, Matt Rothman, and John Hoerr, "A Watershed Strike at Wheeling-Pitt," *BW*, Aug. 5, 1985, p. 26; Cynthia Piechowiak, "W-P Steel's Chief Gets Praise, Brickbats," *PP*, Aug. 11, 1985 (Carney quoted); "W.Va. Shake-Up Aims to Aid W-P Strikers," *PPG*, Oct. 1, 1985 (West Virginia UC decision); John P. Moody, "Resignations Cost Paulson $2.3 Million," *PPG*, Oct. 3, 1985 (Carney's resignation); Gregory L. Miles, Matt Rothman, and William C. Symonds, "Can Wheeling-Pittsburgh Pick Up the Pieces?" *BW*, Oct. 28, 1985, p. 37 (strike settlement).

7. Changes at National Steel: Lee M. Ozley supplied two papers, Howard M. Love, A. Lee Barrett, and Lee M. Ozley, "Creating the Future: The Transformation of National Steel Corporation," and a joint paper prepared by Love and Ozley for presentation at a Conference on Managing Organization-wide Transformations, University of Pittsburgh Graduate School, Oct. 23, 1986. See also Lee M. Ozley and Douglas

White, *Cooperative Partnership: A New Beginning for National Steel Corporation and the USWA* (U.S. Dept. of Labor, Bureau of Labor-Management Relations and Cooperative Programs, 1987).

Primary interviews: Stanley C. Ellspermann, Pittsburgh, Feb. 20 and May 2, 1986; Buddy Davis, phone, Feb. 7, 1984, and Aug. 5, 1987; Bob Pastor, phone, Feb. 7, 1984; Lee M. Ozley, phone, June 24, 1987. Davis's charge in 1984 that U.S. Steel had the "worst labor relations" was disputed by Bruce Johnston, who pointed out that Davis had never negotiated with USS. Davis said he had formed this perception from talking to other USW leaders who did negotiate with USS.

8. Calculations on the cost of shutting down the Weirton plant: William E. Fruhan, Jr., "Management, Labor and the Golden Goose," *Harvard Business Review*, September–October 1985, p. 131. Other background on Weirton Steel: "Making Money—and History—at Weirton," *BW*, Nov. 12, 1984, p. 136. Interviews: R. Alan Prosswimmer, Weirton, June 10, 1987; Alan Gould, manager of Employee Participation Groups, Weirton, June 10, 1987; Irving Bluestone, phone, Apr. 20, 1987.

9. The section on the breakup of coordinated bargaining relies on several sources. Announcement: USS, news release, May 2, 1985; Bruce Johnston, May 1, 1986. Percentage of market represented by major producers: Carol Hymowitz and Thomas F. O'Boyle, "Breakup of Big Steel Bargaining System for Labor Pacts Could Speed Cost War," *WSJ*, May 6, 1985. William C. Symonds and Gregory L. Miles, "Steel's Labor Strategy: Divide and Conquer," *BW*, May 20, 1985, p. 64.

USW abandoned pattern bargaining: Johnston testimony in Industrial Commission of Utah, Department of Employment Security, *In re United Steelworkers*, transcript of hearing, Sept. 24, 1986, Salt Lake City, docket no. 86-A-4765-MC, p. 721; USW, text of remarks by Lynn Williams, Industrial Relations Research Association, Dec. 28, 1985, New York City.

Reasons for pulling out of the CCSC: Anthony St. John, phone, July 21, 1987; A. Cole Tremain, phone, Aug. 3, 1987; James H. Wallace, phone, Nov. 9, 1987. U.S. Steel's labor-relations reputation: Gloria T. LaRue, "Steel Panel's Future Clouded," *American Metal Market*, Aug. 14, 1984.

Chapter 18

1. National plant concessions and buy off: Manuel Stoupis, McKeesport, Feb. 16, 1986; Albert L. Voss, McKeesport, Apr. 28, 1986. Severance pay: U.S. Steel Board of Arbitration, *U.S. Steel, Eastern Steel Div., National Plant and USW Local 1408*, case USS-20, 575, Feb. 1, 1985; Manuel Stoupis, Feb. 16, 1986, and phone, Aug. 30, 1987; Rudolph Milasich, Jr., USW lawyer, Pittsburgh, June 10, 1987, and phone, Nov. 9, 1987.

2. LTV's troubles: Thomas F. O'Boyle and Mark Russell, "Steel Giants' Merger Brings Big Headaches, J&L and Republic Find," *WSJ*, Nov. 30, 1984; William C. Symonds and Gregory L. Miles, "It's Every Man for Himself in the Steel Business," *BW*, June 3, 1985, p.76.

3. Union relations at LTV: A. Cole Tremain, Cleveland, Apr. 15, 1986 (citation covers further quotes by Tremain in this chapter); Bernard Kleiman, Pittsburgh, Apr. 30, 1986. Statistics on LMPTs: Gary L. Wuslich, LTV Director of Participative Programs, phone, Aug. 11, 1987. Regarding George Becker's elevation to top office, David McDonald's downfall in 1965 could be traced in part to festering anger on the executive board that he maneuvered a staff assistant, Howard Hague, into a vice-presidency in 1955. Hague hadn't even worked as a steelworker. Becker, however, had been a local union president and a long-time staff representative.

4. LSE plant: personal tour, Nov. 14–15, 1985. John Hoerr, "LTV Steel Knocks the

Rust off Its Labor Relations," *BW*, Dec. 23, 1985, p. 57. Details of LSE labor agreement: (a) Starting salary of $17,680 per year, about $1,600 less than the average annual income of beginning hourly workers in steel; a basic salary boost of 9 percent over four years without a cost-of-living adjustment. (b) Gain-sharing plan with twice yearly performance bonuses of up to 25 percent of salary. (c) Guaranteed layoff pay which, when combined with unemployment compensation, totals 100 percent of income; 60 percent of former salary guaranteed when UC runs out. The contract stated that labor and management intended to develop "a work life system that will be unique in the domestic steel industry" and "to create a highly efficient facility in which the quality of work life is enhanced by the work environment and embraces human dignity."

5. Developing a strategy with Locker Report: Lynn Williams, Pittsburgh, May 2 and Dec. 9, 1986; phone, May 11 and Aug. 20, 1987; James Smith, Pittsburgh, Apr. 17 and Dec. 9, 1986; Mike Locker, New York, Jan. 6 and July 29, 1986 (these citations cover other quotes in this chapter by these people); Locker/Abrecht Associates, Inc., "Confronting the Crisis: The Challenge for Labor," report to USW, Dec. 16, 1985. Interviews with other sources cited elsewhere, including Bernard Kleiman, Buddy Davis, Sam Camens, Carl Frankel, James McGeehan, and Cole Tremain, also helped me piece together this account.

6. Johnston quote: Johnston testimony, Industrial Commission of Utah, Department of Employment Security, *In re United Steelworkers*, transcript of hearing, Sept. 24, 1986, Salt Lake City, docket no. 86-A-4765-MC, p. 743 (hereafter cited as *Utah Transcript*).

7. BSIC meeting, Dec. 16, 1985: Gregory L. Miles and John Hoerr, "The Steelworkers Are Getting Desperate," *BW*, Dec. 30, 1985, p. 54; report on Williams news conference by *BW* correspondent Gregory Miles. Posco deal: William C. Symonds, "To Beat the Foreign Competition, U.S. Steel Joins it," *BW*, Dec. 30, 1985, p. 55; Mitani quote from reportorial memo by Symonds. Korean wages: Bureau of Labor Statistics, "Hourly Compensation Costs for Iron and Steel Production Workers, 21 Countries," (unpublished data), February 1987.

8. USW, BSIC Statement, Washington, D.C., Jan. 16, 1986.

9. Strategy sessions: Buddy Davis, phone, Aug. 5, 1985; James McGeehan, phone, Aug. 17, 1987; previously cited interviews with Williams, Smith, and Kleiman; testimony by Kleiman, *Utah Transcript*, Sept. 17, 1986, p. 136.

Range of employment costs: The $22 to $27 range for hourly employment costs is based on the "profit and loss" method of calculating these costs. The "cash" method produces a somewhat higher range. The trouble is that employment costs can't be stated with a fail-safe precision, because the calculation depends on how each company factors in pension costs and the cost of paying for health insurance for retirees and laid-off workers. Furthermore, the cost could change substantially in one company from month to month, depending on the number of hours worked. On the P&L basis, the hourly costs at other companies early in 1986 were these, as reported by the union and verified in most cases by the companies: Inland, $22; National, $23.72; USS, $24.21; Bethlehem, $25.05; and LTV, $26.40. However, USS later objected to the P&L method and insisted on using the cash method of calculating its costs (see chap. 19 and notes).

These costs were calculated as if all wage and benefit restorations and COLA required under the USW's 1983 agreement were actually paid. For example, the costs include an anticipated 10¢ COLA payment on May 1, 1986, and a restoration of Sunday premium pay to time-and-a-half, up from time-and-a-quarter. Because the May 1 COLA estimate turned out to be wrong, the costs later were reduced (see n. 17 below).

10. U.S. Steel initiative: Bruce Johnston, Pittsburgh, July 1, 1987. He told the same story in somewhat greater detail while testifying in Salt Lake City: *Utah Transcript,* pp. 723–33. I have supplemented his remarks to me with that testimony. Other members of Johnston's task force included James D. Short, vice-president of benefits administration; Jared H. Meyer, general manager of labor relations, arbitration, and administration; George A. Hensarling, director of employee benefits; James T. Carney, a benefits lawyer; and some industrial engineers. Roderick quote: USS, transcript of Roderick news conference, Feb. 11, 1986.

The 1983 steel labor agreement, signed by USS and six other firms, contained a "Memorandum of Understanding on Joint Efforts" which listed various problems faced by the steel industry and declared: "The parties have concluded from these discussions that many of the critical problems threatening Steelworker jobs and the industry are directly attributable to government action or inaction in these areas. Consequently, the United Steelworkers and the Coordinating Committee Steel Companies have agreed to join mutually in the task of petitioning the federal government to respond to these legitimate concerns, including imports, in a realistic, responsible, and vigorous manner" (*Agreement Between U.S. Steel Corp. and United Steelworkers of America,* 1983, appendix P).

LTV's situation in late 1985: Laurie P. Cohen and Thomas F. O'Boyle, "LTV, Dragged Down by Steel Subsidiary, Struggles to Survive," *WSJ,* Jan. 6, 1986.

11. Donald B. Thompson, "Divided We Fall?" *Industry Week,* Feb. 17, 1986, p. 22.

12. Contracting out: USW, letter to all LTV members, Feb. 28, 1986. My account generally of LTV negotiations is based on these interviews: Sam Camens, Pittsburgh, Apr. 29, 1986; Tremain, Kleiman, Smith (previously cited); Eugene J. Keilin, New York City, Sept. 25, 1986; Joshua Gotbaum and Ronald Bloom, New York City, July 25, 1986.

How LTV saved money under the contracting out ban: It cost LTV $4.50 to support people on layoff who could do work assigned to outside contractors at a cost, perhaps, of $19 per hour, or a total of $24.50. If LTV recalled the laid-off worker to do the work after the new contract reduced labor costs by $3.60 per hour, the company would save the $4.50 burden plus the $3.60 and have the work done for $17.40, or more than $7 less than it cost to bring in a contractor. In the end, LTV saved money, according to Tremain—phone, Aug. 3, 1987.

13. Following is a list of other details of the LTV agreement. Benefit cuts: (a) paid holidays from ten to seven; (b) shift premiums to 20¢ per hour for the second shift and 30 for the third shift, as well as a permanent cut in Sunday premium pay to time-and-a-quarter; (c) elimination of vision care, increase in major medical deductibles to $300 per family, and cut in sickness and accident benefits. Profit-sharing: If the company makes a profit, it must put aside each April 10 percent of the first $100 million in pretax profits and 20 percent above that. If profits are not sufficient to repay the sacrifices, the balance will be made up in the issuance of preferred stock in LTV Steel which can be exchanged for common stock in the parent company. The amount would be deposited in each employee's account in a trust fund.

The company also agreed to try to curb excessive overtime, to put $300,000 annually for three years into a steel crisis action fund to conduct legislative and media campaigns on the import problems, and to provide $975,000 a year for training and development programs for laid-off employees. The USW gained a unique "successorship" provision under which the company guaranteed that it would not sell or transfer a plant unless the union is recognized as the bargaining agent, and the sale cannot be completed until the buyer reaches an agreement acceptable to the union. A special joint committee was named to oversee implementation of twelve new craft jobs, formed by combining the many old crafts into supercrafts. USW, *Summary of Proposed LTV Agreement.*

14. LTV votes: Jim McKay, "USW Approves Wage, Benefit Cuts at LTV," *PPG,* Apr. 5, 1986; Nicholas Knezevich, "USW, LTV Hail $3.15 Giveback," *PP,* Apr. 5, 1986. USW, letter to all LTV members, Mar. 20, 1986. Donald B. Thompson, "No Real Solutions to Crisis Apparent in 1st Steel Pact," *Industry Week,* Apr. 14, 1986, p. 22.

15. Bargaining at National Steel: Stanley Ellspermann, Pittsburgh, May 2, 1986, phone, Feb. 12, 1987; Buddy Davis, phone, Aug. 5, 1987. USW, *Summary of Proposed National Agreement,* The "Japanese-style" brand was affixed by these articles, among others: Jim McKay, "National Workers OK Innovative Contract," *PPG,* Apr. 29, 1986; Peter Behr, "National Steel's Pact with Union Holds Out Hope for Steelmakers," *Washington Post,* Apr. 20, 1986; Gregory L. Miles and Matt Rothman, "The Steel Deal That Everybody's Watching," *BW,* Apr. 21, 1986, p. 32.

16. "Critics Fault Trautlein for Failure to Revive an Ailing Bethlehem," *WSJ,* May 27, 1986 (Trautlein received an 11 percent pay hike in 1985 and was promised a big severance bonus if a takeover occurred); "Bethlehem Workers Approve New Pact," *PPG,* June 17, 1986; USW, *Summary of Proposed Bethlehem Agreement;* "Inland Steel, Union Reach Tentative Pact," *WSJ,* June 23, 1986; "Steelworkers Accept Slight Cutback in New Agreement with Inland Steel," *Daily Labor Report,* July 9, 1986; USW, *Summary of Proposed Inland Agreement;* "Armco Workers Approve Pact," *PP,* Nov. 22, 1986; James Smith, phone, Aug. 30, 1987; James H. Wallace, Armco, phone, Nov. 9, 1987.

17. Value of concessions at LTV: A. Cole Tremain, phone, Aug. 3, 1987 (also data on number of employees and dependents); Joshua Gotbaum, New York, July 25, 1986. The value of the 1986 concessions turned out to be $3.50 per hour, not $3.60. The original estimate contained an anticipated 10¢ per hour in a COLA payment scheduled for May 1, 1986, under the old contract, which the USW gave up. Because the CPI change was less than expected, the company would not have had to pay anything. In the first year under the new agreement, LTV worked about 36 million man-hours, about four million less than anticipated. Thus, the total savings came to $126 million. Gotbaum developed the estimates of what would have been saved at various concession rates. The "full $6 per hour in concessions asked by LTV" includes the Feb. 1, 1986, wage and COLA increase waived by the USW.

18. Cancellation of retiree benefits: J. Ernest Beazley, "Steelworkers Strike LTV Mill in Indiana, Citing Cancellation of Retirees' Benefits," *WSJ,* July 28, 1986. Kevin L. Carter, "Benefits Restored to LTV Retirees, Strike Ends," *PP,* July 31, 1986. Pension plan takeover: "LTV Loses Its Pension Plans," *BW,* Jan. 26, 1987, p. 46. The PBGC refused to pay: (a) any excess over $1,858 a month for regular pensions (by virtue of the USW's traditional criterion that benefits rise with age and years of service, many long-service employees had received more than $1,858); (b) widows' pensions; (c) disability pensions; (d) shutdown pensions.

"LTV Retirees Protest Pension Benefit Cuts at Aliquippa Works," *PP,* Feb. 13, 1987; Jim McKay, "USW Chiefs Vote Down LTV Offer," *PPG,* May 15, 1987, and "Union Presidents Back LTV Contract," June 26, 1987. In the 1987 agreement, the USW reduced LTV's costs by committing active workers and retirees to contribute $26.82 per month to medical and life insurance coverage. The job-cutting plan enabled LTV to "buy out" employees with severance pay of $1,000 per year of service up to twenty-five years.

19. PBGC problems and LTV: A. Cole Tremain, phone, Aug. 3, 1987; James Smith, phone, Aug. 28, 1987; Gregory L. Miles, "Who's Going to Pay Steel's Pensions," and John Hoerr, "Saving the Agency That Saves Workers' Pensions," *BW,* Nov. 2, 1987,

pp. 89–90. Williams quote: Lynn R. Williams, "PBGC Policies Bring Misery and Pain," *PPG*, Aug. 8, 1987.

20. PBGC premium raise: "PBGC Premiums, Funding Standards Highlight Pension Provisions of New Law," *Daily Labor Report*, Jan. 4, 1988, p. A11; Cost of restructuring: Wharton Econometrics, "Restructuring and Revival: The World Steel Industry 1987–2000," vol. 3, part 2, section 1, pp. 1.1–1.15; Ben Fischer, phone, Nov. 9, 1987 ($150,000 cost of retiring high-paid steelworkers). Rejection of bailout plan: J. Ernest Beazley, "Government Bailout of Steel Industry Is Stalled by Divisions in Administration," *WSJ*, Aug. 20, 1987. The PBGC's lack of written policy was brought out in questioning of the agency's director, Kathleen P. Utgoff, by LTV attorney Lewis B. Kaden at a hearing: *In the Matter of Chateaugay, et al., Debtors*, United States Bankruptcy Court, Southern District of New York, transcript of hearing, July 16, 1987.

Chapter 19

1. Sources for steel crisis campaign: USW, copies of ads, pamphlets, news releases. Aliquippa rally: Jim McKay, "Rally's Turnout Sparse at LTV," *PPG*, June 23, 1986. Interviews: Lynn Williams, Pittsburgh, May 2, 1986; James Smith, Pittsburgh, Apr. 17, 1986; John J. Sheehan, USW legislative director, phone, Oct. 16, 1987. Cost of crisis campaign: USW Treasurer James McGeehan, letter to author, Aug. 18, 1987. Example of favorable editorial comment: "The steelworkers have a story to tell the nation about how 11.5 million jobs have been lost to imports," said the *Pittsburgh Press* in "The Steelworkers' Message," June 20, 1986.

2. Bargaining in other industries: Mark Russell, "National Can Contract Is Cleared by Union Officials," *WSJ*, Apr. 6, 1986. From *Daily Labor Report:* "Agreements Reached to End Strike at Alcoa," July 7, 1986, p. A9; "Copper Unions at Asarco Overwhelmingly Approve Pay Reductions of $3.50 per Hour," July 8, 1986, p. A14; and "Copper Workers Agree to Take 20 Percent Cut for Four Years," July 10, 1986, p. A14. Thomas Buell, Jr., "Tale of Sharon Hangs on Steelmaker," *PP*, May 5, 1987; Nicholas Knezevich, "USW Refuses Concessions, Pickets Babcock & Wilcox," *PP*, Sept. 15, 1986. Sharon and Babcock & Wilcox buyout attempts: James Smith, phone, Nov. 6, 1987.

3. My account of Round One bargaining in this chapter is pieced together from many sources. Primary interviews: Sam Camens, Pittsburgh, Aug. 1, 1986; James Smith, Pittsburgh, July 31 and Dec. 9, 1986; Carl Frankel, Pittsburgh, June 10, 1987; Bernard Kleiman, phone, Feb. 11 and Feb. 20, 1987.

A brief interview in Pittsburgh with Bruce Johnston, Dec. 11, 1986, provided some information about the company's point of view, especially with regard to the costing controversy mentioned below. However, Johnston's views on the background of Round One, the issues involved, and offers and counterproposals made by both sides, are given in his extensive testimony at an unemployment compensation hearing in Salt Lake City in September, 1986. Citation: Industrial Commission of Utah, Department of Employment Security, *In re United Steelworkers*, transcript of hearing, Sept. 19 and 24, 1986, docket no. 86-A-4765-MC, pp. 640–861 (hereafter referred to as *Utah Transcript*).

Newspaper and magazine stories appearing during the negotiating period are also sources. Start of USS negotiations: Nicholas Knezevich, "Cut 'Double-talk,' Union Tells USS," *PP*, June 13, 1986. July 4 deadline: Jim McKay, "Customers Push USS to Settle by July," *PPG*, June 14, 1986.

4. USS audit: Price Waterhouse to USS board of directors, "Statement of Internal Cash Generated and Property, Plant and Equipment Expenditures," Steel and Domes-

tic Iron Ore Division, Mar. 18, 1986. Figures also presented in USS, *USS Today*, June, 1986; Johnston testimony, *Utah Transcript*, p. 745. "Artificial Steel" analysis: Mike Locker, New York City, July 29, 1986, also USW pamphlet (see below); Kleiman testimony, *Utah Transcript*, pp. 126–33.

As a result of the bookkeeping organization used in the Price Waterhouse audit, "Artificial Steel" was charged by the corporation for buying limestone, coal, and foreign ore from USS itself. In Locker's analysis, Artificial Steel's bookkeeping losses generated $1.2 billion of income tax savings which were used to shelter gains from the company's sale of $3 billion worth of coal mines, real estate, and other assets. USS also refused to give forecasts of prices, volumes, and plant shutdowns over three years, as had other steelmakers.

USS could perhaps justify the makeup of this entity by showing that it had begun giving some results in the annual report of "steel and related resources" businesses, which did not include coal, iron ore, and the railroads. But, the USW responded, without the incorporation of these assets, as well as other data, the union could not make an accurate comparison between USS's steel operations and those of the other companies to determine how much relief it should grant.

5. Loss of GM business and USS's $18 offer: Johnston testimony, *Utah Transcript*, pp. 762–72; Nicholas Knezevich, "USS Reportedly Seeks $7 Hourly Cost Cut," *PP*, July 1, 1986. Stockpiling: Indiana Employment Security Division, *Decision of Review Board*, July 9, 1987. This decision, based on USS and USW testimony, summarized the production and stockpiling situation at the Gary Works, USS's largest plant. Other data on stockpiling: Jim McKay, "Loss of Sales to GM 'Pressures' USS," *PPG*, July 2, 1986; J. Ernest Beazley, "Moves in Steelworkers' Strike at USX Indicate Walk-out Will Be Costly, Long," *WSJ*, Aug. 4, 1986.

6. USS name change: USX, transcript of Roderick news conference, Pittsburgh, July 8, 1986. *PPG* editorial: "X-ing Out a Name," July 9, 1986. USW propaganda: USW, *USS USX Tomorrow*, July, 1986.

7. USX lowers demands: Johnston testimony, *Utah Transcript*, pp. 775–81. Cost-ing controversy: ibid., pp. 207–12 (Kleiman), pp. 838–40 (Johnston); Bruce Johnston, Pittsburgh, Dec. 11, 1986; James Smith, Pittsburgh, Dec. 9, 1986.

USX used the "cash" method of computing employment costs, under which actual cash payments for pensions in each time period are divided by the number of hours worked. A difference of opinion arose because USX had closed several plants from 1979 to 1983. It had been required under the union contract to provide early pensions to many employees who under normal conditions would not have retired until age sixty-two. On an hourly basis, this cost was figured at $1.14 in 1986, raising total hourly employment costs to $25.35. USX negotiators argued that this amount must be included in the employment cost base every year until the retirees reached the normal retirement age. The cost of funding the pensions from that time on had already been accounted for. Booking a writeoff is a requirement of the Securities and Exchange Commission (SEC), according to Johnston. Accounting for a new pension obligation to the SEC is not the same as measuring the increased hourly cost of paying off that obligation every day and year until the actuarial life is completed.

The union's economist, Jim Smith, argued that the cost of the early pensions should be counted only for the year in which the obligation was taken on. Since no labor was being performed in the closed plants in later years, there were no hours of work to divide into the total cost of providing the early pensions. Smith insisted, instead, on using the "profit and loss" method of determining the employment-cost base. This shows the relationship of assets to liabilities at a specific point in time. In this method, with USX's early pension costs apportioned over several years, the base totaled only $24.21 an hour. Smith also pointed out that the cost base at LTV,

Bethlehem, and National had been determined by the P&L method; to be consistent, it must also be used at USX. However, if the cash method had been used at LTV, its presettlement cost base of $27.60 would be reduced by an undetermined amount, because the company had been permitted to postpone pension fund payments since 1982.

USW strike vote: Nicholas Knezevich, "USX Reportedly Lowers Demands; Union Talks Strike," *PP*, July 15, 1986.

8. LTV bankruptcy: Robert L. Simison and Karen Blumenthal, "LTV Corp. Files for Protection from Creditors Under Chapter 11," *WSJ*, July 18, 1986. Johnston's comments about LTV strategy: Pittsburgh, July 1, 1987; testimony, *Utah Transcript*, pp. 804–07. Denial of intent to deceive: Eugene Keilin, New York City, Mar. 19, 1987; Cole Tremain, phone, Aug. 3, 1987. The three specialty steel plants sold by LTV were at Midland, Pa., Louisville, Ohio, and Detroit.

9. USW view of USX strategy: Sam Camens, Pittsburgh, Aug. 1, 1986. "No political risk": Bruce Johnston, Dec. 11, 1986. Bottomline proposal: *Utah Transcript*, pp. 821–37 (Johnston), pp. 139–53 (Kleiman).

The second disagreement over costing also was a highly technical issue. USX put a value on the requested concessions that was $1.35 per hour lower than the USW's estimate. This is because the company wanted to charge certain costs to employees, thus offsetting $1.35 of the amount they would give up. The $1.35 would come from four sources, the company argued:

(a) To eliminate the 1,105 jobs through the proposed manning agreement, the company would have granted 1,105 early pensions. The company figured this would cost 8¢ an hour in increased pension funding. The USW contended it was ludicrous to charge steelworkers 8¢ an hour for a provision that would save the company an average of $33,000 per year for each job eliminated. (b) In compressing thirty craft classifications into six under its proposal, the company would pay a slightly higher wage to workers who took on added duties in the merger such as millwright and welder. Although the company would save millions of dollars by employing many fewer skilled workers, it wanted to charge the union 10¢ an hour for an average rise of four-tenths of a job class.

(c) By promising to declare idled facilities permanently closed (so laid-off employees could receive benefits), USX would have to provide early pensions and severance pay to eligible workers. This would cost 14¢, the company said. (d) Because of continuing escalation in hospital and doctors' bills, USX estimated that its cost of providing health insurance—without any benefit improvements—would rise by $1.03 over the life of the contract.

10. Final two days and maneuvering over UC pay: *Utah Transcript*, pp. 161–66 (Kleiman), pp. 179–84 (McGeehan), pp. 848–58 (Johnston). Letters: McGeehan to Johnston, July 31, 1986, and Johnston to McGeehan, July 31, 1986 (personal files, copies supplied by USW). Pittsburg local extends contract against orders from Pittsburgh: James McGeehan, phone, Aug. 17, 1987.

11. Strike's effect in Pittsburgh area: Chet Wade, "Even Long Strike Won't Match Impact of USW's '59 Walkout," *PPG*, Aug. 4, 1986.

12. Background for section on UC decisions: James English, phone, Feb. 17, 1987; Carl Frankel, Pittsburgh, June 11, 1987. Referee's decision, Pennsylvania Unemployment Compensation Board of Review, appeal no. 86-1-I-715, Nov. 6, 1986 (reasons for granting UC and good explanation of state supreme court decisions on the issue). UC rulings in five states by Aug. 26: "USX Workers Win Jobless Pay in Ohio," *WSJ*, Aug. 26, 1986.

13. Indiana Employment Security Division, *Decision of Review Board*, July 9, 1987. Andrew Palm, McKeesport, Dec. 11, 1986.

14. Discrimination in distributing strike benefits: Curtis Vosti, "1066 Post Candidate Blasts Use of Funds," *Hammond Times*, August, 1987. Total of strike benefits and consulting services paid by USW: James McGeehan, letter to author, Aug. 18, 1987.

15. USW's lockout activities: George Becker, phone, June 30, 1987. Picket line skirmishes: Don Hopey, "USX Files Complaint to Halt Clairton Plant Blockade," *PP*, Oct. 14, 1986; Jim McKay, "USW Told to Let Trains Leave Irvin," *PPG*, Oct. 17, 1986; Lawrence Walsh, "Judge Restricts Pickets at USX's National Works," *PP*, Dec. 28, 1986; Jim McKay, "No Decision in USX Dispute," *PPG*, Dec. 30, 1986; "14 Arrested as USW Blockades Lorain Mill," *Plain Dealer* (Cleveland), Nov. 27, 1986. USX preparations: Pamela Gaynor, "USX Continues to Ship Steel during Dispute, Analysts Conclude," *PPG*, Sept. 24, 1986. John D. Oravecz, "Costs of Strike Are Still Adding Up, May Hit $500 Million, Analysts Say," *PP*, Jan. 18, 1987; Gloria T. LaRue, "40% USS White Collar Staff Expected to Be Idle by Sept.," *American Metal Market*, Aug. 20, 1986.

Chapter 20

1. Takeovers and the bull market: the financial press was filled with stories on this phenomenon in 1986; for example, see "Deal Mania: The Restructuring of Corporate America," *BW*, Nov. 24, 1986, p. 74.

2. Bidding war for USX: Linda Sandler, "U.S. Steel Catches Eye of Some Asset Players Who See Chance of Big Gains Down the Road," *WSJ*, May 30, 1986; Pamela Gaynor, "Australian Aims to Buy Stake in USX," *PPG*, Aug. 21, 1986; Daniel Hertzberg and J. Ernest Beazley, "USX Rumors Grow as Raiders Purchase Shares," *WSJ*, Sept. 19, 1986; Robert J. Cole, "4 Raiders Said to Buy USX Stock," *NYT*, Sept. 20, 1986 (this article reported "industry sources" as estimating USX pension funds were overfunded by $2.5 billion).

Icahn bid and USX maneuvers: William M. Carley, "USX Might Face TWA-Style Shake-Up Under Icahn," *WSJ*, Oct. 10, 1986 (quotes Icahn describing USX management as "arrogant"). Golden parachutes: Cynthia Piechowiak, "USX Execs Gave Up Pay, Gained 'Parachutes' in '86," *PP*, Mar. 12, 1987. TWA strike: "Flight Attendants at TWA to End Strike, Union Says," *NYT*, May 19, 1986.

3. The nature of the Icahn-USW talks was not disclosed when they took place. Some press stories reported, or speculated, only that Williams and Icahn had met; the possibility of an employee buyout of the steel segment of USX was not public knowledge. As this book went to press, my account of the Icahn meetings with USW officials and the union's rationale for the decision to avoid an Icahn takeover was the first presentation of the full story. It is based on the following interviews, supplemented by followup checks with each source: Lynn Williams, Pittsburgh, Dec. 9, 1986; Eugene Keilin, New York City, Mar. 19, 1987; Mike Locker, New York City, May 6, 1987; James Smith, phone, Mar. 8, 1987.

4. USX pension fund excess only $300,000: David M. Roderick, New York City, Apr. 12, 1987. He disclosed this at a conference with *Business Week* editors. Caliguiri quote: Chuck Hawkins and Gregory L. Miles, "Carl Icahn: Raider or Manager?" *BW*, Oct. 27, 1986, p. 98. Saving or destroying Roderick: Lynn Williams, phone, Aug. 20, 1987.

5. My account of Round Two bargaining is based on the following sources: Lynn Williams, Pittsburgh, Dec. 9, 1986; James Smith, Pittsburgh, Dec. 9, 1986, and phone, Aug. 28, 1987; Bruce Johnston, Pittsburgh, Dec. 11, 1986; Bernard Kleiman, phone, Feb. 11, 1987. Personal reportage: USW and USX news conferences, Pittsburgh, Nov. 21, 1986; USW presidents' meeting, Nov. 21, 1986. USW, "Differences of Economic

Proposals," Nov. 15, 1986 (an analysis distributed to local presidents). The following news stories provided various details: Jonathan P. Hicks, "USX Plans to Raise $1 Billion Cash," *NYT*, Oct. 30, 1986; Don Hopey, " 'Window' Still Open for USX Labor Accord," *PP*, Nov. 8, 1986; James B. Stewart and Daniel Hertzberg, "Fall of Ivan F. Boesky Leads to Broader Probe of Insider Information," *WSJ*, Nov. 17, 1986; Tim Carrington, "Shock Waves of the Boesky Affair Could Bring Shrinkage in Drexel's Merger Investment Lines," *WSJ*, Nov. 21, 1986; Jim McKay, "Union Officials to Get Update on USX Talks," *PPG*, Nov. 21, 1986. Smith on manning: tape cassettes of address to meeting of employees, Geneva Works, late November 1986 (personal files).

6. Johnston letter and McGeehan response: Jim McKay, "USX Takes Its Case to Workers," *PPG*, Dec. 5, 1986. Garrett background: personal reportage; Jim McKay, "USX, USW Choose Referee," *PPG*, Dec. 19, 1986. Garrett's recommendations: They were not disclosed by him or either party before a tentative settlement was reached in mid-January, 1987. For the first public indication that the recommendations might have spurred the bargaining process, see Jim McKay, "USX, Union Involved in 'Serious Bargaining,'" *PPG*, Jan. 6, 1987. The recommendations are listed in USW, "Summary of Tentative Agreement Between the United Steelworkers of America and USX," Pittsburgh, Jan. 18, 1987. This summary quotes Garrett as saying that the parties "might wish to deal" with the bankruptcy issue. It also described the economic and manning proposals as "distasteful."

As a mediator, Garrett himself would not comment publicly on his recommendations. I subsequently confirmed their content in interviews upon which I base my account of Round Three bargaining: Bernard Kleiman, phone, Feb. 11 and Feb. 20, 1987; James Smith, phone, Mar. 8 and Aug. 28, 1987; Carl Frankel, Pittsburgh, June 11, 1987; Lynn Williams, phone, May 11 and Aug. 20, 1987; James McGeehan, phone, Aug. 17, 1987; Richard Fontana, Pittsburgh, June 12, 1987. To a limited degree, my account reflects comments by Bruce Johnston, Pittsburgh, July 1, 1987, although this interview covered only a few aspects of Round Three negotiations.

USX concern about steel orders: J. Ernest Beazley, "USX Corp. Discussed the Possible Sale of Energy Assets to British Petroleum," *WSJ*, Dec. 11, 1986 (this article also contains Roderick's warning about possible plant closings). Icahn withdrawal: Icahn to Roderick, Jan. 8, 1987, contained in Securities and Exchange Commission, schedule 13D, 902905 108, Jan. 8, 1987.

7. Local presidents' vote: USW news conference, Pittsburgh, Jan. 18, 1987. Ratification: Nicholas Knezevich, "Steelworkers Ratify USX contract," *PP*, Feb. 1, 1987. The pact was approved by wide margins in Mon Valley plants: 8 to 1 in Clairton, 4 to 1 in McKeesport, 22 to 1 in Braddock (811 to 37), where the company promised to install a caster. Even Gary Local 1014 voted 2,690 to 717 to ratify, despite the opposition of Local President Larry Regan. Ratification failed in only one local, 1938 at the Minntac, Minn. iron ore operations, which rejected the contract by a 63 percent vote.

8. Settlement terms: *Settlement Agreement Between the United Steelworkers of America and USS, Division of USX Corporation*, Feb. 1, 1987. This document is recognized by both sides as the basic source for the terms of the settlement which will be incorporated in several contract and benefit booklets and distributed to workers and managers. I have also relied on interviews listed above for interpretations of various provisions. Annual cost reductions are enumerated in the profit-sharing provision, pp. 133–45. Economic terms summarized in USW, "Summary of Tentative Agreement," Jan. 18, 1987. They included these changes for the term of the agreement: (a) wage cut of 99¢ per hour, plus 13¢ in incentive earnings, totaling $1.47 including compounding effect; (b) suspension of COLA; (c) reduction of Sunday premium from 1½ to 1¼ times straight pay; (d) deletion of three holidays; (e) cut in vacation pay to eliminate effects of shift differentials and overtime and Sunday pre-

miums. Temporary benefit reductions: (a) one week of vacation eliminated for employees entitled to three or more weeks, restored in second year; (b) shift premiums cut by one-third, restored in third year; (c) one additional holiday eliminated, restored in third year.

Profit-sharing plan: At the end of each year, USS must pay into a profit-sharing pool the greater of (1) an amount calculated by a formula relating LTV and Bethlehem stock prices to worker sacrifices at those companies and USX, or (2) 10 percent of the first $200 million in pre-tax income and 20 percent of additional income. Payments, limited to amount of employee sacrifices, to be made each Apr. 15. Benefit improvements: Increase in medical benefits and rise in monthly pension payment per year of service from $20.50 to $21.50 for thirty years' service, $19 to $20 for fifteen to thirty years, $17.50 to $18.50 for less than fifteen years.

9. Contracting out provisions: *Settlement Agreement*, pp. 291–315. Roderick didn't want to "pay twice": Lynn Williams, Pittsburgh, Dec. 9, 1986. USX didn't accept "boilerplate language": David Roderick, New York City, Apr. 12, 1987. Two thousand jobs: estimated by James Smith, USW news conference, Pittsburgh, Jan. 18, 1987. Checking with locals: Bernard Kleiman, phone, Feb. 11, 1987; Carl Frankel, Pittsburgh, June 11, 1987.

Manning provision: *Settlement Agreement*, pp. 237–81. Entitled "employee protection/job realignment agreement," this provision calls for a fundamental reorganization of work in USS plants. The long-term benefit for USS lay in establishing new, smaller production crews through "one-time remanning," if that is the way it chooses to eliminate the jobs. The reduction also could be accomplished by other means. The company could combine duties of trade and craft workers into the "expanded" crafts of motor inspector, millwright, ironworker, mechanical and hydraulic repairman, and systems repairman. A new job of "equipment tender" was created, enabling craft workers to operate equipment and also repair it to the extent their skills allow. An equipment tender thus would be paid at the rate of his or her skilled craft and also receive production incentive pay for operating equipment. The creation of this job finally recognized the artificiality of the old division between production and maintenance personnel. National USW negotiators committed the locals to reduce jobs by these specific amounts: Fairless, 422; Irvin, 139; Gary Steel, 268; Gary Sheet & Tin, 123; South Works, 83; Lorain, 188; Minntac (Minnesota ore mine), 123.

10. Other provisions contained in *Settlement Agreement:* investment commitments, Edgar Thomson and Irvin plants, p. 180 (in this side letter from Johnston to McGeehan, USX declared it intended "to be fully bound" to commence building the caster and modernizing the Irvin hot strip mill by the end of the 1987 contract), Fairfield caster, p. 219, Fairless Works, p. 220; training and development, pp. 230–31; team leaders, pp. 246–47. Improving relations: David Roderick, New York City, Apr. 12, 1987; Lynn Williams, phone, May 11, 1987.

Not mentioned in my text are several other changes in the noneconomic portions of the USS agreement: (a) expansion of civil rights protections for workers, including prohibitions against sexual harassment and discrimination on the basis of age, handicap, or Vietnam military service; (b) protection of employees and the union under a "successorship" provision covering conditions under which USS may sell a union plant or subdivision; (c) improved safety protections.

11. USX fourth quarter results: Pamela Gaynor, "$1.83 Billion Loss in '86," *PPG*, Jan. 27, 1987. This article stated that some analysts "said USX was being deliberately unclear about how many plants were affected to avoid prejudicing settlement of its six-month labor dispute with the United Steelworkers union." J. Ernest Beazley of *WSJ* quoted analyst Charles Bradford of Merrill Lynch Capital Markets: "Some of

these steelworkers could be voting themselves out of a job." See "USX Posts Loss of $1.42 Billion for 4th Quarter," Jan. 28, 1987. Williams on reaffirming rights: "Workers Called Back to USX Steel Plants as Stoppage Ends," *NYT*, Feb. 2, 1987.

12. Plant closings announced: USX, transcript of Roderick news conference, Pittsburgh, Feb. 4, 1987, pp. 2–7. Roderick indicated that the USX board of directors had approved the plant closings on Jan. 27 (p. 6). He also affirmed that the writeoffs of the four plants were, in fact, charged against the corporation's fourth quarter 1986 earnings (p. 20). USW interpretations of closing threat: James McGeehan, phone, Aug. 17, 1987; James Smith, phone, Aug. 28, 1987.

Letters to employees were signed by V. John Goodwin, general manager of the Mon Valley Works; Don Kistler, manager of the National plant; and Warren Bartel, manager of Geneva. All of the letters were dated Dec. 29, 1986, indicating that top USX or USS officials organized a concerted effort to convince employees that they should return to work under USX's proposed agreement of November 1986. Goodwin stated that the Saxonburg plant (part of the Mon Valley Works because it supplied sinter to Edgar Thomson for use in the blast furnaces) had been on the verge of being closed but added: "*With reduced manning and its development of Synflux* [a type of sinter], that plant now is a vital part of the Mon Valley Works. With these and other efforts, we have earned due respect of executive management and greatly improved the outlook for the future of the Mon Valley" (emphasis in original). Kistler's letter, sent to National employees, indicated that three production operations would be restarted. A USX spokesman acknowledged in early January 1987 that these letters had been sent to employees but did not attempt to modify the implication in the letters that the plants would reopen. See Jim McKay, "USX, Union Involved in 'Serious Bargaining,'" *PPG*, Jan. 6, 1987.

13. Williams on plant closings: phone, May 11, 1987. At a news conference, Williams criticized Roderick's "callous and insensitive" decision; see Nicholas Knezevich, "USW President Rips Roderick for Post-Contract Layoffs," *PP*, Feb. 7, 1987.

"Economic hammer": transcript of Roderick news conference, Feb. 4, 1987, pp. 29–30. In this response, Roderick overstated the warnings he had previously given. In July 1986, he had said a strike shutdown "raises a great risk" as to whether the more vulnerable plants would reopen (USX, transcript of Roderick news conference, Pittsburgh, July 30, 1986, p. 12). In December 1986, he had said that "the longer the strike goes, the greater the chances are" that some mills would not reopen (see n. 6 above). At the Feb. 4 news conference, Roderick referred to these vague warnings as if they had been definite statements of fact. Moreover, the USW disputed the assertion that the company had said "locally" that the more marginal plants would not reopen. If anything, the letters written by local plant managers had strongly implied that they would *resume* operations.

"I never dreamed": Robert Cross, phone, Feb. 5, 1987. Cross said that twenty-seven maintenance workers had been recalled at the National plant before the Feb. 4 closing announcement. When this was mentioned to Roderick at the Feb. 4 news conference by a reporter from the McKeesport *Daily News*, the chairman said he did not know about that, adding: "I don't know what the need would be for maintenance people going back in at this stage" (Feb. 4 transcript, p. 15). See also Jim McKay, "USX Begins Calling Workers Back," *PPG*, Feb. 3, 1987. Declaring plants permanently closed: "USX Layoffs Permanent," *PP*, Apr. 24, 1987, and "USX Closing Plant," May 7, 1987. Utah suit: "Utah Workers Sue USX," *PPG*, Apr. 14, 1987. Sale of Geneva plant: USS public relations, phone, Jan. 20, 1988; Kay B. Mitani, vice-president, USW Local 2701, phone, Feb. 3, 1988.

14. Arbitration decisions: Carl Frankel, phone, Jan. 8, 1988. Many of the union

losses involved heavy construction work that USW members did not normally perform. USX declined to make its tabulation of the decisions available for this book. Very likely, the company would dispute the union count. For example, the USW counted seven cases arising at Clairton as only one loss since it involved the same roofing work but on different days. USX also might contend that more than one USW loss involved exceptions won in negotiations. In two or three cases, the company was permitted to contract out on grounds that the local had previously allowed the practice in return for a benefit granted by the company. The USW, however, argued that USX would have won these cases even under the old agreement.

Roderick savings estimate: USX, transcript of Roderick news conference, Feb. 4, 1987, p. 54. He did not explain how he arrived at $400 million in savings, but it appears that USX could save $100 million in the first year of the contract. If the company employs about 16,000 hourly workers, as Thomas Graham anticipated (see Cynthia Piechowiak, "Graham Says USX Won't Start Price War," *PP*, Feb. 10, 1987), an hourly cost reduction of $2.52 for this many workers, employed an average of 1,900 hours annually, would produce savings of $76.6 million. In addition, the elimination of 1,346 jobs would yield $28.1 million annually, according to an estimate by John Tumazos of Oppenheimer & Company (*Progress Report* no. 87-120, Feb. 2, 1987). Recovery of losses: John Tumazos, phone, Jan. 25, 1988. He and other analysts had anticipated that USS would lower prices after the stoppage to regain customers. Instead, the company raised prices as a weaker dollar and increased steel demand helped the domestic steelmakers in 1987. See J. Ernest Beazley, "USX to Boost Prices 4% to 6% for Sheet Steel," *WSJ*, July 14, 1987. Man-hours per ton and elimination of 2,000 jobs: Bruce Johnston, Pittsburgh, July 1, 1987.

15. Profit-sharing payment: transcript of Roderick news conference, Pittsburgh, Jan. 27, 1988; James McGeehan, phone, Feb. 5, 1988. Mon Valley get-together: Cynthia Piechowiak, "Breaking Bread: USX Touts Future at 'Pep Rally' for Workers," *PP*, Mar. 17, 1987. USX wants cooperation: David Roderick, New York City, Apr. 12, 1987. Arbitration backlog: Bernard Kleiman, phone, Nov. 6, 1987. USX is still USX: Carmen Rettzo, phone, Sept. 14, 1987. Improvement at Clairton: Richard Pastore, phone, Nov. 4, 1987.

Chapter 21

1. My description of plant closedowns and employment in the Mon Valley is based on personal observation and confirmation of various facts in 1987 by phone with following sources: Wheeling-Pittsburgh Steel, July 22; Robert Watson, Monongahela Industrial Development Authority, Nov. 23; Thomas Urbanski, Glassport mayor, Aug. 8; Steelmet, Aug. 4; American Standard, Sept. 9; Westinghouse Electric, Sept. 10; LTV Steel, July 31; USX, Aug. 27.

Estimates of steelworkers in Mon Valley. I start with a 1949 estimate of about 80,000 hourly and salaried workers at steel plants then operating, based partly on a USW survey of its locals in late 1949 and partly on an estimate of salaried employees (see chap. 4, n. 11). Estimate for USX in 1979: Tri-State Conference on Steel, *Steel Valley Authority: A Community Plan to Save Pittsburgh's Steel Industry* (Pittsburgh, c. 1984), p. 3, says USX employed more than 28,000 production and maintenance workers in 1979, according to dues checkoffs of local unions. Employment at J&L and W-P plants in 1979 estimated by public relations departments of the companies. USX in 1987: USS Public Relations, phone, Aug. 27, 1987 (Braddock, 874 hourly, 34 salaried, or 908; Clairton, 1,227 hourly, 39 salaried, or 1,266; Irvin, 1,215 hourly, 92 salaried, or 1,307).

2. Dependent jobs: Morton Coleman, "Decline of the Mon Valley Viewed in a

Global Context," in *Steel People: Survival and Resilience in Pittsburgh's Mon Valley*, ed. James Cunningham and Pamela Martz, a study by the River Communities Project, School of Social Work, University of Pittsburgh, 1986, p. 4. Pittsburgh Countywide Corporation Steel Employment Advisory Committee, *Pittsburgh Steel Production and Employment: Regional Public Policy Initiatives*, p. 12, quoted in *Steel Valley Authority*, p. 1.

Nine communities study: Robert P. Strauss and Beverly Bunch, "The Fiscal Position of Municipalities in the Steel Valley Council of Governments," a study conducted for Steel Valley Council of Governments, 1986, pp. 6–7. Most of the job loss from 1983 to 1987 was concentrated in Clairton (667), Duquesne (1,250), Homestead (2,640), Munhall (2,400)—the Homestead Works was located in both Homestead and Munhall—and West Homestead (840). The 60 percent decline in jobs does not mean the areas involved had a 60 percent unemployment rate; typically in the Mon Valley, workers at any given plant commute from dozens of surrounding towns, including some outside the region. Other communities surveyed were Braddock Hills, Swissvale, West Elizabeth, and Whitaker. McKeesport, Braddock, and West Mifflin do not belong to the Council of Governments.

3. Jobless rates: Michael A. Acquaviva, Pennsylvania Department of Labor and Industry, *Labor Market Report* (Pittsburgh, various issues up to August 1987). Duquesne survey: David Biegel, James Cunningham, and Pamela Martz, "Mon Valley People Speak," *Steel People*, pp. 35–69. White Oak example: Lawrence and Maureen Delo, White Oak, Feb. 21, 1986. Social fabric: James Cunningham, "Conclusion: The Issues That Emerge," *Steel People*, p. 87.

4. Relocating: ibid., pp. 48–55; also quotes Jane Blotzer, "Leaving the Most Livable City: Graduates in Mon, Ohio Valleys Plan to Go Elsewhere for Jobs," *PPG*, June 18, 1986. Psychological distress: Ray M. Milke, "A Survey of Unemployed Steelworkers in the Mon Valley," Ph.D. diss., University of Pittsburgh School of Education, 1984, pp. 115–46. Suicides: Jacqueline Corbett, "A Study on Unemployment in the Mon-Yough Valley and Its Impact on Social and Psychological Functioning," Mon-Yough Community Mental Health and Mental Retardation Center, McKeesport, July 27, 1985, pp. 70–71.

5. McKeesport's abandoned homes and aging population: Christine Seltzer, McKeesport, Feb. 21, 1986. Facts from Strauss and Bunch, "Fiscal Position of Municipalities": USX assessed valuation, p. 13; cutting services, pp. 30–43; deficit projections, p. 48; consolidation, pp. 116–53. Definition of Mon Valley: Mon Valley Commission, "For the Economic Revitalization of the Monongahela, Youghiogheny, and Turtle Creek Valleys," Report to the Allegheny County Board of Commissioners, November 1986, p. 6. McKeesport deficit: Mayor Louis Washowicz, phone, Sept. 2, 1987. "Unique solution": Louis Washowicz, McKeesport, Feb. 21, 1986. People thrown away: Barney Oursler, McKeesport, Feb. 23, 1986. How governments helped steel regions in European countries and Japan is detailed in *Manpower Policies to Facilitate Structural Change in the Steel Industry: Report by a Consultant* (Paris: Organization for Economic Cooperation and Development, June 19, 1984). See also Susan Houseman, "Job Security and Economic Adjustments: Lessons from Steel," *The Brookings Review* (Washington, D.C., Summer 1987), pp. 40–46.

6. Paul Lodico, McKeesport, June 10, 1987.

7. The DMS's doings were chronicled in many news stories, including the following. Shadyside Church: "The 'Mon Valley' Struggles to Find Life After Steel," *BW*, May 21, 1984, p. 61. Mesta campaign: "Is the Mellon Bank Really a 'Good Neighbor'?" *BW*, June 20, 1983, p. 27. Anger at Mellon: Michael Hoyt, "Angry Churches Confront Mellon Bank," *NYT*, Nov. 20, 1983. Honeywell profile: Bohdan Hodiak and Andrew Sheehan, "Strategist for Protesters Keeps Low Profile," *PPG*, May 2, 1984. See also

Michael Hoyt, "How to Make Steel: Agitate and Organize," *Christianity and Crisis,* Mar. 4, 1985, p. 62. DMS support fades: Mike Stout, Munhall, Jan. 28, 1985. Roth story: Carol Waterloo, "Roth Back in Pulpit; No Arrest Attempt; Churchgoers Protest," *MDN,* Nov. 12, 1984, and "Judge Orders Jail Term Start for Contempt," Nov. 13, 1984; "Supporters of Defiant Pastor Found in Contempt," *NYT,* Jan. 1, 1985; Cristina Rouvalis, "Roth Defrocked, Arrested," *PPG,* June 7, 1985. Turning against allies: Matthew Brelis, "USW Local Says DMS Can't Use Facilities," *PP,* May 22, 1986.

8. Chopping ice: "Cutting Ice from Rivers Once Gave Work to Many Here," *MDN,* June 30, 1934. Wearing armbands and playing taps: "The Death of 'Dorothy Six,'" *Steelabor,* June 1984. Origin of Tri-State: Tri-State Conference on Steel, news release. Other organizers in Pittsburgh included Charles McCollester, a United Electrical Workers official at the Union Switch & Signal plant in Swissvale; Frank O'Brien, a former state legislator and USW local president in Pittsburgh; Jay Weinberg, Local 1397; attorney Joseph Hornack; Sr. Ligouri Rossner, Jubilee Soup Kitchen; and Thomas Kerr, a Carnegie-Mellon professor and chairman of the Allegheny County Civil Rights Commission. Also see *Steel Valley Authority,* 1984, prepared by Tri-State.

Story of Dorothy: personal reportage; John Hoerr and William C. Symonds, "A Brash Bid to Keep Steel in the Mon Valley," *BW,* Feb. 11, 1985, p. 30; "Dorothy Six Headed for the Scrap Yard," *PP,* Sept. 9, 1986. American Standard: Matthew Brelis, "Steel Valley Group to Appeal Switch Ruling," *PP,* June 18, 1986. *PPG* editorial: "Dorothy Demolished," Sept. 11, 1986. Roderick's comments: USX, transcript of Roderick news conference, Pittsburgh, Feb. 11, 1986, pp. 21–23.

Chapter 22

1. Analysis of job losses and gains: Provided by Michael Acquaviva, Pennsylvania Department of Labor and Industry, Pittsburgh, Sept. 15, 1987. The five counties involved—Allegheny, Beaver, Fayette, Washington, and Westmoreland—comprise the Consolidated Metropolitan Statistical Area (CMSA). Four of them, excluding Beaver, make up the Standard Metropolitan Statistical Area (SMSA). Population decline: Economics Department, Mellon Bank, Sept. 19, 1987. These figures pertain only to the four-county SMSA to conform to per capita income statistics from the same source. Per capita income figures exist only for the four counties. On per capita income and a discussion of other economic changes, see Jim McKay, "Labor Day, 1987: A Brighter Outlook for Jobs and Personal Income," *PPG,* Sept. 7, 1987.

2. Roger S. Ahlbrandt, Jr., and Morton Coleman, *The Role of the Corporation in Community Economic Development As Viewed by Twenty-One Corporate Executives* (Pittsburgh: University of Pittsburgh, Jan. 5, 1987), pp. 47–48. Those interviewed were nineteen chief executive officers of companies headquartered in Pittsburgh, as well as the presidents of the University of Pittsburgh and Carnegie-Mellon University. The authors agreed to quote the executives anonymously to encourage a candid expression of opinion, and the extended comments provide a valuable insight into the attitudes of the executives. The names of the CEOs interviewed are listed separately. Ahlbrandt and Coleman found "general agreement" among the executives about the seriousness of the middle income decline.

3. Health service and research strengths: Allegheny Conference on Community Development, *A Strategy for Growth: An Economic Development Program for the Pittsburgh Region,* vol. 2 (Pittsburgh, November 1984), pp. III-3, I-7 (Allegheny Conference is hereafter referred to as ACCD). High tech job growth: Charles C. Robb, "Stormy Labor Climate Clouds Pittsburgh's Image," *Executive Report,* August 1987,

p. 28; Cynthia Piechowiak, "High-Tech Firms Like J&L Site, Cyert Says," *PP*, Aug. 26, 1986; Tom Barnes, "Awaiting the Phoenix," *PPG*, Sept. 9, 1986.

4. City rankings: "Dallas Catches Up to City in Fortune 500," *PP*, Apr. 7, 1987. Pittsburgh's fourteen largest companies are USX (ranked twenty-second nationally), Rockwell International, Westinghouse Electric, PPG Industries, Alcoa, H.J. Heinz, Bayer USA, Allegheny International, Koppers, Cyclops, Wheeling-Pittsburgh, Joy Manufacturing, H.H. Robertson, and L.B. Foster. Mellon's troubles: Robert E. Norton, "A Muddled Future for Mellon Bank," *Fortune*, Mar. 30, 1987, p. 68; Carol Hymowitz and Michael McQueen, "Mellon to Tap Frank Cahouet as Chairman," *WSJ*, June 15, 1987; Robert A. Bennett, "The Mellon Bank's Fall from Grace," *NYT*, Mar. 9, 1986. New ventures: Chet Wade, "City Still Lags as New Firms Provide More Jobs," *PPG*, June 23, 1987.

5. Douglas E. Booth, *Regional Long Waves, Uneven Growth, and the Cooperative Alternative* (New York: Praeger, 1987), pp. 13–17. Booth uses his analysis of mature industries to demonstrate why a region with such an industry experiences a lengthy period of relatively poor economic growth after the industry peaks and begins to decline. The bureaucratic form of production organization in mature industries impedes innovation. Booth hypothesizes that the organizational flexibility needed for new businesses to grow in this atmosphere can best be provided by producer cooperatives, in which all employee-members vote for a supervisory board, share in the profits, and therefore have an incentive to adapt to changing markets (pp. 38–41). Although Booth does not examine the steel industry, his theory is applicable to it.

Heirs not eager to stake innovators: Edgar M. Hoover, "Pittsburgh Takes Stock of Itself," *Pennsylvania Business Survey*, January 1964, pp. 1–6. When this article was written, Hoover was director of the Center for Regional Economic Studies at the University of Pittsburgh. He also served as study director of the three-volume study of the Pittsburgh region mentioned later in my text. See Pittsburgh Regional Planning Association, Vol. 1, *Region in Transition*; Vol. 2, *Portrait of a Region*; Vol. 3, *Region with a Future* (Pittsburgh: University of Pittsburgh Press, 1963).

Analysts of bureaucracies: Warren Bennis, *Beyond Bureaucracy: Essays on the Development and Evolution of Human Organization* (New York: McGraw-Hill, 1973; originally published as *Changing Organizations*, 1966); Michel Crozier, *The Bureaucratic Phenomenon* (Chicago: University of Chicago Press, 1964); Richard Edwards, *Contested Terrain: The Transformation of the Workplace in the Twentieth Century* (New York: Basic Books, 1979); Larry Hirschhorn, *Beyond Mechanization: Work and Technology in a Postindustrial Age* (Cambridge: MIT Press, 1986); Michael Maccoby, "The Problem of Work: Technology, Competition, Employment, Meaning," paper delivered at a meeting of the American Sociological Association, Washington, D.C., Aug. 26, 1985.

6. Bureaucracy prevents pioneering: Booth, *Regional Long Waves*, p. 38. Refusal to believe jobs are gone: Matt Hawkins, "Homestead Community Credit Union: Come Back Mechanism for a Depressed River Town," in *Steel People: Survival and Resilience in Pittsburgh's Mon Valley*, ed. James Cunningham and Pamela Martz, a report prepared by the River Communities Project, School of Social Work, University of Pittsburgh, 1986, p. 84. "Many young people, in their late thirties with families to support, still expect the mills to reopen and provide jobs for the valley in the near future." The difficulty in accepting that jobs are gone for good has been noted in many industrial towns since the 1940s, particularly where plants had operated for decades before closing.

7. Pittsburgh's adversary labor relations: ACCD, *A Strategy for Growth*, pp. VIII–11 (emphasis in original). The ACCD's ambivalence on the labor issue did not go undetected. The conference is seldom criticized in Pittsburgh's dailies, but two *Post-*

Gazette writers raised questions about the ACCD's view of organized labor. In a critical column on the future of the conference, Associate Editor Joseph Plummer wrote: "Labor is the widely acknowledged, rarely-addressed wild card in Pittsburgh's future" ("An Agenda for the Conference," Nov. 12, 1986). Clarke Thomas, the paper's chief editorialist, noted in another column that the issue had been "conspicuously fuzzed" in the ACCD study and that as of early 1987 the problem "continues to stay on the Conference's back burner" ("The Problem No One Can Solve," Apr. 14, 1987).

8. Unionization estimate: Morton Coleman, "Decline of the Mon Valley Viewed in a Global Context," in *Steel People*, p. 5. Comments by Cyert and consulting firm official: Robb, "Stormy Labor Climate," pp. 27–28. The consulting firm was Fantus Company, which has advised 200 of the *Fortune* 500 on location decisions. A Fantus vice-president, Saul Grohs, told Robb that western Pennsylvania is normally not even in the search area for his firm's clients. "Grohs concedes that perceptions of Pittsburgh may not conform to reality, and that a close examination might present the area's labor climate in a more positive light. But a company looking for a new location will make a close examination only in an area it is seriously considering. Pittsburgh, Grohs says, rarely makes the first cutoff. 'We haven't looked at Pittsburgh at all in the last several years,' Grohs says. 'That's true for the whole western half of western Pennsylvania.'"

9. CEO comments: Ahlbrandt and Coleman, *Role of the Corporation:* automatic unionization, p. 50; extraordinary truculence, p. 48; authors' conclusion, p. 50. Mon-Yough survey: Carmen J. Lee, "Bad Image of Labor a Myth, Poll Says," *PPG*, Nov. 18, 1986. The survey questioners contacted 250 companies in 44 communities and received answers from 173. Of course, these firms had an interest in *not* furthering the bad-climate image. Even so, the survey is instructive.

10. Enlightened management practices: Robb, "Stormy Labor Climate," p. 31. Regarding the old-line companies that have changed management style, National Steel and PPG do not have plants in the Pittsburgh area. Make Pittsburgh a central focus: Ben Fischer, phone, Sept. 30, 1987. Unfair to hard-working people: Ahlbrandt and Coleman, *Role of the Corporation*, p. 56. "Greater Pittsburgh mentality": ibid., p. 32. Exacerbated mill town problems: ibid., p. 22. Lawrence Delo, White Oak, Feb. 21, 1986.

11. Planning for Mon Valley: Mon Valley Commission, "For the Economic Revitalization of the Monongahela, Youghiogheny, and Turtle Creek Valleys," report to the Allegheny County Board of Commissioners, February 1987. People are important: Hoover, "Pittsburgh Takes Stock," p. 6.

Chapter 23

1. For more on associational unionism, see Charles C. Heckscher, *The New Unionism* (New York: Basic Books, 1988). Gregory L. Miles, "Cancel the Funeral—Steel Is on the Mend," *BW*, Oct. 5, 1987, p. 74. Mission Statement: Gary Hubbard, phone, Jan. 15, 1988. Loss of steel jobs: AISI, *Annual Statistical Report* (Washington, D.C., various years). The AISI companies employed an average of 512,395 hourly and salaried workers in 1974, a high-volume year, and 174,783 in 1986. Steel Output in West: AISI, *Annual Statistical Report*, 1986.

2. Profit-sharing at Ford and GM: Jacob M. Schlesinger and Paul Ingrassia, "Ford Schedules Profit-Sharing Payments for '86, Averaging Over $2,100 a Worker," *WSJ*, Feb. 19, 1987.

3. For more on American workers' attitudes toward a working class, see David Halle, *America's Working Man* (Chicago: University of Chicago Press, 1984). Halle studied a group of workers at a chemical plant in New Jersey, exploring their lives on

and off the job. He concludes that while these blue-collar workers were very much aware of their common interests in the workplace, the feeling of group identity dissipated outside of work. Many of the workers lived in new suburban areas along with white-collar and managerial employees and identified their political and community interests with those of their neighbors. "The Road Not Taken": *The Poetry of Robert Frost*, ed. Edward Connery Lathem (New York: Henry Holt, 1969), p. 105.

4. Thomas A. Kochan, Harry C. Katz, and Robert B. McKersie, *The Transformation of American Industrial Relations* (New York: Basic Books, 1986), chap. 2.

5. Ibid., p. 236.

6. Companies 25 percent to 50 percent more productive, Walton quote: John Hoerr and Michael A. Pollock, "Management Discovers the Human Side of Automation," *BW*, Sept. 29, 1986, p. 70. Michael J. Piore and Charles F. Sabel, *The Second Industrial Divide: Possibilities for Prosperity* (New York: Basic Books, 1984).

7. Packard Electric: personal reportage; see also Kochan et al., *Transformation*, pp. 199–200. Fraser on Fremont plant: address at Columbia University, Oct. 7, 1987. Warren Bennis, *Beyond Bureaucracy: Essays on the Development and Evolution of Human Organization* (New York: McGraw-Hill, 1973), p. 21.

8. For more on conflicts with labor law, see Stephen I. Schlossberg and Steven M. Fetter, "U.S. Labor Law and the Future of Labor-Management Cooperation," U.S. Department of Labor, June 1986. "We're never going back": Fraser, Columbia address.

9. "A great way to run the business": James Howell, Ecorse, Michigan, June 9, 1987. For more on the productivity-enhancing benefits of employment security, see Jerome M. Rosow and Robert Zager, eds., *Employment Security in a Free Economy* (New York: Pergamon Press, 1984). For more on participation as the foundation of the economy, see Ray Marshall, *Unheard Voices: Labor and Economic Policy in a Competitive World* (New York: Basic Books, 1987).

Index

Pittsburgh Series in Social and Labor History

Maurine Weiner Greenwald, Editor